T0201916

RANDOM PROCESS ANALYSIS WITH R

Random Process Analysis with R

Marco Bittelli

University of Bologna

Roberto Olmi

National Research Council, Italy

Rodolfo Rosa

National Research Council, Italy

OXFORD
UNIVERSITY PRESS

Great Clarendon Street, Oxford, OX2 6DP,
United Kingdom

Oxford University Press is a department of the University of Oxford.
It furthers the University's objective of excellence in research, scholarship,
and education by publishing worldwide. Oxford is a registered trade mark of
Oxford University Press in the UK and in certain other countries

© Marco Bittelli, Roberto Olmi, and Rodolfo Rosa 2022

The moral rights of the authors have been asserted

Impression: 1

All rights reserved. No part of this publication may be reproduced, stored in
a retrieval system, or transmitted, in any form or by any means, without the
prior permission in writing of Oxford University Press, or as expressly permitted
by law, by licence or under terms agreed with the appropriate reprographics
rights organization. Enquiries concerning reproduction outside the scope of the
above should be sent to the Rights Department, Oxford University Press, at the
address above

You must not circulate this work in any other form
and you must impose this same condition on any acquirer

Published in the United States of America by Oxford University Press
198 Madison Avenue, New York, NY 10016, United States of America

British Library Cataloguing in Publication Data
Data available

Library of Congress Control Number: 2022935991

ISBN 978–0–19–886251–2 (hbk)
ISBN 978–0–19–886252–9 (pbk)

DOI: 10.1093/oso/9780198862512.001.0001

Printed and bound by
CPI Group (UK) Ltd, Croydon, CR0 4YY

Links to third party websites are provided by Oxford in good faith and
for information only. Oxford disclaims any responsibility for the materials
contained in any third party website referenced in this work.

What men really want is not knowledge but certainty
Bertrand Russell

Preface

A random or stochastic process is a process that can be defined by random variables. In other words, it is a process involving observations whose outcome at every time instant is not certain. Mathematically, this reflects in working with functions whose arguments are characterized by a probability distribution instead of having a certain value.

The choice of describing a phenomenon by deterministic or probabilistic laws may depend upon several reasons. The phenomenon appears to be conceptualized only as a random process. On the other hand, we know that it can be described by deterministic laws, but due to lack of knowledge about system parameters or high complexity, we decide to model it as a random process.

Overall, what is a random process? Is randomness an inherent feature of nature or simply our inability to describe it with deterministic laws? Therefore, is there an inherent randomness in nature or is it our lack of knowledge that bring us to describe it as a random process?

These differences can be referred to the 'objective' and 'subjective' viewpoints of randomness. The first viewpoint considers randomness an inherent feature of nature, while the second conceptualizes it as an 'anthropomorphic'–'subjective' interpretation of nature due to lack of knowledge. Is there a true distinction between these two viewpoints? These ideas have been the subject of scientific and philosophical discussion for centuries and a brief discussion is presented in the final chapter of this book.

However, when the phenomena we are studying appear as stochastic processes and have been subjected to rigorous mathematical tests and do not reveal a fully deterministic framework, probabilistic tools must be employed. In this book we present concepts, theory and computer code written in R for random process analysis.

Acknowledgements

We are grateful to Dr. Sonke Adlung for being most cooperative, considerate and helpful during the publication process.

Contents

1
Introduction

1.1 Introduction

The subject of random or stochastic process analysis is a very important part of scientific inquiry. The terms stochastic and random process are used interchangeably. Random processes are used as mathematical models for a large number of phenomena in physics, chemistry, biology, computer science, information theory, economics, environmental science and others. Many books about random processes have been published over the years. Over time, it become more and more important to provide not only the theory and examples regarding a specific processes, but also the computer code and example data. Therefore, this book is intended to present concepts, theory and computer code written in R, that helps readers with limited initial knowledge of random processes to become operational with the material. Each subject is described and problems are implemented in R code, with real data collected in experiments performed by the authors or taken from the literature. With this intent, the reader can promptly apply the analysis to her or his own data, making the subject operational. Consistent with modern trends in university instruction, this book make readers active learners, with hands-on computer experiments directing readers through applications of random process analysis (RPA). Each chapter is also introduced with a brief historical background, with specific references, for further readings about each subject.

Chapter 2 provides a brief historical background about the origin of random processes theory. In Chapter 3, the reader will find an in-depth description of the fundamental theory of stochastic processes. The chapter introduces concepts of stationarity, ergodicity, Markov processes and Markov chains. Examples from mathematics and physics are presented to exemplify random processes such as the Buffon's needle and the Ehrenfest Urn Model. In Chapter 4, Poisson's processes are presented. Derivation of the well-known distribution is presented as well as homogeneous and non–homogeneous Poisson processes are discussed. One the cornerstones of random processes, random walk, is described in Chapter 5. The concepts of absorbing and reflecting barriers are presented along with the gambler's ruin example, as well as a two-dimensional random walk code and discussion about random walk applied to the process of Brownian motion.

Chapter 6 enters into stochastic time series analysis with the description of moving average, autoregressive and autoregressive moving average processes. Seasonal time series analysis is introduced with examples applied to measures of temperature and water budget. The analysis of random processes requires a thorough understanding of spectrum and noise analysis. In Chapter 7 Fourier transforms for deterministic and stochastic time series are presented, with application to spectrum analysis. The singular spectrum analysis technique is also presented, for analysis and removal of trends.

Chapter 8 presents the Markov Chain Monte Carlo method with a description of probably the most famous algorithm in stochastic theory, the Metropolis algorithm. After a through description of the theory and code, the travelling salesman problem is presented, with the simulated annealing approach. The concept of Bayesian analysis is here briefly introduced and leading into the next chapter. Chapter 9 focus on a cornerstone of modern statistics, Bayesian inference, which is applied to a description of autoregressive processes. After introducing the main concepts, examples applied to real data of temperature and CO_2 concentration in Antarctica, as well as radar detection, are presented. Bayesian analysis of the Poisson process is presented with the waiting-time paradox. The chapter ends with an application to lighthouse detection as a remarkable example of Bayesian inference.

Random processes are used as tools for random search in minimization algorithms, as an alternative to gradient-based search algorithms used for instance in least square optimization. Genetic algorithms are presented in Chapter 10, with application to non–linear fitting, autoregressive moving average models. As an example of improved optimization with respect to other approaches, the travelling salesman problem is here solved with genetic algorithms. The modelling of stochastic processes depends on the accuracy of the estimators derived in the process analysis. The problem of accuracy is discussed in Chapter 11, with examples on averaging of time series, batch means methods, moving bootstrap and other techniques to improve accuracy in random processes modelling.

Chapter 12 addresses a topic that is not traditionally described in books about random processes: spatial analysis. It is nevertheless an important subject dealing with the application of statistical concepts to properties varying in space. The chapter provides an introduction to geostatistical concepts and then present a novel approach, where spatial and temporal analysis are combined into a stochastic analysis of spatio-temporal processes. At the end of the chapter, the optimization procedure for spatial parameters is computed also with genetic algorithm, showing the possibility of connecting and applying various techniques presented in the book.

The book ends with Chapter 13, which discusses the very definition of a random process, the mathematical definition of randomness and a discussion of the definition of entropies. This discussion is developed into a general framework and its implications for scientific inquiry. The book also has two appendices providing additional tools presented in the main part of the book.

The codes presented in this book are written using the RStudio integrated development environment (IDE). RStudio includes a console, an editor that supports direct code execution, as well as tools for plotting, debugging and workspace management. There are many books about programming in R that can be used as reference and in particular publications and links presented in the official Comprehensive R Archive Network (CRAN) available at: `https://cran.r-project.org/`. The codes and example data written in this book can be dowloaded from the website `http://www.marcobittelli.it` under the section *Computer codes for books*. Exercises are presented at the end of each chapter and solutions are downloadable on the book's website.

Open source languages and related libraries are subject to changes, updates and modifications, therefore the packages presented here may undergo changes in the future. To obtain specific information and documentation about a library, the following instruction should be used: `library(help=GA)`, where for example the library `(GA)` for genetic algorithms can be explored. Here we list the libraries necessary to run the examples in different chapters:

Chapter 7 requires the library `lubridate`; Chapter 9 the library `rjags`, described in detail in the Appendix B; Chapter 12 requires `ggplot2`, `gstat`, `lattice`, `mapview`, `GA`, `quantmod`, `reshape`, `sf`, `sp`, `stars`, `tidyverse`, `xts` and `zoo` and Chapter 13 requires `entropy`, `tseriesEntropy`.

2
Historical Background

It is not certain that everything is uncertain.

Blaise Pascal

2.1   The Philosopher and the Gambler

To introduce the role of computer studies in stochastic processes analysis, we will go back a few centuries to the invention of probability theory. It is the year 1654, according to a familiar story (Hacking, 1975). Antoine Gombaud Chevalier de Méré, Sieur de Baussay (1607−1684), asks some questions on a game of chance to Blaise Pascal (1623−1662). Later, Siméon−Denis Poisson (1781−1840), calls Antoine Gombaud 'man of the world' and Blaise Pascal 'austere Jansenist': *Un problème relatif aux jeux de hasard, proposé à un austère janséniste par un homme du monde, a été l'origine du calcul des probabilités* (A problem about games of chance proposed to an austere Jansenist by a man of the world was the origin of the calculus of probabilities) (Poisson, 1837). We know that Pascal was not only a philosopher, but also a physicist, a mathematician, a writer, a theologian. Antoine Gambaud was a writer and a philosopher, not only a gambler.

We now discuss one of the questions that our Chevalier asked of Pascal concerning the throws of two dice. We throw two dice and bet on the double six. How many throws do we need to have a change of winning?

Antoine Gombaud said that a gambling rule, based on the mathematical analogy between the probabilities of obtaining six with a single die or double six with a couple of dice, indicates that you need at least 24 throws, but from his personal gambling experiences the throws must be at least 25. Pascal, after discussing the problem with Pierre de Fermat (1601−1665), answered that mathematics is not contrary to experience. Let us briefly discuss the topic.

Let A_1 be the event $\{6, 6\}$ at the first throw, so $\mathsf{P}\{A_1\} = 1/36$ (the symbol $\mathsf{P}\{.\}$ means 'probability'). Then, the probability of *not* obtaining two six is that of the *complementary* event $\overline{A}_1$: $\mathsf{P}\{\overline{A}_1\} = 1 - 1/36 = 35/36$. At the second throw, the event $\overline{A}$: none double six at the first throw *and* none double six at the second throw has probability $\mathsf{P}\{\overline{A}\} = \mathsf{P}\{\overline{A}_1\}\,\mathsf{P}\{\overline{A}_2\} = (35/36)^2$, and so on. The probability of not winning in 24 throws is:

$$\mathsf{P}\{\overline{A}\}^{[24]} = \mathsf{P}\{\overline{A}_1\}\,\mathsf{P}\{\overline{A}_2\}\cdots\mathsf{P}\{\overline{A}_{24}\} = (35/36)^{24} = 0.5086$$

so the probability of winning is $P\{A\}^{[24]} = 1 - 0.5086 = 0.4914$. While in 25 throws it is $P\{\overline{A}\}^{[25]} = 0.4945$ and $P\{A\}^{[25]} = 0.5055$. Notice that the difference is very small, and this honours the power of observation of our Chevalier. But there are doubts about the truth of this story (Ore, 1960).

Let us imagine being the Chevalier de Méré who, for 30 nights, goes to the game table to throw two dice. Every night, we play 20 games, with 25 and 24 throws each. If in a game two sixes appears, we win the game. At the end of the night, that is after 20 games, if the victories are more than 10, we had a lucky night.

We can describe the throw of a die as a *stochastic process*. In the next chapter we will rigorously define 'stochastic process', but here we simply say that stochastic processes are mathematical models of dynamical systems that evolve over time or space in a probabilistic manner.

In our case, the dynamical system is the die, that at each throw shows a face with probability $1/6$. The code below is the 'transcription' in R of the dice game above.

```
## Code_2_1.R Throw of two dice
#   25 throws
#   p_25<- 0.5055: probability of getting two sixes in 25 throws

n.nights <- 30          # number of nights
n.games <- 20           # number of games
n.throws <- 25          # number of throws
spot <- c(1:6)          # spots of a 6-sided die
p_fair <- rep(1/6,6)    # probabilities of a "fair" die
d6 <- numeric()
d6T <- numeric()
nseed <- 50
for (j in 1:n.nights)
{          # loop on the nights
   nseed <- nseed+1
   set.seed(nseed)
   for (l in 1:n.games)
   {   # loop on the games
      d6[l] <- 0
      for (i in 1:n.throws)
      {    # loop on the throws
         die.1 <- sample(spot,1,p_fair,replace=T)  # i-th throw with the die 1
         die.2 <- sample(spot,1,p_fair,replace=T)  # i-th throw with the die 2
         s.points <- die.1+die.2
         if (s.points == 12) d6[l] <- 1
      }       # end loop on the throws
   }        # end loop on the games
   d6T[j] <- sum(d6)
} # end loop on the nights

d6T

###  24 throws
##   p_24<- 0.4914    # probability of getting two sixes in 24 throws

n.nights <- 30          # number of nights
n.games <- 20           # number of games
n.throws <- 24          # number of throws
spot <- c(1:6)          # spots of a 6-sided die
```

```
p_fair <- rep(1/6,6)        # probabilities of a "fair" die
d6 <- numeric()
d6T <- numeric()
nseed <- 500
for (j in 1:n.nights)
{       # loop on the nights
   nseed <- nseed+1
   set.seed(nseed)
   for (l in 1:n.games)
   {       # loop on the games
      d6[l] <- 0
      for (i in 1:n.throws)
      {       # loop on the throws
         die.1 <- sample(spot,1,p_fair,replace=T)  # i-th throw with the die 1
         die.2 <- sample(spot,1,p_fair,replace=T)  # i-th throw with the die 2
         s.points <- die.1+die.2
         if (s.points == 12) d6[l] <- 1
      }       # end loop on the throws
   }       # end loop on the games
   d6T[j] <- sum(d6)
} # end loop on the nights
d6T
```

For the sake of clarity, we repeat for 24 throws the instructions for 25 throws, changing only the lines n.throws <- 24 and nseed <- 500. The vector d6 at the beginning of each game is 0, if a double six is obtained its value becomes 1, by summing the n.nights components of d6, we know if we won or lost. Notice the R function **set.seed(.)** is used to set the initial seed of the (pseudo) random number generator (RNG). RNGs are in fact fully deterministic algorithms, so the same seed generates the same sequences, changing the seed, we get different sequences. The result of the code above for 25 throws is:

```
d6T:
10 10 12  9 11 13 12 10 12 10 13  9  9  8 13  9 12 11  9 11 11 11 15 12 12
11 10  8 11 13
```

We see that the two first games were tied, we won the third and lost the fourth, and so on. We won 18 games out of 30, lost 7 games and tied 5 games. The result of the code above for 24 throws is:

```
d6T:
11  7  6 11  8 17 11  9 12 10 13 10  7  7 11 16  6 12  9  7  9 13 12 11 13
13  9  7 14 11
```

In this case, we won 16 games, so that after 30 nights we are again win-making. These results appear not to support the Chevalier's claim that 24 throws are not enough to hope to win. However, such a statement is not correct. For instance, if we put nseed <- 100 with n.throws <- 25 we have:

```
d6T:
12 13 10 10 10 12  9 10  8 12 12  9  9 10 11  9 12 10  9  9  8  9  6  9 12
13  9  9 11 10
```

We won only 10 games out of 30. The reason for this variability is that the sample size is too small, that is the number of games is not enough to give reliable results. Let us increase the number of games. For each night we play 100 games and the nights are 180. In the `Code_2_1.R`, it is now: `n.nights <- 180` and `n.games <- 100`. The result is for 25 throws:

```
d6T:
[1] 59 44 57 53 54 49 51 54 50 45 43 50 50 45 41 ...
.....................................................
[176] 59 55 50 58 45
```

that is, the first night we won 59 games out of 100, the second only 44, and so on. The result is for 24 throws:

```
d6T:
[1] 46 49 52 48 52 41 47 53 40 46 55 50 51 42 55 ...
.....................................................
[176] 53 48 52 49 54
```

We can show the results both with 25 and 24 throws as in Fig. 2.1. The figure is obtained adding the following lines after `d6T`, both for 25 and 24 throws.

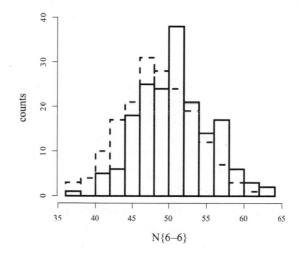

Fig. 2.1 Number of times for obtaining double six has the value in the abscissa. Solid line: 25 throws, dashed line: 24 throws.

For `n.throws <- 25` and `nseed <- 100`:

```
lbin <- 2
par(lwd=3)
hist(d6T,main=" ",freq=T,xlab="N{6-6}",ylab="counts",cex.lab=1.3,lty=1,
border="black",ylim=c(0,40),br=seq(36,64,by=lbin),font.lab=3)
```

For `n.throws <- 24` and `nseed <- 301`:

```
hist(d6T,lty=2,br=seq(36,64,by=lbin),add=T)
```

Note that `freq=T` means that the histogram reports the counts component of the result, if `freq=F` the histogram reports probability densities, in this case the total area of the plot is 1.

In the abscissa it reports the number of times the double six was obtained in 100 games. The symbol `N{6-6}` indicates the double six. In the ordinate it reports the number of times this event occurred in 180 nights. For instance, the bin $(50, 52]$ is 38 for the games with 25 throws, meaning that in 38 times out of 180 the double six occurred 51 or 52 times in 100 games.

In the histogram each bin is closed on the right and open on the left, therefore the 50 occurrences of the double six are not counted in the $(50, 52]$ bin, but rather in the $(48, 50]$ bin. The results show that in the total number of games, which is $180 \times 100 = 18000$, the double six occurred in 9204 games, then the probability of a double six estimated in 18000 games is $\widehat{\mathsf{P}}\{A\}^{[25]} = 9204/18000 \approx 0.5113$ (the hat stands for estimate), very close to the 'theoretical' probability. Here 'theoretical' means the probability of perfect dice, that is that expected assuming equiprobability for each face. Rigorously speaking we should not define probability by counting on 'equiprobable' events, because that makes the definition recursive. However, putting aside philosophy, if the die is fair, its faces are 'equally' likely to occur and the probability of outcomes can be computed as we did.

For games with 24 throws, a double six occurred in 8797 games, then the estimated probability of the double six is $\widehat{\mathsf{P}}\{A\}^{[24]} \approx 0.4887$, in agreement with the 'theoretical' one $\mathsf{P}\{A\}^{[24]} = 0.4914$.

Let us consider the 18000 games as N independent trials, each with probability p of success, and let n be the number of successes. The standard error of the estimate of the proportion p of 1's in the N long sequence is $\hat{\sigma} = \sqrt{\hat{p}(1 - \hat{p})/N}$, where $\hat{p}$ is an estimate of p, denoted above as $\widehat{\mathsf{P}}\{A\}^{[25]}$. In our case $\hat{\sigma}^{[25]} = 0.0037$, practically equal to the theoretical one. Obviously also $\hat{\sigma}^{[24]}$ results in the same.

We have seen that $\widehat{\mathsf{P}}\{A\}^{[25]} = 0.5113$ and $\widehat{\mathsf{P}}\{A\}^{[24]} = 0.4887$, we could ask ourselves if the difference between the two means $\hat{d} = 0.5113 - 0.4887 = 0.0226$ is significant. We can test for the significance of the difference between two population means using the Student's t, which can be done in R by the line:

```
t.test(z,y,alt="greater",var.equal=TRUE)
```

where z are the winnings in the 180 nights with 25 throws in each game (59, 54, ..., 58, 45) and y with 24 throws in each game (46, 49, ..., 49, 54). The option `alt="greater"` is to specify a one–tailed test and `var.equal=TRUE` to specify equal variances. The significance level (p value) is `p-value = 5.159e-06`, that is highly statistically significant. In passing, 20 throws for 30 nights yield no significant difference, confirming what we said above about the small number of games.

We can test the difference of the means also by the bootstrap method. We will discuss this method in Appendix A, here we limit ourselves to show the result in Fig. 2.2, which is also presented as Exercise 2.1. As can be seen in Fig. 2.2, the difference $\hat{d}$ is significant, since out of $B = 1000$ replications $\hat{d}^*$, none of them is less than 0.

2.2 Comments

It is difficult to believe that a real gentleman such as Antoine Gombaud went to play with dice for about three months playing 100 games each night. Regardless of whether the story is true or false, it teaches us something. In the doubt expressed by the Chevalier, different concepts of probability are involved. Doubtless, the term

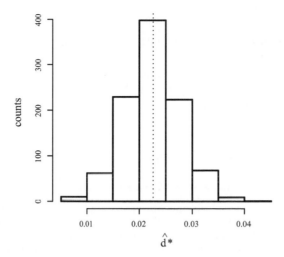

Fig. 2.2 Distribution of 1000 bootstrap replications $\hat{d}^*$ of the difference of the means of the two samples obtained with 25 and 24 throws (sample size = 180). The dotted line locates the observed difference $\hat{d} = 0.0226$. There are no replications less than 0.

'probability' had not yet appeared, but from one side he speaks about a theoretical mathematical argument to calculate the number of chances to get two sixes. From the other side he relies on his experience as a gambler to evaluate the frequencies of the results. It is already recognizable the tensions between theory and experience, between probability as subject of study of a purely mathematical discipline and probability as a property of a real physical random process evolving over time.

We could ask ourselves why the Chevalier affirmed that mathematics was wrong in alleging 24 throws. According to some historians, perhaps he believed that. If the probability of success in one throw is $1/n$, in m throws it is m/n, that is $24/36 = 0.667$. Of course, this reasoning is wrong: probabilities have to be multiplied, not summed.

Historians say that similar problems about games of chance were present before Pascal and Fermat. Gerolamo Cardano, for instance, wrote *Liber de Ludo Aleae* ('The Book on Games of Chance'), written about in 1560 and published posthumously in 1663, in which many results in various games with dice are discussed. In particular, the chance of various combinations of points in games with three dice are presented. The same problems about three dice were studied also by Galileo Galilei (Todhunter, 1865), in about 1610−1620 in his *Sopra le scoperte de i dadi*, translated in different ways, for instance, 'Analysis of Dice Games', 'On a Discovery Concerning Dice', 'Concerning an Investigation on Dice', and so on. Actually the word '*scoperte*' means the faces of the dice that appear, so we could translate it simply as 'On the Outcomes of Dice'.

Galileo was asked why playing with three dice the sum of points 10 or 11 are observed more frequently than the sum of points 9 or 12. Galileo's answer was:

Che nel giuoco de dadi alcuni punti sieno più vantaggiosi di altri, vi ha la sua ragione assai manifesta, la quale è il poter quelli più facilmente e più frequentemente scoprirsi che questi (The fact that in dice games certain outcomes are more advantageous than others has a very clear reason, which is that certain outcomes can appear more easily and more frequently than others.)

For instance, the sum 9 is obtained with the following six triple number (*triplicità*), that is the *scoperte* of the three dice:
1.2.6.; 1.3.5.; 1.4.4.; 2.2.5.; 2.3.4.; 3.3.3.
Six triple number are also necessary to get the sum 10:
1.3.6.; 1.4.5.; 2.2.6.; 2.3.5.; 2.4.4.; 3.3.4.
However the sum 3.3.3, for instance, can be produced by only one throw, while the sum 3.3.4. by three throws: 3.3.4., 3.4.3., 4.3.3. In conclusion, the sum of points 10 can be produced by 27 different throws, while the sum of points 9 by 25 only.

Let us do a forwards time warp and read the following quotation:

From an urn, in which many black and an equal number of white but otherwise identical spheres are placed, let 20 purely random drawings be made. The case that only black balls are drawn is not a hair less probable than the case that on the first draw one gets a black sphere, on the second a white, on the third a black, etc. The fact that one is more likely to get 10 black spheres and 10 white spheres in 20 drawings than one is to get 20 black spheres is due to the fact that the former event can come about in many more ways than the latter. The relative probability of the former event as compared to the latter is the number 20!/10!10!, which indicates how many permutations one can make of the terms in the series of 10 white and 10 black spheres [...]. Each one of these permutations represents an event that has the same probability as the event of all black spheres.

That is what Ludwig Boltzmann wrote in 1896 in his *Vorlesungen über Gastheorie* translated by Brush (1964). It is *not impossible* to draw 20 black balls, since this balls extraction has the same probability as any other one, but the number of ways to draw 10 black balls and 10 white balls is far greater than that of all black balls.

In his Foundations of Statistical Mechanics, Boltzmann explicitly introduces the postulate of equal a priori probability of *microstates* being compatible with a given *macroscopic* state. On this *Ansatz*, Boltzmann explains why the 'arrow of time' points to the more probable macrostate. We will – so to speak – hold in our hand these concepts by studying the stochastic process 'Ehrenfest's urn model' in the next chapter.

Both Cardano and Galileo found the solution of the problems of three dice, assuming that the possible outcomes are *equally* possible and counts the chance of compound events. In statistical mechanics, each microstate describes the position and velocity of each molecule. A macrostate is a state description of the macroscopic properties of the system: for instance its pressure, volume and such. Each macrostate is made up of many microstates. To have an idea of microstates and macrostates, let us think of macrostates as the sum of the points of the three dice and of microstates as the number of favourable outcomes. So we say that the 'system' (the system is formed by the three dice) is in the macrostate 10 (that is, the sum of the points is 10) which is realized by 27 microstates. In the Boltzmann's example of the 20 balls, each extracted sequence is a microstate, while the number of white (or black) balls is a macrostate.

The hypothesis of equal a priori probability (explicitly stated or implicitly assumed), in its turn, rests on the principle of indifference: equal probabilities have to be assigned to each occurrence if there is no reason to think otherwise. With a small time leap, we learn from Einstein (1925) that, in what will be called 'Bose-Einstein Statistics', the microstates are not equally possible, even though there is no reason to consider any one of these microstates either more or less likely to occur than any other (on this subject, see Rosa (1993)).

To conclude this introductory chapter, we notice that in the computer experiments performed with two dice (`Code_2_1.R`) (the term most used is *simulation* and more exactly *Monte Carlo simulation*), it is possible to experience the notion of probability. In other words, the computer is regarded as something like a 'statistical laboratory' with which probabilistic experiments can be performed, experiments not quite feasible in practice. We will encounter expressions like 'measurements', 'statistical and systematic errors', 'error propagation', and so on, just as in a laboratory experiment.

2.3 Exercises

Exercise 2.1 We have seen in `Code_2_1.R`, relative to the throw of two dice, that the estimated probability of the double six in 18000 games with 24 throws is $\widehat{P}\{A\}^{[24]} \approx 0.4887$, while with 25 throws it is $\widehat{P}\{A\}^{[25]} = 0.5113$, result practically equal to the theoretical one. The difference $\hat{d} = 0.5113 - 0.4887 = 0.0226$ resulted significant. Write a code to obtain Fig. 2.2, showing if the difference of the means is significant with the bootstrap method. Before you have to read *Appendix A* if you are not familiar with the bootstrap method.

Exercise 2.2 In the dice game Unders and Overs (U&O), two dice are rolled. Players bet on one of the following alternatives: (1) The result (sum of the dice faces) is below 7, (2) the result is 7, (3) the result is above 7.

In cases (1) and (3) the pay off odds are 1:1, i.e. if you bet £1 the house gives you back your money plus an additional £1. In case (2) the odds are 4:1, i.e. betting £1 you gain £4 (you get £5).

Suppose you bet at £1 on the outcome (1). What is your expected average win/loss (i.e. in an infinite number of throws)?

Exercise 2.3 Referring to the previous exercise, write a code to simulate a finite game consisting of 10, 100 or 1000 throws. Discuss the result of the simulations, compared to the theoretical win/loss expectation.
Hint: Use the R function `sample` *for sampling a die face, i.e. an integer number from 1:6.*

Exercise 2.4 A variant of the U&O game allows the player to bet up to two alternatives (placing £1 over each one). How does the win/loss expectation change?

Exercise 2.5 Justify the following assertion: 'the house always wins'. Hint: If you can bet £1 on each of the three alternatives, what is the expected outcome?

Exercise 2.6 With reference to exercise 2.5, compare the theoretical result for an infinite number of throws with those obtained in a small number of them, e.g. 10. Simulating the problem in R, in 100 repetitions how many times do you win and how many do you lose your money?

3
Introduction to Stochastic Processes

Noi corriamo sempre in una direzione,
ma qual sia e che senso abbia chi lo sa...

We are all headed in one direction,
but which it is and what sense it makes, who knows...

Francesco Guccini, Incontro

Probability theory is essential to the understanding of many processes (physical, chemical, biological, economic, etc.). By means of random variables, we build models of such processes, that is of systems that evolve over time. We are interested in what happens in the future. If we know the probability distribution until now, how will it be modified if carried forwards, through time? The answer is a matter of stochastic processes. For further reading many books are available on the subject (Feller, 1970; Lawler, 2006; Yates and Goodman, 2015; Jones and Smith, 2018; Grimmett, 2018).

3.1   Basic notion

In dictionaries of classical Greek, the word $\sigma\tau o\chi\acute{a}\zeta\varepsilon\sigma\vartheta\alpha\iota$ (*stochazesthai*) means 'to aim at something', 'to aim at a target, at a goal', at a $\sigma\tau o\kappa\acute{o}\varsigma$ (*stóchos*). Later, figuratively, 'to aim at something' becomes 'to have something in view', or 'to conjecture', from which $\sigma\tau o\chi\alpha\sigma\tau\iota\kappa\acute{o}\varsigma$ (*stochastikós*), 'skilful in aiming at', 'able to conjecture'. So the 'target' becomes the 'conjecture'. Conjecture of what? Of something below the apparent chance? Of undisclosed causes? Is there a hidden 'determinism' even in (seemingly) random phenomena? That is the question. The interested reader can refer to Chapter 2, where some historical answers we recalled succinctly.

A stochastic (or random) process is defined as a family of random variables:

$$X_1, X_2, \ldots, X_t, \ldots,$$

indexed by a parameter t, and defined on the same probability space $(\Omega, \mathcal{F}, \mathsf{P})$ formally defined as follows: Ω is the sample space, i.e. the space of all possible outcomes, $\mathcal{F}$ is a family of subsets of Ω, mathematically defined a σ-algebra, with particular properties (for example, that of including the whole sample space and all possible unions of subsets) that make Ω a measurable space. P is a probability measure function operating on $\mathcal{F}$, such that $\mathsf{P}(\Omega) = 1$ and $\mathsf{P}(\Phi) = 0$, Φ being the empty set. The index t often, but not always, stands for a time (days, years, seconds, nanoseconds, etc.). The 'time' can also be a non-physical time, as for instance 'Monte Carlo steps'.

A stochastic process is written as $\{X_t; t \in \mathbb{T}\}$. The set $\mathbb{T}$ is the *parametric space*, it can be a subset of natural numbers or integers, that is $\mathbb{T} = \{0, 1, 2, \dots\}$, or $\mathbb{T} = \{\dots, -2, -1, 0, -1, -2, \dots\}$, or $\mathbb{T} = \{0, 1, 2, \dots, n\}$. In these cases $\{X_t\}$ is said to be a *discrete*-time stochastic process. If $\mathbb{T}$ is the real line $\mathbb{R}$ or its subset, for instance $\mathbb{T} = (-\infty, \infty)$, or $\mathbb{T} = [0, \infty)$, or $\mathbb{T} = [a, b)$, or $\mathbb{T} = [a, b]$, $\{X_t\}$ is said to be a *continuous*-time stochastic process.

Discrete-time and continuous-time processes essentially differ in the time scale: in the former case events occur in a predetermined succession of time points $t_1, t_2, \dots$, in the latter events can occur at each time point t of a continuous range of possible values.

The name *stochastic process* refers therefore to two inherent aspects: the term 'process' refers to a time function; the adjective 'stochastic' refers to randomness, in the sense that a random variable is associated to each event in the time scale. In some cases the *stochastic process* can also be associated to space and not just time.

Time has an arrow. The process has therefore a before and an after, a past and a future. The realization x_t at time t of the random variable X_t is supposed to be closer to observations x_{t-1} and x_{t+1}, rather than to those farther in time. This means that the *chronological order* of observations plays an essential role.

We said that the X_t's are random variables. They are defined on the same probability space $(\Omega, \mathcal{F}, \mathsf{P})$. They take values in a measurable space, whose values are called *states*. We say that 'the process at time t is in the state x_i', or more simply, 'the process at t is in i', to mean that the random variable X_t has taken the value x_i. The set of all values taken by the variables of the process is called the *state space* and it will be denoted as $\mathcal{S}$. We can say that 'the *system* at time t is in the state i', or that it 'occupies' or 'visits' the state i, if $X_t = x_i$ for $x_i \in \mathcal{S}$.

A process is *discrete*, or is in discrete values, if $\mathcal{S}$ is discrete, that is if $\mathcal{S}$ is countable (finite or infinite): $\mathcal{S} \subseteq \mathbb{N}$ or $\mathcal{S} \subseteq \mathbb{Z}$. The process is *continuous*, or is in continuous values, if $\mathcal{S} \subseteq \mathbb{R}$. So that, stochastic processes may be classified into four types:

1. discrete-time and discrete state space,

2. discrete-time and continuous state space,

3. continuous-time and discrete state space,

4. continuous-time and continuous state space.

In other words, discrete or continuous time concerns the domain of the time variable t, while discrete or continuous state concerns the domain of X_t for a given t.

Let us take a practical example. We are interested in *continuously* recording the temporal variations of the temperature of a device. For technical reasons, the temperature does not remain constant, but floats within a certain range. Suppose the measurements are read on an analogue scale and, to be specific, suppose also that measurements are down to thousandths of a degree, with precision of the order of 1%. We can consider the measurements expressed in real numbers, even though, obviously, any measurement has a finite number of digits. With such premises, the sequence in time of the random variable 'temperature' can be represented as a continuous-time stochastic process in continuous values. If we decide to group the measurements within a range, say, of tenths of a degree, the process is still a continuous-type process, but in discrete values. If we group the data readings within a predefined range, for instance every 20 or 60 seconds, the process will be a discrete-time process, in (approximately) continuous values (thousandths of a degree) or discrete values (tenth of a degree)

Further classification concerns the dimension of Ω and $\mathbb{T}$. If the space Ω has dimension greater than 1, we refor to a *multivariate* stochastic process, as they are, for instance, space-time processes. All the variables X_t are defined in the same space Ω, then each X_t is a random function of two arguments of different nature: the variable of probabilistic nature $\omega \in \Omega$ indicates the event, the variable of mathematical nature $t \in \mathbb{T}$ creates an order in the random variables family. The stochastic process $\{X_t; t \in \mathbb{T}\}$, in more complete manner, should be written as:

$$\{X(\omega, t); \omega \in \Omega, t \in \mathbb{T}\}$$

to highlight that the particular realization of the stochastic process at time t depends on the particular event $\omega \in \Omega$.

Let us fix t, $t = \bar{t}$. Then $X_{\bar{t}}(\omega) \equiv X(\omega, \bar{t})$ is a random variable and, if the possible outcomes of the 'trial' are $\omega_1, \omega_2, \ldots$, the possible realizations of $X_{\bar{t}}(\omega)$ are given by:

$$X_{\bar{t}}(\omega_1) = x_1, X_{\bar{t}}(\omega_2) = x_2, \ldots,$$

where the subscript i ($i = 1, 2, \ldots$) of the x_i's numbers are the different possible realizations of the same random variable $X_t(\omega)$ at time $t = \bar{t}$.

It is possible to regard $X(t, \omega)$ as a function of t, fixed $\omega = \bar{\omega}$ in Ω, so we have $X_t(\bar{\omega})$. In this case, we consider a particular outcome at times $t = t_1, t = t_2, \ldots$, that is:

$$X_{t_1}(\bar{\omega}_1) = x_1, X_{t_2}(\bar{\omega}_2) = x_2, \ldots, \tag{3.1}$$

where the subscripts i ($i = 1, 2, \ldots$) of the x_i's number are the time points t_1 at which the event $\bar{\omega}_1$ has occurred, t_2 at which the event $\bar{\omega}_2$, has occurred, etc. In this case, for each fixed $\bar{\omega}$, the sequence $(x_1, x_2, \ldots)$ is called *realization* or *history* or *sample path* or *trajectory* of the process. All the possible sample paths resulting from an experiment constitute an *ensemble*. Therefore a stochastic process can be regarded as formed by the wholeness of all its possible realisations. A finite portion of a realization is called *time series*:

$$(\ldots, \underbrace{x_{k+1}, x_{k+2}, \ldots, x_{k+t}, \ldots, x_{k+n}}_{\text{time series}}, \ldots) \tag{3.2}$$

So, to recap, $X_t(\omega)$ means, depending on the context:

1. $X_t(\omega)$ (t and ω variables): a family of time dependent real-valued functions, that is a stochastic process.
2. $X_{\bar{t}}(\omega)$ (t constant and ω variable): a random variable, that is, for definition, a measurable function defined on a probability space.
3. $X_t(\bar{\omega})$ (t variable and ω constant): a single mathematical function depending on time.
4. $X_{\bar{t}}(\bar{\omega})$ (t constant and ω constant): a real number.

Coming back to the example of the temperature measurement of a device, suppose that the measurements are taken every minute and we round up the records to tenths of a degree. The stochastic process, modelling the time variation of the temperature, is therefore a discrete-valued process in discrete time. Suppose we have carried out the measurements for one hour on 30 similar devices ($k = 1, \ldots, 30$).

The records of the measurements at every minute will not be the same for the 30 devices, therefore the records will be displayed as in Table 3.1.

Table 3.1 Temperature vs time dependence of 30 devices.

k \ t	1	2	3	4	...	59	60
1	17.4	21.1	13.9	19.4	...	18.7	19.6
2	18.4	16.3	17.1	17.8	...	14.4	15.0
3	20.7	16.0	19.5	12.1	...	15.6	19.3
...	...	...	...	...	...	...	...
29	16.3	14.8	11.5	17.2	...	19.7	19.4
30	21.1	18.1	20.1	26.7	...	14.7	15.1

If t is fixed, for instance, $t = \bar{t} = 3$, the corresponding column shows the temperature of the 30 devices at time $\bar{t}$. The values $x_1 = 13.9$, $x_2 = 17.1, \ldots$, $x_{30} = 20.1$ are then 30 realizations of the random variable $X_{\bar{t}}(\omega)$. If ω is fixed, it means picking a device and following its change in temperature over time. For instance, the ensemble $\{18.4, 16.3, 17.1, 17.8 \ldots, 14.4, 15.0\}$ shows a possible sample path corresponding to the device $k = 2$. Figure 3.1 shows the temperature as a function of time for the first four devices and it represents four realizations, or time series, of a discrete-valued stochastic process in discrete time.

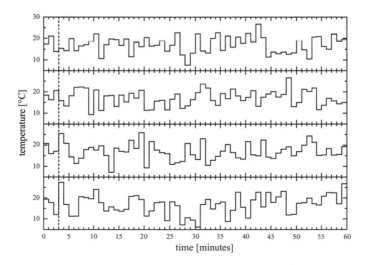

Fig. 3.1 Realizations of a discrete-valued stochastic process in discrete time. Dashed line: devices temperature at time $\bar{t} = 3$. The lines between points are to guide the eye.

Let us consider some well-known stochastic processes, based on the *Bernoulli process*, i.e. on a sequence of independent and identically distributed, *i.i.d.*, random variables:

$$X_1, X_2, \ldots, X_i, \ldots, \text{ with } X_i \sim Bern(p)$$

This sequences is called a *Bernoulli process* $\{X_t; t \in \mathbb{N}\}$ if the X_i's are independent of each other and $\forall i \in \mathbb{N}, \mathsf{P}\{X_i = 1\} = p$ and $\mathsf{P}\{X_i = 0\} = 1 - p$. It is a discrete-valued stochastic process in discrete time. The index set is the set of natural numbers $\mathbb{T} = \mathbb{N}$, or $\mathbb{T} = \mathbb{N} \smallsetminus 0)$, and the state space $\mathcal{S}$ is the set $\{0, 1\}$.

Define the random variable sequence:

$$S_t = \sum_{i=1}^{t} X_i$$

where $\{X_t; t \in [1, \infty)\}$ is the Bernoulli process. The sequence $\{S_t; t \in [1, \infty)\}$ is also a discrete-valued stochastic process in discrete time, with $\mathcal{S} = \mathbb{N}$.

Consider the random variable sequences:

$$Y_t = \frac{S_t}{t} = \frac{1}{t} \sum_{i=1}^{t} X_i$$

The sequence $\{Y_t; t \in [1, \infty)\}$ is a discrete-time stochastic process, but with continuous values, that is $\mathcal{S} = \mathbb{R}$. We also know, from the law of large numbers for binomial random variables, that the sequence $\{Y_t\}$ converges in probability to $\mathrm{E}\,[X_i] = p$.

Some general considerations about stochastic processes are now defined. A first question is: what does it mean to have a complete knowledge of a process $\{X_t; t \in \mathbb{T}\}$? We have a complete knowledge of a random variable X, and we say that X is known when we know its repartition (or cumulative distribution) function $F(x) = \mathsf{P}\{X \leqslant x\}, \forall x \in \mathbb{R}$.

As a consequence, a stochastic process is known when the repartition function of each variable of the family is known, that is all $F_t(x) = \mathsf{P}\{X_t \leqslant x\}, \forall x \in \mathbb{R}$ and $\forall t \in \mathbb{T}$. But this is not enough. We have to know all the joint probability distributions of the variables X_t, that is all the double cumulative functions:

$$F_{t_1, t_2}(x_1, x_2) = \mathsf{P}\{X_{t_1} \leqslant x_1 \cap X_{t_2} \leqslant x_2\}, \forall x_1, x_2 \in \mathbb{R} \text{ and } \forall t_1, t_2 \in \mathbb{T}$$

and, in general, all the $n-$tuple:

$$F_{t_1, t_2, \ldots, t_n}(x_1, x_2, \ldots, x_n) =$$
$$\mathsf{P}\{X_{t_1} \leqslant x_1 \cap X_{t_2} \leqslant x_2, \ldots, \cap X_{t_n} \leqslant x_n\},$$
$$\forall x_1, x_2, \ldots, x_n \in \mathbb{R} \text{ and } \forall t_1, t_2, \ldots, \in \mathbb{T} \quad (3.3)$$

The family of functions $F_{t_1, t_2, \ldots, t_n}(x_1, x_2, \ldots, x_n)$ is called the *temporal law* of the process. For a finite dimension family and under certain precise conditions the probabilistic structure of the complete process may be specified. Such a formidable theoretical problem was faced by Kolmogorov in the first half of the 20th century (Kolmogorov, 1950).

For practical purposes, it is advantageous to search for values summarizing the main properties of the process. Such summaries are the finite–order moments, in particular the first- and second-order moments. Suppose that such moments exist for each X_t. The first-order moment, the *expected value*, is the mean of each X_t, defined as:

$$\mu_t = \mathrm{E}\left[X_t\right]$$

which, in general, is different for each t. The second central moment, or autocovariance function, at the lag k, is defined as:

$$\gamma_{t,t-k} = \mathrm{Cov}\left[X_t, X_{t-k}\right] = \mathrm{E}\left[(X_t - \mu_t)(X_{t-k} - \mu_{t-k})\right] \tag{3.4}$$

The autocovariance function represents the covariance between the random variable X_t and X_{t-k}, $k = 0, 1, 2, \ldots$, that is the covariance of the process with itself at pairs of time points. This function measures the joint variation of X_t and X_{t-k}, either in the same direction (positive values of $\gamma_{t,t-k}$) or in the opposite direction (negative values of $\gamma_{t,t-k}$), at the time points $k = 0, 1, 2, \ldots$.

Remark 3.1 *In bivariate, or generally multivariate processes, covariances are named 'cross-covariances', when the dependence of one process over another (or more than one) is investigated. For instance, for two processes $\{X_t\}$ and $\{Y_t\}$, the cross-covariance is given by:*

$$\gamma_{xy}(t, t - k) = \mathrm{E}\left[(X_t - \mu_x(t))\left(Y_{t-k} - \mu_y(t - k)\right)\right] \tag{3.5}$$

with a slight change of symbols to indicate a bivariate process.

If $k = 0$, from eqn (3.4):

$$\gamma_t = \mathrm{Var}\left[X_t\right] = \mathrm{E}\left[(X_t - \mu_t)^2\right] \tag{3.6}$$

is the variance of the process, also denoted by σ_t^2.

The *(linear) autocorrelation function* $\rho_{t,t-k}$ at the lag k, is given by normalizing the autocovariance $\gamma_{t,t-k}$, eqn (3.4), to the variance (3.6):

$$\rho_{t,t-k} = \frac{\mathrm{Cov}\left[X_t, X_{t-k}\right]}{\sqrt{\mathrm{Var}\left[X_t\right]\mathrm{Var}\left[X_{t-k}\right]}} = \frac{\gamma_{t,t-k}}{\sqrt{\gamma_t\,\gamma_{t-k}}} \tag{3.7}$$

$$= \frac{\mathrm{E}\left[(X_t - \mu_t)\left(X_{t-k} - \mu_{t-k}\right)\right]}{\sqrt{\mathrm{E}\left[(X_t - \mu_t)\right]^2\,\mathrm{E}\left[(X_{t-k} - \mu_{t-k})\right]^2}} \tag{3.8}$$

The function $\rho_{t,t-k}$ is dimensionless and does not vary by interchanging X_t and X_{t-k}. It reaches its maximum when $k = 0$. If X_t and X_{t-k} are linearly independent $\rho_{t,t-k}$ is equal to 0, while it is $+1$ or -1, in presence of a perfect correlation (there is an exact overlapping when time is shifted by k) or a perfect anti-correlation, respectively. The quantities $\rho_{t,t-k}$ and $\gamma_{t,t-k}$ are essential to characterize stochastic processes; indeed, they measure the internal structure of processes and their memory.

3.1.1   Stationary processes

It was shown that moments depend on time and on the joint distribution of X_t and X_{t-k}. This makes inferential applications very limited because, usually, only a time series is available, that is a single and limited realization of the process: at each t, only one realization of X_t is available. At every t, X_t varies, causing μ_t and the correlation between X_t and X_{t-k} to vary as well. One question is whether the available series properties persist through time, or they belong only to the time interval where the series has been observed. Furthermore, are the observed characteristics specific to a particular realization or do they belong also to other realizations of the same process?

If we have to make valid inferences on the properties of the X_t's, for instance on their moments and their transformations, but with only one available observation, we have, so to speak, to 'approach' the most usual statistical situation: the analysis of a sequence of *i.i.d.* random variables.

The requirement that the X_t distributions have to be 'not too much' different in time, leads to specify the conditions to define a *stationary process*. This is a clear restriction on the heterogeneity of series and conforms to the condition of identically distributed random variables. The requirement that the X_t have to be 'not too much' dependent on one another, when they are quite far in time, leads to the notion of *ergodicity*. In this case, the restriction is on the memory of the series and conforms to the condition of independent random variables. Figure 3.2 schematically summarizes the above reasoning.

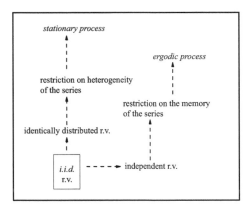

Fig. 3.2 From *i.i.d.* random variables (*r.v.*) to stationarity and ergodicity properties.

We can write schematically:

$$\text{identical distributed r.v.} \quad \xrightarrow{\text{restriction on the heterogeneity}} \quad \textit{stationarity}$$

and

$$\text{independent r.v.} \quad \xrightarrow{\text{restriction on the memory}} \quad \textit{ergodicity}$$

Stationarity is defined at different degrees. The process $\{X_t\}$ is *strictly stationary, strongly stationary* or *strict-sense stationary* when the joint distribution of

$$X_t, X_{t+1}, \ldots, X_{t+h} \tag{3.9}$$

is the same for every $t \in \mathbb{T}$ and for every time shift $h = 0, 1, 2, \ldots$. That is:

$$\begin{aligned}\mathsf{P}\left\{X_{t_1} \leqslant x_1, X_{t_2} \leqslant x_2, \ldots, X_{t_n} \leqslant x_n\right\} = \\ \mathsf{P}\left\{X_{t_1+h} \leqslant x_1, X_{t_2+h} \leqslant x_2, \ldots, X_{t_n+h} \leqslant x_n\right\}\end{aligned} \tag{3.10}$$

or in different notation:

$$F_{t_1,t_2,\ldots,t_n}(x_1, x_2, \ldots, x_n) = F_{t_1+h,t_2+h,\ldots,t_n+h}(x_1, x_2, \ldots, x_n)$$

for every $t_1, t_2, \ldots, t_n$, for every $x_1, x_2, \ldots, x_n$, $\forall n$ and $\forall h$.

Intuitively, if we imagine dividing the process into time chunks, all the pieces are 'statistically similar', in the sense that statistical properties do not vary over time, and there is no preferred choice of start time. Under such conditions, all the moments, note: *if they exist*, do not vary over time. This condition, as we said above, reflects that of identical distributions. Indeed, if $n = 1$ in eqn (3.10), all the one–dimensional repartition functions become identical to each other: $F_t(x) = \mathsf{P}\left\{X_t \leqslant x\right\} = F_{t+h}(x) = \mathsf{P}\left\{X_{t+h} \leqslant x\right\}$, for all t and h. If $n = 2$, all the two-dimensional repartition functions $F_{t_1,t_2}(x_1, x_2) = \mathsf{P}\left\{X_{t_1} \leqslant x_1, X_{t_2} \leqslant x_2\right\}$ are dependent only on $(t_2 - t_1)$, but not specifically on t_1 and t_2. Stationarity up to order m is also defined, if the joint probability distribution of the $X_t, X_{t+1}, \ldots, X_{t+h}$ has finite and equal moments up to order m.

Consider $m = 2$. In that case the stationarity is named *weak stationarity*, or *covariance-stationarity*, or *wide-sense stationary*, or simply *stationary*. If a stochastic process is weakly stationary, the first and second moments exist and do not vary throughout the time. The second cross moment $\mathrm{E}\left[X_t \cdot X_{t+k}\right]$, $\forall k$ may depend on k. The expected value is finite and constant, at all time points:

$$\mu_t = \mathrm{E}\left[X_t\right] = \mu < \infty, \ \forall t$$

Variance and autocovariance are finite and constant, at all time points:

$$\mathrm{Var}\left[(X_t)\right] = \sigma_t^2 \equiv \gamma_t = \sigma^2 \equiv \gamma < \infty, \ \forall t$$

and

$$\gamma_{t,t-k} = \mathrm{E}\left[(X_t - \mu)(X_{t-k} - \mu)\right] = \gamma_k < \infty, \ \forall t$$

Note that γ_k does not depend on t, but only on k, i.e. $\gamma(k)$ is a function of one variable, the time difference k.

To give greater emphasis to this property, the autocovariance may also be written:

$$\gamma(|t - s|) = \mathrm{E}\left[(X_t - \mu)(X_s - \mu)\right]$$

k being the time lag between t and s. If $k = 0$, $\gamma_0 = \sigma^2$. Note that $\gamma_k = \gamma_{-k}$.

The autocorrelation function becomes:

$$\rho_k = \frac{\gamma_k}{\gamma_0} = \frac{\gamma_k}{\sigma^2}$$

clearly $\rho_0 = 1$.

Strong stationarity does not necessarily imply weak stationarity, since the definition of strong stationarity does not assume the existence of finite second moments. However, if the process has finite second moments, then the implication is valid. On the contrary, the time constancy up to the second moment, does not imply that the distribution of marginal random variables X_t is the same for all t. In one case the two definitions, strong and weak, overlap: when the the process is normal (Gaussian), that is when for every n-tuple $(t_1, t_2, \ldots, t_n)$ the joint distribution of $(X_{t_1}, X_{t_2}, \ldots, X_{t_n})$ is n-tuple variate, for every n. Not-stationary processes (in any sense) are not 'less important' than the stationary ones. The Poisson process, for instance (see next section) continuous time process in discrete values, is not-stationary in any sense, since $\mu(t) = \lambda t$, (λ is a constant).

3.1.2   Ergodic processes

We have that stationarity conditions contribute to reducing the heterogeneity degree of the process, but are they enough to estimate moments of interest? Can we estimate μ, σ, γ_k and ρ_k from the available observations, i.e. from time series? We are interested, for example, in obtaining an estimate $\hat{\mu}$ of the mean value μ of the stochastic process $\{X_t; t \in \mathbb{T}\}$.

Suppose we have a time−discrete process in discrete values that is also weak stationary. One could reasonably assume that, as μ is the same for every X_t, the average can be computed at time t' by considering m realizations of $X_{t'}$. Unfortunately, only one realization of $X_{t'}$ is available at t', the observation $x_{t'}$. But in return, n observations taken in a time window of the process are available. Call $t = 1$ the time point at the starting time, so we have the time series $x_1, x_2, \ldots, x_n$, where x_1 is the realization of the variable X_1, at time point $t = 1$, x_2 is the realization of the variable X_2, at time point $t = 2$, $\ldots$, x_n is the realization of the variable X_n, at time point $t = n$. Compute the arithmetic mean $\bar{x}$, namely the average of the sample of n observations $x_1, x_2, \ldots, x_n$ of the same trajectory:

$$\bar{x} = \frac{1}{n} \sum_{i=1}^{n} x_i$$

What can we say about $\bar{x}$? This value is the realization of the random variable $\overline{X}_n$, named *time average*, defined as:

$$\overline{X}_n = \frac{1}{n} \sum_{i=1}^{n} X_i \leftarrow \text{time}$$

Can we use the random variable $\overline{X}_n$, time average, to estimate the process mean μ? It depends. First of all, we require that such random variable $\overline{X}_n$ must be an *unbiased* (or correct) estimator of μ, that is $\mathrm{E}\left[\overline{X}_n\right] = \mu$. Remember that an estimator is unbiased if its expected value is equal to the value of the quantity (parameter) to be estimated. The answer is affirmative Indeed, from stationarity:

$$\mathrm{E}\left[\overline{X}_n\right] = \mathrm{E}\left[\frac{1}{n} \sum_{i=1}^{n} X_i\right] = \frac{1}{n} \sum_{i=1}^{n} \mathrm{E}\left[X_i\right] = \mu$$

Unbiased is a required property of estimators, as is consistency.

Recall that an estimator Θ_n (the subscript n to indicate its dependence on sample size n) is *consistent* of the parameter θ if, as the sample size n increases, it converges in probability to the value of the parameter to be estimated, that is:

$$\Theta_n \xrightarrow[n\to\infty]{p} \theta$$

More precisely, the above is the definition of *weak consistency*, the *strong consistency* requiring the *almost sure convergence* to the parameter θ. If the estimator is unbiased, namely $\mathrm{E}\big[\Theta_n\big] = \theta$, it is more convenient to study its convergence in quadratic mean (*qm*). We have:

$$\mathrm{E}\left[(\Theta_n - \theta)^2\right] = \mathrm{E}\left[(\Theta_n - \mathrm{E}\left[\Theta_n\right])^2\right] = \mathrm{Var}\left[\Theta_n\right]$$

If $\mathrm{Var}\left[\Theta_n\right] \to 0$, it follows:

$$\Theta_n \xrightarrow[n\to\infty]{qm} \theta \qquad \text{and then also} \qquad \Theta_n \xrightarrow[n\to\infty]{p} \theta$$

that is the estimator Θ_n is consistent.

Remembering the asymptotic unbiased of estimates:

$$\mathrm{E}\left[\Theta_n\right] \xrightarrow[n\to\infty]{p} \theta$$

an estimator is weakly consistent if it is asymptotic unbiased, namely when it converges in distribution to the parameter to be estimated. If it convergences to a constant, as in the present case, the two notions of 'convergence in probability' and 'convergence in distribution' are equivalent.

In our case, it must be:

$$\overline{X}_n = \frac{1}{n}\sum_{i=1}^{n} X_i \xrightarrow[n\to\infty]{p} \mu \tag{3.11}$$

We have to ensure if:

$$\mathrm{Var}\left[\overline{X}_n\right] = \mathrm{E}\left[(\overline{X}_n - \mu)^2\right] \xrightarrow[n\to\infty]{} 0$$

Now:

$$\mathrm{Var}\left[\overline{X}_n\right] = \mathrm{E}\left[\left(\frac{1}{n}\sum_{i=1}^{n} X_i - \mu\right)^2\right] = \frac{\sigma^2}{n} + \frac{2}{n^2}\sum_{k=1}^{n-1}(n-k)\gamma_k \tag{3.12}$$

Note that $\mathrm{Var}\left[\overline{X}_n\right]$ is enhanced by the second term. From eqn (3.12), then also:

$$\frac{1}{n^2}\sum_{k=1}^{n-1}(n-k) = \sum_{k=1}^{n-1}\frac{n-k}{n^2} = \frac{1}{2} - \frac{1}{2n}$$

For the last sum in eqn (3.12) to converge to 0, with increasing n, it is necessary that autocovariances have to converge to 0, with increasing time interval between the X_i's.

Autocovariance, we said, is a measure of the interdependence between time points of elements of the process. Therefore, if such a sum does not converge to 0, it means that memory persists for very long times. As a consequence, even though the process is stationary, the mean estimator computed by observed trajectories is not consistent, since its mean square error does not converge to 0 with increasing sample size. This behaviour, in the context of the Monte Carlo method is what once was called 'statistical inefficiency' (Friedberg and Cameron, 1970), namely the variance inflation due to time correlations. It will be discussed again in Chapter 8 for the Monte Carlo method.

On the contrary, as time grows − i.e. with increasing the number of observations − we require the time average to converge to the so-called *ensemble average*. In other words, it is as if we were observing a large number of trajectories at a given time instant, although we actually have only a single trajectory at hand. In the following a step-by-step derivation of eqn (3.12) is reported.

Proof Consider in general the random variable X, linear combination of n random variables: $X = a_1 X_1 + a_2 X_2 + \cdots + a_n X_n$. The variance of X is given by:

$$\text{Var}\,[X] = \text{Var}\left[\sum_{i=1}^{n} a_i X_i\right] = \sum_{i=1}^{n} a_i^2 \text{Var}\,[X_i] + 2\sum_{i=1}^{n}\sum_{j=i+1}^{n} a_i a_j \text{Cov}\,[X_i, X_j] \tag{3.13}$$

By putting:

$$\sigma_{ij} = \text{Cov}\,[X_i, X_j], \quad (\text{for } i = j \text{ it is } \sigma_{ii} = \text{Var}\,[X_i])$$

it results in:

$$\text{Var}\,[X] = \text{Var}\left[\sum_{i=1}^{n} a_i X_i\right] = \sum_{i=1}^{n} a_i^2 \sigma_{ii} + 2\sum_{i=1}^{n}\sum_{j=i+1}^{n} a_i a_j \sigma_{ij} \tag{3.14}$$

Let us consider $\text{Var}\left[\overline{X}_n\right]$ and, to follow more easily the reasoning, let us take $n = 3$. So we have (dropping the subscript n for convenience):

$$\overline{X} = \frac{1}{3}\sum_{i=1}^{3} X_i \quad \text{written as} \quad = \frac{X_1}{3} + \frac{X_2}{3} + \frac{X_3}{3}$$

so that $\text{Var}\left[\overline{X}\right]$ becomes for eqn (3.13):

$$\text{Var}\left[\overline{X}\right] = \frac{\text{Var}\,[X_1]}{3^2} + \frac{\text{Var}\,[X_2]}{3^2} + \frac{\text{Var}\,[X_3]}{3^2} +$$

$$+ 2\left[\frac{1}{3}\frac{1}{3}\text{Cov}\,[X_1, X_2] + \frac{1}{3}\frac{1}{3}\text{Cov}\,[X_1, X_3] + \frac{1}{3}\frac{1}{3}\text{Cov}\,[X_2, X_3]\right]$$

Since it is $\sigma_{ij} = \text{Cov}\,[X_i, X_j]$ and $\sigma_{ii} = \text{Var}\,[X_i]$, we can write:

$$\text{Var}\left[\overline{X}\right] = \frac{1}{3^2}\,\sigma_{11} + \frac{1}{3^2}\,\sigma_{22} + \frac{1}{3^2}\,\sigma_{33} +$$

$$+ \frac{1}{3^2}\,\sigma_{12} + \frac{1}{3^2}\,\sigma_{12} + \frac{1}{3^2}\,\sigma_{13} + \frac{1}{3^2}\,\sigma_{13} + \frac{1}{3^2}\,\sigma_{23} + \frac{1}{3^2}\,\sigma_{23}$$

We have written in this way to allow us to understand the term '2' in eqn (3.14). In a compact form, we write:

$$\text{Var}\left[\overline{X}\right] = \frac{1}{3^2}\sum_{i=1}^{3}\sum_{j=1}^{3}\sigma_{ij}$$

Indeed:

for $i=1$ we have $\sigma_{11}, \sigma_{12}, \sigma_{13}$

for $i=2$ we have $\sigma_{21}, \sigma_{22}, \sigma_{23}$,   but $\sigma_{21} = \sigma_{12}$

for $i=3$ we have $\sigma_{31}, \sigma_{32}, \sigma_{33}$,   but $\sigma_{31} = \sigma_{13}$, $\sigma_{32} = \sigma_{23}$

Then, in general:

$$\text{Var}\left[\overline{X}\right] = \frac{1}{n^2}\sum_{i=1}^{n}\sigma_{ii} + \frac{2}{n^2}\sum_{i=1}^{n}\sum_{j=i+1}^{n}\sigma_{ij} = \frac{1}{n^2}\sum_{i=1}^{n}\sum_{j=1}^{n}\sigma_{ij} \tag{3.15}$$

Since the process is stationary, we have $\text{Var}\left[X_t\right] = \sigma^2$, $\forall t$, that is to say $\sigma_{ii} = \sigma^2$, $\forall i$ ($\sigma_{11} = \sigma_{22} = \cdots = \sigma_{nn} = \sigma^2$). So the part with σ_{ii} becomes:

$$\frac{1}{n^2}\sum_{i=1}^{n}\sigma_{ii} = \frac{1}{n^2}n\sigma^2 = \frac{\sigma^2}{n}$$

Let us look at the covariance term σ_{ij}. The autocorrelation function in the stationary case is $\rho_k = \gamma_k/\sigma^2$. For a continuous process, formally we can write $\rho(k) = \gamma(k)/\sigma^2$ or also $\gamma(k) = \sigma^2\,\rho(k)$. Now $\gamma(k)$ is the autocovariance at lag k, that is the covariance between times i and j, before written as σ_{ij}. The autocorrelation function $\rho(k)$ can be written as $\rho(|i-j|)$. Still remaining in the example with $n=3$, if $t=1,2,3$, it is $k=0,1,2$. Consider the terms with σ_{ij}:

$$\frac{2}{n^2}\sigma_{12} + \frac{2}{n^2}\sigma_{13} + \frac{2}{n^2}\sigma_{23} = \frac{2}{n^2}\rho(2-1) + \frac{2}{n^2}\rho(3-1) + \frac{2}{n^2}\rho(3-2)$$

which can be summarized as follows:

$$\sum_{j=2}^{3}\sum_{i=1}^{j-1}\rho(j-i)\quad\text{and in general}\quad\sum_{j=2}^{n}\sum_{i=1}^{j-1}\rho(j-i),\ j>i$$

From eqn (3.15), we can write $\text{Var}\left[\overline{X}\right]$ as:

$$\text{Var}\left[\overline{X}\right] = \frac{1}{n^2}\sum_{i=1}^{n}\sum_{j=1}^{n}\sigma_{ij} = \frac{\sigma^2}{n^2}\sum_{i=1}^{n}\sum_{j=1}^{n}\rho(j-i)$$

which always holds, even though observations are taken at not-regular time intervals. Rewrite the summations with $n=3$. Let $j-i=k$ be, it is:

j	i	k		
2	1	1	$\longrightarrow$	$\rho(1)$
3	1	2	$\longrightarrow$	$\rho(2)$
3	2	1	$\longrightarrow$	$\rho(1)$

The sum of the ρ's can be written as $2\rho(1) + 1\rho(2)$. In general:

$$\sum_{i=1}^{n}\sum_{j=1}^{n}\rho(j-i) = \sum_{k=1}^{n-1}(n-k)\rho(k)$$

Eventually $\text{Var}\left[\overline{X}_n\right]$ (retaining the n index) is the sum of two terms: the first is σ^2/n, the second is $2\sigma^2/n^2$, multiplied by the sums on the $\rho(k)$'s:

$$\text{Var}\left[\overline{X}_n\right] = \frac{\sigma^2}{n}\left[1 + \underbrace{\frac{2}{n}\sum_{k=1}^{n-1}(n-k)\rho(k)}_{\text{inflation term}}\right]$$

This equation is the same as eqn (3.12), in which $\gamma_k = \rho_k\gamma_0$, with $\gamma_0 = \sigma^2$, and, for a continuous-time process, we have written $\rho(k)$ instead of ρ_k. □

We have to take into consideration the notion of *ergodicity*, whose discussion can be introduced from various points of view.

To maintain the exposition at a rather intuitive level, let us return to Table 3.1. If we consider the arithmetic average of the values in the $k = 2$ raw, that is $(18.4+16.3+17.1+17.8+\cdots+14.4+15.0)/60$ we get the time average $\bar{x} = 17.19$. If we consider the arithmetic average of the values in the $t = 3$ column (dashed line in Fig. 3.1), that is $13.9+17.1+19.5+\cdots+11.5+20.1)/30$ we get an estimate $\hat{\mu} = 16.96$ of the expected value μ of the process. Note we say 'estimate' since the 30 considered values are supposed to be a sample extracted from all the possible realizations of $X_{t=3}$. To ascertain if the two averages are not significantly different, it is necessary to know their variability – not a simple task.

Usually, in real situations, only the time average $\bar{x}$ is available, but we expect to estimate μ from it, without computing ensemble averages. If, for a long enough time, the process takes a large number of possible values, or, in other words, it visits a large number of possible states, a 'time sample' of these observed values is equivalent to a sample extracted from all the possible available values.

So far we have considered the mean, but analogous reasoning holds also for the second moments. In the following, we report some results. In general, a process is ergodic with respect to the mean value $\text{E}\left[f(X_t)\right]$ of a function $f(X_t)$ (also mean-ergodic), if the time average of the $f(X_t)$'s converges in probability to the mean value of the function, namely:

$$\underbrace{\frac{1}{n}\sum_{i=1}^{n}f(X_i)}_{\text{time average}} \xrightarrow[n\to\infty]{p} \underbrace{\text{E}\left[f(X_t)\right]}_{\text{ensemble average}}$$

The left-hand side term is a time average, while the right-hand side is an ensemble average, or 'spatial' average, with reference to the state space.

In particular, for the mean, eqn (3.11) holds:

$$\overline{X}_n = \frac{1}{n}\sum_{i=1}^{n} X_i \xrightarrow[n\to\infty]{p} \mu$$

while for the variance, it must be:

$$\frac{1}{n}\sum_{i=1}^{n}(X_i - \mu)^2 \xrightarrow[n\to\infty]{p} \sigma^2$$

and for the covariance:

$$\frac{1}{n}\sum_{i=j+1}^{n}(X_i - \mu)(X_{i-j} - \mu) \xrightarrow[n\to\infty]{p} \gamma_j$$

Note that we have to specify which particular moment (mean, variance, etc.) ergodicity refers to. Clearly, if a process is ergodic, it is also stationary, but not vice versa. The Slutski ergodic theorem (Grimmett and Stirzaker, 2001) states a necessary and sufficient condition for the mean-ergodicity, namely the validity of eqn (3.11). The theorem says that a stationary stochastic process $\{X_t\}$ is ergodic with respect to the mean if and only if:

$$\lim_{n\to\infty} \frac{\sum_{i=1}^{n}\gamma(t, t-k)}{n} = 0$$

which holds (sufficient, but not necessary, condition) if $\gamma(t, t-k) \to 0$, for $k \to \infty$. The underlying idea is that, with the time interval increasing, the information contained in x_t is not connected to that contained in x_{t+k}.

A brief account of ergodic theory is reported in Appendix A of Huffaker *et al.* (2017), here we recall that the word 'ergodic' derives from the Greek $\epsilon\rho\gamma o\nu$, that is *ergon*, work, energy, and *οδόσ*, that is *odon*, route, path. The term was introduced by Boltzmann in statistical mechanics to describe the 'work trajectory' covered in the phase space by the representative point of the system. The system is ergodic if its work trajectory visits, sooner or later, all the points of the phase space allowed by energy constraints. This hypothesis was later called the *ergodic hypothesis*.

3.2 Markov processes

A *Markov process* can be defined as a stochastic process whose state space has the property of being 'past-forgetting', one of the possible practical definitions of the 'markovian property'. Conversationally, we say that the process is a Markov process if the future does not depend on the past, but only on the present instant. The verb 'to depend' has to be interpreted in probabilistic terms.

Let us consider the process $\{X(t)\}$. We know that in this instant t the process is in the state e, that is we know the event 'Present': $\{e\} = \{X(t) = e\}$. In the past, namely at times $s_1, s_2, \ldots, s_n$, the process was in states $d_1, d_2, \ldots, d_n$. Let us call 'Past' the set of events $\mathscr{P}$, defined as:

$$\mathscr{P} = \{X(s_1) = d_1, X(s_2) = d_2, \ldots X(s_n) = d_n\}$$

As we said, it doesn't matter if the 'Past' is known or not. In a similar way, we call 'Future' the set of events $\mathscr{F}$:

$$\mathscr{F} = \{X(u_1) = f_1, X(u_2) = f_2, \ldots X(u_m) = f_m\}$$

then, in the future, that is at times $u_1, u_2, \ldots, u_m$, the process is in states $f_1, f_2, \ldots, f_m$. The Markovian condition states that the *probability* that the process, at a certain time $u_i > t, u_i \in \mathbb{T}$ ($\mathbb{T}$ is the parametric space), will be in a state f_i (with $i = 1, \ldots, m$) – and such an event is the realization of an event belonging to $\mathscr{F}$ – does not depend on (it is not conditioned by) where the process was in the past, namely it does not depend on the events $\{X(s_i) = d_i\} \in \mathscr{P}$ (con $i = 1, \ldots, n$) which occurred before t. This probability depends only on (is conditioned by) the state in which the process is at the instant t, that is on the precise event $\{e\}$ that occurs *hic et nunc*, at the time t. In short:

$$\mathsf{P}\{\mathscr{F}|e, \mathscr{P}\} = \mathsf{P}\{\mathscr{F}|e\} \tag{3.16}$$

Since there is no possibility of confusion, the braces in $\{e\}$ have been removed. The above relation defines the Markov process. If a stochastic process meets this relation, then the process is a Markov process. If a stochastic process is a Markov process, then it must meet the above relation.

Markov processes are memoryless, or better: they have no ancestral memory, but preserve a memory only of the state where they are at the present time t. It is worth noting that eqn (3.16) is the probabilistic analogue of a deterministic law of mechanics. For instance, the future space coordinates of a satellite are completely determined by the equations of classical physics. If I know its state at time t, I can predict exactly (better: to an appropriate accuracy) its time evolution, no matter what its state was before t. Analogously, for a Markovian process, if we know the state of the system at time t, we can predict, not with certainty, but with perfectly computable probability, the future visited states, no matter of its past history.

In the following, as is often used in Markov chains literature, we write $\mathsf{P}\{A, B\}$, with the comma in place of the symbol $\cap$. Recall that, given the events A, B, C, if $\mathsf{P}\{B, C\} > 0$, we have:

$$\mathsf{P}\{A|B, C\} = \frac{\mathsf{P}\{A, B, C\}}{\mathsf{P}\{B, C\}} = \frac{\mathsf{P}\{A, B, C\}}{\mathsf{P}\{B|C\}\,\mathsf{P}\{C\}} = \frac{\mathsf{P}\{A, B|C\}}{\mathsf{P}\{B|C\}}$$

We aim to show that:

$$\mathsf{P}\{\mathscr{P}|\mathscr{F}, e\} = \mathsf{P}\{\mathscr{P}|e\} \tag{3.17}$$

Starting with:

$$\underset{A}{\mathsf{P}\{\mathscr{F}}, \underset{B}{\mathscr{P}}|\underset{C}{e}\} = \mathsf{P}\{\mathscr{F}|e, \mathscr{P}\}\,\mathsf{P}\{\mathscr{P}|e\}$$

and for eqn (3.16):

$$P\left\{\mathscr{F}, \mathscr{P} | e\right\} = P\left\{\mathscr{F} | e\right\} P\left\{\mathscr{P} | e\right\}$$

now:

$$P\left\{\mathscr{P} | \mathscr{F}, e\right\} = \frac{P\left\{\mathscr{F}, \mathscr{P} | e\right\}}{P\left\{\mathscr{F} | e\right\}} = P\left\{\mathscr{P} | e\right\}$$

That is eqn (3.17). Equation (3.17), symmetric with respect to eqn (3.16), says that we can know the past history of the system, regardless of how it will evolve (probabilistically) in the future, provided the present state is known. It is obvious enough that not all stochastic processes have the Markovian property; let us think of a system that cannot return to states previously visited. In this case the process must remember all the visited states, since the probability that the system may return in the future, to an already visited state, must be zero.

State spaces of Markov processes can be discrete or continuous. First we will study the *discrete–time Markov chains*, for which the state space $\mathcal{S}$ is discrete. In this case the parametric space $\mathbb{T}$ is a subset of natural or integer numbers, for example $\mathbb{T} = \{0, 1, 2, \ldots, n, \ldots\}$. More specifically, the Markovian property can be written as follows. For every state $i_0, i_1, \ldots, i_{n-1}, i, j \in \mathcal{S}$ and for each $n \in \mathbb{T}$, it is:

$$P\left\{X_{n+1} = j | X_n = i, X_{n-1} = i_{n-1}, \ldots, X_0 = i_0\right\}$$
$$= P\left\{X_{n+1} = j | X_n = i\right\} \quad (3.18)$$

which can be read: the probability that the process at time $n + 1$ is in the state j is conditioned only by the knowledge of the state i in which it stays at the present time n, it being not crucial to know the state visited at past times $i_{n-1}, \ldots, i_0$.

The probability $P\left\{X_{n+1} = j | X_n = i\right\}$ is called *transition probability* from the state i to the state j. We write $p_{ij}(n)$ the transition probability, at the time n, from the state i to the state j *in one step*, that is from the instant n to the following instant $n + 1$.

$$p_{ij}(n) = P\left\{\underset{\text{time}}{X_{n+1}} = \underset{\text{state}}{j} \mid \underset{\text{time}}{X_n} = \underset{\text{state}}{i}\right\}$$

Such probability $p_{ij}(n)$ can be different if one considers a transition at a time $m \neq n$, $p_{ij}(n) \neq p_{ij}(m)$. The probability that, after a rainy night, the following day is a sunny day depends on the season. Weather forecasts are obviously not the same in summer or winter. We have assumed that there exists some model based on Markov chain to describe changes in weather, as it occurs in the Land of Oz, as we will see below.

There are Markov chains, called *time homogeneous* or having *stationary transition probabilities*. For these chains the transition probability from i to j is always the same, at any instant, independent of the time t, namely such probability does not depend on $n, \forall n \in \mathbb{T}$. In the following we will only deal with homogeneous chains, so we will write simply p_{ij}, without (n). Symbolically, a Markov chain is homogeneous if:

$$p_{ij} = P\left\{X_{n+1} = j | X_n = i\right\}, \quad \forall n \in \mathbb{T}, \forall i, j \in \mathcal{S}$$

and it is also:

$$p_{ij} = P\left\{X_1 = j | X_0 = i\right\}$$

It is important not to confuse stationary, or homogeneous, transition probability with stationary (or invariant or equilibrium) probability distribution.

This distribution plays an essential role in the theory of Markov chains. To say that a chain is in (or has achieved) a stationary distribution means that the stochastic process is stationary as discussed in Section 3.1.1. In short, if a chain has a stationary distribution, then $X_0 \sim \pi$, and also $X_n \sim \pi$, $\forall n \in \mathbb{T}$.

Let us first consider the case of a *finite* number of states, equal to N. In this case $\mathcal{S}$ can be the set of integers $\{1, 2, \ldots, N\}$, or $\{0, 1, \ldots, N-1\}$. In the following, both numerations will be adopted. Transition probabilities p_{ij} are collected in a matrix $N \times N$, called transition probability matrix, or the state transition matrix or simply *transition matrix* in one step:

$$\mathbf{P} = \begin{pmatrix} p_{11} & p_{12} & \cdots & p_{1N} \\ p_{21} & p_{22} & \cdots & p_{2N} \\ \cdots\cdots\cdots\cdots\cdots \\ p_{N1} & p_{N2} & \cdots & p_{NN} \end{pmatrix} \tag{3.19}$$

with the conditions:

$$p_{ij} \geqslant 0, \ \forall i, j \in \mathcal{S} \qquad \text{and} \qquad \sum_{j \in \mathcal{S}} p_{ij}(t) = 1, \ \forall i \in \mathcal{S} \tag{3.20}$$

The first condition simply ensures that p_{ij} are probabilities, so they cannot be negative. The second condition means that the rows sum of the matrix must be all 1; indeed, if the system is in the state i, either it stays in i, or it goes to some other state j. Note: the columns of the matrix do not in general sum to 1.

The sum $\sum_{j \in \mathcal{S}}$ means the sum of states in the space $\mathcal{S}$, either as a finite number N set, or if it is countably infinite.

If eqns (3.20) hold for a matrix, then such a matrix is called a *stochastic matrix*. For convention we can write:

$$p_{ij}^{(0)} = \delta_{ij} = \begin{cases} 1, & \text{if } i = j \\ 0, & \text{if } i \neq j \end{cases} \tag{3.21}$$

and in matrix form $\mathbf{P}^{(0)} = \mathbf{I}$, where $\mathbf{I}$ is the identity matrix. The term (0) on the exponent will be clear in the following.

We ask ourselves: knowing the transition matrix, can we claim to know the whole process? The answer is negative. Indeed, if the transition matrix is known, we can say what is the probability that the system will be in the state j, at time $n+1$, if at time n it is in i. But I have to know what is this state i at the time n, otherwise I have to know where the system was at the time $n-1$, and so on at the time $n-2$ etc. In short, I have to know the state from which the process starts, that is the *initial state*. Starting from an initial state, I am able to trace the whole process step by step, by means of the transition matrix. Therefore transitions probabilities define the probabilistic law of the chain, *conditioned on the initial state*. However, the initial states are usually not known with certainty, but only probabilistically: for instance, we could know that all the states are equally probable. To sum up, Markov chains are defined by:

1. the transition matrix

2. the initial state distribution

Consider a system with four states, say *1, 2, 3, 4*. At each time instant $t = \tau$, $t = \tau + 1$, $t = \tau + 2$, etc., the system goes to another state, or stays in the same state, with a particular transition probability. State transition diagrams can be useful to describe Markov chains, as the example in Fig. 3.3 shows. In the diagram, nodes represent states, the arrows from state i to state j represent the transition probabilities p_{ij}. If the system stays in the state i, there is a cycle around the state, this means that $p_{ij} = 0$.

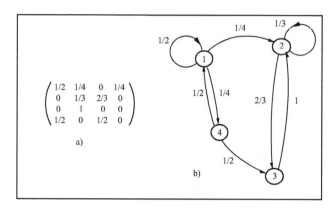

Fig. 3.3 Transition matrix (a) and states transition diagram (b) for the system with four states.

In the example of Fig. 3.3, if the system is in the state *2*, it cannot jump either to *1*, or to *4*; it either stays in *2* with probability 1/3, or moves to *3*, with probability 2/3. Clearly, the time evolution of the system is not 'deterministic': if the system is in *2*, it may stay there or not, but this option does not depend on the past visited state. If the system obeyed a deterministic law, we could state with certainty that either it goes to *3*, or it stays in *2*. But also in this case, the option does not depend on its past place.

The transition matrix of the example, also reported in Fig. 3.3, is:

$$
\begin{array}{c c c c c}
 & 1 & 2 & 3 & 4 \\
\begin{matrix} 1 \\ 2 \\ 3 \\ 4 \end{matrix} &
\left(\begin{matrix}
1/2 & 1/4 & 0 & 1/4 \\
0 & 1/3 & 2/3 & 0 \\
0 & 1 & 0 & 0 \\
1/2 & 0 & 1/2 & 0
\end{matrix} \right)
\end{array}
\tag{3.22}
$$

Notice that if the system visits (or starts from) the states *2* or *3*, it 'gets trapped', namely it cannot reach any other state. We will look at this in more detail. Now we would wonder: is it possible to model the time evolution of the system, and in this case why do we do it? The reason is that by means of simulations we are able to answer questions such as, for instance, how many steps are necessary to reach the state *3*, starting from *1* ? Actually, this problem is also solvable analytically, but there are situations which can be dealt with only by means of simulations.

To model the motion of the system, let us consider the transition matrix eqn (3.19), and suppose that the system starts from the state 1. The vector components, call it, row_1 are the transition probabilities $p_{11}, p_{12}, \ldots, p_{1N}$, according to which the system moves to states $1, 2, \ldots N$. The states are chosen through random numbers and, one after the other, the trajectory is constructed. In R this procedure can be written as the function below:

```
markov<- function(x0,n,x,P) {    # starting function
s<- numeric()
s[1]<- x0
row<- which(x==x0)
for(i in 2:n) {
s[i]<- sample(x,1,P[row,],replace=T)
row<- which(x==s[i])
                }
return(s)
                         }  # ending function
```

The initial state is x0, and n is the number of transitions executed by the process. States can be enumerated as $1, 2, \ldots, N$, or named, for instance, 'north', 'northeast', 'east', and so on. x is the vector containing the state names. The transition matrix is P[.]. The vector s[.] collects the states visited at each step. Code running translations between states, as the above function, is called an *update function*. The whole code to perform the system trajectory is of the type reported in the following:

```
## Code_3_1.R
#Example of code to simulate the system motion
set.seed(2)
n<- 20                              # number of steps
P<- matrix(c(1/2,1/4,0,1/4,0,1/3,2/3,0,0,1,0,0,1/2,0,1/2,0)
          ,nrow=4,ncol=4,byrow=T)     # transition matrix
P
x<- c(1,2,3,4)
x0<- 1                              # initial condition
s<- markov(x0,n,x,P)               # call function Markov
t<- c(0:(n-1))
t
s
```

In R a matrix is constructed using the function **matrix(.)**. To directly create a matrix as in the above code, the matrix entries are placed along the columns by default. If **byrow=T**, the matrix is filled by rows. Matrices are written in R as below:

```
> P
      [,1]      [,2]        [,3] [,4]
[1,]   0.5 0.2500000 0.0000000 0.25
[2,]   0.0 0.3333333 0.6666667 0.00
[3,]   0.0 1.0000000 0.0000000 0.00
[4,]   0.5 0.0000000 0.5000000 0.00
```

where t is the vector of steps. The final result is reported below

```
> t
 [1] 0 1 2 3 4 5 6 7 8 9 10 11 12 13 14 15 16 17 18 19 20
> s
 [1] 1 1 2 3 2 2 2 3 2 3 2 3 2 2 3 2 2 2 3 2 3

> t
```

```
[1]  0  1  2  3  4  5  6  7  8  9  10  11  12  13  14  15  16  17  18  19  20
>  s
[1]  1  1  4  1  1  2  3  2  3  2  3  2  3  2  3  2  2  3  2  2  3
```

The two trajectories are reported in Fig. 3.4. Dashed and dotted lines between states are to guide the eye.

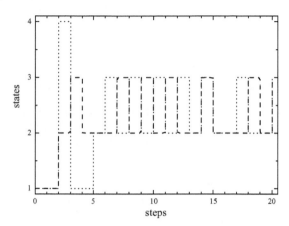

Fig. 3.4 Two trajectories with the same initial state *1* of the process with transition matrix eqn (3.22).

We see, for instance, that the first trajectory (dashed line) visits the state *3* for the first time after thee steps, the second trajectory (dotted line), after six steps. It could happen that the initial state x0 is not known with certainty, but only the initial distribution of states is available. In this case, an *initiation function* chooses the initial state by a random number. So the line x0<- 1 becomes:

```
v0<-c(0.2,0.1,0.4,0.3)      # initial distribution
x0<-sample(x,1,v0,replace=T) # initiation function
```

In this example, the system starts from state *1*, with probability 0.2, from state *2*, with probability 0.1, etc.

Sketching something like a philosophical background of computer experiments on Markov chains, the basic idea is the following. If the estimate of mean values of parameters (the number of visits, the first visit to a state, etc.) are required, and those quantities are representable as random variables, a suitable number of trajectories are simulated and averages of interested quantities are computed. The term *average* means *ensemble average*, but we must be aware that this kind of averages is not always computable. This is a delicate matter: there are situations in which only one trajectory can be computed. In this case the 'average' is a *time average*, with all problems of stationarity, ergodicity, and so on.

3.3 Predicting the future

Through a simple example, we will introduce basic concepts, such as visit probability after n steps, n-step transition matrix, Chapman–Kolmogorov equation, and more.

The system is the weather over multiple days. Suppose that the weather each day can be rainy or sunny. Suppose also that the event {tomorrow is rainy} depends only on the weather today, not on how it was in past days. In the literature, similar models are usually defined as 'describing the weather in the Land of Oz'. The random variables X_n, therefore, can assume only two values, say 0 and 1. If at time n, X_n is 0, the system is in the state 'rainy', and if X_n is 1, the system is in the state 'sunny'. Index n marks the days, so we have:

$$\underbrace{X_0}_{\substack{0 \quad 1 \\ \text{today}}} \qquad \underbrace{X_1}_{\substack{0 \quad 1 \\ \text{tomorrow}}} \qquad \underbrace{X_2}_{\substack{0 \quad 1 \\ \text{after 2 days}}} \qquad \underbrace{X_3}_{\substack{0 \quad 1 \\ \text{after 3 days}}} \qquad \underbrace{X_4}_{\substack{0 \quad 1 \\ \text{after 4 days}}} \quad \cdots \quad \underbrace{X_n}_{\substack{0 \quad 1 \\ \text{after } n \text{ days}}} \quad \cdots$$

Suppose the transition probabilities are known. If today is raining, tomorrow will be raining with probability 0.7, and then it will be sunny with probability $1 - 0.7 = 0.3$. A similar reasoning applies if today is sunny. The transition matrix is then:

$$
\begin{array}{cc}
 & \begin{array}{cc} 0 & \qquad\qquad\qquad 1 \end{array} \\
\begin{array}{c} 0 \\ \\ 1 \end{array} &
\left(
\begin{array}{cc}
\underset{\text{today rain, tomorrow rain}}{0.7} & \underset{\text{today rain, tomorrow sun}}{0.3} \\
\underset{\text{today sun, tomorrow rain}}{0.4} & \underset{\text{today sun, tomorrow sun}}{0.6}
\end{array}
\right)
\end{array}
\tag{3.23}
$$

In terms of random variables, if I know that today is rainy, that is $\mathsf{P}\{X_0 = 0\} = 1$, the probability that tomorrow is rainy, that is $\mathsf{P}\{X_1 = 0 | X_0 = 0\}$ is just $p_{00} = 0.7$. If I know that today is sunny, then $\mathsf{P}\{X_1 = 0 | X_0 = 1\}$ is $p_{10} = 0.4$. Suppose now we know only the probability that today is rainy equal to 0.2, then $\mathsf{P}\{X_0 = 1\} = 0.8$. The probability that tomorrow is rainy $\mathsf{P}\{X_1 = 0\}$, is given by the Total Probability Theorem, recalled here.

Let $B_1, B_2, \ldots, B_n$ be mutually exclusive (also called disjoint) events, forming a partition of the sample space Ω, all with a non-zero probability of occurrence. Let A be an event $\subset \Omega$, we can write $A = \bigcup_{i=1}^{n}\{A, B_i\}$, then:

$$\mathsf{P}\{A\} = \sum_{i=1}^{n} \mathsf{P}\{A, B_i\} = \sum_{i=1}^{n} \mathsf{P}\{A | B_i\}\,\mathsf{P}\{B_i\}$$

Turning to our problem, we have:

$$
\begin{aligned}
\mathsf{P}\{X_1 = 0\} &= \underbrace{\mathsf{P}\{X_1 = 0 | X_0 = 0\}}_{p_{00}}\, \underbrace{\mathsf{P}\{X_0 = 0\}}_{v_0^{(0)}} \\
&+ \underbrace{\mathsf{P}\{X_1 = 0 | X_0 = 1\}}_{p_{10}}\, \underbrace{\mathsf{P}\{X_0 = 1\}}_{v_1^{(0)}}
\end{aligned}
\tag{3.24}
$$

where $v_0^{(0)}$ and $v_1^{(0)}$ are unconditional probabilities $\mathsf{P}\{X_0 = 0\}$ and $\mathsf{P}\{X_1 = 0\}$, respectively. In general, $\mathsf{P}\{X_n = i\}$ or also $v_i^{(n)}$, means the probability that the system at time n, after n steps, is in the state i:

$$P\{X_n = i\} \equiv v_i^{(n)} \begin{array}{l} \rightarrow \text{ number of steps} \\ \rightarrow \text{ state after } n \text{ steps} \end{array}$$

Let N be the number of states, then after n steps the system can be in the states $1, 2, \ldots, N$, with probability $v_1^{(n)}, v_2^{(n)}, \ldots, v_N^{(n)}$. The vector

$$\mathbf{v}^{(n)} = (v_1^{(n)}, v_2^{(n)}, \ldots, v_N^{(n)}) \tag{3.25}$$

is called *state vector at n-th step* or *stochastic vector*, or *probability vector*, being a vector with non–negative entries which add up to 1:

$$v_i^{(n)} \geqslant 0 \quad \forall i \in \mathcal{S}, \ \forall n \in \mathbb{T} \qquad \text{and} \qquad \sum_{i \in \mathcal{S}} v_i^{(n)} = 1$$

In the example, the vector $\mathbf{v}^{(0)} = (v_0^{(0)}, v_1^{(0)})$ is the *initial probabilities vector* (or *initial state vector*). Note that in the following, the same symbol $\mathbf{v}$ will be used both to denote a row vector, as in $\mathbf{v}\mathbf{P}$, or a column vector, as in $\mathbf{P}\mathbf{v}$.

Turning to eqn (3.24), it is written in general as:

$$P\{X_1 = j\} = \sum_{i \in \mathcal{S}} P\{X_1 = j | X_0 = i\} P\{X_0 = i\}, \qquad j \in \mathcal{S}$$

or in different notation:

$$v_j^{(1)} = \sum_{i \in \mathcal{S}} p_{ij} v_i^{(0)}, \qquad j \in \mathcal{S}$$

and in matrix notation:

$$\mathbf{v}^{(1)} = \mathbf{v}^{(0)} \mathbf{P}$$

We ask for the probability that the day after tomorrow is raining, knowing that today is raining. We have to compute the two-step transition probability. Tomorrow the state can be either rainy or sunny, therefore the requested probability is the sum of the probability that tomorrow is rainy again with the probability that tomorrow is sunny and after tomorrow is rainy:

$$P\{X_2 = 0 | X_0 = 0\} = P\{X_2 = 0, \underbrace{X_1 = 0}_{\text{tomorrow rain}} | X_0 = 0\}$$

$$+ P\{X_2 = 0, \underbrace{X_1 = 1}_{\text{tomorrow sun}} | X_0 = 0\}$$

Recall that $P\{A, B | C\} = P\{A, B, C\}/P\{C\}$, consider the first addend of the right-hand side:

$$P\{X_2 = 0, X_1 = 0 | X_0 = 0\} = \frac{P\{X_2 = 0, X_1 = 0, X_0 = 0\}}{P\{X_0 = 0\}} =$$

multiplying numerator and denominator for $P\{X_1 = 0, X_0 = 0\}$ and recalling that $P\{B|A\} = P\{B, A\} / P\{A\}$:

$$= \frac{P\{X_2 = 0, X_1 = 0, X_0 = 0\}}{P\{X_1 = 0, X_0 = 0\}} \times \frac{P\{X_1 = 0, X_0 = 0\}}{P\{X_0 = 0\}}$$

$$= P\{X_2 = 0 | X_1 = 0, \underset{\text{Markov property}}{X_0 = 0}\} \times P\{X_1 = 0 | X_0 = 0\} =$$

for the Markov property the term $X_0 = 0$ in the first brackets is unnecessary, then:

$$= P\{X_2 = 0 | X_1 = 0\} \times P\{X_1 = 0 | X_0 = 0\}$$

The same procedure is applied to the second addend, so the final result is:

$$P\{X_2 = 0 | X_0 = 0\} = \underbrace{P\{X_2 = 0 | X_1 = 0\}}_{p_{00}} \times \underbrace{P\{X_1 = 0 | X_0 = 0\}}_{p_{00}}$$

$$+ \underbrace{P\{X_2 = 0 | X_1 = 1\}}_{p_{10}} \times \underbrace{P\{X_1 = 1 | X_0 = 0\}}_{p_{01}}$$

Shortly, putting $P\{X_2 = 0 | X_0 = 0\}$ as $p_{00}^{(2)}$, we have:

$$p_{00}^{(2)} = p_{00}p_{00} + p_{01}p_{10} = \sum_{k=0}^{1} p_{0k}p_{k0}$$

We now introduce a third state in the Land of Oz: a snowy day. Let the transition matrix be:

$$
\mathbf{P} = \begin{array}{c} \\ 0 \\ 1 \\ 2 \end{array}
\begin{array}{ccc}
0 & 1 & 2 \\
\end{array}
\left(\begin{array}{ccc}
0.6 & 0.1 & 0.3 \\
0.3 & 0.5 & 0.2 \\
0.5 & 0.4 & 0.1
\end{array} \right)
\tag{3.26}
$$

To compute the probability that past tomorrow is sunny, if today is snowy, we have to take into account that tomorrow possible states are rain, sun and snow, denoted as $k = 0, 1, 2$, so we have:

$$P\{X_2 = 1 | X_0 = 2\} = p_{21}^{(2)} = \sum_{k=0}^{2} p_{2k}p_{k1}$$

thus $p_{21}^{(2)} = 0.5 \times 0.1 + 0.4 \times 0.5 + 0.1 \times 0.4 = 0.29$.

In general, the two-step transition probability can be written as:

$$P\left\{X_{n+2} = j \middle| X_n = i\right\} = p_{ij}^{(2)} = \sum_{k \in \mathcal{S}} p_{ik} p_{kj} \tag{3.27}$$

The quantities $p_{ij}^{(2)}$ can be considered as entries in a matrix $\mathbf{P}^{(2)}$, named a two-step transition matrix, that is the one-step transition matrix multiplied by itself:

$$\mathbf{P}^{(2)} = \mathbf{P} \cdot \mathbf{P} = \mathbf{P}^2$$

note the difference between $\mathbf{P}^{(2)}$ and $\mathbf{P}^2$.

Considering only two states *0* and *1*, the state vector after two steps is:

$$\mathbf{v}^{(2)} = (v_0^{(2)}, v_1^{(2)})$$

with:

$$v_0^{(2)} = P\left\{X_2 = 0\right\} = p_{00}^{(2)} v_0^{(0)} + p_{10}^{(2)} v_1^{(0)}$$
$$v_1^{(2)} = P\left\{X_2 = 1\right\} = p_{01}^{(2)} v_0^{(0)} + p_{11}^{(2)} v_1^{(0)}$$

with N states:

$$v_j^{(2)} = \sum_{i \in \mathcal{S}} p_{ij}^{(2)} v_i^{(0)}$$

and in matrix notation:

$$\mathbf{v}^{(2)} = \mathbf{v}^{(0)} \, \mathbf{P}^{(2)} = \mathbf{v}^{(0)} \, \mathbf{P}^2$$

From the Chapman–Kolmogorov equation discussed below, the n-step transition matrix $\mathbf{P}^{(n)}$ can be obtained by multiplying n times the one-step transition matrix:

$$\mathbf{P}^{(n)} = \underbrace{\mathbf{P} \cdot \mathbf{P} \cdots \mathbf{P}}_{n \text{ times}} = \mathbf{P}^n \tag{3.28}$$

and the state vector at n-th step:

$$\mathbf{v}^{(n)} = \mathbf{v}^{(0)} \, \mathbf{P}^{(n)} = \mathbf{v}^{(0)} \, \mathbf{P}^n \tag{3.29}$$

So in conclusion, by knowing the initial probabilities vector $\mathbf{v}^{(0)}$ and the one-step transition matrix $\mathbf{P}$, we are able through eqn (3.29) to predict the future of the system after n steps, for all large n, namely we achieve a *complete statistical description* of the process, as we stated before: Markov chains are defined by a transition matrix and an initial distribution over states.

We will prove an equation relating the transition probabilities at different steps, known as the Chapman–Kolmogorov equation, which is written as:

$$p_{ij}^{(m+n)} = \sum_{k \in \mathcal{S}} p_{ik}^{(m)} \, p_{kj}^{(n)} \tag{3.30}$$

Before proving eqn (3.30), note than eqn (3.27) is a special case of eqn (3.30), with $n = m = 1$ (of course, $p_{ij}^{(1)}$ is p_{ij}). In matrix form eqn (3.30) is:

$$\mathbf{P}^{(n+m)} = \mathbf{P}^{(n)} \cdot \mathbf{P}^{(m)}$$

and from eqn (3.30), it follows eqn (3.28). Indeed, if $m = 1$ and $n = k - 1$, for any $k \geqslant 1$, it is: $\mathbf{P}^{(k)} = \mathbf{P}^{(k-1)} \cdot \mathbf{P}$, and by induction:

$$\mathbf{P}^{(n)} = \mathbf{P}^{(n-1)} \cdot \mathbf{P} = \mathbf{P}^{(n-2)} \cdot \mathbf{P}^2 = \cdots = \mathbf{P} \cdot \mathbf{P}^{n-1} = \mathbf{P}^n$$

Equation (3.30) says that the $m+n$-step transition probabilities can be expressed as the sum of the products between the m-step and the n-step transition probability. In eqn (3.30), the factor $p_{ik}^{(m)}$ represents the probability that the process, starting from i, visits an intermediate state k in m steps, while the factor $p_{kj}^{(n)}$ represents the probability that the process, from k arrives in j in n steps, as shown in Fig. 3.5, as an example. By adding all intermediate states k, we obtain all possible different trajectories from i to j in $m + n$ steps. The Markov property ensures that the probability of going from i to j in $m + n$ steps is independent of the steps the process visited k.

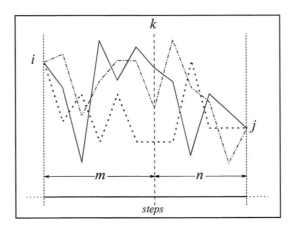

Fig. 3.5 Example of application of the Chapman–Kolmogorov eqn (3.30). Three trajectories, starting from the state i, end in the state j, in $m+n$ steps, each visiting a different intermediate state k.

To prove eqn (3.30), the Markov property is exploited. Transition probability $p_{ij}^{(m+n)}$ is given by:

$$p_{ij}^{(m+n)} = \underbrace{\mathsf{P}\{X_{m+n} = j|}_{\text{the process}} \underbrace{X_0 = i}_{\text{starts from } i} \ \} = \sum_{\underbrace{k \in \mathcal{S}}} \mathsf{P}\{\underbrace{X_{m+n} = j, X_m = k}_{\text{visits } k}|X_0 = i\}$$

Recalling that $\mathsf{P}\left\{A, B|C\right\} = \mathsf{P}\left\{A, B, C\right\}/\mathsf{P}\left\{C\right\}$, within the summation, we have:

$$P\left\{X_{m+n}=j, X_m=k \middle| X_0=i\right\} = \frac{P\left\{X_{m+n}=j, X_m=k, X_0=i\right\}}{P\left\{X_0=i\right\}} =$$

multiplying numerator and denominator for $P\left\{X_m=k, X_0=i\right\}$:

$$=\frac{P\left\{X_{m+n}=j, X_m=k, X_0=i\right\}}{P\left\{X_m=k, X_0=1\right\}} \times \frac{P\left\{X_m=k, X_0=i\right\}}{P\left\{X_0=i\right\}}$$

$$= P\{X_{m+n}=j|X_m=k, \underbrace{X_0=i}_{\text{Markov property}}\} \times P\left\{X_m=k|X_0=i\right\}$$

$$= \underbrace{P\left\{X_{m+n}=j|X_m=k\right\}}_{p_{kj}^n} \times \underbrace{P\left\{X_m=k|X_0=i\right\}}_{p_{ik}^m}$$

In passing, notice that if a process is a Markov process, Chapman–Kolmogorov eqn (3.30) holds, as we have seen; nevertheless there are discrete-time processes, that are *not* Markov processes, for which eqn (3.30) still holds.

Let us turn to the example of the weather in the Land of Oz. The knowledge of the transition matrix eqn (3.23) and the initial state distribution enable to forecast the weather in two, four or eight days. Very naively, in R we have to multiply the initial probability vector v0 by the power of the transition matrix P, as in the short code below.

```
##
mult <- function(A,n){
M <- A
if(n>1){
for(i in 2:n) {M <- M %*% A}  }
M
                          }
nm<-2        # for instance
P<-matrix(c(0.7,0.3,0.4,0.6),2,2,byrow=T)
# P   # comment if you do not wish the initial matrix printed
for(i in 1:nm){
Pm<-mult(P,i) }
# Pm        # reported below
v0<-c(0.2,0.8)
vm<-v0%*%Pm
# vm        # reported below
```

The output is:

```
> P2
        [,1] [,2]
   [1,] 0.61 0.39
   [2,] 0.52 0.48
> v2
          [,1]   [,2]
   [1,] 0.538 0.462
```

Four and eight days forecasts are reported below:

```
> P4                    |   > P8
      [,1]    [,2]      |            [,1]        [,2]
[1,] 0.5749 0.4251      |   [1,] 0.5714567 0.4285433
[2,] 0.5668 0.4332      |   [2,] 0.5713911 0.4286089
> v4                    |   > v8
        [,1]     [,2]   |            [,1]        [,2]
[1,] 0.56842 0.43158    |   [1,] 0.5714042 0.4285958
```

We see that:

$$v_0^{(8)} \approx p_{00}^{(8)} \approx p_{10}^{(8)}$$
$$v_1^{(8)} \approx p_{01}^{(8)} \approx p_{11}^{(8)}$$

That is, as steps increase, transition matrix rows tend to become the same. In fact, there exists a limit behaviour of the transition matrix, such that it does not change any further, even if time increases. This limiting distribution does not depend on the initial probabilities; the system has completely forgotten the initial state and has stabilized in a stationary condition. Such *asymptotic transition matrices* $\mathbf{P}^{(\infty)}$, are defined as:

$$\mathbf{P}^{(\infty)} = \lim_{n \to \infty} \mathbf{P}^n$$

3.3.1 Stationarity

We have introduced the asymptotic transition matrix $\mathbf{P}^{(\infty)}$, given by:

$$\mathbf{P}^{(\infty)} = \begin{pmatrix} v_1^{(\infty)} & v_2^{(\infty)} & \cdots & v_N^{(\infty)} \\ v_1^{(\infty)} & v_2^{(\infty)} & \cdots & v_N^{(\infty)} \\ \cdots\cdots\cdots\cdots\cdots \\ v_1^{(\infty)} & v_2^{(\infty)} & \cdots & v_N^{(\infty)} \end{pmatrix} = \begin{pmatrix} \mathbf{v}^{(\infty)} \\ \mathbf{v}^{(\infty)} \\ \cdots \\ \mathbf{v}^{(\infty)} \end{pmatrix}$$

where the row is the convergence limit of the state vector $\mathbf{v}^{(n)}$ for $n \to \infty$. Then:

$$v_j^{(\infty)} = \lim_{n \to \infty} p_{ij}^{(n)}$$

is the limiting probability, for $n \to \infty$, that the system is in the state j.

Let N be the number of states. The vector, called the *vector of asymptotic probabilities*,

$$\mathbf{v}^{(\infty)} = (v_1^{(\infty)}, v_2^{(\infty)}, \dots, v_N^{(\infty)}), \qquad \text{with} \quad \sum_{i \in \mathcal{S}} v_i^{(\infty)} = 1$$

represents the limiting probability that the system is in the states $1, 2, \dots, N$.

Let us introduce another probability vector. Let $\boldsymbol{\pi} = (\pi_1, \pi_2, \dots, \pi_N)$ be a probability vector. We say that the probability vector $\boldsymbol{\pi}$ is a *stationary distribution*, or *invariant distribution*, or *equilibrium distribution* for the transition matrix $\mathbf{P}$, if:

$$\begin{aligned} &1) \quad \pi_i \geqslant 0, \forall i \in \mathcal{S} \quad \text{and} \quad \sum_{i \in \mathcal{S}} \_i = 1 \\ \\ &2) \quad \boldsymbol{\pi} = \boldsymbol{\pi}\mathbf{P} \quad \text{that is} \quad \sum_{i \in \mathcal{S}} \pi_i p_{ij} = \pi_j, \forall j \in \mathcal{S} \end{aligned} \tag{3.31}$$

The Property 1) simply says that π describes the probability distribution over states $1, 2, \ldots, N$. Property 2) entails the following. Suppose that the initial distribution $\mathbf{v}^{(0)}$ is equal to π: $\mathbf{v}^{(0)} = \pi$. Then:

$$\mathbf{v}^{(1)} = \mathbf{v}^{(0)}\mathbf{P} = \pi\mathbf{P} = \pi$$
$$\mathbf{v}^{(2)} = \mathbf{v}^{(1)}\mathbf{P} = \pi\mathbf{P} = \pi$$
$$\ldots\ldots\ldots\ldots\ldots\ldots\ldots\ldots$$

and by iteration $\forall n$:

$$\mathbf{v}^{(n)} = \pi$$

which means that the distribution over states remains the same, the initial one. That is the reason for the term 'invariant' distribution.

Let us now compare stationary distributions and asymptotic distributions given by the asymptotic probability vector. In the example with two state, rain and sun, the stationary distribution is:

$$\pi = (0.57143, 0.42857)$$

which is also the asymptotic distribution. In this case, we can write:

$$\lim_{n\to\infty} \mathbf{P}^n = \begin{pmatrix} 0.7 & 0.3 \\ 0.4 & 0.6 \end{pmatrix}^n = \begin{pmatrix} v^{(\infty)} \\ v^{(\infty)} \end{pmatrix} = \begin{pmatrix} \pi \\ \pi \end{pmatrix}$$

$$= \begin{pmatrix} 0.57143 & 0.42857 \\ 0.57143 & 0.42857 \end{pmatrix}$$

It is also:

$$\pi_i = \lim_{n\to\infty} v_i^{(n)}$$

which we found is not always true: the stationary distribution might not be the asymptotic one.

Further comments about stationary distributions are now in order. In our example, we know that, if today is rainy, the probability that tomorrow is rainy is 0.7, or is 0.4 if today is sunny. At stationarity, practically after eight days, the probability that it is rainy is 0.57, no matter what the weather is today, and this probability always remains the same. Then, after eight days, choosing any day, we are able to say that the probability that it will be a rainy day is 0.57, not depending on initial probabilities.

In essence: at stationarity, the probability of observing the system in a certain state, does not depend on the chosen time instant. From another point of view, we can say that 57 days out of 100 will be rainy, and 43 will be sunny. Such a reasoning rests on frequency interpretation (better: conception) of probability. To say that π_i is the probability that the system at stationarity is in the state i means to say that π_i is the limiting frequency distribution of the occupation time of the state i at equilibrium.

The *occupation time* is the time spent by the process in a state, or in a given subset, of the state space $\mathcal{S}$. Let us see a couple of examples of a chain with only two sates *0* and *1*. Let the transition matrix be:

$$\mathbf{P} = \begin{pmatrix} 1-a & a \\ b & 1-b \end{pmatrix} \tag{3.32}$$

suppose $0 < a, b < 1$. Equations (3.31) are written as:

$$\pi_0 = (1-a)\pi_0 + b\pi_1$$
$$\pi_1 = a\pi_0 + (1-b)\pi_1 \tag{3.33}$$
$$\pi_0 + \pi_1 = 1$$

The solution is immediate:

$$\pi_0 = \frac{b}{a+b} \quad e \quad \pi_1 = \frac{a}{a+b} \tag{3.34}$$

In conclusion, for $0 < a, b < 1$, stationary and asymptotic distributions exist and coincide. Indeed, with $a = 0.3$ and $b = 0.4$ of the above example, we find just $\boldsymbol{\pi} = (0.5714285714, 0.4285714286)$, (see also Fig. 3.6a). With $a = b = 1$, we have:

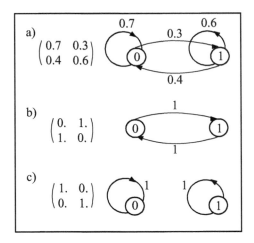

Fig. 3.6 Examples of systems with two states and different transition matrices.

$$\mathbf{P} = \begin{pmatrix} 0 & 1 \\ 1 & 0 \end{pmatrix}, \quad \mathbf{P}^2 = \begin{pmatrix} 1 & 0 \\ 0 & 1 \end{pmatrix}, \quad \mathbf{P}^3 = \begin{pmatrix} 0 & 1 \\ 1 & 0 \end{pmatrix}, \quad \mathbf{P}^4 = \begin{pmatrix} 1 & 0 \\ 0 & 1 \end{pmatrix}, \quad \dots$$

It is clear that here no asymptotic matrix exists, since there is no $\lim_{n \to \infty} p_{ij}^{(n)}$. Indeed, $p_{00}^{(n)} = 1$, for $n > 1$ even, and $= 0$, for $n > 1$ odd. That is to say, the states 0 and 1 are periodic, 0 is visited at steps $2, 4, 6, \dots$, and 1 at steps $1, 3, 5, \dots$, (see also Fig. 3.6b). Nevertheless the stationary distribution exists and is, by eqn (3.34), $\boldsymbol{\pi} = (0.5, 0.5)$. This means that the system spends half the time in state 0, and half the time in state 1.

Lastly, take $a = b = 0$ (see Fig. 3.6c). The system remains forever in the initial state, and obviously it is $\lim_{n \to \infty} \mathbf{P}^{(n)} = \mathbf{P}^n = \mathbf{P}, \forall n$. Stationary distributions are infinite, being always $\boldsymbol{\pi} = \boldsymbol{\pi}\mathbf{P}$, for any component π_0 and π_1 of $\boldsymbol{\pi}$. In similar cases, the transition matrix is the *identity matrix*.

The question is: under which conditions is the stationary distribution unique and equal to the asymptotic distribution? Or in other words, when can we write, for any initial distribution $\mathbf{v}^{(0)}$?

$$\lim_{n\to\infty} \mathbf{v}^{(0)}\mathbf{P}^n = \boldsymbol{\pi}, \ \forall \mathbf{v}^{(0)}$$

A general approach to deriving the stationary distribution (if it exists) is to view equation $\boldsymbol{\pi} = \boldsymbol{\pi}\mathbf{P}$ as an eigenvalue equation. By means of its eigenvalues and eigenvectors, the matrix $\mathbf{P}$ can be diagonalized, that is $\mathbf{D} = \mathbf{A}^{-1}\mathbf{PA}$, where $\mathbf{A}$ columns are the right eigenvectors of $\mathbf{P}$, and $\mathbf{A}^{-1}$ rows are the left eigenvectors. Then the diagonal matrix is:

$$\mathbf{D} = \begin{pmatrix} \lambda_1 & 0 & \cdots & 0 \\ 0 & \lambda_2 & \cdots & 0 \\ \vdots & \vdots & \cdots & \vdots \\ 0 & 0 & \cdots & \lambda_N \end{pmatrix} \tag{3.35}$$

Given the diagonal matrix, it is not difficult, in general, to raise $\mathbf{P}$ to n, and to take the limit $n \to \infty$. The method based on eigenvalue analysis, while certainly elegant, is not applicable if the number of states is infinite. In the following part of this section, examples in R will be provided.

To wonder if a stationary distribution exists is to wonder if a non-negative left eigenvector of $\mathbf{P}$ exists, corresponding to an eigenvalue equal to 1, all others have modulus $\leqslant 1$. $\mathbf{P}$ is a stochastic matrix (rows add up to 1), it always has the eigenvector $\mathbf{u} = (1,\dots,1)$ corresponding to the eigenvalue 1. In conclusion, *for a finite state chain, there exists always at least one stationary distribution*. To wonder if the stationary distribution is unique, is to wonder if the chain is *ergodic*.

$$\mathbf{P} \text{ is ergodic if } \exists m \text{ such that for } \forall n \geqslant m, p_{ij}^{(n)} > 0, \forall i,j \in \mathcal{S}$$

This means that, starting from a certain n, matrices $\mathbf{P}^n$ (n–step transition matrix) have all elements strictly greater than zero, no matter if $\mathbf{P}$ (1-step transition matrix) has some entry equal to zero. It seems that, at least *prima facie*, if the chain is periodic, then it cannot converge to equilibrium, but, if after a certain time n, all the states communicate to each other, there cannot be periodicity, so the chain converges to equilibrium.

If $\mathbf{P}$ is ergodic, then:

a) 1 is an eigenvalue of $\mathbf{P}$. It is single and all the other eigenvalues are less than 1 in absolute value.

b) a state vector $\boldsymbol{\pi}$ exists and is unique, which is also an eigenvector of $\mathbf{P}$ corresponding to eigenvalue 1

c) let $\mathbf{v}$ be any state vector, even of the initial state, then $\mathbf{v}\mathbf{P}^n \to \boldsymbol{\pi}$, for $n \to \infty$.

So to recap: stationary distribution exists, it is unique and coincides with the asymptotic distribution if the transition matrix is ergodic. In this case, the chain is said to be *asymptotically stationary*.

An ergodic Markov chain is sometimes called 'regular' in the literature. Ergodic chains can be regular or periodic, as we will see in the next section.

3.3.2 Classification of states

States are named according to some specific features. We have seen, for instance, an example of periodic states, eqn (3.32) with $a = b = 1$, while with $a = b = 0$, it is $p_{00} = p_{11} = 1$. In general, if $p_{ii} = 1$, the state i is called *absorbing*, and the system stays at the state forever, once it enters the state. This does not mean that the process has stopped: time keeps running and the system continues to remain in the same state.

We say that a state j is accessible from state i if the system starting from i has the possibility of reaching j in some number $n > 0$ of steps. Note $n > 0$, since for $n = 0$ it results $p_{ij}^{(0)} = \delta_{ij}$ (see eqn (3.21)). In symbols:

$$j \leftarrow i : \exists n > 0 \quad \text{such that} \quad p_{ij}^{(n)} > 0$$

Obviously, $j \leftarrow i$ and $i \rightarrow j$ are equivalent. The accessibility relation has the following properties ($\forall i, j, k \in \mathcal{S}$):

reflexive: any state is accessible from itself, in fact $p_{ii}^{(0)} = 1$
transitive: if $i \rightarrow j$ and $j \rightarrow k$, then $i \rightarrow k$

Indeed, the relation $k \leftarrow i$ means $\exists r > 0$ such that $p_{ik}^{(r)} > 0$. If $r = n + m$ from Chapman–Kolmogorov equation (3.30), since $p_{ij}^{(n)}$ and $p_{jk}^{(m)} > 0$, it results in:

$$p_{ik}^{(n+m)} = \sum_{q \in \mathcal{S}} p_{iq}^{(n)} p_{qk}^{(m)} \geqslant p_{ij}^{(n)} p_{jk}^{(m)} > 0$$

Note that the property of accessibility is *not* symmetric. For instance, in the matrix eqn (3.22) (see also Fig. 3.3) it is $2 \leftarrow 1$, but $1 \not\leftarrow 2$.

Two states accessible from each other ($i \rightarrow j$ and $j \rightarrow i$) are said to *communicate*.

$$j \leftrightarrow i : \exists n, m > 0 \quad \text{such that} \quad p_{ij}^{(n)} > 0 \text{ and } p_{ji}^{(m)} > 0$$

Communicability is an *equivalence relation* ($\forall i, j, k \in \mathcal{S}$), i.e. it has the following properties:

reflexive: any state communicates with itself, indeed $p_{ii}^{(0)} = 1$.
symmetric: if $i \leftrightarrow j$ then $j \leftrightarrow i$, by definition.
transitive: if $i \leftrightarrow j$ and $j \leftrightarrow k$ then $i \leftrightarrow k$. The above is a proof for $k \leftarrow i$, for $i \leftarrow k$.

The communicability relation divides the state space $\mathcal{S}$ into disjoint communicating classes (a class can include only one state), such that all states belonging to the same class communicate with each other, but not if they are in different classes. A chain is *irreducible* if there is only one class, that is all states communicate with each other. As a slogan, we could say: *from anywhere to anywhere*, and more formally a chain is irreducible if:

$$\forall i, j, \quad \exists n, m > 0 \quad \text{such that} \quad p_{ij}^{(n)} > 0 \text{ and } p_{ji}^{(m)} > 0$$

Example 3.1 A chain having the transition matrix:

$$
\mathbf{P} = \begin{array}{c} \\ 0 \\ 1 \\ 2 \end{array}
\begin{array}{ccc}
0 & 1 & 2 \\
\begin{pmatrix} 1/2 & 1/2 & 0 \\ 1/2 & 1/4 & 1/4 \\ 0 & 1/3 & 2/3 \end{pmatrix}
\end{array}
\tag{3.36}
$$

is irreducible.

Note that the state *0* and the state *2* communicate through a two-step transition:

$$0 \xrightarrow{1/2} 1 \xrightarrow{1/4} 2 \text{ and } 2 \xrightarrow{1/3} 1 \xrightarrow{1/2} 0.$$

Example 3.2 The chain with transition matrix:

$$
\mathbf{P} = \begin{array}{c} \\ 0 \\ 1 \\ 2 \\ 3 \end{array}
\begin{array}{cccc}
0 & 1 & 2 & 3 \\
\begin{pmatrix} 1/3 & 2/3 & 0 & 0 \\ 2/3 & 1/3 & 0 & 0 \\ 1/4 & 1/4 & 1/4 & 1/4 \\ 0 & 0 & 0 & 1 \end{pmatrix}
\end{array}
\tag{3.37}
$$

it is not irreducible, that is *reducible* (see Fig. 3.7).

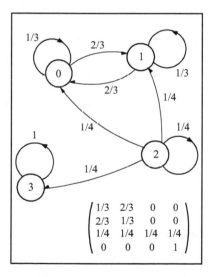

Fig. 3.7 States transition diagram of the reducible chain eqn (3.37).

We have three classes: $\{0, 1\}$, $\{2\}$, $\{3\}$. Observe that if the states *0* or *1* are accessible from *2*, this state is accessible from no other states. Then, the system may be in *2*, only if it starts from this state; once it get out, it never comes back. The state *3* is absorbing, that is $p_{ii} = 1$; an irreducible chain can have no absorbing state.

Irreducibility can be defined via the notion of *closed class*. A set of states $\mathcal{C} \subset \mathcal{S}$ is *closed* if it is impossible to exit, that is if $i \in \mathcal{C}$ and $j \notin \mathcal{C}$, then $p_{ij}^{(n)} = 0$, $\forall i \in \mathcal{C}$, $\forall j \notin \mathcal{C}$ and $\forall n$. A chain is irreducible if there exists only one closed set, namely the state space $\mathcal{S}$. In the above example (matrix eqn (3.37)), the subset $\mathcal{C} = \{0, 1\}$ is a closed irreducible class. Also the subset $\{0, 1, 3\}$ is a closed class, but not irreducible.

Consider the transition matrix:

$$
\mathbf{P} = \begin{array}{c} \\ 0 \\ 1 \\ 2 \\ 3 \\ 4 \end{array} \begin{array}{ccccc} 0 & 1 & 2 & 3 & 4 \\ \left(\begin{array}{ccccc} 0.7 & 0.3 & 0 & 0 & 0 \\ 0.4 & 0.6 & 0 & 0 & 0 \\ 0 & 0 & 0.6 & 0.1 & 0.3 \\ 0 & 0 & 0.3 & 0.5 & 0.2 \\ 0 & 0 & 0.5 & 0.4 & 0.1 \end{array} \right) \end{array}
$$

If the process starts from states 0 or 1 belonging to the subset $\mathcal{C}_1 = \{0, 1\}$ of $\mathcal{S}$, never leaves these states, and similarly if it starts from one of the states belonging to the subset $\mathcal{C}_2 = \{2, 3, 4\}$. The sets $\mathcal{C}_1$ and $\mathcal{C}_2$ are closed and irreducible classes.

If the process starts from $\mathcal{C}_1$, it can be described by the transition matrix with only states 0 and 1 (we have chosen the matrix eqn (3.23)), with its asymptotic stationary distribution, is, as we know, $\pi = (0.5714, 0.4286)$. It is the same reasoning, if the process starts from $\mathcal{C}_2$ (the matrix is eqn (3.26). In this case, its asymptotic stationary distribution is (to four decimal places) $\pi = (0.4933, 0.2800, 0.2267)$. In this sense, we speak of an 'reducible' matrix, since the state space $\mathcal{S}$ can be 'reduced' to one or more subsets of $\mathcal{S}$.

Clearly, if the chain is reducible, there cannot be a unique asymptotic distribution, as it depends on the initial state. But for the existence of the asymptotic distribution, the chain needs to be aperiodic, not just irreducible.

The occupation times of the states at stationarity can be obtained with the R code `Code_3_2.R` below. We will see later how to derive it analytically in a simple way.

```
## Code_3_2.R
# Weather in the Land of Oz
# States: rain (0), sun (1), snow (2)

markov<- function(x0,n,x,P) {   # starting function
s<- numeric()
s[1]<- x0              # initial state
raw<- which(x==x0)
for(i in 2:n) {
s[i]<- sample(x,1,P[raw,],replace=T)
raw<- which(x==s[i])
                }
return(s)
                          } # ending function
set.seed(5)          #  reset random numbers
n<- 1000            # number of steps
# transition matrix:
P<-matrix(c(0.6,0.1,0.3,0.3,0.5,0.2,0.5,0.4,0.1),3,3,byrow=T)
P
x<-c(0,1,2)        # x is the state space
x0<-0
s<-markov(x0,n,x,P)     # visited states
```

```
t<-c(0:(n-1))
        # comment if you do not wish t and s values printed
# t
# s
# for clarity, stop the plot at npl
npl<-50
par(mai=c(1.02,1.,0.82,0.42)+0.1)     # to control the margin size
plot(t[0:npl],s[0:npl],type="s",col="black",xlim=c(0,npl),
ylim=c(0,2),xlab="days",ylab="states",
cex.lab=1.3,font.lab=3,lwd=1.7,yaxt="none")
axis(2,seq(0,2,1))
# alternative representation: plot(t,s,type="p", ...
# ........................................................

# number of visits to states during the time-interval [0,n-1],
# given that the initial state is 0 (rain).
# number of raining, sunny and snowing days
n_rain<- numeric()
n_sun<-  numeric()
n_snow<- numeric()
# visiting probability
P_rain<- numeric()
P_sun<-  numeric()
P_snow<- numeric()
# The initial state (x0<- 0) is known
n_rain[1]<- 1
n_sun[1]<-  0
n_snow[1]<- 0
P_rain[1]<- 1
P_sun[1]<-  0
P_snow[1]<- 0
# stationary probability distribution
for(i in 2:n) {              # starting loop on steps
n_rain[i]<- n_rain[i-1]
n_sun[i]<-  n_sun[i-1]
n_snow[i]<- n_snow[i-1]
if(s[i]==0){n_rain[i]<- n_rain[i-1]+1 }
if(s[i]==1){n_sun[i]<-  n_sun[i-1]+1 }
if(s[i]==2){n_snow[i]<- n_snow[i-1]+1 }
P_rain[i]<- n_rain[i]/i
P_sun[i]<-  n_sun[i]/i
P_snow[i]<- n_snow[i]/i
                }                 # ending loop on steps
# number of total visits
n_rain[n]
n_sun[n]
n_snow[n]
# stationary probability vector
par(mai=c(1.02,1.,0.82,0.42)+0.1)     # to control the margin size
plot(P_rain,type="s",col="black",xlim=c(0,n),
ylim=c(0,1),xlab="days",ylab="prob. vector",
cex.lab=1.3,font.lab=3,lwd=1.7)
lines(P_sun,type="s",lty=2,lwd=1.7)
lines(P_snow,type="s",lty=3,lwd=2)

# ........... time average
ta<- 600
tb<- 1000
l_Int<- length(ta:(tb-1))
timeAv_rain<- mean(P_rain[ta:(tb-1)] )
```

```
timeAv_rain
se.rain<- sd(P_rain[ta:(tb-1)])/sqrt(l_Int)
se.rain
timeAv_sun<- mean(P_sun[ta:(tb-1)] )
timeAv_sun
se.sun<- sd(P_sun[ta:tb])/sqrt(l_Int)
se.sun
timeAv_snow<- mean(P_snow[ta:(tb-1)] )
timeAv_snow
se.snow<- sd(P_snow[ta:tb])/sqrt(l_Int)
se.snow
```

The first ten steps are:

```
> t
 [1]  0  1  2  3  4  5  6  7  8  9 10 ...
> s
 [1] 0 0 2 2 0 0 2 1 2 2 ...
```

As is also shown in Fig. 3.8, we have 'rain' for the first two days and the third day it is snowy (the initial time, that is first day, is $t = 0$). The first sunny day is the eighth day.

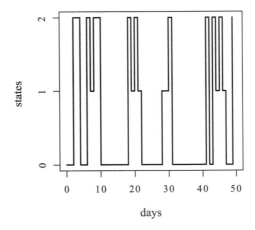

states

days

Fig. 3.8 First 50 steps of the trajectory of the process with transition matrix eqn (3.26).

The numbers identifying the rainy days are 1 (initial day), 2 $(1 + 1)$, up to the rainy fifth day. The first numbers of sunny and snowy days are also reported.

```
> n_rain
 [1]  1  2  2  2  3  4  4  4  4  4 ...
> n_sun
 [1] 0 0 0 0 0 0 0 1 1 1 ...
> n_snow
 [1] 0 0 1 2 2 2 3 3 4 5 ...
```

The components of the probability vector π describing the probability distribution over states, at *stationarity*, are computed:

```
> P_rain
 [1] 1.0000000 1.0000000 0.6666667 0.5000000 0.6000000 0.6666667 0.5714286
 [8] 0.5000000 0.4444444 0.4000000  ...
> P_sun
 [1] 0.00000000 0.00000000 0.00000000 0.00000000 0.00000000 0.00000000
 [7] 0.00000000 0.12500000 0.11111111 0.10000000  ...
> P_snow
 [1] 0.0000000 0.0000000 0.3333333 0.5000000 0.4000000 0.3333333 0.4285714
 [8] 0.3750000 0.4444444 0.5000000  ...
```

For instance, for the rainy day the probability of the system of staying in the state 'rain' is 1 for the first two days, since we know that 'rain' is the starting state. The third day, this probability becomes $2/3 \sim 0.6666667$, and the day after, it is $2/4 = 0.5$, and so on, and the same for the other states. However, we have to keep in mind that the probability distribution refers to stationarity, and stationarity is reached after a certain number of steps. Figure 3.9 shows the time evolution of the components of the probability vector π.

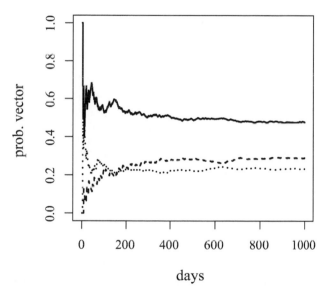

Fig. 3.9 Time evolution of the components of the probability vector π of the process with transition matrix eqn (3.26). Rain: continuous line, sun: dashed line, snow: dotted line.

We see that the stationarity is reached after, more or less, 600 steps. We can perform the time average in the interval $[600, 1000]$, obtaining:

```
timeAv_rain: 0.4820,   timeAv_sun: 0.2837, timeAv_snow: 0.2343
```

in good agreement with the results from the short code presented above.

Adding to the code `Code_3_2.R` the following instructions, we can estimate the *ensemble average*. For the sake of brevity only the state 'rain' is considered.

```
## Code_3_2.R continues

# .......... ensemble average
set.seed(5)      # reset random numbers if wished
n<- 1000         # number of steps
nhist<- 100      # number of histories
mstep<- 900    # compute the ensemble average at step = mstep
nen<-numeric()
PM_rain<-matrix(,nhist,n)  # to save all histories
par(mai=c(1.02,1.,0.82,0.42)+0.1)     # to control the margin size
plot(P_rain,type="n",col="black",xlim=c(0,n),
ylim=c(0,1),xlab="days",ylab="P_rain",
cex.lab=1.2,font.lab=3,lwd=1.7)

for(l in 1:nhist){     # starting loop on histories
x0<-0
s<-markov(x0,n,x,P)
n_rain[1]<- 1
P_rain[1]<- 1
PM_rain[l,1]<- 1
for(i in 2:n) {
n_rain[i]<- n_rain[i-1]
P_rain[i]<- P_rain[i]/i
PM_rain[l,i]<- P_rain[i]
if(s[i]==0){n_rain[i]<- n_rain[i]+1 }
P_rain[i]<- n_rain[i]/i
PM_rain[l,i]<- P_rain[i]
            }
# for clarity, stop the histories plot at hpl
hpl<-20
if(l<=hpl)lines(PM_rain[l, ],type="s",lty=3,lwd=1)
            }    # ending loop on histories
for(i in 1:n){
nen[i]<- mean(PM_rain[,i]) }               # "mean history"
lines(nen,lwd=2,type="s",col="black",lty=1) # better: col="red"
abline(v=mstep,lty=4,lwd=2,col="black") # here ensemble average
m_ens<- mean(PM_rain[,mstep])
m_ens
se.ens<- sd(PM_rain[,mstep])/sqrt(nhist)
se.ens
```

In the code above, nhist<- 100 different trajectories are computed, and the ensemble averages are performed at step mstep<- 900. Figure 3.10 shows the time evolution of 20 trajectories out of 100 of the component of the probability vector π referring to the state 'rain'. The continuous line depicts the 'mean trajectory' on nhist<- 100 trajectories. The step at which the ensemble average is computed mstep<- 900 is also shown (dot-dashed line).

The two averages, time and ensemble, are quite similar: time average = 0.4820 and ensemble average = 0.4928. The differences is the standard errors of the mean: they are 0.0003218 and 0.002196, respectively. The reason is that the time series of the components of the probability vector π are highly correlated.

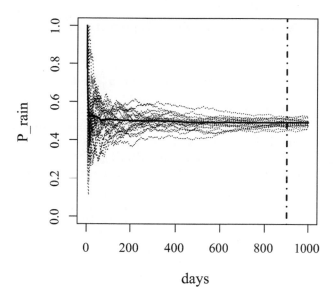

Fig. 3.10 Time evolution of 20 trajectories out of 100 of the component of the probability vector π referring to the state 'rain' of the process with transition matrix eqn (3.26). Continuous line: 'mean trajectory' on `nhist<- 100` trajectories. Vertical dot-dashed line: time (days = 900) at which the average is performed.

3.3.3 Periodic and aperiodic states

Consider the following transition matrix:

$$
\mathbf{P} = \begin{array}{c}
\\ 0 \\ 1 \\ 2 \\ 3
\end{array}
\begin{array}{cccc}
0 & 1 & 2 & 3 \\
\left(\begin{array}{cccc}
0 & 0 & 0.6 & 0.4 \\
1 & 0 & 0 & 0 \\
0 & 1 & 0 & 0 \\
0 & 1 & 0 & 0
\end{array}\right)
\end{array}
\tag{3.38}
$$

and follow the trajectory of the system, assuming that it starts from state *2*. Possible trajectories are:

$$2 \to 1 \to 0 \to 2 \to 1 \to 0 \to 2 \to 1 \to 0 \to 2 \to 1 \cdots$$
$$2 \to 1 \to 0 \to 3 \to 1 \to 0 \to 2 \to 1 \to 0 \to 2 \to 1 \cdots$$
$$2 \to 1 \to 0 \to 2 \to 1 \to 0 \to 3 \to 1 \to 0 \to 3 \to 1 \cdots$$

We immediately see that the chain is irreducible, and also that states *0* and *1* occur every three steps. State *2* does not occur every three steps, but when it occurs, it is every three steps. The same holds for state *3*. The states of this chain are called periodic with period 3. A state i is *periodic* with period $d^{(i)}$ if the chain starting from i can return to i only at a multiple of some n steps, where $n = d^{(i)}, 2d^{(i)}, 3d^{(i)}, \ldots$, and $d^{(i)}$ is the greatest integer with this property. More formally, a state i is *periodic* with period $d^{(i)}$ if $p_{ii}^{(n)} = 0$, *except* for the n values that are multiples of some integer

$d^{(i)}$, and if there exist more integers with this property, $d_1^{(i)}, d_2^{(i)}, \ldots, d_m^{(i)}$, then $d^{(i)} = \max(d_1^{(i)}, d_2^{(i)}, \ldots, d_m^{(i)})$.

Formally:

$$d^{(i)} = \gcd\{n \text{ such that } p_{ii}^{(n)} > 0\}$$

where 'gcd' is the greatest common denominator of all integer $n > 0$, for which $p_{ii}^{(n)} > 0$. Periodicity is a class property. If for the state i, $d^{(i)} = 1$, the state i is called *aperiodic*. A Markov chain is aperiodic if every state is aperiodic, otherwise it is periodic.

If the chain is periodic, an asymptotic distribution $\mathbf{v}^{(\infty)}$ does not exist (as we have seen for matrix eqn (3.32) with $a = b = 1$). For the matrix eqn (3.38), it is indeed:

$$\mathbf{P}^2 = \begin{pmatrix} 0 & 1 & 0 & 0 \\ 0 & 0 & 0.6 & 0.4 \\ 1 & 0 & 0 & 0 \\ 1 & 0 & 0 & 0 \end{pmatrix}, \quad \mathbf{P}^3 = \begin{pmatrix} 1 & 0 & 0 & 0 \\ 0 & 1 & 0 & 0 \\ 0 & 0 & 0.6 & 0.4 \\ 0 & 0 & 0.6 & 0.4 \end{pmatrix},$$

$$\mathbf{P}^4 \equiv \mathbf{P} = \begin{pmatrix} 0 & 0 & 0.6 & 0.4 \\ 1 & 0 & 0 & 0 \\ 0 & 1 & 0 & 0 \\ 0 & 1 & 0 & 0 \end{pmatrix}, \ldots$$

However, periodic chains are allowed to possess an invariant distribution $\boldsymbol{\pi}$, independent of the initial distribution. In the example above, we see that the system stays for a third of the time in the states *0* and *1*, then $\pi_0 = \pi_1 = 1/3$, in the remaining third in the states $\{2, 3\}$, then $\pi_2 = 1/3 \times 0.6 = 0.2$ and $\pi_3 = 1/3 \times 0.4 = 0.133$.

In general, we can say that the succession of transition matrices $\mathbf{P}^{(n)}$ of an irreducible and periodic chain does not converge to the asymptotic matrix $\mathbf{P}^{(\infty)}$. If, however, the sequence of arithmetic means:

$$\frac{1}{n}\left[\mathbf{P} + \mathbf{P}^{(2)} + \cdots + \mathbf{P}^{(n)}\right]$$

is considered, then for $n \to \infty$, such a sequence converges to a limiting matrix $\mathbf{P}^*$ with all positive entries and all equal rows. In the above example, the rows are the vector $\boldsymbol{\pi} = (0.333, 0.333, 0.2, 0.133)$. When the convergence of arithmetic mean succession is taken into account, it is referred as 'convergence in the Cesaro sense'.

Some theorems exist characterizing irreducible chains. One of them states that *all states in an irreducible chain are of the same type*. This means that, for instance, states are all aperiodic or all periodic, and in this case they all have the same period. Then, in an irreducible chain, if $\exists i$ such that $p_{ii} > 0$, we can state that $\forall i \in \mathcal{S}$ it is $p_{ii} > 0$ and, therefore, that the chain is aperiodic. Sufficient, but not necessary, condition.

A theorem states that: for irreducible and aperiodic chain, there exists a number m such that for $\forall n \geqslant m$, all entries of the n-step transition matrix $\mathbf{P}^n$ are strictly positive, namely, as previously stated, the chain is ergodic. It follows that there exists a unique stationary distribution coinciding with the asymptotic distribution. It should be noted, however, that if the chain has a countably infinite set of states, for ergodicity it is not enough that the chain is irreducible and aperiodic, as we will see later.

The proof is rather simple. If i and j are any two states in $\mathcal{S}$, since the chain is irreducible, there exists q such that $p_{ij}^{(q)} > 0$. The chain is also aperiodic, therefore there exists r such that from a certain $n > r$ onwards it gives $p_{ii}^{(r)} > 0$. If m is the greatest value between q and r, for all pair i and j, the theorem holds.

As an example, turn to the irreducible chain with transition matrix eqn (3.36). It is $p_{02} = p_{20} = 0$. Moreover, matrix $\mathbf{P}^{(2)}$ has all entries strictly positive:

$$
\mathbf{P}^{(2)} = \begin{array}{c} \\ 0 \\ 1 \\ 2 \end{array} \begin{pmatrix} \begin{array}{ccc} 0 & 1 & 2 \end{array} \\ 0.500 & 0.375 & 0.125 \\ 0.375 & 0.396 & 0.229 \\ 0.167 & 0.306 & 0.528 \end{pmatrix} \tag{3.39}
$$

The chain is ergodic and there exists a unique stationary distribution coinciding with the asymptotic distribution. It is:

$$
\boldsymbol{\pi} = (0.364, 0.364, 0.273)
$$

3.3.4 Stopping time and other relevant random times

Turn again to the matrix eqn (3.22). We see that if the system starts from state *1* or *4*, it can visit these states a certain number of times, until it enters state *2* or *3*, after which these states are visited infinite times. There are states, called *recurrent*, that are revisited infinite times, and fleeting states, called *transient*, which are visited only a finite number of times.

To characterize the two types of state, let us introduce new random variables, the first of which is the *random time*. A random time T is a discrete random variable which takes values in the parameter space $\mathbb{T} = \{0, 1, 2, \dots\}$. Then, the event $\{X_T = i\}$ means that the process at the random time T is in the state i. A particular random time is the *stopping time*, a name which may be a bit misleading, since it does not mean that the process stops for certain values of T.

We say that T is a *stopping time* with respect to the stochastic process $\{X_t\}$ if the occurrence, or the non-occurrence, of event $\{T = n\}$ can be exhaustively determined only by observing the values assumed by the variables $\{X_0, X_1, \dots X_n\}$. It is 'stopping time' since if we want to stop at time T, we know when to do it by looking at the process until T, with no need to know the future. In the heads and tails game, a stopping time could be the moment to stop gambling after, for instance, ten heads:

$$
T = \min\{n \geqslant 0 : X_0 + \dots + X_n = 10\}
$$
$$
(\text{if } X_n = 0 : \text{tail, if } X_n = 1 : \text{head})
$$

A time which is *not* a stopping time is the *last exit time*, the time of the last visit of a certain state. To say that the process never comes back to this state, you have to know not only the time at which the system leaves the state, but also its future evolution.

Another relevant random time is the time of the first visit of the process (or of the first passage) into the state i, also known as the *hitting time* to state i. The random variable T_i is a *first visit time* to state i if:

$$T_i = \min\{n \geqslant 1 : X_n = i\}$$

that is T_i takes the value n when no variable $X_1, X_2, \ldots, X_{n-1}$ takes the value i, but i is assumed by X_n. Therefore, the event:

the process visits i for the first time at time n}

can be expressed as:

$$\{T_i = n\} = \{X_1 \neq i, X_2 \neq i, \ldots, X_{n-1} \neq i, X_n = i\}$$

Note in T_i, the subscript i refers to that particular state i, while n is a time, that is the value taken by the random variable T_i.

A couple of observations. If the initial value is $n = 1$, in 0 step the process does not enter i, unless it is not already there. The random variable T_i can assume the value $+\infty$ with probability > 0, therefore it is a special random variable. Lastly, note that in the T_i definition the initial state is not mentioned. More generally, we can speak of the time of the first visit into the set of states $\{A\}$, for instance $\{A\} = \{0, 2, 4\}$, and write

$$T_{\{A\}} = \min\{n \geqslant 1 : X_n \in \{A\}\}$$

A gambler decides to stop playing when his pockets (the system) are in the state *0* (he lost all) or when the state is *1000* (he won 1000 euro). In this case $\{A\} = \{0, 1000\}$.

3.3.5 Strong Markov property

We have defined a Markov process as one having the property of being memoryless, i.e. if the system at time n (present) is in the state i, $\{X_n = i\}$, the future $\{X_{n+1}, X_{n+2}, \ldots\}$ does not depend on the past $\{X_{n-1}, \ldots, X_0\}$. If, instead of the 'deterministic' time n, we consider the 'random' time T, stopping time, we obtain the definition of a *strong Markov property*.

We can rewrite the above definition of a Markov process with the stopping time T in place of n to define the strong Markov property: if the system at time T (present) is in the state i, $\{X_T = i\}$, the future $\{X_{T+1}, X_{T+2}, \ldots\}$ does not depend on the past $\{X_{T-1}, \ldots, X_0\}$. The meaning of this definition is the following. I know that T is a stopping time, and that $\{X_T = i\}$. Then any other information is irrelevant to the process $\{X_{T+k}\}$, $k \geqslant 0$. In other words, I have the Markov chain starting from a certain $t = 0$. Let T be a stopping time with $\{X_T = i\}$. It follows that the process $\{X_{T+k}\}$, $k \geqslant 0$ is the same original Markov chain $\{X_t\}$, but with the initial state i. We want to stress here that in the definition of the strong Markov property, the essential point is that T is a stopping time.

3.3.6 Recurrent and transient states

Turn to the first visit time in terms of which accessibility can be defined. Consider a pair of states i and j, $\forall i, j \in \mathcal{S}$ and $i \neq j$. We say that j is accessible from i if:

$$f_{ij} = \mathsf{P}\{T_j < \infty | X_0 = i\} > 0 \qquad \text{note: it is not} = 0$$

We have introduced the probability f_{ij}, *not null*, that the system starting from i hits for the first time j in a finite time. To be more specific, define the probability $f_{ij}^{(n)}$ that the system starting from i, takes the time n for visiting j for the first time:

$$f_{ij}^{(n)} = \mathsf{P}\left\{T_j = n | X_0 = i\right\}\left\{X_1 \neq j, \ldots, X_{n-1} \neq j, X_n = j | X_0 = i\right\} \qquad (3.40)$$

Note that $f_{ij}^{(0)} = 0$, since the process cannot go from i to j in zero steps, if it is not $i = j$, but in this case $f_{ii}^{(0)} = 1$. We can write f_{ij} as a function of $f_{ij}^{(n)}$, by summing on n:

$$f_{ij} = \sum_{n=1}^{\infty} \mathsf{P}\left\{T_j = n | X_0 = i\right\} = \sum_{n=1}^{\infty} f_{ij}^{(n)}$$

Let us introduce the probability f_{ii} that the system starting from i returns in i:

$$f_{ii} = \mathsf{P}\left\{T_i < \infty | X_0 = i\right\} = \mathsf{P}_i\left\{T_i < \infty\right\} \qquad (3.41)$$

where the probability of event $\{A\}$ with the process starting from i is written in short as:

$$\mathsf{P}\left\{A | X_0 = i\right\} = \mathsf{P}_i\left\{A\right\}$$

By analogy with eqn (3.40), define the probability $f_{ii}^{(n)}$ that the process, starting from i, returns there after n steps:

$$f_{ii}^{(n)} = \mathsf{P}\left\{T_i = n | X_0 = i\right\} \qquad (3.42)$$

From this:

$$f_{ii} = \sum_{n=1}^{\infty} f_{ii}^{(n)}$$

On the basis of eqn (3.41), the notion of *recurrence* can be introduced. The state i is *recurrent* if:

$$\mathsf{P}\left\{T_i < \infty | X_0 = i\right\} = \mathsf{P}_i\left\{T_i < \infty\right\} = 1 \qquad (3.43)$$

A recurrent state is often called *persistent*, because a system starting from such a state returns to it with certainty. By contrast, the state i is *transient* if:

$$\mathsf{P}\left\{T_i < \infty | X_0 = i\right\} = \mathsf{P}_i\left\{T_i < \infty\right\} < 1$$

In general, $f_{ii}^{(n)}$ is the probability that the process, being in i at any time m, returns to i after n steps. Recurrence and transience are written:

$$\mathsf{P}\left\{X_{n+m} = i | X_m = i\right\} \xrightarrow[n \to \infty]{} \begin{cases} 1 & i \text{ is recurrent} \\ 0 & i \text{ is transient} \end{cases}$$

With f_{ii} given by

$$f_{ii} = \begin{cases} 1 \text{ or } & f_{ii}^{(n)} \xrightarrow[n \to \infty]{} 1 & i \text{ is recurrent} \\ < 1 \text{ or } & f_{ii}^{(n)} \xrightarrow[n \to \infty]{} 0 & i \text{ is transient} \end{cases}$$

If $f_{ii} = 1$ the process will return to state i any time it has visited that state. But thanks to the strong Markov property, as it is recurrent, it will be visited an infinite number of times. Clearly, the absorbing state, $p_{ii} = 1$, is also recurrent, since the process stays forever in i. If i is recurrent, the transition probability $p^{(n)}$ has to be greater than zero for some n value.

If state i is not recurrent, but transient, then the process can return to i only with a certain probability, so that there is a non-zero probability that the process will never return to i after leaving it. This means that a transient state after a certain time will never occur in the chain.

Generalize T_i, first visit time to i, in T_i^k, time of the k-th visit to i:

$$T_i^k = \min\{n > T_i^{k-1} | X_n = i\}$$

We have defined f_{ij}, the probability that the system starting from i hits for the first time j in a finite time, similarly f_{ji} for j to i. Then, we start from j and go to i for the first time, then we will visit it again $k - 1$ times. This is expressed by the following events:

$$\{j \to i \quad \text{first time}\} \quad \text{and} \quad \{i \to i \quad k - 1 \text{ times}\}$$

For the strong Markov property the two events are independent, therefore:

$$\mathsf{P}\left\{T_i^k < \infty | X_0 = j\right\} = f_{ji}\, f_{ii}^{k-1}$$

An important notion is the *mean recurrence time* μ_i, defined as the expected value of the random variable T_i, conditioned on the event that the process starts from i, that is:

$$\mu_i = \mathsf{E}\left[T_i | X_0 = i\right] \equiv \mathsf{E}_i\left[T_i\right] \tag{3.44}$$

Let i be recurrent, then if the states of the chain are *infinite*, μ_i can be finite or infinite, If μ_i is finite the state i is said to be *positive recurrent*, while if μ_i is infinite the state i is said to be *null recurrent*. Clearly, if the states of the chain are finite, the state i is either recurrent or transient, without further specification.

If i is recurrent, $f_{ii}^{(n)}$ is a probability distribution; indeed, in this case $\sum_{n=1}^{\infty} f_{ii}^{(n)} = 1$. As further consequence, we can define the mean recurrence time through $f_{ii}^{(n)}$ in eqn (3.42):

$$\mu_i = \mathsf{E}\left[T_i | X_0 = i\right] = \sum_{n=1}^{\infty} n\, f_{ii}^{(n)} \quad \text{if} \quad \begin{cases} < \infty & i \text{ is positive recurrent} \\ +\infty & i \text{ is null recurrent} \end{cases}$$

Recurrence has been characterized through the random variable T_i (and its generalizations, as T_i^k) which informs about the *time* taken by the process to reach a certain state i. Now, recurrence is characterized through the *number of times* that a state i is visited. Let V_i be the random variable counting the number of times that the process visits i, defined as:

$$V_i = \sum_{n=0}^{\infty} \mathbb{1}_{(X_n = i)}$$

where $\mathbb{1}_{(\cdot)}$ is the *indicator function* taking value 1, if the condition $(\cdot)$ is verified, and 0 otherwise. In our case, $\mathbb{1}_{(X_n = i)} = 1$, if $X_n = i$, while $\mathbb{1}_{(X_n = i)} = 0$, if $X_n \neq i$.

The expected value of V_i is:

$$\mathrm{E}_i\left[V_i\right] \equiv \mathrm{E}\left[V_i \middle| X_0 = i\right] = \sum_{n=0}^{\infty} p_{ii}^{(n)}$$

Indeed:

$$\mathrm{E}_i\left[V_i\right] = \mathrm{E}\left[\sum_{n=0}^{\infty} \mathbb{1}_{(X_n=i)} \middle| X_0 = i\right] = \sum_{n=0}^{\infty} \mathrm{E}\left[\mathbb{1}_{(X_n=i)} \middle| X_0 = i\right] \qquad (3.45)$$

since the expected value of a sum of random variables is equal to the sum of their expected values. We have:

$$\mathrm{E}\left[\mathbb{1}_{(X_n=i)}\right] = \mathsf{P}\left\{X_n = i\right\}$$

because by definition the indicator function of an event takes the value 1 when the event occurs and 0 when the event does not occur. Then eqn (3.45) is equal to:

$$\sum_{n=0}^{\infty} \mathsf{P}\left\{X_n = i \middle| X_0 = i\right\} = \sum_{n=0}^{\infty} p_{ii}^{(n)}$$

A theorem can be proved, which says:

state i is recurrent if and only if $\mathrm{E}_i\left[V_i\right] = \infty$

If i is recurrent, the number of visits in i is infinite. If i is transient the number of visits in i is finite and the expected value of V_i is finite:

$$\mathrm{E}\left[V_i \middle| X_0 = i\right] = \frac{f_{ii}}{1 - f_{ii}}$$

Also $\sum_{n=0}^{\infty} p_{ii}^{(n)}$ is finite.

We said that the states of an irreducible chain are all of the same type. Therefore if a state is recurrent and the chain is irreducible, all the state are recurrent. But in a *finite-state* Markov chain, not all states can be transient, since after a certain time there are no more states to be reached. Only if the process has infinite states, can they be all transient states.

In summary, for a finite-state chain, to say 'irreducible' or to say that 'all states are recurrent' is the same thing. A finite-chain with all states recurrent (that is irreducible) and aperiodic is *ergodic*. If the number of states is infinite, the chain is ergodic only if it is aperiodic and it has *positive* recurrent states.

If both types of state belong to the chain, the state space can be partitioned in a disjoint set of states: $\mathcal{T}, \mathcal{R}_1, \mathcal{R}_2, \ldots$, such that:

1. $\mathcal{T}$ it is the set of all transient states.

2.   if $i \in \mathcal{R}_l$,   then: $\begin{cases} f_{ik} = 1, \forall k \in \mathcal{R}_l \\ f_{ik} = 0, \forall k \notin \mathcal{R}_l \end{cases}$

Therefore $\mathcal{R}_l$ is a irreducible set of states with only recurrent states.

Example 3.3 Let us consider the following transition matrix:

$$
\mathbf{P} = \begin{array}{c} \\ 0 \\ 1 \\ 2 \\ 3 \\ 4 \end{array}
\begin{array}{ccccc}
0 & 1 & 2 & 3 & 4 \\
\begin{pmatrix}
1/2 & 1/2 & 0 & 0 & 0 \\
1/2 & 1/2 & 0 & 0 & 0 \\
0 & 0 & 1/2 & 1/2 & 0 \\
0 & 0 & 1/2 & 1/2 & 0 \\
1/4 & 1/2 & 0 & 0 & 1/2
\end{pmatrix}
\end{array}
$$

States belonging to the set $\mathcal{R}_1 = \{0, 1\}$ communicate with others allowing to visit again the states in $\mathcal{R}_1$ and are recurrent, and so are those belonging to the set $\mathcal{R}_2 = \{2, 3\}$. The set $\mathcal{T} = \{4\}$ is composed of only one transient state. In fact, if the process starts from the state *4*, it cannot return there, once it entered the closed set $\mathcal{R}_1 = \{0, 1\}$.

Figure 3.11 schematically shows the states classification discussed so far.

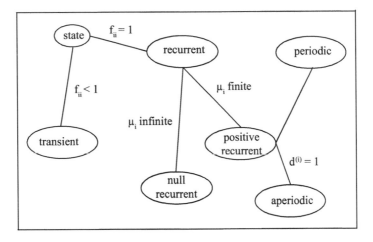

Fig. 3.11 Scheme of the state classification in Markov chain.

3.3.7 Mean recurrence time and stationary distribution

We are interested in estimating the mean recurrence time μ_i of the state i defined in eqn (3.44). To do so we have to count the number of steps to return to i after it was left. Consider the following example. Let i be the starting state, for short we write T^k instead of T_i^k to denote the time of the k-th visit to i. Suppose the random variable T^1 assumes, for instance, the value $\tau^1 = 5$. This means that the process starts from i and after five steps comes back to i for the first time. The process goes on as follows:

The proess starts from i, then:

It comes back to i after 5 steps, then: $\tau^1 = 5$ first return

It comes back to i after 2 steps, then: $\tau^2 = 2$ second return

It comes back to i after 3 steps, then: $\tau^3 = 3$ third return

It comes back to i after 2 steps, then: $\tau^4 = 2$ fourth return

The time spent by the process until the fourth return to i is $5 + 2 + 3 + 2 = 12$ steps. The mean is $(5 + 2 + 3 + 2)/4 = 3$, therefore there are on average 3 steps between one visit and the next, which is an estimate of the mean recurrence time of the state i.

In general, the time spent by the process until the k-th return to i is given by the sum of the random variables T^i, all with the same distribution:

$$T^1 + T^2 + \cdots + T^j + \cdots + T^k$$

where T^j is the time between the $(j-1)$-th and the j-th return. In the example, T^3 has the value 3.

The weak law of large numbers states that the average of the sequence converges in probability to the ensemble average of the variables T^i, $\forall i \in \mathcal{S}$:

$$\frac{1}{k}\left(T^1 + T^2 + \cdots + T^k\right) \xrightarrow[k\to\infty]{p} \mathrm{E}\left[T_i \middle| X_0 = i\right] \equiv \mathrm{E}_i\left[T_i\right]$$

(omitting the subscript i) $\equiv \mathrm{E}\left[T\right]$

This means that there are about k visits in $k\,\mathrm{E}\left[T\right]$ steps (with k large, of course).

We said that π_i is the mean time spent in i, if the state i is visited $n\pi_i$ times in n steps. Then we can write (omitting the subscript i):

k visits in $k\,\mathrm{E}\left[T\right]$ steps, and also:

$n\pi$ visits in n steps

if $n = k\,\mathrm{E}\left[T\right]$, it follows:

$k\,\mathrm{E}\left[T\right]\pi$ visits in $k\,\mathrm{E}\left[T\right]$ steps

From which:

$$\pi = \frac{1}{\mathrm{E}\left[T\right]}$$

Now $\mathrm{E}\left[T\right]$ is the average recurrence time, usually written for the state i as μ_i, then:

$$\mu_i = \frac{1}{\pi_i}$$

Remembering the example relative to the weather of the Land of Oz, notably eqn (3.23), the recurrence time μ_0, average number of days between two rainy days, is $1/0.57 = 1.75$, while μ_1, average number of days between two sunny days, is $1/0.43 = 2.33$.

Turn to the example involving three states: rain, sun and snow, whose transition matrix is eqn (3.26). The stationary distribution is given by the system:

$$\pi_0 = 0.6\pi_0 + 0.3\pi_1 + 0.5\pi_2$$
$$\pi_1 = 0.1\pi_0 + 0.5\pi_1 + 0.4\pi_2$$
$$\pi_1 = 0.5\pi_0 + 0.4\pi_1 + 0.1\pi_2$$
$$\pi_0 + \pi_1 + \pi_2 = 1$$

The solution is the stationary distribution $\pi = (0.4933, 0.2800, 0.2267)$. The recurrence time of a rainy day (between two rainy days, there can be sunny or snowy days) is $1/0.4933 = 2.03$.

We could be interested in the average number v_{ij} of visits to the state i between two visits to the states j. This is given by:

$$v_{ij} = \frac{\pi_i}{\pi_j} = \frac{\mu_j}{\mu_i}$$

In the example above, the average number of rainy days between two snowy days is given by:

$$v_{02} = \frac{\pi_0}{\pi_2} = \frac{\mu_2}{\mu_0} = \frac{0.4933}{0.2267} = 2.18$$

With a small code `Code_3_3.R`, we can estimate via Monte Carlo the recurrence time in Markov chain. The example is the weather in the Land of Oz, and part of the code is the same as in `Code_3_2.R`, repeated here for convenience. The new instruction is the **diff** function which computes the difference between pairs of successive elements of a vector.

```
## Code_3_3.R
# mean recurrence time
# stationary distribution = (0.4933,0.2800,0.2267)
# Weather in the Land of Oz
# States: rain (0), sun (1), snow (2)
markov<- function(x0,n,x,P) {   # starting function
s<- numeric()
s[1]<- x0               # initial state
raw<- which(x==x0)
for(i in 2:n) {
s[i]<- sample(x,1,P[raw,],replace=T)
raw<- which(x==s[i])
            }
return(s)
                           } # ending function
set.seed(5)           #  reset random numbers
n<- 200               # number of steps
# transition matrix:
P<-matrix(c(0.6,0.1,0.3,0.3,0.5,0.2,0.5,0.4,0.1),3,3,byrow=T)
x<-c(1,2,3)
x0<- 1
s<-markov(x0,n,x,P)
t<-c(0:(n-1))
r<-which(s==1)
d<-diff(r)  # difference between successive values
# for instance: r = 3,7,12,13, diff(r) = 4,5,1
```

```
m<-mean(d)
m                   # theory 1/0.4933 = 2.03
r<-which(s==2)
d<-diff(r)
m<-mean(d)
m                   # theory 1/0.2800 = 3.57
r<-which(s==3)
d<-diff(r)
m<-mean(d)
m                   # theory 1/0.2267 = 4.41
```

With the stationary distribution $\pi = (0.4933, 0.2800, 0.2267)$, the recurrence times of rainy, sunny and snowy days are 2.03, 3.57 and 4.41, respectively. The results of the code (number of steps = 200) are 1.895, 3.957 and 4.355, respectively, in good agreement, as expected.

3.3.8 Sojourn time

Today is raining, so how many days (excluding today) will be rain days? To answer this question, let us introduce a new random variable S_i, called *sojourn time* in the state i.

Suppose that at initial time $t = 0$, the process is in the state i. It may remain in i with probability p_{ii} for a further step ($t = 1$), or may leave i with probability $(1 - p_{ii})$. If it remained in i, suppose it leaves i at the next step ($t = 2$), that is the events are:

$$\{\text{the process stays in } i \text{ still one step } (t - 1)\}$$
$$\{\text{the process leaves } i \text{ at the next step } (t = 2)\}$$

Thanks to the strong Markov property, the two events are independent, therefore:

$$\mathsf{P}\left\{X_2 = j \middle| X_1 = i, X_0 = i\right\} = p_{ii}\left(1 - p_{ii}\right), \qquad (j \neq i)$$

is the probability that the process at $t = 2$ is in the state j, given at $t = 0$ was in i, and at $t = 1$ was still in i. Instead, if at $t = 1$, it still stays in i, the probability that it leaves at $t = 3$ is given by:

$$\mathsf{P}\left\{X_3 = j \middle| X_2 = i, X_1 = i, X_0 = i\right\} = p_{ii}\, p_{ii}\left(1 - p_{ii}\right) \qquad (j \neq i)$$

We introduce the random variable S_i as the time during which the system stays in i, once inside. For instance, the process enters i at step 4, and stays there at steps $5, 6, 7, 8, 9$, and at the step 10 is no longer in i, but it is in j, which means that the process remained in i for five steps. So in general:

$$\mathsf{P}\left\{S_i = k\right\} = p_{ii}^{k-1}\left(1 - p_{ii}\right), \qquad k = 1, 2, \ldots \quad \text{and} \quad \forall i \in \mathcal{S}$$

where S_i is the random variable *sojourn time*, with *geometric* distribution, and represents the probability that the process stays in i for further $k - 1$ steps, once entered there.

To estimate the average number of steps spent in i, once the process is entered and before leaving it, we have to compute the mean value of the variable S_i, given by:

$$\mathrm{E}\left[S_i\right] = \sum_{k=1}^{\infty} k p_{ii}^{k-1} \left(1 - p_{ii}\right) = \frac{1}{1 - p_{ii}}, \qquad \forall i \in \mathcal{S}$$

Therefore, the answer to the initial question is: if today it is raining, we have to wait on average for

$$\mathrm{E}\left[S_i\right] = \frac{1}{1 - \pi_0} = \frac{1}{\pi_1} = \frac{1}{0.429} = 2.33$$

days, before the weather changes.

3.3.9   Summing up

There are chains with a finite number of states (*finite-chains*), and chains with a infinite, but numerable, number of states (*infinite-chains*)

$$\text{chains} \begin{cases} \text{finite} \\ \text{infinite} \end{cases}$$

Finite-chains can be irreducible or not irreducible (namely reducible).
If a chain is irreducible, all states are recurrent, i.e. to say irreducible or recurrent is the same thing. If it is reducible there can be recurrent and transient states, but they cannot to be all transient.

$$\text{finite-chains} \begin{cases} \text{irreducible: all recurrent states} \Rightarrow \text{irreducible} \equiv \text{recurrent} \\ \text{reducible} \begin{cases} \text{recurrent states} \\ \text{transient states, but } not\ all \end{cases} \end{cases}$$

A chain can possess the following combinations of attributes:

- An irreducible and aperiodic finite-chain ($\equiv$ all states recurrent) is called ergodic.
- An irreducible, aperiodic and positive recurrent infinite-chain is called ergodic.
- An irreducible, aperiodic and positive recurrent infinite-chain is *not* ergodic if recurrent states are null recurrent.
- An irreducible and aperiodic infinite-chain is *not* ergodic if states are transient.

3.4   Continuous-time Markov chain

In a continuous-time stochastic process, events can occur at any moment t of a continuous set of values, not at fixed times as in the time-discrete case. In the Poisson process, described in the next chapter, we will see that events can happen at any point of the time axis, unlike what happens, for example, in a Bernoulli process where events can only occur at certain times. Of course, also for continuous-time processes there exists a timescale, but the time units, that is the time interval in which the system changes state (or stays in the same state), is not 'implicit', as in discrete-time processes. The concept of 'step' has no more reason to exist, just as it makes no sense to speak of 'one-step transition', or of 'n steps'. If time is continuous, the 'step' can be as small as we want. In any case, the transition probability in 'time period t' remains an essential concept. A further important difference is that continuous-time Markov chains are *aperiodic* by definition. We will see, indeed, that the random variable representing state changes has an exponential distribution, so states cannot follow one another at regular intervals.

3.4.1 Matrix transition probability function

Let us study continuous-time Markov chains in discrete state space, that is $t \in \mathbb{T} \subseteq \mathbb{R}[0, \infty)$, but the values of random variables X_t are discrete. For discrete-time chains, we wrote the Markovian property:

$$\mathsf{P}\left\{X_{n+1} = j \big| X_n = i, X_{n-1} = i_{n-1}, \ldots, X_0 = i_0\right\}$$
$$= \mathsf{P}\left\{X_{n+1} = j \big| X_n = i\right\}$$

Now we write:

$$\mathsf{P}\left\{X(s+t) = j \big| X(s) = i, X(u) = x(u), 0 \leqslant u < s\right\}$$
$$= \mathsf{P}\left\{X(s+t) = j \big| X(s) = i\right\} \tag{3.46}$$

where $x(u), 0 \leqslant u < s$ are the states visited in the past, before time s. The chain is *time homogeneous* if the probability:

$$\mathsf{P}\left\{X(s+t) = j \big| X(s) = i\right\} \quad \forall s \leqslant t, \ \forall t \geqslant 0, \ \forall i, j \in \mathcal{S}$$

depends only on the interval amplitude $[t-s]$, not on the particular values of extremes at s and t. In other words, whatever moment the system enters the state i, its probabilistic time evolution is the same, as the system should start ($t = 0$) from the state i. In the following we will deal with homogeneous and regular chains. A continuous-time Markov chain is called *regular* if the number of transitions in a finite time interval is finite. Define the *transition probability function*, or simply the *transition function*, from state i to state j in the time interval $[s, t]$ (not in n steps as in the discrete case):

$$p_{ij}(s,t) = \mathsf{P}\left\{X(s+t) = j \big| X(s) = i\right\}, \quad s \leqslant t$$

and for homogeneous chains:

$$p_{ij}(s, s+t) = p_{ij}(0, t) \equiv p_{ij}(t)$$

Notice that for each state i and j, the transition function $p_{ij}(s,t)$ is a continuous function of t. Also for continuous-time Markov chains, we introduce a matrix $\mathbf{P}(t)$, called the *matrix transition probability function*, or simply the *matrix transition function*, which can be regarded as the continuous time counterpart of the transition matrix in n steps at discrete time. With finite number N of states, it is given by:

$$\mathbf{P}(t) = \begin{pmatrix} p_{11}(t) & p_{12}(t) & \ldots & p_{1N}(t) \\ p_{21}(t) & p_{22}(t) & \ldots & p_{2N}(t) \\ \multicolumn{4}{c}{\ldots\ldots\ldots\ldots\ldots\ldots\ldots} \\ p_{N1}(t) & p_{N2}(t) & \ldots & p_{NN}(t) \end{pmatrix} \tag{3.47}$$

where $p_{ij}(t) = \mathsf{P}\left\{X(t) = j \big| X(0) = i\right\}$, transition probability from state i to state j in time t. 'In time t' does not mean, of course, in t steps, but it means in a time interval $\Delta t = [0, t]$, with amplitude t. Transition probabilities $p_{ij}(t)$ exhibit the same properties of p_{ij} in the discrete case, that is $\forall t$:

$$p_{ij}(t) \in [0, 1], \ \forall i, j \in \mathcal{S} \qquad \text{and} \qquad \sum_{j \in \mathcal{S}} p_{ij}(t) = 1, \ \forall i \in \mathcal{S} \qquad (3.48)$$

As in the discrete case, $\sum_{j \in \mathcal{S}}$ means summation on all the states belonging to the space $\mathcal{S}$, both with finite number N, and in the case of numerable infinity. The matrix $\mathbf{P}(t)$, for eqns (3.48), is therefore a stochastic matrix. Similarly to the discrete case, we have:

$$p_{ij}(0) = \delta_{ij} = \begin{cases} 1, & \text{if } i = j \\ 0, & \text{if } i \neq j \end{cases}$$

and in matrix form $\mathbf{P}(0) = \mathbf{I}$, where $\mathbf{I}$ is the identity matrix.

The Chapman−Kolmogorov equation is now written:

$$p_{ij}(s + t) = \sum_{k \in \mathcal{S}} p_{ik}(s) \, p_{kj}(t), \qquad 0 \leqslant s \leqslant k \leqslant t \qquad (3.49)$$

and in matrix form:

$$\mathbf{P}(s + t) = \mathbf{P}(s) \, \mathbf{P}(t), \qquad s, t > 0$$

Due to this last property, together with eqns (3.48) and because $\mathbf{P}(0) = \mathbf{I}$, the family $\{\mathbf{P}(t)\}$ is defined as a *stochastic semigroup*.

The process is completely described by the matrix transition function $\mathbf{P}(t)$, and as in the discrete case, by the initial distribution over the N states, that is by the probability vector at time 0: $\mathbf{v}(0) = \big(v_1(0), v_2(0), \ldots, v_N(0)\big)$, with $v_i(0) = \mathsf{P}\{X(0) = i\}$, $i = 1, 2, \ldots, N$.

Supposing that $p_{ij}(t)$ are derivable functions $\forall t \geqslant 0$, define ($\forall i, j \in \mathcal{S}, i \neq j$) and $\Delta t \geqslant 0$:

$$q_{ij} = \left. \frac{p_{ij}(t)}{t} \right|_{t=0} = +p'_{ij}(0) = \lim_{\Delta t \to 0} \frac{p_{ij}(\Delta t)}{\Delta t} \qquad (3.50)$$

and for $i = j$:

$$q_{ii} = \left. \frac{p_{ij}(t)}{t} \right|_{t=0} = -p'_{ii}(0) = - \lim_{\Delta t \to 0} \frac{1 - p_{ii}(\Delta t)}{\Delta t} \qquad (3.51)$$

Note that it is $q_{ii} < 0$.

Consider eqn (3.50). The quantity q_{ij} is defined as the prime derivative with respect to time of the transition function $p_{ij}(t)$ at $t = 0$. Therefore, q_{ij} is not a probability; at the most we might say that it is a 'probability' less than infinitesimal of order higher than the first with respect to Δt, a 'probability per time unit'. In fact, q_{ij} represents the 'instantaneous transition speed' of the system of moving from state i to state j; q_{ij} is called also the *transition intensity*

Let us think of a procedure of this type, to understand the reason that Δt must tend to 0. We begin the experiment by resetting the clock, namely $t = 0$. At $t = 0$, the system is in state i. We wait for a time Δt. If at this time, the system is in state j, we say we have achieved success. Of course, the system might not reach state j, in which case there is no success. If the system went to j, we would bring back it again in i, and would repeat the experiment. Let us give some numbers. We take a time period of 1 minute, and execute six experiments, that is one every 10 s, which means $\Delta t = 10$ s. We say that the *average number* of successes in 1 minute is given by the executed number of experiment in 1 minute, times the success probability $p_{ij}(\Delta t)$ in each experiment. Then, if $p_{ij}(\Delta t) = 0.5$, the average number of successes in 1 minute is $6 \times 0.5 = 3$. Quite correct! But if Δt is finite, we cannot affirm with certainty that the system moved directly from i to j. In 10 s, it could have arrived at j, starting from i and visiting states $k_1, k_2, \ldots$. However, if $\Delta t \to 0$, practically if Δt is small enough for the realization of a unique experiment (in the example, one second), the success guarantees exactly a direct transition $i \to j$, therefore q_{ij} can be legitimately regarded as the average number of transitions $i \to j$ per time unit, namely as *transition frequency*, and this is just its physical meaning. Turning to eqn (3.51), the quantity q_{ii}, always $\leqslant 0$, represents the frequency with which the system in state i, leaving it to reach another state.

Suppose that the system at time t is in state i. In the interval Δt, the system leaves i, namely it is no more in i at the instant $t + \Delta t$. If accessible states are $h, k_1, k_2, \ldots, k_h$ (see Fig. 3.12), the frequency q_{ii} has to be distributed over such states, that is, the following relation holds:

$$q_{ii} = -\sum_{k \neq i} q_{ik}, \tag{3.52}$$

where k stays for states $k_1, k_2, \ldots, k_h$.

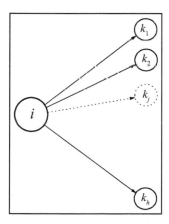

Fig. 3.12 Transition from state i to another possible state.

For instance, suppose there are four states and that $q_{33} = -2$, that is the system left two times (in suitable time units) the state *3*. A possible scenario is that the system once went from *3* to *2* and once from *3* to *4*. Since it is $q_{31} = 0$, $q_{32} = 1$ and $q_{34} = 1$, it results in $-2 = -(0 + 1 + 1)$.

To derive the relation eqn (3.52), the starting point is the certain event:

$$1 = \sum_{k \in \mathcal{S}} p_{ik}(\Delta t), \quad \forall i \in \mathcal{S}$$

and separating the term $p_{ii}(\Delta t)$ from the others with $k \neq i$, it is:

$$1 = \sum_{k \in \mathcal{S}} p_{ik}(\Delta t) = p_{ii}(\Delta t) + \sum_{k \neq i \in \mathcal{S}} p_{ik}(\Delta t) \tag{3.53}$$

From eqn (3.51), it is: $-q_{ii}(\Delta t) = 1 - p_{ii}(\Delta t)$, so that, for the first term of the right-hand side, it is $p_{ii}(\Delta t) = 1 + q_{ii}(\Delta t)$. Using eqn (3.50), $q_{ik}(\Delta t) = p_{ik}(\Delta t)$ is replaced in the summation in eqn (3.53), giving:

$$1 = \sum_{k \in \mathcal{S}} p_{ik}(\Delta t) = 1 + q_{ii}(\Delta t) + \sum_{k \neq i \in \mathcal{S}} q_{ik}(\Delta t)$$

from which eqn (3.52). It may also be $q_{ii} = 0$, meaning that i is an *absorbing* state (in discrete-time case it is $p_{ii} = 1$).

The quantities q_{ij} are important, since they represent observed data. We will see now how, by knowing q_{ij} and the initial distribution $\mathbf{p}(0)$, the matrix transition function $\mathbf{P}(t)$ and consequently the probability distribution over states as a time function can be obtained.

3.4.2   Transition intensity matrix

Turning to the Chapman−Kolmogorov equation, eqn (3.49) is written a bit differently, substituting $(t + \Delta t)$ for $(s + t)$. Therefore eqn (3.49) becomes:

$$p_{ij}(t + \Delta t) = \sum_{k \in \mathcal{S}} p_{ik}(t) \, p_{kj}(\Delta t)$$

Summing only the terms $k \neq j$ we obtain:

$$p_{ij}(t + \Delta t) = \sum_{k \neq j \in \mathcal{S}} p_{ik}(t) \, p_{kj}(\Delta t) + p_{ij}(t) \, p_{jj}(\Delta t)$$

Subtracting $p_{ij}(t)$ from both members:

$$p_{ij}(t + \Delta t) - p_{ij}(t) = p_{ij}(t) \, p_{jj}(\Delta t) - p_{ij}(t) + \sum_{k \neq j \in \mathcal{S}} p_{ik}(t) \, p_{kj}(\Delta t)$$

$$= p_{ij}(t)[p_{jj}(\Delta t) - 1] + \sum_{k \neq j \in \mathcal{S}} p_{ik}(t) \, p_{kj}(\Delta t)$$

Dividing both members by Δt:

$$\frac{p_{ij}(t + \Delta t) - p_{ij}(t)}{\Delta t} = p_{ij}(t) \frac{p_{jj}(\Delta t) - 1}{\Delta t} + \sum_{k \neq j \in \mathcal{S}} p_{ik}(t) \frac{p_{kj}(\Delta t)}{\Delta t}$$

If Δt tends to 0, the ratio at the first member is the time derivative $p'_{ij}(t)$, and by eqn (3.51), the first term of the second member is equal to $q_{jj} p_{ij}(t)$. So we can write:

$$p'_{ij}(t) = q_{jj}p_{ij}(t) + \sum_{k \neq j \in S} p_{ik}(t)q_{kj} \qquad (3.54)$$

Assigning the initial conditions $p_{ij}(0)$, the system eqn (3.54) can be resolved, then functions $p_{ij}(t)$ are determined. In matrix notation, it can be written:

$$\mathbf{Q} = \begin{pmatrix} q_{11} & q_{12} & \cdots & q_{1N} \\ q_{21} & q_{22} & \cdots & p_{2N} \\ \cdots & \cdots & \cdots & \cdots \\ q_{N1} & q_{N2} & \cdots & q_{NN} \end{pmatrix}$$

or defining $q_{ii} = -q_i$

$$= \begin{pmatrix} -q_1 & q_{12} & \cdots & q_{1N} \\ q_{21} & -q_2 & \cdots & p_{2N} \\ \cdots & \cdots & \cdots & \cdots \\ q_{N1} & q_{N2} & \cdots & -q_N \end{pmatrix}$$

$$(3.55)$$

The matrix $\mathbf{Q}$ is called the *intensity matrix* or *infinitesimal generator matrix* or *rate matrix*. Notice that:

a) non-diagonal elements are non-negative: $q_{ij} \geqslant 0$, $\forall i \neq j$.
b) diagonal elements are non-positive: $0 \leqslant -q_{ii} < \infty$, $\forall i$. Note: $(-q_{ii}) > 0$.
c) the sum of each row is $= 0$: $\sum_{j \in S} q_{ij} = 0$, $\forall i$.

For example, matrix $\mathbf{Q}$ can be as ($S = \{1, 2, 3, 4\}$):

$$\mathbf{Q} = \begin{pmatrix} -1 & 1 & 0 & 0 \\ 1 & -3 & 1 & 1 \\ 0 & 1 & -2 & 1 \\ 0 & 1 & 1 & -2 \end{pmatrix}$$

where: -1 is q_{11} or $-q_1$, -3 is q_{22} or $-q_2$, and so on. As we said, q_{ij} are not probabilities, so they can assume values greater than one. Matrix entries are to be read adding the time unit, for instance: "the system left three times the state *2 in 1 hour*, (or in one minute, one second, etc.) and visited once the state *1*, once the state *3*, and once the state *4*".

The system eqn (3.54) can be written in terms of transition functions:

$$\mathbf{P}'(t) = \frac{d\,\mathbf{P}(t)}{dt} = \mathbf{P}(t)\,\mathbf{Q} \quad (\textit{forward equation})$$

or:

$$\mathbf{P}'(t) = \frac{d\,\mathbf{P}(t)}{dt} = \mathbf{Q}\,\mathbf{P}(t) \quad (\textit{backward equation})$$

with initial conditions: $\mathbf{P}(0) = \mathbf{I}$. The *unique solution* is given by:

$$\mathbf{P}(t) = e^{\mathbf{Q}t} \qquad (3.56)$$

where the exponential can be written as:

$$e^{\mathbf{Q}t} = \sum_{j=0}^{\infty} \frac{\mathbf{Q}^j t^j}{j!} = \mathbf{I} + \mathbf{Q}t + \frac{\mathbf{Q}^2 t^2}{2} + \frac{\mathbf{Q}^3 t^3}{3} + \cdots$$

and eqn (3.56) can be also written:

$$\mathbf{P}(t) = \mathbf{I} + \sum_{j=1}^{\infty} \frac{\mathbf{Q}^j t^j}{j!} \qquad (3.57)$$

We said that if time is discrete, the probability vector (or state vector) at the n-th step is (N number of states) (see eqn (3.25)):

$$\mathbf{v}^{(n)} = (v_1^{(n)}, v_2^{(n)}, \ldots, v_N^{(n)}),$$
$$\text{with} \quad \forall i \in \mathcal{S}, \, \forall n \in \mathbb{T}, \, v_i^{(n)} \geqslant 0 \ \text{and} \ \sum_{i \in \mathcal{S}} v_i^{(n)} = 1$$

In this case, $v_i^{(n)} = \mathsf{P}\{X_n = i\}$ is the probability that the system is in state i after n steps.

With continuous time, we have the probability vector:

$$\mathbf{v}(t) = \big(v_0(t), v_1(t), \ldots, v_N(t)\big),$$
$$\text{with} \quad \forall i \in \mathcal{S}, \, v_i(t) \geqslant 0 \ \text{and} \ \sum_{i \in \mathcal{S}} v_i(t) = 1$$

Now $v_i(t) = \mathsf{P}\{X(t) = i\}$ is the probability that the system is in state i at time t. In general, $\mathbf{v}(t)$ says that the system at time t is in state *0* with probability $v_0(t)$, in state *1* with probability $v_1(t)$, etc.

Let $\mathbf{v}(0)$ be the probability vector at time 0. It is:

$$\mathbf{v}(t) = \mathbf{v}(0)\,\mathbf{P}(t)$$

by deriving both members with respect to time:

$$\mathbf{v}'(t) = \mathbf{v}(0)\,\mathbf{P}'(t) = \mathbf{v}(0)\,\mathbf{P}(t)\,\mathbf{Q}$$

from which the equation describing the variation over time of the states probability, given the initial conditions $\mathbf{v}(0)$, is:

$$\mathbf{v}'(t) = \mathbf{v}(t)\,\mathbf{Q} \qquad (3.58)$$

or also:

$$\frac{dv_i(t)}{dt} = \sum_{j \in \mathcal{S}} q_{ji} v_j(t), \quad \forall i \in \mathcal{S} \tag{3.59}$$

and in explicit form:

$$
\begin{aligned}
v_0'(t) &= q_{00} v_0(t) + q_{10} v_1(t) + \cdots + q_{N0} v_N(t) \\
v_1'(t) &= q_{01} v_0(t) + q_{11} v_1(t) + \cdots + q_{N1} v_N(t) \\
&\cdots\cdots\cdots\cdots\cdots\cdots\cdots\cdots\cdots\cdots\cdots\cdots\cdots\cdots \\
v_N'(t) &= q_{0N} v_1(t) + q_{1N} v_1(t) + \cdots + q_{NN} v_N(t)
\end{aligned}
\tag{3.60}
$$

and also for eqn (3.56):

$$\mathbf{v}(t) = \mathbf{v}(0)\, e^{\mathbf{Q}t}$$

3.4.3 Embedded matrix

We have seen that a continuous-time Markov process is defined if, given the intensity matrix $\mathbf{Q}$ (whose entries are observable quantities) and the initial conditions $\mathbf{v}(0)$, we are able to determine the matrix transition function $\mathbf{P}(t)$. Either we resolve the differential equation system eqn (3.58), or we diagonalize $\mathbf{Q}$ and compute $\mathbf{P}(t)$.

We will now describe a further way to describe the process, by means of the random variable S_i *sojourn time*. The sojourn time was defined in Section 3.3.8 for discrete-time Markov chains. The sojourn time in state i is the time spent by the system in state i before leaving it, or in other words the time instant at which the system in i changes states. It is defined as *the* distribution of the sojourn time in i since it is the same every time the system enters state i (recall the assumption of homogeneity of the chain). Supposing that the system is in state i at time 0, the the sojourn time in i is given by:

$$S_i = \min\{t \geqslant 0, X(t) \neq i \,|\, X(0) = i\} \tag{3.61}$$

Therefore the value j, given by $X(S_i) = j$, denotes the state visited the first time by the process when it leaves the initial state i.

A simple example can illustrate the meaning of S_i, in practice. Consider a process with only two states *0* and *1*, and suppose that only one transition from *0* to *1* occurs. Let λ be the *transition rate* (or transition intensity) q_{01}. From the above assumptions, it follows: $q_{00} = -\lambda$ and $q_{10} = q_{11} = 0$. The intensity matrix results:

$$\mathbf{Q} = \begin{pmatrix} -\lambda & \lambda \\ 0 & 0 \end{pmatrix} = -\lambda \begin{pmatrix} 1 & -1 \\ 0 & 0 \end{pmatrix} = -\lambda\, \overline{\mathbf{Q}}$$

Since $\mathbf{Q}^k = (-\lambda)^k\, \overline{\mathbf{Q}}$, $k = 1, 2, \ldots$, from eqn (3.57) it is:

$$\mathbf{P}(t) = \mathbf{I} + \sum_{k \geqslant 1} \frac{\mathbf{Q}^k t^k}{k!} = \mathbf{I} + \overline{\mathbf{Q}} \sum_{k \geqslant 1} \frac{(-\lambda)^k t^k}{k!}$$

The series above is equal to:

$$\sum_{k \geqslant 1} \frac{(-\lambda)^k t^k}{k!} = e^{-\lambda t} - 1$$

Therefore:

$$\mathbf{P}(t) = \mathbf{I} + \overline{\mathbf{Q}}\left(e^{-\lambda t} - 1\right)$$

More explicitly:

$$\mathbf{P}(t) = \begin{pmatrix} 1 & 0 \\ 0 & 1 \end{pmatrix} + \begin{pmatrix} 1 & -1 \\ 0 & 0 \end{pmatrix} \left(e^{-\lambda t} - 1\right)$$

that is:

$$\mathbf{P}(t) = \begin{pmatrix} 1 & 0 \\ 0 & 1 \end{pmatrix} + \begin{pmatrix} e^{-\lambda t} - 1 & -e^{-\lambda t} + 1 \\ 0 & 0 \end{pmatrix} = \begin{pmatrix} e^{-\lambda t} & 1 - e^{-\lambda t} \\ 0 & 1 \end{pmatrix} \tag{3.62}$$

Therefore, the transition probability function $p_{00}(t)$, namely the probability that the system still stays in state *0* at time t, has an exponential form:

$$\mathsf{P}\left\{X(t) = 0 \middle| X(0) = 0\right\} = p_{00}(t) = e^{-\lambda t} = e^{q_{00}t} = e^{-q_0 t}$$

At time t, the process goes into state *1* with transition probability function $p_{01}(t)$:

$$\mathsf{P}\left\{X(t) = 1 \middle| X(0) = 0\right\} = p_{01}(t) = 1 - e^{-\lambda t} = 1 - e^{-q_0 t}$$

which can be regarded as the distribution of the random variable S_0, "sojourn time in state *0*", that is $\mathsf{P}\left\{S_0 \leqslant t\right\}$. Take, for instance, $\lambda = 3$. As saying:

$$q_{01} = 3, \ q_{00} = -3, \ -q_0 = -3$$

Then:

$$p_{01}(t) = 1 - e^{-3t}$$
$$p_{00}(t) = e^{-3t}$$

At time $t = 0$, we know $\mathsf{P}\left\{X(0) = 0\right\} = 1$, namely: $v_0(0) = 1$ and $v_1(0) = 0$. At time $t = 1$, *e.g.* after one second, we have:

$$t = 1: \begin{cases} p_{01}(t) = 0.9502 \\ p_{00}(t) = 0.0498 \end{cases}$$

We see that the probability $\mathsf{P}\left\{X(t) = 1 \middle| X(0) = 0\right\}$ that the system at time $t = 1$ enters the state *1* is given by $p_{01}(t) = 0.9502$. Similarly, 0.0498 is the probability that the system still stays in state *0* after one second. After a further second:

$$t = 2: \begin{cases} p_{01}(t) = 0.9975 \\ p_{00}(t) = 0.0025 \end{cases}$$

the probability to remain in state *0* diminishes, and as a consequence the probability of leaving it increases.

The reasoning holds for any time t. Therefore, if the system is in i at time $t = 0$ it stays there for a time given by the exponential random variable with intensity q_i:

$$\mathsf{P}\left\{S_i > t\right\} = e^{-q_i t}, \quad t \geqslant 0 \tag{3.63}$$

The system enters state j with probability given by:

$$\mathsf{P}\left\{X(S_i) = j\right\} = \frac{q_{ij}}{q_i}$$

as we see later.

The same results can also be obtained without resorting to the random variable S_i. Consider a generic state i, to which there are no transitions, that is transitions towards i are excluded. Equation (3.60) reduces to:

$$\frac{v_i(t)}{t} = -q_i v_i(t), \quad \text{con } v_i(0) = 1$$

The solution is:

$$v_i(t) = 1 - e^{-q_i t} \tag{3.64}$$

which says that the probability that the system at time t is in state i decreases exponentially with time, the frequency being equal to the number of times that the system, in time units, left state i.

This process is *memoryless*. We have already seen what this means: the transition probability at a time after a certain instant s is the same regardless of the time spent before s, that is of the time considered as an initial instant, which may be 0, or any t. Suppose that the system has been in state i for 60 minutes. We want the probability that it will stay there for another 30 minutes, that is $\mathsf{P}\{S_i < 90 | S_i > 60\}$. If the process is memoryless, the system behaves as if it had left i, that is:

$$\mathsf{P}\{S_i \leqslant 90 | S_i > 60\} = \mathsf{P}\{S_i \leqslant (60 + 30) | S_i > 60\} = \mathsf{P}\{S_i \leqslant 30\}$$

We have seen that the process stays in state i for some time, described by an exponential random variable, before changing state. Figure 3.12 suggests that the system will visit states $k_1, k_2, \ldots, k_h$ with probability proportional to the transition probabilities q_{ik}. By normalizing with respect to the transition intensity $q_i = -q_{ii}$ with which the system leaves i, we can define a new type of transition probability $\tilde{p}_{ij}$:

$$\tilde{p}_{ij} = \frac{q_{ij}}{\displaystyle\sum_{k \neq i} q_{ik}} = \frac{q_{ij}}{q_i}, \quad i \neq j$$

Such probabilities are not a transition function, and they do not depend on time. Of course, $\tilde{p}_{ii} = 0$, since the process cannot stay in i when it leaves it. That is also justified by an empirical point of view, because what is observable is a state change. It must be $\sum_{j \neq i} \tilde{p}_{ij} = 1$, since the system must reach some state in any case. For the sake of completeness, we have to take into account that if i is absorbing, then $q_i = 0$. In this case $\tilde{p}_{ii} = 1$ and $\tilde{p}_{ij} = 0$, for $i \neq j$.

In conclusion, we know *when* the system moves and *where* it goes, that is *we know the process*. Then, if the process at time 0 is in state i_0, it will stay in such a state for a time described by an exponential distribution with intensity q_{i_0}, after it jumps in state i_1 with probability $\tilde{p}_{i_0 i_1}$, and it stays there for a time q_{i_1}, after which it jumps to state i_2 with probability $\tilde{p}_{i_1 i_2}$, and so on.

Turning to eqn (3.63), representing the distribution of the sojourn time S_i, it gives the time spent in i by the system, but it does not give any information about where it goes after leaving i. To know the probability that the sojourn time in i ends with a transition to state j, we have to know the transition function $p_{ij}(t)$, given by:

$$p_{ij}(t) = \frac{q_{ij}}{q_i} e^{-q_i t} = \tilde{p}_{ij} e^{-q_i t}, \quad i \neq j$$

The quantities $\tilde{p}_{ij}$ form a stochastic matrix $\widetilde{\mathbf{P}}$ called the *embedded* transition matrix. This matrix completely describes the behaviour of a discrete-time Markov chain. To every continuous-time Markov chain $\{X(t)\}$ it is possible to associate a discrete-time *embedded* Markov Chain $\{\widetilde{X}(t)\}$. The embedded chain is also called the 'skeleton' of the continuous-time chain. The matrix $\widetilde{\mathbf{P}}$ only accounts for transition probabilities from state to state, but not for transition rates, sojourn times, etc. We can form a continuous-time chain from a discrete-time embedded chain, but we need q_i in addition to $\tilde{p}_{ij}$.

Example 3.4 Let us write an intensity matrix $\mathbf{Q}$, and the corresponding embedded matrix $\widetilde{\mathbf{P}}$, with $\mathcal{S} = \{1, 2, 3\}$.

$$\mathbf{Q} = \begin{pmatrix} -1 & 1 & 0 \\ 1 & -2 & 1 \\ 0 & 1 & -1 \end{pmatrix} \qquad \widetilde{\mathbf{P}} = \begin{pmatrix} 0 & 1 & 0 \\ 0.5 & 0 & 0.5 \\ 0 & 1 & 0 \end{pmatrix}$$

For example:

$$\tilde{p}_{12} = \frac{q_{12}}{q_{12} + q_{13}} = \frac{1}{1+0}, \quad \cdots \quad \tilde{p}_{32} = \frac{q_{32}}{q_{31} + q_{32}} = \frac{1}{0+1}$$

Figure 3.13 provides a visual synthesis of the meaning of the embedded chain of a Markov process. Let $\{X(t)\}$ be a continuous-time Markov chain and let $\{\widetilde{X}(t)\}$ be the corresponding embedded chain. The process at time $t = 0$ is in state $0 = 1$, where it stays for a time $[0, t_1)$, determined by the exponential distribution with intensity q_0. At t_1, with probability $\tilde{p}_{01} = q_{01}/q_0$, it jumps into state $1 = 5$, and so on. The realization of $\widetilde{X}(0)$ is denoted as $\tilde{x}_0 = 1$.

3.4.4 Poisson *retrouvé*

The Poisson process is described in the next chapter, here we describe the Poisson process from the point of view of a Markov process. Let's go back to reconsider the process involving two states 0 and 1, and a unique transition from 0 to 1 with frequency $\lambda = q_{01}$, $q_{00} = -\lambda$ and $q_{10} = q_{11} = 0$. Consider now a countable infinity of states: $0, 1, 2, \ldots$, as sketched in Fig. 3.14. In this case, the transition function matrix eqn (3.47) has infinite rows and infinite columns. At the beginning, the process is in state 0, and at a certain instant it jumps to state 1, leaves 1 and enters 2, forgetting what it did before, and so on. Let us suppose that all the random variables S_i, sojourn time in states $i, i = 0, 1, \ldots$, have the same exponential distribution with intensity λ. Therefore:

$$q_{ij} = \begin{cases} \lambda, & \text{if } j = i + 1 \\ 0, & \text{if } j \neq i + 1, \end{cases} \qquad i = 0, 1, \ldots$$

with $q_{ii} = -\lambda$ and initial condition $\mathsf{P}\{X_0 = 0\} = 1$. The intensity matrix is:

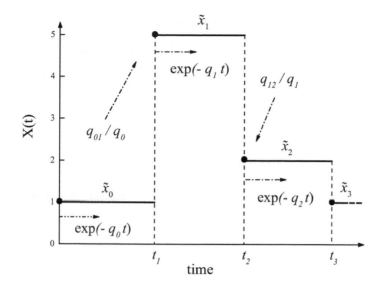

Fig. 3.13 Scheme of realizations of embedded chain $\{\widetilde{X}(t)\}$ of a continuous-time Markov chain $\{X(t)\}$.

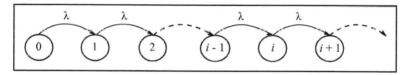

Fig. 3.14 Sketch of the Poisson process with a countable infinity of states, and with a transition frequency from state to state equal to λ.

$$
\mathbf{Q} = \begin{pmatrix}
\underset{q_{00}}{-\lambda} & \underset{q_{01}}{\lambda} & \underset{q_{02}}{0} & \underset{q_{03}}{0} & \dots \\[4pt]
\underset{q_{10}}{0} & \underset{q_{11}}{-\lambda} & \underset{q_{12}}{\lambda} & \underset{q_{13}}{0} & \dots \\[4pt]
\underset{q_{20}}{0} & \underset{q_{21}}{0} & \underset{q_{22}}{-\lambda} & \underset{q_{23}}{\lambda} & \dots \\[4pt]
\multicolumn{5}{c}{\dots\dots\dots\dots\dots\dots\dots}
\end{pmatrix}
\tag{3.65}
$$

The corresponding embedded matrix ('skeleton') is:

$$
\widetilde{\mathbf{P}} = \begin{pmatrix}
0 & 1 & 0 & 0 & \dots \\
0 & 0 & 1 & 0 & \dots \\
0 & 0 & 0 & 1 & \dots \\
\multicolumn{5}{c}{\dots\dots\dots\dots\dots\dots}
\end{pmatrix}
$$

Let us turn to eqn (3.58), namely:

$$
\mathbf{v}'(t) = \mathbf{v}(t)\,\mathbf{Q}
$$

With the aim to determine how the probability vector $\mathbf{v}$ changes with time, consider that we now have a countable infinity of states, so we write:

$$\mathbf{v}(t) = \big(v_0(t), v_1(t), \dots, v_j(t), \dots\big)$$

recalling that $v_i(t) = \mathsf{P}\{X(t) = i\}$ is the probability that the process is in state i at time t.

The above equation is written in explicit form (see also eqn (3.60)):

$$
\begin{aligned}
v_0'(t) &= q_{00}v_0(t) + q_{10}v_1(t) + q_{20}v_2(t) + \cdots + q_{j0}v_j(t) + \cdots \\
v_1'(t) &= q_{01}v_0(t) + q_{11}v_1(t) + q_{21}v_2(t) + \cdots + q_{j1}v_j(t) + \cdots \\
&\cdots\cdots\cdots\cdots\cdots\cdots\cdots\cdots\cdots\cdots\cdots\cdots\cdots\cdots\cdots \\
v_i'(t) &= q_{0i}v_0(t) + q_{1i}v_1(t) + q_{2i}v_2(t) + \cdots + q_{ji}v_j(t) + \cdots \\
&\cdots\cdots\cdots\cdots\cdots\cdots\cdots\cdots\cdots\cdots\cdots\cdots\cdots
\end{aligned}
\tag{3.66}
$$

Equation (3.66) greatly simplifies considering that the only non-zero entries q_{ij} are those with $j = i+1$, equal to λ, and the entries q_{ii} are equal to $-\lambda$. For instance, for $v_0'(t), v_1'(t)$ and $v_2'(t)$, it is:

$$v_0'(t) = \underbrace{(-\lambda)}_{q_{00}} v_0(t) \quad (q_{10}, q_{20}, \text{ etc. are all } = 0)$$

$$v_1'(t) = \underbrace{\lambda}_{q_{01}} v_0(t) + \underbrace{(-\lambda)}_{q_{11}} v_1(t) \quad (q_{21}, \text{ etc. are all } = 0)$$

$$v_2'(t) = \underbrace{q_{02} v_0(t)}_{} + \underbrace{\lambda}_{q_{12}} v_1(t) + \underbrace{(-\lambda)}_{q_{22}} v_2(t) + \underbrace{q_{32} v_3(t)}_{}, \text{ etc.}$$

$$= \lambda v_1(t) - \lambda v_2(t)$$

$$\cdots\cdots\cdots\cdots\cdots\cdots\cdots\cdots\cdots\cdots\cdots\cdots\cdots$$

In general:

$$v_j'(t) = \lambda v_{j-1}(t) - \lambda v_j(t) \tag{3.67}$$

The integral of $v_0'(t) = -\lambda v_0(t)$ is:

$$v_0(t) = e^{-\lambda t}$$

Equation (3.67) is integrated by multiplying both members by $\exp(\lambda t)$, obtaining:

$$e^{\lambda t} v_j'(t) = e^{\lambda t} \lambda v_{j-1}(t) - \lambda v_j(t)$$

or also:

$$\frac{d}{dt}\left[e^{\lambda t} v_j(t)\right] = e^{\lambda t} v_{j-1}(t)$$

By integrating both members:

$$e^{\lambda t} v_j(t) = \int_0^t \lambda e^{\lambda y} v_{j-1}(y)\, dy$$

from which:

$$v_j(t) = e^{\lambda t} \lambda \int_0^t e^{\lambda y} v_{j-1}(y)\, dy$$

Let us compute $v_1(t)$. In the place of v_0, that is v_{j-1}, we take its value $v_0(t) = \exp(-\lambda t)$:

$$v_1(t) = \lambda e^{\lambda t} \int_0^t e^{\lambda y} \underbrace{e^{-\lambda y}}_{v_0(y)} \, dy = \lambda e^{\lambda t} \int_0^t dy = \lambda t \, e^{-\lambda t}$$

By solving recursively the remaining equations, that is: $v_2(t) = \ldots, v_3(t) = \ldots$, etc., the general solution is obtained:

$$v_i(t) = \mathsf{P}\{X(t) = i\} = \frac{(\lambda t)^i}{i!} \, e^{-\lambda t}, \quad t \geqslant 0, \ \ i \in \mathcal{S} = \{0, 1, 2 \ldots\}$$

which is just the Poisson distribution with parameter λ. From the point of view of Markov processes, the Poisson process can be defined as a continuous-time Markov process, in discrete state space, $\mathcal{S} = \{0, 1, 2 \ldots\}$. Every transition of the system corresponds to an arrival at time t. We know that the system stays in a state during a time given by the exponential random variable, therefore we find again that in the Poisson process, waiting times between arrivals are described by independent random variables, all with the same exponential distribution.

3.4.5 Birth-death process

We can look at Fig. 3.14 as a depiction of a *pure birth process*. Let us imagine observing an animal population which reproduces according to the following rule. When a new individual comes to light, we say that a transition occurs and the state of the system (number of animals) is increased by one. In theory, all animal species are good, except rabbits, since we know that from early 1200 that Leonardo da Pisa, who in the 18th century was called Fibonacci, precisely modelled the growth of a rabbit population by a nonlinear time series, later known as the Fibonacci's.

We wish now to make the model more 'realistic', so first the intensities λ can change from state to state, so they become λ_i. Moreover, we admit also transitions from i to $i - 1$, not only from i to $i + 1$, that is transitions can only be to a neighbouring state. Intensities given by sojourn times from i to $i - 1$ are denoted as μ_i. In this model, transitions from state i to state $i + 1$ are called as births, and those from i to $i - 1$ as deaths.

The matrix eqn (3.65) becomes:

$$\begin{array}{c@{\quad}c} & \begin{array}{cccccccc} 0 & 1 & 2 & 3 & \ldots & i-1 & i & i+1 & \ldots \end{array} \\ \begin{array}{c} 0 \\ 1 \\ 2 \\ \ldots \\ i-1 \\ i \\ i+1 \\ \ldots \end{array} & \left(\begin{array}{ccccccccc} -\lambda_0 & \lambda_0 & & & & & & & \\ \mu_1 & -(\lambda_1 + \mu_1) & \lambda_1 & & & & & & \\ 0 & \mu_2 & -(\lambda_2 + \mu_2) & \lambda_2 & & & & & \\ \ldots & \ldots & \ldots & \ldots & \ldots & \ldots & \ldots & \ldots & \ldots \\ & & & & & & & & \\ & & & & & \mu_i & -(\lambda_i + \mu_i) & \lambda_i & \\ & & & & & & & & \\ \ldots & \ldots & \ldots & \ldots & \ldots & \ldots & \ldots & \ldots & \ldots \end{array} \right) \end{array} \quad (3.68)$$

where $\lambda_i \geqslant 0$, while $\mu_i > 0$, with $\mu_0 = 0$ and $q_i = -q_{ii} = \lambda_i + \mu_i$.

Figure 3.14 is generalized as Fig. 3.15.

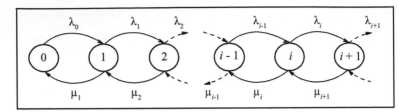

Fig. 3.15 Sketch of the birth-death process with a countable infinity of states, and with a transition frequency from state to state equal to λ_i (birth) or μ_i (death).

To summarize:

$$q_{ij} = \begin{cases} \lambda_i, & \text{if } j = i+1 \quad & \text{(rate of birth)} \\ \mu_i, & \text{if } j = i-1 \quad & \text{(rate of death)} \\ 0, & \text{if } j \neq i+1,\, i-1, \quad & i = 0, 1, \ldots \\ q_i = -q_{ii} = \lambda_i + \mu_i, & \forall i \end{cases}$$

Stationary distribution does not depend on time, as if to say that at stationarity the probability vector $\mathbf{v}(t)$ does not depend on time, as if to say that its derivative with respect to time must be zero:

$$\lim_{t \to \infty} \frac{dv_i(t)}{dt} = 0, \quad i = 0, 1, \ldots$$

The system of equations eqns (3.60) becomes in the present case:

$$\begin{aligned} v_0'(t) &= q_{00}v_0(t) + q_{10}v_1(t) + \cdots + q_{N0}v_N(t) \\ v_1'(t) &= q_{01}v_0(t) + q_{11}v_1(t) + \cdots + q_{N1}v_N(t) \\ &\cdots\cdots\cdots\cdots\cdots\cdots\cdots\cdots\cdots\cdots\cdots\cdots\cdots \\ v_N'(t) &= q_{0N}v_1(t) + q_{1N}v_1(t) + \cdots + q_{NN}v_N(t) \end{aligned}$$

In the present case:

$$v_0'(t) = -\lambda_0 v_0(t) + \mu_1 v_1(t)$$

and in general:

$$v_k'(t) = -\lambda_{k-1}v_{k-1}(t) + \mu_{k+1}v_{k+1}(t) - (\lambda_k + \mu_k)v_k(t)$$

The solution is obtained by knowing the initial probability distribution $\mathbf{v}^{(0)}$. Another way to derive the stationary distribution is through the condition *detailed balance* for continuous-time processes. The *global* detailed balance is written as:

$$\pi_i \sum_{j \neq i} q_{ij} = \sum_{j \neq i} \pi_j q_{ij}$$

where the left-hand side represents the total flow outgoing from state i into states $j \neq i$, while the right-hand side represents the total flow entering state i outgoing from all states $j \neq i$.

The *local* detailed balance holds if the probability fluxes from i to j and vice versa are equal:

$$\pi_i q_{ij} = \pi_j q_{ji}, \qquad \forall i, j$$

In discrete time, the detailed balance is (π equilibrium distribution):

$$\pi_k p_{ij} = \pi_j p_{ji}, \qquad \forall i, j$$

In our case, at stationary the probability vector $\mathbf{v}(t) = \big(v_0(t), v_1(t), \ldots, v_j(t), \ldots\big)$, can be derived by writing the balance equations:

$$
\begin{aligned}
\mu_1 v_1 &= \lambda_0 v_0 \\
\lambda_0 v_0 \mid \mu_2 v_2 &= (\lambda_1 + \mu_1) v_1 \\
\lambda_1 v_1 + \mu_3 v_3 &= (\lambda_2 + \mu_2) v_2 \\
&\cdots\cdots\cdots\cdots\cdots\cdots
\end{aligned}
$$

from which:

$$v_1 = v_0 \frac{\lambda_0}{\mu_1}$$

$$v_2 = v_1 \frac{\lambda_1}{\mu_2} = v_0 \frac{\lambda_0 \lambda_1}{\mu_1 \mu_2}$$

$$v_3 = v_2 \frac{\lambda_2}{\mu_3} = v_0 \frac{\lambda_0 \lambda_1 \lambda_2}{\mu_1 \mu_2 \mu_3}$$

$$\cdots\cdots\cdots\cdots\cdots\cdots$$

In general:

$$v_n = v_0 \left(\frac{\lambda_0 \lambda_1 \ldots \lambda_{n-1}}{\mu_1 \mu_2 \ldots \mu_n} \right), \, n \geqslant 1$$

From the normalization condition $\sum_{i=0}^{\infty} v_i = 1$, the existence of stationary probability implies for all $n \geqslant 0$ the following *ergodicity condition*:

$$\sum_{n=1}^{\infty} \left(\frac{\lambda_0 \lambda_1 \ldots \lambda_{n-1}}{\mu_1 \mu_2 \ldots \mu_n} \right) < \infty$$

Then all v_n can be determined:

$$v_n = \frac{\prod_{i=0}^{n-1} \left(\dfrac{\lambda_i}{\mu_{i+1}} \right)}{1 + \sum_{n=1}^{\infty} \prod_{i=0}^{n-1} \left(\dfrac{\lambda_i}{\mu_{i+1}} \right)}, \, n \geqslant 1 \tag{3.69}$$

where v_0 is:

$$v_0 = \left(1 + \sum_{n=1}^{\infty} \prod_{i=0}^{n-1} \frac{\lambda_i}{\mu_{i+1}} \right)^{-1} \tag{3.70}$$

or more explicitly:

$$v_0 = \left(1 + \sum_{n=1}^{\infty} \frac{\lambda_0 \lambda_1 \ldots \lambda_{n-1}}{\mu_1 \mu_2 \ldots \mu_n}\right)^{-1}$$

The following code can be seen somewhat as a generalization of the `Code_3_2` in which exponential generation of arrival times are simulated. Now we have to take into account transitions from state i either to state $i+1$, or to state $i-1$. The probability of moving to state $i+1$ is $\lambda_i/(\lambda_i + \mu_i)$, while that of moving to state $i-1$ is $\mu_i/(\lambda_i + \mu_i)$. The process spends a length of time in the new state according to an exponential distribution with parameter $(\lambda_i + \mu_i)$, and the process continues with choosing the subsequent state to be visited. The binomial random variable decides whether there is a birth (`bin==1`) or a death (`bin==0`). The process is followed until a fixed value `Deltat`, in some time units, unless the population becomes extinct before.

```
## Code_3_4.R
#  Birth-death process simulation
set.seed(2)         # reset random numbers
nhistories<-  4     # number of simulated histories
Deltat<- 3          # the process is followed no more than Deltat
lambda<- 0.5     # rate of birth
mu<-       0.9     # rate of death
n.init <- 20     # initial population size
time<-matrix(,nhistories,200)
population<-matrix(,nhistories,200)
ls<- numeric()
lp<- numeric()
### 2 loops: most external loop on histories,
#            inner loop generates births/deaths
for(l in 1:nhistories)      {       # starting loop on histories
ev<- 0
np<- n.init
n<- np
t<- 0
while(t<Deltat & np>0) {    # population may become extinct
t<- t + rexp(1,(lambda+mu)*n)
# if 1: birth, if 0: death
bin<-rbinom(1,1,(lambda/(lambda+mu)))
np<-ifelse (bin==1, np<- np+1, np<- np-1)
ev<- c(ev,t)
n <- c(n,np)
}               # ending while loop
ls[l]<-length(ev)
time[l,1:ls[l]] <- ev
lp[l]<-length(n)
population[l,1:ls[l]] <-n
}               # ending loop on histories
## # uncomment the following lines to print the transition instant
# and the number of individuals over time for each history
# time[1,1:ls[1]]
# time[2,1:ls[2]]
# time[3,1:ls[3]]
# time[nhistories,1:ls[nhistories]]
# population[1,1:ls[1]]
# population[2,1:ls[2]]
# population[3,1:ls[3]]
# population[nhistories,1:ls[nhistories]]
par(mfrow=c(2,2))
```

```
plot(time[1,1:ls[1]],population[1,1:ls[1]],type="s"
,xlab="time",ylab="population")
abline(v=Deltat,lty=3)
plot(time[2,1:ls[2]],population[2,1:ls[2]],type="s"
,xlab="time",ylab="population")
abline(v=Deltat,lty=3)
plot(time[3,1:ls[3]],population[3,1:ls[3]],type="s",
xlab="time",ylab="population")
abline(v=Deltat,lty=3)
plot(time[nhistories,1:ls[nhistories]],
population[nhistories,1:ls[nhistories]],type="s",
xlab="time",ylab="population")
abline(v=Deltat,lty=3)
```

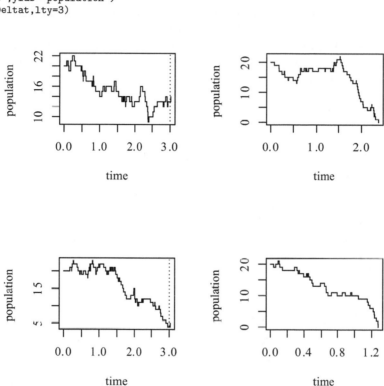

Fig. 3.16 Number of individuals as a function of time for four histories. Transition frequency from state to state are equal to λ_i (birth) or μ_i (death). The dotted line is at $\Delta t = 3$.

Figure 3.16 shows four time histories of the birth-death process. In the histories 1 and 3, the process is stopped after 71 and 90 steps, respectively, while in the histories 2 and 4, the process becomes extinct after 69 and 41 steps, respectively, as reported below:

```
> population[1,1:ls[1]]
[1]   20 21 20 21 20 19 20 21 22 21 ...
[51] 14 15 16 15 14 13 12 11 10  9 10 11 12 13 12 13 14 13 12 13 14
> population[2,1:ls[2]]
[1]   20 19 18 17 16 17 16 15 14 15 ...
[51] 10  9  8  7  6  5  4  5  4  5  6  5  4  3  2  3  2  1  0
```

```
> population[3,1:ls[3]]
 [1]   20 21 20 21 22 23 22 21 20 19 ...
[76]  11 12 11 10 11 10  9 10  9  8  7  6  5  4  5
> population[4,1:ls[4]]
 [1]   20 19 20 21 20 19 18 19 18 17 ...
[26]  11 10 11 10 11 10  9  8  7  6  5  4  3  2  1  0
```

In any case, if the rate of death μ is greater than the rate of birth λ, the population sooner or later becomes extinct; on the contrary, if $\lambda > \mu$, the population increases without bounds. When all individuals have disappeared, the size of the population remains zero from there on; that is 0 is an absorbing state. By increasing the number of histories and following each of them up to extinction, we can estimate the average survival time of the population, about 75 steps.

The birth-death process has wide application in queueing models, in which 'animals' are now customers in a queue. In a simple queue, the process is in state i if there are i costumers in the queue. The costumers are served in order of arrival, the first is receiving the service and the others (if any) are waiting in the queue. In this model, the arrivals (births) occur in a Poisson process with rate λ, then here $\lambda_i = \lambda, \forall i$. The *service rate* (or processing time) is μ ($\mu_i = \mu, \forall i$), so the mean processing time is $1/\mu$. Once the customer has been served, they leave the queue and the number of customers decreases by one (death). For instance, the queue is empty, so the waiting time for the first customer is zero. Suppose that the waiting time for the second customer is 20 minutes, and the waiting time for the third customers is $20 + 25 = 45$ minutes. Then the average waiting time is $(0 + 20 + 25)/3 = 15$ minutes.

We have described a type of queue named $M/M/1$. The two Ms stand for 'Markov', since both the arrival and the service rate are exponential, therefore they have Markovian property. The '1' indicates that there is one server, so customers are served once at a time. Equation (3.69) becomes with $\lambda_i = \lambda$ and $\mu_i = \mu$:

$$v_n = \frac{\left(\frac{\lambda}{\mu}\right)^n}{1 + \sum_{n=1}^{\infty} \left(\frac{\lambda}{\mu}\right)^n} = \left(\frac{\lambda}{\mu}\right)^n \left(1 - \frac{\lambda}{\mu}\right), \; n \geqslant 0$$

The quantity $\rho = \lambda/\mu$ is named the 'load of the queue', or 'traffic intensity', the mean number of arrivals per unit of time. Clearly, ρ must be less than 1, if the customers arrive with a rate λ greater than the service rate μ, the queue tends to become infinitely long. In other words, stationary probability exists if and only if:

$$\sum_{n=1}^{\infty} \frac{\lambda_n}{\mu_n} = \sum_{n=1}^{\infty} \rho^n < \infty$$

Then, eqn (3.70) becomes:

$$v_0 = \left(1 + \sum_{n=1}^{\infty} \frac{\lambda_n}{\mu_n}\right)^{-1} = \left(\sum_{n=0}^{\infty} \rho^n\right)^{-1} = \left(\frac{1}{1-\rho}\right)^{-1} = (1 - \rho)$$

So, if $\mu < \rho$, the process is positive recurrent and the stationarity distribution is:

$$v_n = \left(\frac{\lambda_n}{\mu_n}\right) v_0 = (1 - \rho)\, \rho^n$$

Notice that v_0 represents the probability that the queue has no customers, that is the system is in the state 0, so the next lucky customer is promptly served.

Some formulas characterizing the queueing process are reported below. We said that $v_n(t) = \mathsf{P}\{X(t) = n\}$ is the probability that, at time t, the process is in the state n, that is the probability that the system contains n customers. Therefore, the average number of customers in the system N, that is the average length of the queue, at stationarity is (Ross, 2014):

$$N = \sum_{n=0}^{\infty} n v_n = (1 - \rho) \sum_{n=0}^{\infty} n \rho_n = \frac{\rho}{1 - \rho} = \frac{\lambda}{\mu - \rho}$$

and the variance of the number of customers is:

$$\mathrm{Var}\,[N] = \sum_{n=1}^{\infty} n^2 v_n - N^2 = \frac{\rho}{(1 - \rho)^2}$$

The average amount of time spent by a customer in the system (note: in the system) is:

$$T = \frac{N}{\lambda} = \frac{1}{\mu - \lambda} = \frac{1/\mu}{1 - \lambda/\mu}$$

The formula $N = \lambda T$ holds and is known as *Little's law*, in words: *the average number of customers present in the queue is equal to the arrival rate of customers times the average amount of time spent by customers in the queue.*

Notice that T includes the *queueing delay* plus the mean *service time* $T_S = 1/\mu$. So, to be precise, N refers to the number of customers either in the queue or in service. The average amount of time that a customer spends in the queue is given by:

$$W = T - T_S = T - \frac{1}{\mu} = \frac{1}{\mu - \lambda} - \frac{1}{\mu}$$

However, T and W are not always separately examined, but only a unique average time spent in the system (*in queue or process*) is considered.

Turning to the example of customers going to a pub, modelled by the `Code_3_1.R`. Suppose now that the average number of arrivals is 50 per hour. The time between successive arrivals (inter-arrival time) is exponentially distributed with mean $1/\lambda$:

$$\frac{1}{\lambda} = \text{(in seconds)}\ \frac{60 \times 60}{50} = 72\,\mathrm{s} = 1.2\ \text{minutes}$$

and the average frequency is then $\lambda = 0.0139\,\mathrm{s}^{-1}$. The waiting time W of a costumer in line depends on the length of the queue, that is the number of customers before him in the line, at time t. The bartender takes 1 minute to serve a costumer, then the service rate is $\mu = 60$ customers per hour. From Little's Law, the average amount of time spent by a customer *in the system* is

$$T = \frac{1}{\mu - \lambda} = \frac{1}{60 - 50} = \frac{1}{10} \text{ hrs} = 6.0 \text{ minutes}$$

The waiting time W of a costumer in line is:

$$W = T - \frac{1}{\mu} = 6 - 1 = 5 \text{ minutes}$$

and there are on average:

$$N = \lambda T = 0.8333 \times 6.0 = 5$$

customers in the pub. The load of the queue is then:

$$\rho = \lambda/\mu = 50/60$$

The paper *The Theory of Probabilities and Telephone Conversations* published in 1909 by A. K. Erlang is referred to as the first paper on queuing theory. Many excellent books have been written on this subject, including (Medhi, 2003; Ross, 2014).

3.4.6 Probability and determinism: the Buffon's needle

Before introducing other techniques based on the Monte Carlo method, let us discuss, perhaps, the first Monte Carlo application. This example is extensively covered in the literature on Monte Carlo and has a certain charm, probably because it shows how it is possible to estimate a universal constant, π, probably the most universally known constant, with a very simple game. Also it is an interesting example of a problem that can be solved with either a deterministic approach or a probabilistic one. Buffon's needle was one the earliest problems in geometric probability to be solved.

In 1777 Georges–Louis Leclerc, Count of Buffon, an illustrious naturalist, described in his treatise *Essai d'Arithmétique morale* (Buffon, 1777) a method of estimating π using a random process. Obviously he did not use computers nor did it have sequences of random numbers. The fact remains, however, that the basic idea was precisely that of evaluating an unknown quantity by assuming it as an average value of a suitable random variable and then trying to estimate the latter by examining an appropriate sample.

Suppose we 'randomly' throw a needle of length a on a plane intersected by parallel lines, each d apart on the other and let $d > a$. With N the total number of tosses and n the number of times the needle crosses a line, Buffon showed that:

$$\pi \approx \frac{2aN}{dn} \tag{3.71}$$

Let's see the reason for the formula (3.71). The first step is to formulate the problem in a probabilistic context. In order for the problem to be determined, it must explicitly state what is meant by the expression 'random throw'. In reality, a good 'pitcher' could certainly make the needle intersect all, or almost, the times she or he wants, but here

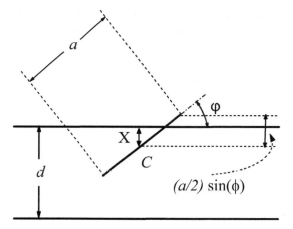

Fig. 3.17 Buffon's needle for the calculation of π. Geometry scheme: a is the length of the needle, C its centre point and d the distance between the lines.

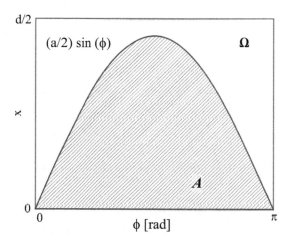

Fig. 3.18 Buffon's needle for the calculation of π. Graphical representation of the change of x (distance of C from the nearest row) as a function of ϕ (angle between the needle and the lines).

we have specified that the launches must be at random. It is about imposing certain conditions − entirely intuitive − on the position of the needle centre and direction. We introduce two random variables: X and Φ such that: (see Fig. 3.17 and 3.18):

- (A) X represents the values that at the distance can take centre C of the needle from the nearest row. X is uniformly distributed in $[0, d/2]$, so that the probability density function is $f_X(x) = 2/d$, for $0 \leqslant x \leqslant d/2$. This is equivalent to assigning equal probability to the ordinate of C.
- (B) The variable Φ represents the values that the angle can take formed by the needle and parallel lines. Φ is uniformly distributed in $[0, \pi]$, so that the probability density function is $f_\Phi(\phi) = 1/\pi$, for $0 \leqslant \phi \leqslant \pi$.

This means assigning equal probability to the direction of the needle. In the simple fact that we have defined the range $[0, \pi]$, it is implied that we know how much is π and, with the procedure that we are going to explain, we also suppose it unknown. The purpose of all this is obviously didactic.

- (C) X and Φ are independent, so $f(x, \phi) = f_X(x) f_\Phi(\phi)$, that is, the joint density function is equal to the product of marginal densities.

The density function $f(x, \phi)$ of the double random variables (X, Φ) is given by:

$$f(x, \phi) = \frac{2}{d} \times \frac{1}{\pi}, \quad \text{if} \quad 0 \leqslant x \leqslant d/2, \quad 0 \leqslant \phi \leqslant \pi$$

Geometrically, therefore, the needle positions are defined by the points of a rectangle of sides $d/2$ and π (see Fig. 3.18). From Fig. 3.18 it can be deduced that, in order for the needle to cross one of the lines, the realizations of the random variables (X, Φ) must belong to A, where A is given by:

$$A = \left\{ (x, \phi) : x \leqslant \frac{a}{2} \sin \phi \right\} \tag{3.72}$$

Recall now the so-called 'geometrical definition' of probability. R is the region of the plane (or of a space or of an hyperspace) and we consider another region r contained in R, as shown in Fig. 3.19.

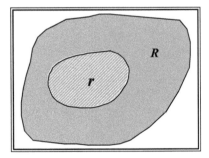

Fig. 3.19 Geometric definition of probability.

The probability P that a randomly chosen point in R falls into r is proportional to the measure (area, volume, hypervolume) of r and it does not depend on either its shape or position within R. We will therefore write:

$$P = \text{mis}\, r \,/\, \text{mis}\, R$$

In agreement with the above definition, in our case, the probability that the needle intersects a line, that is the probability $P\,\{\text{intersection}\}$ of the 'intersection' event, is given by the ratio between the area dashed A and the area of the whole rectangle $\Omega = [0, d/2] \times [0, \pi]$

$$P\,\{\text{intersection}\} = \frac{A}{\Omega} = \int_0^\pi (a/2) \sin \phi \, d\phi \, \times \frac{2}{d\pi} \tag{3.73}$$

In particular, if you choose a needle that is half the length of the width between the parallel lines, i.e. with $a = d/2$, we obtain:

$$P\{\text{intersection}\} = 1/\pi \tag{3.74}$$

Let us dwell on the equation written above. On the one hand we have a probability, and on the other the unknown number. Probability is also a number and it is also unknown, but we can estimate it. Each throw of the needle, in fact, can be regarded as a test with probability p of success (intersection with the line) and $1 - p$ of failure, that it can be described by the Bernoulli (*Bern*) random variable; in N repeated and independent tests

$$S = \sum_{i=1}^{N} Bern(i)$$

where S is the binomial r.v. *Bin* of parameters N and p, or by writing it more extensively $Bin(n|p, N)$, where n represents the number of successes. The first moment (average) $E[Bin]$ and the second moment (variance) $\text{Var}[Bin]$ are:

$$E[Bin] = Np \qquad \text{Var}[Bin] = Np(1 - p)$$

What is observed are the $f = n/N$ of the r.v. with frequency $Fr = Bin/N$, which represents the ratio of the number n of successes and the number N of trials, and has average:

$$E[Fr] = \frac{E[Bin]}{N} = \frac{Np}{N} = p$$

and variance:

$$\text{Var}[Fr] = \frac{\text{Var}[Bin]}{N^2} = \frac{Np(1-p)}{N^2} = \frac{p(1-p)}{N}$$

f_r is a unbiased estimator of p. In other words, in a series of tests, frequencies have mean p, hence eqn (3.71). The standard deviation σ_f of f_n − the so-called 'standard error of the average' − is given, from:

$$\sigma_f = \sqrt{\frac{p(1-p)}{N}} \tag{3.75}$$

where, commonly, in the calculation of p its estimate f_n is used.

The procedure suggested by Buffon can be reproduced by computers, once a suitable program has been developed. This essentially consists of determining N values of x and ϕ, in the appropriate ranges of variation, through sequences of random numbers provided by the computer itself. The calculation program will contain one statement to perform the eqn (3.72) inequality, and another to count the times it is verified. Finally, eqn (3.71) ratio is calculated.

The code to simulate needle throws is `Code_3_5.R`, where `pi` is the 'known' π, `pig` is the 'unknown' one. The meaning of `n` and `N` is that described in the text and the `runif (k, a, b)` function generates k random numbers uniformly distributed between a and b.

Table 3.2 shows the results of calculations with $a = d/2$ and in which the number of 'throws' has changed, that is to say the number N of the assumed values randomly from x and from ϕ. The first column shows N, the second the number of intersections obtained, i.e. how many times the inequality eqn (3.72) was found to be valid, and in the third the value of π given by eqn (3.74) and in the fourth the percentage error ϵ compared to the true value of π (3.141592654 ...), showing that as N increases the dispersion decreases, as eqn (3.75) also indicates.

Table 3.2 Buffon needle from computations.

throws	intersections	π computed	$\epsilon(\%)$
100	40	2.50000	20
1000	286	3.49650	11
10000	3086	3.24044	3.1
100000	31652	3.15936	0.57
1000000	318319	3,14150	0.17

In this example, the value of π is known, but of course the Monte Carlo method is used when the value to be estimated is unknown; in this case, the error will have to be evaluated by running simulations with increasing values of N. If we didn't know the value of π, all that we could get from the calculator's answers would be a *confidence interval* which the value sought belongs. If, for example, we had chosen $N = 1000$ and a level equal to 95% for the confidence interval, we would have final results:

$$1/\pi = f \pm 1,96 \sqrt{\frac{f(1-f)}{N}} = 0.286 \pm 0,028$$

where f is the proportion of successes in 1000 launches (see Table 3.2 second line). Results of the computation are shown in Fig. 3.20

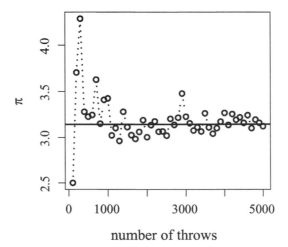

number of throws

Fig. 3.20 Buffon needle: computation of π as function of the number of throws N.

The above procedure for calculating π can also be regarded as a general method of calculating through random numbers a definite integral. Consider, in fact, Fig. 3.18. Suppose we want to evaluate the dotted area A. It holds for eqn (3.73) written differently:

$$A = \underset{\text{number}}{\mathsf{P}} \underset{\text{probability}}{\{\text{intersection}\}} \ \Omega \approx \underset{\text{estimate}}{(n/N) \ \Omega}$$

Therefore, the computation of a definite integral is reduced to the computation of how many times by 'randomly pulling' on the rectangle of area Ω, the area A is hit. Such a method for calculating integrals defined is called Monte Carlo 'win or lose' (*hit-or-miss*). The `Code_3_5.R` computes the solution as presented below.

```
#Code_3_5.R
## Buffon needle problem
# needle length a = half the distance d between the lines.
# d = 1
# pig is the known mathematical constant pi

set.seed(2)
pigl<- numeric()
Nin<- 100
Nfin<- 5000
N<- seq(Nin,Nfin,by=100)
Nl<-length(N)
for(l in 1:Nl){          # external loop: changes number of throws
n<-0
for(i in 1:N[l]){        # internal loop on throws
x<-runif(1,0,1/2)
phi<-runif(1,0,pi)
if( x <=(1/4)*sin(phi))n<- n+1   # formula: x <= a/2 sin(phi) a=d/2 con d=1
p<- n/N[l]
pig<- 1/p
}
pigl[l]<- pig
]
N
pigl
par(mai=c(1.02,1.,0.82,0.42)+0.1)
plot(N,pigl,type="b",xlab="number of throws",
ylab=expression(pi),cex.lab=1.3,cex.axis=1.2,
lwd=2,lty=3,col="black",font.lab=3)
abline(h=pi,lty=1,lwd=2,col="black")
```

3.5   Ehrenfest urn model

We conclude this chapter describing a famous stochastic problem whose roots deeply penetrate into the ground of physics. There are two urns, A and B, and 100 balls enumerated from 1 to 100. We put, for instance, 80 ball in A and 20 in B. It does not matter how we choose them. We can put in A the balls $1, 2, \ldots, 80$ and the remaining ones in B, or we can put them at random. We also have a third urn with 100 cards, each with a number from 1 to 100. We draw a card and pick up the ball with this number. If it is in A, we move it in B, if it is in B, it goes in A. The drawn card is put again in the card-urn. We continue the drawing, and follow for some time the balls going from one urn to the other one. This 'game' was conceived in 1907 by Paul and Tatiana Ehrenfest (Ehrenfest and Ehrenfest, 1907), and it is known as the 'Ehrenfest urn model', but also 'Ehrenfest dog-flea model'. Costantini and Garibaldi (2004) quote the 1907 article in which the Ehrenfest say:

Jedesmal, wenn eine Nummer gezogen wird, huepft die Kugel mit dieser Nummer aus der Urne [...] in die andere Urne
[Every time a number is drawn, the ball with that number bounces out of it]

Perhaps the Ehrenfest had in mind just fleas, or anyway jumping bugs, since they used the verb *hüpfen*, which means 'to jump', more suitable to fleas rather than balls.

It was not an easy life, that of Tatiana Alexeyevna Afanassjewa and Paul Ehrenfest. She was an Orthodox Russian, while he was Jewish. They both had to declare that they were without religion to get married. Tatiana was born in 1876 in Kiev in the Ukraine, Paul in 1880 in Vienna. The interested reader can find many articles about the Ehrenfest, for instance in Klein (1985).

In the years 1899−1900, in Vienna, Paul followed Boltzmann's lectures on thermodynamics. In 1902 he went to Göttingen and in 1904 he received his doctorate with the dissertation (English translation) *On the extension of Hertz's mechanics to problems in hydrodynamics*, with Boltzmann's supervision. In 1906, he came back to Göttingen. Ehrenfest and Boltzmann would never see each other again. Boltzmann killed himself in that year in Duino near Trieste and Ehrenfest did the same in Leiden in 1933. Tatiana and Paul got married in Vienna on December 21, 1904, so Tatiana became Tatiana Ehrenfest.

Difficult years followed, without any regular employment, but they continued studying and writing. The 'Urn model' was published in 1907 and in 1906 they had begun to write the famous monograph (English translation) *The Conceptual Foundations of the Statistical Approach in Mechanics* for the *Encyklopaedie der mathematischen Wissenschriften* (published in 1912). For five years (1907−1912) they lived in St. Petersburg and eventually, in October 1912, they moved to Leiden, where Paul became professor of theoretical physics, succeeding H.A. Lorentz. But depression dug deep into Paul's soul. Their third son Wassik was born with Down's syndrome. On September 25, 1933 Paul committed suicide by shooting himself.

Tatiana continued her life in Leiden, taking care of teaching mathematics up to her death in 1964. One of her final writings, published in the *American Journal of Physics* in 1958, is *On the use of the notion of 'probability' in physics*. We quote here the abstract.

An analysis of the meaning of such terms as 'at random', 'equal chances', 'probability calculus', 'laws of probability' as used in physics. The author treated this subject for the first time in a lecture at St. Petersburg in 1911, after which it was published in Russian in the *Journal of the Physical-Chemical Association of St. Petersburg*. Later, Paul Ehrenfest made it the topic of several lectures.

To get an idea of the underlying problems addressed by the Ehrenfest we have to go back in time, some years before the 'Urn model', to 1872 when Boltzmann pursued the intent of *deducing* from the laws of Newtonian mechanics the thermodynamic properties of a system formed by an extremely large number of particles (atoms or molecules), given constraints and initial conditions. The development of such a programme culminated in the famous '*H*-theorem'. In this theorem Boltzmann aimed to prove that for a perfect (or ideal) gas with no external forces acting on the particles, and under the assumption of absence of correlation in the initial state between the positions and the velocities of colliding particles, there is a function, called the *H* function, that may never grow over time and tends to decrease as equilibrium is approached up to a limiting value corresponding to Maxwell's distribution of molecule velocities. Moreover, the *H* function is linked to the thermodynamic entropy S through the relation:

$$S = -k_B H$$

with k_B the Boltzmann's constant. This equation is, in the intentions of Boltzmann, the proof of the second law of thermodynamics from principles of mechanics and it shows that the time evolution of the macroscopic state of a gas is irreversible. Note that the assumption of absence of correlation (*Stosszahlansatz*, or *molecular chaos assumption*) is purely statistical in character. That introduces an element of statistical nature in Boltzmann's 'deduction'.

The theorem was strongly criticized. In 1876, Loschmidt maintained that if at equilibrium the direction of all particles is reversed, since the microscopic laws are reversible, the system should return to its initial state with a decrease of *H* (*reversibility objection*). Later, Poincaré (1890) and Zermelo (1896) argued that, waiting for a long enough time, the distribution of particles in the state space becomes indistinguishable from the original distribution (*recurrent objection*). Therefore the *H* function should be periodic, contrary to *H*-theorem.

These and similar objections led Boltzmann to revise his theory published in 1877. There are no more references to molecules collisions, and the key point is the explicit introduction of the *statistical hypothesis* about microstate equiprobability. For the first time a statistical hypothesis enters, so to speak, in a pair with deterministic laws, those of Newtonian mechanics. Boltzmann relates the entropy S to the logarithm of the probability W, that is to the number of microstates corresponding to the same macrostate, through the well-known formula:

$$S = k_B \log W$$

We return to the Ehrenfest model to read how Paul and Tatiana attempted to explain as clearly as possible Boltzmann's use of probability inside the conceptual framework of physics. We will describe the model by means of a discrete-time Markov chain in a discrete state space. The time steps are the draws of the cards and the consequent transfers of the balls from one urn to the other. The state of the system at time i, that is after the i–th draw of the card, is defined by the number of balls in the urns. If N is the number of balls (100 in our example) the possible states are 0 balls in urn A and 100 in urn B, 1 ball in A and 99 in B, ... 100 in A and 0 in B. Obviously if there are j balls in A, there will be $N - j$ in B.

As before in this chapter, $X_i = j$ means that there are j balls in A at time i. At the next time $i + 1$, either a ball has been drawn in A, so that $X_{i+1} = j - 1$, or in B, then $X_{i+1} = j + 1$. It is clear that this chain is a Markov chain, $X_{i+1} = j \pm 1$ depends only on how many balls there were in the urns at the previous time. If j balls are in A at time i, the probability of drawing a ball from A is j/N, therefore the transition probability from state j to state $j - 1$ is given by:

$$p_{j,j-1} = \frac{j}{N}$$

and of course:

$$p_{j,j+1} = \frac{N-j}{N} = 1 - \frac{j}{N}$$

Let us remark on a point, that might appear obvious but is very important. The motion of the single ball is *time symmetric*, meaning that each ball at each time instant (i.e. at each card draw) has the same probability equal to $1/N$ of moving from A to B or from B to A. The ball number 4, for instance, always has the same probability of being raffled off, either it is in A or in B. Here there is an underlying hypothesis of equiprobability: the cards continue to have the same odds of being drawn. This is not so obvious as perhaps it might seems.

Other notations can be used to define the state of the system. For example, instead of $X_i = j$, we can write $(N_A)_i = n_A$ or $(N_B)_i = n_B$, to indicate the number of balls in A or in B at time i. We write also $N = n_T$ for the total number of balls, and by putting $n_e = n_T/2$, we can consider the variable $\Delta n = N_A - n_e$. The state of the system at time i is defined by $(\Delta n)_i = (N_A - n_e)_i = \delta$. In any case, regardless the chosen notation, the state of the system at time i is defined by means of only one variable: N_A or Δn or $|N_A - N_B|$, and so on.

We now study the Markov chain. The transition matrix $\mathbf{E}$ is given by:

$$
\mathbf{E} = \begin{array}{c} \\ 0 \\ 1 \\ 2 \\ \vdots \\ n_T-1 \\ n_T \end{array}
\begin{array}{c}
\begin{array}{ccccccc} 0 & 1 & 2 & 3 & \cdots & n_T-1 & n_T \end{array} \\
\left(\begin{array}{ccccccc}
0 & 1 & 0 & 0 & \cdots & 0 & 0 \\
n_T^{-1} & 0 & 1-n_T^{-1} & 0 & \cdots & 0 & 0 \\
0 & 2n_T^{-1} & 0 & 1-2n_T^{-1} & \cdots & 0 & 0 \\
\vdots & \vdots & \vdots & \vdots & \vdots & \vdots & \vdots \\
0 & 0 & 0 & 0 & \cdots & 0 & n_T^{-1} \\
0 & 0 & 0 & 0 & \cdots & 1 & 0
\end{array}\right)
\end{array}
\tag{3.76}
$$

For instance, with only 4 balls ($n_T = 4$) the states are 0, 1, 2, 3, 4 and the transition matrix is (the state of the system is defined by the variable N_A):

$$\mathbf{E} = \begin{pmatrix} 0 & 1 & 0 & 0 & 0 \\ 1/4 & 0 & 3/4 & 0 & 0 \\ 0 & 1/2 & 0 & 1/2 & 0 \\ 0 & 0 & 3/4 & 0 & 1/4 \\ 0 & 0 & 0 & 1 & 0 \end{pmatrix} \qquad (3.77)$$

Notice that there are no absorbing states, that is $p_{jj} = 0$, $\forall j$, $j = 0, \ldots, n_T$, even though at a certain time there may be zero balls in an urn, the process continues since, at the next time, there will be certainly one ball in the same urn.

Let us note also that the chain eqn (3.76) is irreducible and periodic with period 2, i.e. every state has period 2. Recall that a state j is *periodic* with period $d^{(j)}$ if starting from j it is possible (even if it is not sure) to return to j only after n steps, with $n = d^{(j)}, 2d^{(j)}, 3d^{(j)}, \ldots$, and $d^{(j)}$ is the largest of such integer values. With $n_T = 4$, if the system is, for example, in the state $n_A = 1$, two possible trajectories are:

$$1 \to 0 \to 1 \to 0 \to 1 \to 0 \to 1 \cdots$$
$$1 \to 2 \to 1 \to 2 \to 3 \to 2 \to 1 \cdots$$

We can transform the chain from periodic to aperiodic, if the flea has the chance to repent. The flea, while it is in flight, can stop on a shrub and toss a coin. If it gets heads it continues the hop, if tails it comes back on the starting dog. Both chains (periodic and aperiodic) have the same stationary (or invariant) distribution. For an aperiodic chain, remember, the stationary distribution is also the asymptotic distribution.

The stationary distribution is given by:

$$\pi_k = \frac{1}{2^{n_T}} \frac{n_T!}{k!(n_T - k)!}, \qquad k = 0, 1, \ldots, n_T \qquad (3.78)$$

The above eqn (3.78) is nothing different from a binomial distribution $Bin(p)$, giving the probability of observing k successes in n trials, when the probability of success in a single trial is p:

$$Bin(p) = \binom{n}{k} p^k (1 - p)^{n-k}, \qquad k = 0, 1, \ldots, n$$

If we put $n = n_T$ and $p = \frac{1}{2}$, we find again eqn (3.78). The stationary distribution π_k is the probability of the process being in state k in the long run, and such a probability does not change anymore. Therefore we can write:

$$\pi_k = \mathsf{P}\{X_{i\to\infty} = k\}, \qquad k = 0, 1, \ldots, n_T \qquad (3.79)$$

or with another notation, by referring to the number of balls in the urns A and B, eqn (3.78) can be written as:

$$\pi_{n_A} = \frac{1}{2^{n_T}} \frac{n_T!}{n_A! n_B!}, \qquad n_A = 0, 1, \ldots, n_T$$

and so eqn (3.79) becomes:

$$\pi_{n_A} = P\left\{(N_A)_{i \to \infty} = n_A\right\}, \qquad n_A = 0, 1, \ldots, n_T$$

The result is the same, whatever the notation is: at the starting time we can have any number of balls in A, also $n_A = n_T$ or $n_A = 0$, but for $i \to \infty$, that is after many draws of the numbered cards, the probability π_k of k balls in A does not change anymore. The situation is similar to the previously discussed example of the Land of Oz. After a certain number of days, the probability of raining is equal to 57%, whatever the previous time state has been. Saying that the system tends to a stationary state, regardless of the starting state, implicitly indicates a clear and precise time direction. Therefore, while the move of a single ball from one urn to another determines no time direction, a time evolution emerges from the initial to the stationary state, by the way the states of all balls following each other, that is the system as a whole. Once stationarity is reached, the time direction disappears.

Recall that the succession of transition matrices $\mathbf{P}^{(n)}$ of an irreducible and periodic chain does not converge to the asymptotic matrix $\mathbf{P}^{(\infty)}$. If however the sequence of arithmetic means:

$$\frac{1}{n}\left[\mathbf{P} + \mathbf{P}^{(2)} + \cdots + \mathbf{P}^{(n)}\right] \tag{3.80}$$

is considered, then for $n \to \infty$, such a sequence converges ('in the Cesaro sense') to a limiting matrix $\mathbf{P}^*$ with all positive entries and all rows identical. In the example of matrix eqn (3.77) and with $n = 1000$ in eqn (3.80), the limiting matrix $\mathbf{E}^*$ is obtained:

$$\mathbf{E}^* = \begin{pmatrix} 0.06266667 & 0.2506667 & 0.375 & 0.2493333 & 0.06233333 \\ 0.06266667 & 0.2501667 & 0.375 & 0.2498333 & 0.06233333 \\ 0.06250000 & 0.2500000 & 0.375 & 0.2500000 & 0.06250000 \\ 0.06233333 & 0.2498333 & 0.375 & 0.2501667 & 0.06266667 \\ 0.06233333 & 0.2493333 & 0.375 & 0.2506667 & 0.06266667 \end{pmatrix}$$

Through eqn (3.78), the stationary probabilities vector is:

$$\boldsymbol{\pi} = (0.0625, 0.25, 0.375, 0.25, 0.0625) \tag{3.81}$$

So with $n = 1000$, a good approximation to the exact result is already obtained.

Let us still remain on eqn (3.78). We can look at the ratio:

$$\frac{n_T!}{k!(n_T - k)!} \tag{3.82}$$

as giving all the possible ways of distributing k balls in A and $n_T - k$ in B, regardless of the number they have written on. Let us suppose to have six balls 1, 2, 3, 4, 5, 6. If we have $k = 4$ balls in A (2 in B) at time i, we may have the balls 2, 4, 5, 6 are in A and the 1, 3 in B, or balls 1, 2, 3, 4 in A and 5, 6 in B, and so on. There are 15 ways of placing four balls in A and 2 in B, that is there are 15 possible combinations. In formulae, the number of combinations is given by:

$$\frac{6!}{4!(6-4)!} = 15$$

Let us use, from now on, the language of statistical mechanics. To say that there are four balls in A means to specify the *macrostate*, or 'macroscopic state', of the system. To say that the ball number 1 is in A, the number 2 is in B, and so on, means to specify the *microstate*, or 'microscopic state', of the system. In this terminology eqn (3.82) gives the number of microstates corresponding to the same macrostate with k balls in A and π_k in eqn (3.78) gives the *probability* that the system will be in the macrostate with k balls in A in the long run.

Look at Table 3.3. It reports all the possible microstates compatible with the *macroscopic constraints*. The only macroscopic constraint is the number of balls $n_T = 6$. Denote by μ_l the microstates. In total there are $2^6 = 64$ microstates, so $\mu_1 = (1, 2, 3, 4, 5, 6; -), \mu_2 = (1, 2, 3, 4, 5; 6), \mu_3 = (1, 2, 3, 4, 6; 5), \ldots \mu_{64} = (-; 1, 2, 3, 4, 5, 6)$, the semicolon divides the balls in A from those in B. All the microstates have the same occurrence probability, i.e. we make the hypothesis that each ball has probability $1/n_T$ of being drawn, as a consequence of the assumption of equiprobability of the card draws. The equiprobability of microstates is the basic hypothesis of Boltzmann's foundation of statistical mechanics.

If the microstates have the same probability, this is not true for the macrostates. The macrostate composed of 20 microstates is more probable than that composed by six microstates. For instance, the probability of four balls in A is $15/64 \approx 0.23$, that of three balls is $20/64 \approx 0.31$, that of five is $6/64 \approx 0.094$, that of six is $1/64 \approx 0.016$. When the system is in the stationary state it is said that it is 'at equilibrium', a property of the macrostate. Suppose now we have 1000 pairs of urns and 1000 persons drawing the balls at the same time. After stationarity is reached, we take a picture of the 1000 pairs of urns. It results that about 230 urns have four balls in A, about 310 have three balls, ..., about 16 have six balls. Turning to the statistical mechanics language, we can state that the evolution of a system at the microscopic level is symmetric with respect to time, but a precise temporal direction appears if the system is considered at the macroscopic level. Past and future only exist at the macroscopic level. It is like saying that we can talk about a 'before' and an 'after', only if the system is dealt with 'statistically'.

The stationary probability vector comes from eqn (3.78):

$$\boldsymbol{\pi} = (0.016, 0.094, 0.23, 0.31, 0.23, 0.094, 0.016) \tag{3.83}$$

We will shortly deal with the random variable $\Delta n = N_A - n_e$ where $n_e = n_T/2$. Δn has values:

$$\delta_1 = n_e, \delta_2 = n_e + 1, \ldots, \delta_{n_e+1} = 0, \ldots,$$
$$\delta_{n_T} = -n_e - 1, \delta_{n_T+1} = -n_e$$

Of course, the enumeration can be reversed by putting $\delta_1 = -n_e$ and $\delta_{n_T+1} = n_e$.

To summarize, Δn is the random variable, and the δ_n's are its realizations. The expression:

$$\pi_{\delta_n} = \mathsf{P}\left\{(\Delta n)_{i\to\infty} = \delta_n\right\}, \qquad \delta_n = n_e, n_e + 1, \ldots, -n_e - 1, -n_e$$

is the analogue of:

$$\pi_{n_A} = \mathsf{P}\left\{(N_A)_{i\to\infty} = n_A\right\}, \qquad n_A = 0, 1, \ldots, n_T$$

For instance, with $n_T = 6$, the values assumed by the random variable Δn are reported in Table 3.4.

Table 3.3 Macrostates and corresponding number of microstates with six balls.

macrostate	balls in A	balls in B	# microstates
$6 - 0$	$1, 2, 3, 4, 5, 6$	$-$	1
$5 - 1$	$1, 2, 3, 4, 5$ $1, 2, 3, 4, 6$ $\ldots$ $2, 3, 4, 5, 6$ $\ldots$	6 5 $\ldots$ 1 $\ldots$	6
$4 - 2$	$2, 3, 4, 5$ $1, 2, 5, 6$ $\ldots$ $2, 4, 5, 6$ $\ldots$	$1, 6$ $3, 4$ $\ldots$ $1, 3$ $\ldots$	15
$3 - 3$	$2, 3, 4$ $2, 5, 6$ $\ldots$ $1, 4, 6$ $\ldots$	$1, 5, 6$ $1, 3, 4$ $\ldots$ $2, 3, 5$ $\ldots$	20
$2 - 4$	$2, 5$ $1, 2$ $\ldots$ $1, 6$ $\ldots$	$1, 3, 4, 6$ $3, 4, 5, 6$ $\ldots$ $2, 3, 4, 5$ $\ldots$	15
$1 - 5$	2 1 $\ldots$ 5 $\ldots$	$1, 3, 4, 5, 6$ $2, 3, 4, 5, 6$ $\ldots$ $1, 2, 3, 4, 6$ $\ldots$	6
$0 - 6$	$-$	$1, 2, 3, 4, 5, 6$	1

Table 3.4 Macrostates, number of corresponding microstates and values of $\Delta n = N_A - n_T/2$ with $n_T = 6$.

macrostate	# microstates	δ_n
$6 - 0$	1	3
$5 - 1$	6	2
$4 - 2$	15	1
$3 - 3$	20	0
$2 - 4$	15	-1
$1 - 5$	6	-2
$0 - 6$	1	-3

In a further notation the invariant distribution π_k in eqn (3.78) is written as:

$$\gamma g(k) \quad \text{or also} \quad \gamma g(n_A) \quad \text{or also} \quad \gamma g(\delta_n)$$

where $\gamma = 1/2^{n_T}$ and $g(\cdot)$, called *multiplicity function*, is the ratio eqn (3.82), that written as a function of δ_n is:

$$g(\delta_n) = \frac{n_T!}{(n_e + \delta_n)!(n_e - \delta_n)!}, \qquad \delta_n = n_e, \dots, -n_e$$

Note that γ, or more precisely $\gamma(n_T)$, is the probability of realization of any microstate l of the set of all the 2^{n_T} possible microstates $\mu_l, l = 1, \dots, 2^{n_T}$:

$$\gamma(n_T) \equiv \mathsf{P}\{\mu_l\} = \frac{1}{2^{n_T}} \tag{3.84}$$

As said before, $g(\cdot)$ represents the number of microstates corresponding to the same macrostate with $n_A - n_e$ balls in A. The total number of microstates 2^{n_T} can be written as a function of $g(\cdot)$:

$$2^{n_T} = \sum_{r=1}^{n_T+1} g(\delta_r) \tag{3.85}$$

For instance, with $n_T = 6$, it is from Table 3.4: $g(\delta_1) = 1$, $g(\delta_2) = 6, \dots$, therefore:

$$2^{n_T} = 1 + 6 + 15 + 20 + 15 + 6 + 1 = 64$$

Let us slightly increase the number of balls, by putting $n_T = 10$. Table 3.4 becomes Table 3.5. The stationary probabilities vector is now given by:

Table 3.5 Macrostate, number of corresponding microstates and values of $\Delta n = N_A - n_T/2$ with $n_T = 10$.

macrostate	# microstates	δ_n
10 − 0	1	5
9 − 1	10	4
8 − 2	45	3
7 − 3	120	2
6 − 4	210	1
5 − 5	252	0
4 − 6	210	−1
3 − 7	120	−2
2 − 8	45	−3
1 − 9	10	−4
0 − 10	1	−5

$$\pi = (0.00098, 0.0098, 0.044, 0.12, 0.21, 0.25,$$
$$0.21, 0.12, 0.044, 0.0098, 0.00098) \tag{3.86}$$

and the number of microstates is now $2^{10} = 1024$.

The macrostate with $n_A = n_B = 5$ is the realization of 252 microstates, while it is 20 with $n_T = 6$. Therefore if the number of balls increases by a factor of 2, the probability of realization of the macrostate $n_A = n_B$, with respect to the probability of a macrostate with no ball in A (or in B), increases by one order of magnitude. Therefore, the equilibrium situation having $n_A = n_B$ is more and more probable with respect to the other ones, with the increase of the number of balls. In terms of fleas and famous cartoon dogs, even though all the fleas are on the Tramp at the beginning, after a little time about one half of them have hopped onto the Lady.

If $n_T \to \infty$, the binomial distribution tends to a sharply peaked normal distribution. How much peaked? The density function of the binomial random variable

$$Bin(p) = \frac{N!}{k!(N-k)!} p^k (1-p)^{N-k}, \qquad k = 0, 1, \dots, N$$

for very large N tends to the density function of the normal random variable

$$\mathcal{N}(\mu, \sigma) = \frac{1}{\sqrt{2\pi}\sigma} \exp\left[-\frac{1}{2}\frac{(x-\mu)^2}{\sigma^2}\right]$$

where $\mu = Np$ and $\sigma = \sqrt{Np(1-p)}$.

We can see that:

$$g(\delta_n) \approx g(0) \times \exp\left[-\frac{2\,\delta_n^2}{n_T}\right] \tag{3.87}$$

where $g(0)$ is:

$$g(0) = [2/(\pi n_T)]^{1/2}\, 2^{n_T}$$

$g(0)$ is $g(\delta_n)$, when $\delta_n = 0$, that is in the 'continuous' approximation of eqn (3.82) with $k = (n_T - k)$, that is to say $n_A = n_B$.

Proof Compute $g(n_A)$ and make n_T tend to infinity. Compute first the logarithm of $g(n_A)$.

$$\log g(n_A) = \log \frac{n_T!}{n_A! n_B!} = \log n_T! - \log n_A! - \log n_B!$$

and by Stirling's approximation:

$$n! \approx (2\pi n)^{1/2} n^n \exp\left(-n + \frac{1}{12n} + \dots\right) \xrightarrow{n \to \infty} (2\pi n)^{1/2} n^n e^{-n}$$

Then:

$$\log n_T! \approx \frac{1}{2}\log 2\pi + \frac{1}{2}\log n_T + n_T \log n_T - n_T$$
$$= \frac{1}{2}\log 2\pi + \left(n_T + \frac{1}{2}\right)\log n_T - n_T$$

Similarly for n_A and n_B:

$$\log n_A! \approx \frac{1}{2}\log 2\pi + \left(n_A + \frac{1}{2}\right)\log n_A - n_A$$
$$\log n_B! \approx \frac{1}{2}\log 2\pi + \left(n_B + \frac{1}{2}\right)\log n_B - n_B$$

Turning to $\log n_T!$, we put $n_A + n_B$ instead of n_T in some place and add and remove $\frac{1}{2}\log n_T$. So we have:

$$\log n_T! \approx \frac{1}{2} \log \frac{2\pi}{n_T} + \left(n_A + \frac{1}{2} + n_B + \frac{1}{2}\right) \log n_T - (n_A + n_B)$$

and subtracting $\log n_A!$ and $\log n_B!$, we have:

$$\log g(n_A) = \log \frac{n_T!}{n_A! n_B!} \approx \frac{1}{2} \log \frac{1}{2\pi n_T} - \left(n_A + \frac{1}{2}\right) \log \frac{n_A}{n_T} - \left(n_B + \frac{1}{2}\right) \log \frac{n_B}{n_T}$$

Consider that:

$$\log(1 + x) \approx x - \frac{1}{2} x^2 + \dots, \quad \text{se } x \ll 1$$

Now:

$$\frac{n_A}{n_T} = \frac{\delta_n + n_e}{n_T} = \frac{\delta_n}{n_T} + \frac{n_T/2}{n_T} = \frac{1}{2} + \frac{\delta_n}{n_T} = \frac{1}{2}\left(1 + \frac{2\delta_n}{n_T}\right)$$

Similarly:

$$\frac{n_B}{n_T} = \frac{1}{2}\left(1 - \frac{2\delta_n}{n_T}\right)$$

And in logarithm form :

$$\log \frac{n_A}{n_T} = \log\left[\frac{1}{2}\left(1 + \frac{2\delta_n}{n_T}\right)\right] = -\log 2 + \log\left(1 + \frac{2\delta_n}{n_T}\right)$$

$$\approx -\log 2 + \frac{2\delta_n}{n_T} - \left(\frac{2\delta_n}{n_T}\right)^2$$

and

$$\log \frac{n_B}{n_T} \approx -\log 2 - \frac{2\delta_n}{n_T} + \left(\frac{2\delta_n}{n_T}\right)^2$$

Neglecting the terms $(2\delta_n/n_T)^2$:

$$\log g(\delta_n) \approx \frac{1}{2} \log \frac{2}{\pi n_T} + n_T \log 2 - \frac{2\,\delta_n^2}{n_T}$$

Finally the logarithms are removed.

$\square$

So we find that $g(\delta_n)$ for $n_T \to \infty$ is normal, practically a Dirac delta function. At stationarity the peak is on the maximum value corresponding to $\delta_n = 0$. If we put $\delta_n = (n_T/2)^{1/2}$, we have:

$$\frac{\delta_n}{n_T} = \frac{1}{n_T}\left(\frac{n_T}{2}\right)^{1/2} = \left(\frac{1}{2n_T}\right)^{1/2}$$

and from eqn (3.87), it results $g(\delta_n) \approx g(0) \times e^{-1}$, that is $g(\delta_n)$ has decreased by a factor $1/e \approx 0.37$ with respect to its maximum value. Therefore the quantity $1/(2n_T^{1/2})$ can be assumed to be a measure of the relative width, that is in relation to n_T, in $g(\delta_n)$. Figure 3.21 shows the above quantities with $n_T = 100$. With this value, we have $\delta_n = (100/2)^{1/2} \approx 7.071$ and since $g(0) \approx 1 \times 10^{29}$, $g(\delta_n) \approx 1 \times 10^{29} \exp[-2 \times 7.071^2/100] \approx 3.7 \times 10^{28}$, that is $g(\delta_n) \approx g(0) \times e^{-1}$.

To give an idea of how narrow the peak is, let n_T be equal to one million. The probability of finding 505000 balls in the urn A, with respect to that of finding 500000, that is a fluctuation about 1% with respect to n_e, is given from the ratio $g(5000)/g(0)$ that is of the order of 10^{-22}. If the balls are 10^{24}, the peak width is of the order of 10^{-12}, almost 0. Such extreme numbers are those of statistical mechanics, the number of molecules of a gas, for instance. So the lesson is: the process is a Markov chain

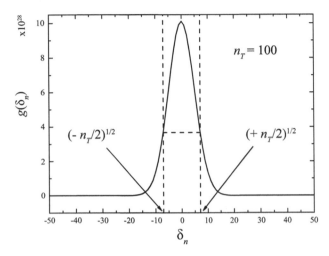

Fig. 3.21 Gaussian approximation of $g(\delta_n)$ with $n_T = 100$. The dashed lines, starting from the marked points, cross the curve at $1/e$ of its maximum value.

converging to an invariant distribution $\boldsymbol{\pi}$. When n_T is large, the probabilities:

$$\pi_{\delta_n} = \mathsf{P}\left\{(\Delta n)_{i\to\infty} = \delta_n\right\}, \qquad \delta_1 = n_e, \ldots, \delta_{n_T+1} = -n_e$$

are all ≈ 0, with the exception of $\delta_n \approx 0$, that is when $n \approx n_e + 1$, for which:

$$\pi_{n_e+1} = \mathsf{P}\left\{(\Delta n)_{i\to\infty} = \delta_{n_e+1} = 0\right\} \approx 1$$

Remember the definition of expected value. Let X be a discrete or continuous random variable and Y a random variable function of X, that is $Y = f(X)$. The expected value $\mathrm{E}\,[Y]$ of Y is given by:

$$\mathrm{E}\,[Y] = \mathrm{E}\,[f(X)] = \begin{cases} \displaystyle\int_{-\infty}^{+\infty} f(x)p(x)\,dx & \text{continuous case} \\ \displaystyle\sum_r f(x_r)p(x_r) & \text{discrete case} \end{cases}$$

where, if X is discrete $p(x_r) = \mathsf{P}\left\{X = x_r\right\}$ is the probability mass function and r covers the support of X, finite or numerable subset of $\mathbb{R}$.

For example, let us take $n_T = 6$ and $X = \Delta n$, and choose $f(\Delta n) = (\Delta n)^2$. In the above sum, x_r is δ_r and $f(x_r) = \delta_r^2$. We are interested in evaluating the expected value of Y, that is $\mathrm{E}\,[Y] = \mathrm{E}\,[f(X)] = \mathrm{E}\,\left[(\Delta n)^2\right]$. The total number of microstates is given by eqn (3.85) for which, if all the microstates are equiprobable, $p(x_r)$ is:

$$p(\delta_r) = \frac{1}{2^{n_T}}\, g(\delta_r) = \frac{1}{\displaystyle\sum_{r=1}^{n_T+1} g(\delta_r)}\, g(\delta_r) \qquad (3.88)$$

We have seen in Table 3.4 that:

$$\delta_r\ (r = 1, \ldots, 7)\ = 3, 2, 1, 0, -1, -2, -3$$

then the values assumed by the random variable $(\Delta n)^2$ are:

$$\delta_r^2\ (r = 1, \ldots, 7)\ = 9, 4, 1, 0, 1, 4, 9$$

Again from the Table 3.4:

$$g(\delta_r)\ (r = 1, \ldots, 7)\ = 1, 6, 15, 20, 15, 6, 1$$

and:

$$\sum_{r=1}^{7} g(\delta_r) = 64$$

The general formula:

$$\mathrm{E}\,[Y] = \mathrm{E}\,\left[(\Delta n)^2\right] = \sum_{r=1}^{n_T+1} \delta_r^2\, p(\delta_r) \qquad (3.89)$$

becomes in the present example:

$$\mathrm{E}\,\left[(\Delta n)^2\right] = \sum_{r=1}^{7} \delta_r^2\, p(\delta_r) = 9 \times \frac{1}{64} + 4 \times \frac{6}{64} + 1 \times \frac{15}{64}$$
$$+ 0 \times \frac{20}{64} + 1 \times \frac{15}{64} + 4 \times \frac{6}{64} + 9 \times \frac{1}{64} = 96/64 = 1.5 = n_T/4 \qquad (3.90)$$

We have reported all the steps for the sake of clarity, but recall that the probabilities $p(\delta_r)$ are already known, indeed they are the components of the probability vector $\boldsymbol{\pi}$ in the eqn (3.83).

When n_T is large enough we can exploit the approximation eqn (3.87) and compute the expected value by replacing the sum with an integral between $-\infty$ and $+\infty$. Then:

$$\mathrm{E}\,\left[(\Delta n)^2\right] = \frac{1}{2^{n_T}} \int_{-\infty}^{+\infty} \delta_n^2\, g(\delta_n)\, d\delta_n$$
$$= \frac{1}{2^{n_T}}\, [2/(\pi n_T)]^{1/2}\, 2^{n_T} \int_{-\infty}^{+\infty} \delta_n^2\, \exp(-2\, \delta_n^2/n_T)\, d\delta_n \qquad (3.91)$$

Changing variables, by putting $\delta_n^2 = x^2$, we obtain:

$$\mathrm{E}\left[(\Delta n)^2\right] = [2/(\pi n_T)]^{1/2}\,(n_T/2)^{3/2} \int_{-\infty}^{+\infty} x^2 e^{-x^2}\,dx$$

$$= [2/(\pi n_T)]^{1/2}\,(n_T/2)^{3/2}(\pi/4)^{1/2} = n_T/4$$

Here we have computed exactly $\mathrm{E}\,[Y]$. We ask ourselves: is there a way to *estimate* this expected value, if we are not able to solve the integral eqn (3.91) or compute the sum eqn (3.89)? One way is to use the Markov Chain Monte Carlo method, by constructing a Markov chain such that its invariant distribution π is the interested distribution.

In the example above with $n_T = 6$, if we do not know the probabilities $p(\delta_r)$ in eqn (3.88), we estimate them through the frequencies with which the system occupies the states corresponding to $\delta_r = 3, \ldots, -3$ at stationarity. In the following we say only 'state' instead of 'microstate', if there is no possibility of misunderstanding.

The R code `Ehrenfest_model` simulates the urns model. Given an initial configuration, chosen at random or fixed, the code computes the number of balls in the urns at each card draw.

```
### Code_3_6.R
## Ehrenfest_model
## Ehrenfest urn model  6 balls 200 steps
## expected value of (Delta n)^2

set.seed(1)
diffe<-numeric()
f<-numeric()
nt<- 6
ne<-nt/2;
nsteps<- 200
numberA <- 4
ta<- 100
tb<- 200
nA<- numberA
diffe0<- nA-ne    # starting state
i<-0
while(i<nsteps){    #  starting loop on steps
i<- i+1
u<- sample(nt,1,1/nt)
if(u<=nA) nA<-nA-1 else nA<-nA+1
diffe[i]<-nA-ne
f[i]<- diffe[i]^2    # function of (Delta n), now = (Delta n)^2
}      # ending loop on steps
t<-c(0:nsteps)        # adding t=0
diffe<- c(diffe0,diffe)  # adding starting  diffe0
# t                # uncomment to print t and/or diffe
# diffe
plot(t,diffe, type="l",ylim=c(-3,3),xlim=c(0,nsteps),main="",
ylab=expression(italic(Delta*n)),xlab="time",cex.lab=1.3,font.lab=3)
abline(h=0,lty=3,col="black",lwd=2)
lt<- length(diffe[(ta+1):tb] )
mtemp<-  mean(diffe[(ta+1):tb] )
mtemp                                   # time average
std.mtemp<-  sqrt(var(diffe[(ta+1):tb])/lt) # std of the mean
std.mtemp
mtempf<-  mean(f[(ta+1):tb] )
mtempf                                  # time average
std.mtempf<-  sqrt(var(f[(ta+1):tb])/lt)   # std of the mean
std.mtempf
```

Figure 3.22 shows the time evolution of the values of the variable Δn, with $n_T = 6$. The initial state is fixed with four balls in A, so $(\Delta n)_0 = 1$. In words: at time 0 the random variable Δn assumes with probability 1 the value $\delta_3 = 1$. We performed the

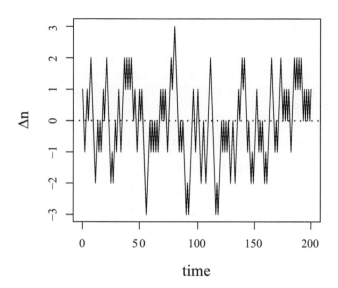

Fig. 3.22 Time evolution of the values assumed by the random variable Δn with $n_T = 6$.

time average of Δn in the interval $[101, 200]$. The time average is given by:

$$\frac{1}{100} \sum_{i=101}^{200} (\Delta n)_{i \underset{\text{time}}{\nwarrow}} = [1 + 0 + (-1) + (-2) + \ldots, +1 + 0]/100 = -0.12 \qquad (3.92)$$

Therefore, at the 101-th step it is: $(\Delta n)_{101} = \delta_3 = 1$, at the 102-th step it is: $(\Delta n)_{102} = \delta_4 = 0$, etc. By saving the number of visits to states, we have the vector $(2, 11, 25, 30, 23, 9, 0)$ to be compared with the second column, '# microstates', of Table 3.4 (the symbol '#' means 'number'). The *estimate* $\widehat{\pi}$ of the stationary probabilities vector π, and π, are then:

$$\widehat{\pi} = (0.020, 0.110, 0.25, 0.30, 0.23, 0.090, 0.00)$$
$$\pi = (0.016, 0.094, 0.23, 0.31, 0.23, 0.094, 0.016)$$

The estimate $\widehat{(\Delta n)^2}$ of $\mathrm{E}\left[(\Delta n)^2\right]$ is obtained by inserting the components

$$(\hat{p}(\delta_1), \ldots, \hat{p}(\delta_1)) = (0.020, \ldots, 0.00)$$

of the vector $\widehat{\pi}$ in eqn (3.90), obtaining:

$$\widehat{Y} = \widehat{(\Delta n)^2} = \sum_{r=1}^{7} \delta_r^2 \, \hat{p}(\delta_r) = 1.46$$

The exact value is $\mathrm{E}\left[(\Delta n)^2\right] = n_T/4 = 1.5$.

More simply, we compute the time average of the values of $f(\Delta n) = (\Delta n)^2$, that is we put in eqn (3.92) $(\Delta n)_i^2$ in the place of $(\Delta n)_i$:

$$\overline{(\Delta n)^2} = \frac{1}{100} \sum_{i=101}^{200} (\Delta n)_i^2 = (1 + 0 + 1 + 4 + \ldots, +1 + 0)/100 = 1.46$$

In this way we avoid doing the sum from $r = 1$ to $r = n_T + 1$ in eqn (3.89), but if n_T were of the order of 10^{20} it would be prohibitive. Moreover, this way of proceeding considers only those states that contribute to the average, that is only those states visited with high probability.

The code computes directly the time averages and their standard error:

```
mtemp             time average of diffe (Delta n = nA - ne)
-0.12
std.mtemp         standard error of the mtemp
0.1208388
mtempf            time average of f[i]<- diffe[i]^2  (Delta n)^2
1.46
std.mtempf        standard error of the mtempf
0.1788967
```

Let us run again the above code **Ehrenfest_model** with $n_T = 1000$. The input data are now:

```
set.seed(2)
nt<- 1000
ne<-nt/2;
nsteps<- 5000
numberA <- 600
ta<- 3000
tb<- 5000
```

in this case, the results are:

```
mtemp             time average of diffe (Delta n = nA - ne)
11.187
std.mtemp         standard error of the mtemp
0.2511813
mtempf            time average of f[i]<- diffe[i]^2  (Delta n)^2
251.054
std.mtempf        standard error of the mtempf
5.731442
```

The time average of the $(\Delta n)_i^2$ is given by:

$$\overline{(\Delta n)^2} = \frac{1}{5000} \sum_{i=3000}^{5000} (\Delta n)_{i \searrow \text{time}}^2 = 245 \tag{3.93}$$

The exact value is 250. Figure 3.23 shows the time evolution of the values of the variable Δn, with $n_T = 1000$. The initial state is fixed with 600 balls in A.

From that figure, we notice that about 2000 steps are necessary to reach the stationarity, but even once it is reached, the system shows a poor mixing around the value $\Delta n = 0$. The positive values assumed by the random variable Δn are more than the negative ones.

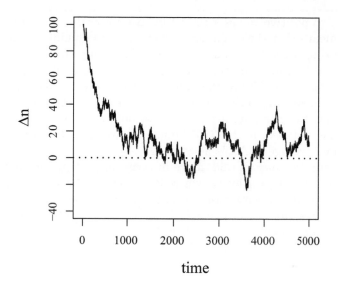

Fig. 3.23 Time evolution of the values assumed by the random variable Δn wit $n_T = 1000$.

Figure 3.24 shows the number of states visited by the system. It is obtained by the commands:

```
y<- diffe[(ta+1):tb]
hist(y,main="",xlab=expression(delta[n]),ylab="# visits",
cex.lab=1.5,font.lab=3)
```

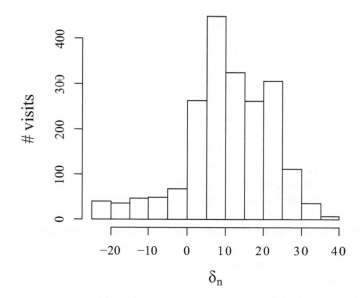

Fig. 3.24 Number of visits in the states δ_n, $n = -n_T/2, \cdots + n_T/2$, with $n_T = 1000$.

Figure 3.24 shows two things. Bad news, it is not symmetric. As we said, the states corresponding to positive values assumed by Δn are visited more frequently than the others. This means that the chain needs more steps to eliminate such a distortion.

But there is also good news. The code considers only those states that contribute substantially to the time average. With $n_T = 1000$, the range of possible states (microstates) is from $\delta_1 = 500$ (all the balls in A) to $\delta_{1001} = -500$ (all the balls in B). Figure 3.24 includes only states from $\delta_n = -30$ to $\delta_n = 40$, because the probability of visiting other states is quite negligible. Just out of curiosity, if the number of steps is increased, for instance, up to 40,000 and the average is taken between 20,000 and 40,000, the time average of Δn is now -1.329.

We can rephrase what we said above in terms of entropy. In Chapter 13 we write Boltzmann's formula $S_B = k \log W$, where W represents the number of real microstates corresponding to the same macrostate, under the hypothesis of equiprobability of the microstates. In the present urns case, W is the number of all the possible ways of distributing n_A balls in A and n_B balls in B:

$$W = \frac{n_T!}{n_A! n_B!}$$

So the Boltzmann's equation can be written:

$$S = k_B \log \left(\frac{n_T!}{n_A! n_B!} \right)$$

In this equation S represents the entropy of the system. One sees clearly that S increases as W increases and reaches its maximum when the balls are equally distributed between the two urns. If the system is far from the equilibrium state, it will evolve in one precise direction, the one leading to the maximum S. Thus it is easy to see, if $n_A > n_B$ it is more probable the transition $A \rightarrow B$ and the opposite if $n_B > n_A$. This can be seen as a model of the spontaneous expansion of gas particles. It is not impossible *de iure* that S decreases, but *de facto* it is extremely improbable.

Are the Ehrenfests urns the answer to the *Wiederkehreinwand (recurrent objection)* and to the *Umkehreinwand (reversibility objection)*? The answer is affirmative, surely in their intention, but also for other readers. In any case, Tatiana and Paul have left a fine lesson to make us understand clearly how statistics enters in physics. Mark Kac refers to the Ehrenfest's model as 'probably one of the most instructive models in the whole of physics.' (Kac, 1959). About lecturer qualities, Delft (2014) writes: '[Paul] Ehrenfest [...] was a gifted lecturer who seemed to do magic tricks at the blackboard.' Albert Einstein once called him 'the best teacher in our profession whom I have ever known'.

Let us recall the concept of ergodicity, by modifying the previous example. Instead of the time average, we estimate the *ensemble average*, executing a number of different trajectories (histories) and performing the average at a certain time. In the R code `Code_3_7.R` below, the number of steps is 5000 and the number of trajectories is 1000. The initial number of balls in the urns is chosen at random, that is at starting time either all the 1000 balls or no balls can be in the urn A.

```
### Code_3_7.R
## Ehrenfest_many_hist
```

```
## Ehrenfest urn model  1000 trajectories, 5000 steps
#           ensemble average of (Delta n)^2

set.seed(1)
nt<- 1000
ne<-nt/2
numberA <- seq(0,nt)
nsteps<- 5000         # number of steps
nhist<-  1000         # number of histories
splot<-  20           # to plot only "splot" histories
diffe0<- numeric()    # see comments below
diffe<-matrix(,nhist,nsteps)  # to save each history
f<-     matrix(,nhist,nsteps) # to save (Delta n)^2
mstep<- 4000   # compute the ensemble average at step = mstep
numberA <- seq(0,nt)  # starting with any number of balls in A
# (0<= numberA <= nt)
plot(t, type="n",ylim=c(-500,500),xlim=c(0,nsteps),main="",
ylab=expression(Delta*n),xlab="time",cex.lab=1.3,font.lab=3)
for(l in 1:nhist)       {    # starting loop on histories
i<-0
nA<- sample(numberA,1)      # nA different for each history
diffe0[l]<- nA-ne
while(i<nsteps)    {         # starting loop on steps
i<-i+1
u<- sample(nt,1,1/nt)
if(u<=nA) nA<- nA-1 else nA<- nA+1
diffe[l,i]<- nA-ne
f[l,i]<- diffe[l,i]^2    # function of (Delta n), now = (Delta n)^2
}      # ending loop on steps
}      # ending loop on histories

t<-c(0:nsteps)           # adding t=0
# adding to the matrix diffe a first column with the starting diffe
diffe<- cbind(diffe0,diffe)
for(l in 1:nhist)    {
# to plot the histories from 1 to splot
if(l<=splot)lines(t,diffe[l,],lty=3)
}
segments(mstep,-500,mstep,+500,lty=4,lwd=2,col="black")
mdi<- mean(diffe[,mstep])      # ensemble average
mdi
se.mdi<- sqrt(var(diffe[,mstep]))/sqrt(nhist)
se.mdi
mdif<-  mean(f[,mstep])        # ensemble average
mdif
se.mdif<-  sqrt(var(f[,mstep]))/sqrt(nhist)
se.mdif
```

The result is shown in Fig. 3.25. At the time 4000 the values of $(\Delta n)_{4000}$ and $(\Delta n)^2_{4000}$ are saved for all the 1000 histories. The ensemble average of the $(\Delta n)^2_i$ (here i numbers the histories: $1 \leqslant i \leqslant 1000$) gives the result:

$$\widehat{(\Delta n)^2} = \frac{1}{1000} \sum_{i=1}^{1000} (\Delta n)^2_{i \, \searrow \text{ history}} = 245.566 \qquad (3.94)$$

Time average and ensemble average are in good agreement: the system is ergodic. A theorem due to Birkhoff, known as *ergodic theorem*, states (in a discursive way)

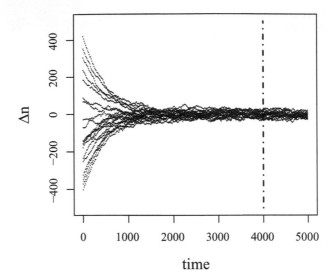

Fig. 3.25 Time evolution of the values assumed by the random variable Δn with $n_T = 1000$ of 20 histories out of 1000. The trajectories length is 5000 steps. The vertical dot-dashed line indicates the time point ($t = 4000$) at which the average is performed.

that because in the long run the system visits a large number of possible states, a time sample of these values is equivalent to a sample extracted from all possible system values. In other words, a single trajectory is representative of all the possible trajectories allowed to the system.

Note the conceptual difference between eqns (3.93) and (3.94). The former is a time average: only one trajectory and average on 2000 values (from 3000 to 5000 steps) assumed by the variable $(\Delta n)^2$. The latter is an ensemble average: 1000 trajectories and the average is on values of $(\Delta n)^2$ that each of them assumes at a precise time instant.

Figure 3.25 shows that all the trajectories, although starting from different initial values, once they have reached stationarity (at about 2000 steps) assume values in a narrow interval around $\Delta n = 0$, as expected. Things are different for the standard error of $\Delta n = 0$, the ensemble average $\widehat{(\Delta n)^2}$ and the time average $\overline{(\Delta n)^2}$, computed through the sample variance `sqrt(var(f[,mstep]))/sqrt(nhist)`, in the code `Code_3_7.R`. For $\overline{(\Delta n)^2}$, `std.mtempf` $= 5.731442$, while for $\widehat{(\Delta n)^2}$, `se.mdif` $= 10.98741$, a little bit more than 50% with respect to the standard error of the time average. Here we are in a situation in which the values assumed by the variable $(\Delta n)^2$ cannot be considered the realization of mutually independent random variables with the same distribution. If only one trajectory is available, we have to address, for instance, to the moving block bootstrap, discussed in Chapter 11 to derive a correct estimate of the standard error of the time average.

3.6 Exercises

Exercise 3.1 Consider a simple game. In a track like the following.

1	2	3

You start on position 1, the goal is to reach position 3. You flip a coin. If you get heads, stay where you are. If you get tails, move forward in the next position. Each player counts the number of coin flips. The winner is who reaches 3 in the lowest number of moves.

Write the transition probability matrix and the state transition diagram for this game.

Exercise 3.2 Write an R code to simulate the game of exercise 3.1.

Exercise 3.3 Consider a simplified version of the Game of The Goose, a very popular old game attributed to Francesco de Medici (1541−1587). The simplified version consists of a path having only 10 positions (instead of 64), starting from 0 and ending at 9, with a single player and one die only (instead of two dice).

The path is like the following:

0	1	2	3	4	5	6	7	8	9

The rules of the game are the following:

- The player starts on position 0. He rolls a dye and moves forwards the number of steps shown by the die face.

- Position 6 is a goose. If the player lands on it he moves forwards the same number of positions. For example, if he was in 5 and the dye shows 1, he goes in 6 and repeats the move, i.e. he goes in position 7.

- Position 4 is the death. The player falling there comes back to position 0.

- The goal is to reach position 9, exactly. If the number of moves is such as to exceed that position, the surplus is counted backwards from position 9.

Represent the problem in terms of its transition matrix. In particular, while in general the transition probabilities are 1/6 (due to the die), rows 4, 7 and 9 of the matrix has peculiar characteristics. Which ones?

Exercise 3.4 Justify the following assertion: an independent and identically distributed sequence of random variables is a Markov chain.
Hint: write the transition matrix.

Exercise 3.5 Assume the following transition probability matrix for a Markov process describing weather, with possible states:

```
1 = clear sky
2 = cloud covered sky
3 = rain
```

Let P be:

$$\mathbf{P} = \begin{pmatrix} 0.4 & 0.3 & 0.3 \\ 0.2 & 0.4 & 0.4 \\ 0.1 & 0.6 & 0.3 \end{pmatrix}$$

A meteorologist predicts for tomorrow, Friday, 40% rain probability and 60% clear sky. What is the probability of rain on Saturday?

Hint: consider the initial vector probability and refer to Section 3.3.

Exercise 3.6 Compute the eigenvectors and eigenvalues of the matrix **P** given by eqn (3.32).

4
Poisson Processes

But if we want to know what to think about such irregular and extraordinary actions,
we might consider the view that is commonly taken of irregular events
that appear in the course of nature and in the operations of external objects.
All causes are not conjoined to their usual effects with the same uniformity.

David Hume, Enquiry Concerning Human Understanding

It was Siméon—Denis Poisson himself who introduced in the work *Recherches sur la probabilité des jugements en matière criminelle et en matière civile, précédées des règles générales du calcul des probabilités*, published in 1837, the distribution that later will bear his name. His intent was to introduce the judiciary statistics into juridical questions, so as to estimate the probability of an incorrect judgement, problems already faced by Nicolas de Condorcet in 1785 and by Pierre Simon Laplace in 1825 (Laplace, 1825).

On pages 205-206 of his *Recherches* one can find the emergence of that probability distribution, which is now called Poisson distribution. He also proved - as he called it — the '*loi générale des grands nombres*', extending the well-known theorem of Jacques Bernoulli, and applied this law in a variety of examples '*pris dans l'ordre physique et dans l'ordre moral*'.

As we shall see below, the Poisson distribution is used to compute how many times a random event will occur in a time or space interval, when the probability of its happening is very small but the number of trials is very large. Historically, one of the most famous applications of the Poisson distribution was made by Ladislaus Bortkewitsch in *Das Gesetz der Kleinen Zahlen* (1898) (*Law of Small Numbers*), who estimated the probability of a Prussian cavalryman being killed by a horse-kick (Preece *et al.*, 1988). Such accidents were very rare, but the number of 'trials' (observations) was very large, recorded over 20 years (1875—1894), from 14 military corps: a paradigmatic scenario for applying Poisson distribution.

Poisson processes are generally associated with the occurrence of some type of events within a given period of time: customers arrivals, particle emissions, telephone calls, bus arrival times, etc., but they also find essential applications in *spatial statistics*, to analyse the distribution of points in plane or space: locations of beeches in a certain area of a forest, galaxy distribution, heavy metal concentrations, etc. In the *Encyclopædia Britannica* it is reported that, during World War II, British statistician R. D. Clarke analysed the distribution of hits of V-1 and V-2 flying bombs in London, finding a Poisson pattern.

In the next section the Poisson process will be regarded as a continuous-time Markov chain on the discrete state space of non-negative integers and here in this section, the aim is to introduce some basic techniques to perform computer experiments on stochastic processes. The Poisson process is chosen as a worked example. After reviewing fundamental concepts, some strategies to build up Monte Carlo simulations are illustrated. Parameters estimates are 'ensemble estimates', since different realizations of the same process are performed, through different sequences of random numbers.

4.1   Counting process

Consider the stochastic process $\{X(t), t \geqslant 0\}$. The process is called a *counting process* if $X(t)$ stands for the total number of 'events' occurred (or 'counted') up to a particular time t. By 'up to time t' we mean 'before and *at* time t'. If the event occurs exactly at time t, it is counted. This process is a continuous-time process in discrete state space $\mathcal{S} = \mathbb{N}$. The number of customers entering a store from opening hours 8.00 up to 10.00 may be described by a counting process, but not the number of customers inside the store at 10 o'clock. A counting process might model the number of births of an animal species up to a certain day, or the number of pieces of music written by a composer up to a certain year.

The counting process $\{X(t), t \geqslant 0\}$ is defined on the basis of the following properties:

1. $\{X(t)\} \geqslant 0$.
2. $\{X(t)\}$ is integer-valued for all $t \geqslant 0$.
3. if $s < t$ then $X(s) \leqslant X(t)$.
4. if $s < t$ then the difference $X(t) - X(s)$ counts the number of events that occurred in the time interval $(s, t]$.

Therefore, the differences $X(t) - X(s)$ takes only non-negative integer values. In particular, if $X(0) = 0$, $\{X(t)\}$ denotes the number of events counted in the time interval $[0, t]$. Such processes are also called *purely discontinuous* processes. The occurrence of each event is referred to also as an 'arrival', or an 'emission', for instance, of radioactive particles from a substance.

The random variables $X(t) - X(s)$ $\forall t, s$ positive are named *increments* of the process and are of major importance when they are *independent* and *stationary*. A counting process has *independent increments* if the numbers of events in disjoint intervals are independent. This means that the number of events, the realization of the random variable $X(t) - X(s)$, that occurred in any interval $(s, t]$, is independent of the number of events, realizations of the random variable $X(v) - X(u)$, occurred in any other *disjoint* interval $(u, v]$, for instance in $[0, s]$. With this definition, the random variables, describing the number of events in disjoint intervals, are independent of each other. That is the probability of n events in $(s, t]$ does not change even though it is conditioned on information on events in other intervals different from $(s, t]$. This independence property of increments has its counterpart in the independence of trials in the discrete-time: the discrete values Bernoulli process.

Examples of independent increment processes are those describing the number of customers entering a store, or the number of pieces written by a composer. Although that assumption could appear reasonable also for the number of births, actually it is not. Indeed, if the number of births is large enough, it is not plausible that the number of future births does not depend on births in the past times.

A counting process has *stationary increments* (or *homogeneous increments*) if the number of events in any interval $[t, t + \Delta t]$ only depends on the length Δt of that interval, and not on t, that is on its particular position in time. In other words, the random variable 'number of events' in Δt interval has the same distribution of the random variable 'number of events' in $t + \Delta t$ interval. So, the stationarity property is written as:

$$X(z + t + \Delta t) - X(z + t) \sim X(t + \Delta t) - X(t)$$

or, with $\Delta t = t - s$, $t > s$:

$$X(t + z) - X(s + z) \sim X(t) - X(s), \qquad \forall z > 0, \ \forall t > s \tag{4.1}$$

This means that events are 'equally probable' for any t.

The stationary increments property has its counterpart in the constancy of probability of success in the Bernoulli process. In the store example, the stationarity assumption is not reasonable, since almost certainly there are periods of time with greater access of customers. Also in the other examples, the increments are not stationary. For the births, the stationary increments assumption would be reasonable if the population was constant. For the composer, it is probable that inspiration changes over time. Lastly, note that if a process has stationary increments, that does not imply its stationarity. An example is the Poisson process.

Further random variables are associated to $X(t)$, the number of counted arrivals in $[0, t]$: the instant of the n-th arrival and the time between each arrival and the next. In the following, we usually refer to the occurrence of each event also as an 'arrival'. Let T_n be the random variable describing the n-th arrival, or also the time to wait the n-th arrival, or the *waiting time*, so defined:

$$T_n = \inf\{t \geqslant 0, X(t) = n\} \tag{4.2}$$

where inf is the mathematical *infimum*. At time t at least n arrivals are counted if and only if $T_n \leqslant t$. This means that the following events are equivalent:

$$\{X(t) \geqslant n\} \qquad \text{and} \qquad \{T_n \leqslant t\} \tag{4.3}$$

It follows that a counting process may be defined on the basis of the waiting time:

$$\{X(t)\} = \max\{n : T_n \leqslant t\}$$

For instance, in a store the event {at 11 o'clock there were *at least* 6 arrivals} (they might have been also 7, 8, etc.), that is $X(11) \geqslant 6$, is equivalent to the event {the waiting time of the 6-th arrival does not exceed 11 o'clock}, that is $T_6 \leqslant 11$. The following events are equivalent:

$$\{X(t) > n\} = \{T_{n+1} \leqslant t\} \qquad \text{and} \qquad \{X(t) = n\} = \{T_n \leqslant t < T_{n+1}\}$$

The time between any two consecutive arrivals, that is between the event $(n-1)$-th and the event n-th, (*inter-arrival time*), is described by the random variable $A_n =$

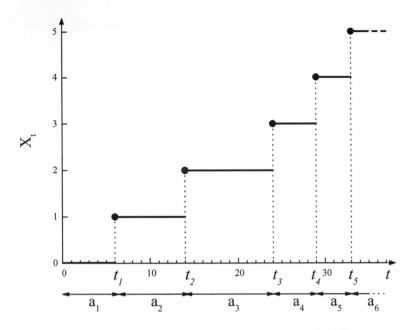

Fig. 4.1 Sample path of the counting process $\{X(t)\}$.

$T_n - T_{n-1}$. Figure 4.1 shows a sample path of the counting process $\{X(t), t \geqslant 0\}$. Let us suppose that the first arrival occurs at the instant $t_1 = 6$ (in some unit of time, for instance seconds). There is no arrival up to the time $t_2 = 14$. The third arrival occurs after 10 further seconds, at $t_3 = 24$, and so on. We have denoted with t_1 a realization of the variable T_1, with t_2 a realization of the variable T_2, etc. Therefore, the realizations of the random variable A_n are $a_1 = 6$, $a_2 = 8$, $a_3 = 10$, etc. Figure 4.1 clearly shows the equivalence of the events reported in eqn (4.3). For instance, the third arrival occurs at a time $t \leqslant 30$ (before the time 30 or at the time 30) if and only if the number of arrivals in $[0, 30]$ is $\geqslant 3$. It may even be 4, but in any case $\geqslant 3$. A small note: in this case the time has been discretized, supposing for recording arrivals, for instance, with a timer accurate to one second.

4.2   Poisson process from counting process

The Poisson process is a counting process with stationary and independent increments, defined in the following. The counting process $\{X(t)\}$ is a Poisson process if:

a) $X(0) = 0$. The counting process begins at time $t = 0$.

b) The process has independent increments.

c) The probability of n arrivals in any interval of finite length Δt is given by the random variable, $X(\Delta t)$ with distribution, named *Poisson distribution*:

$$P\{X(\Delta t) = n\} = \frac{(\lambda \Delta t)^n}{n!} e^{-\lambda \Delta t}, \qquad n = 0, 1, \ldots \qquad (4.4)$$

The interval Δt may be written also as $(t - s), s < t$ or, for the independence of increments, $(t - 0) = t$. So the above equation may be rewritten as:

$$P\{X(t + s) - X(s) = n\}$$
$$= P\{X(t) = n\} = \frac{(\lambda t)^n}{n!}\, e^{-\lambda t}, \qquad n = 0, 1, \dots$$

The state space is clearly $\mathcal{S} = \{0, 1, 2, \dots\}$. In that counting process, $X(t) - X(s) \geq 0, \ \forall t > s$, all the trajectories of the Poisson process are non-decreasing in t. The average number of arrivals in $[0, t]$ is given by:

$$E[(X(t)] = \sum_{1}^{\infty} n \frac{(\lambda t)^n}{n!}\, e^{-\lambda t}$$

(sum from 1, with 0 the term cancels out)

$$= \sum_{1}^{\infty} \frac{(\lambda t)^n}{(n-1)!}\, e^{-\lambda t} = \lambda t$$

It is clear that the Poisson process is not stationary in any sense.

The parameter λ, also called the *intensity* or *rate* of the process, represents the number of arrivals per unit of time, that is the *frequency*, so it has the dimensions of inverse time, i.e. s^{-1}. If λ is the frequency, then $1/\lambda$ represents the average time between arrivals. Then, the larger λ the smaller the average waiting time is.

Condition *c*) also implies that the process has stationary increments. From eqn (4.1), we have to prove that:

$$P\{X(t + z) - X(s + z) = n\} = P\{X(t) - X(s) = n\}$$

Now:

$$P\{X(t + z) - X(s + z) = n\} = \frac{[\lambda(t - s)]^n}{n!}\, \exp[-\lambda(t - s)]$$
$$= P\{X(t) - X(s) = n\}$$

Therefore the process *is not stationary, but has stationary increments*.

4.3 Poisson process from Bernoulli process

Let us start from different conditions and derive that the number of arrivals in a time interval Δt has a Poisson distribution.

The counting process $\{X(t), t \geqslant 0\}$ is a Poisson process with intensity λ $(\lambda > 0)$ if:

a) $X(0) = 0$.

b) The following conditions hold:

$$P\{X(t + dt) - X(t) = 0\} = 1 - \lambda dt + o(dt)$$
$$P\{X(t + dt) - X(t) = 1\} = \lambda dt + o(dt)$$
$$P\{X(t + dt) - X(t) \geqslant 2\} = o(dt)$$

c) The process has independent and stationary increments.

The notation $o(dt)$ (the Landau symbol "little-o") means $\lim_{dt \to 0} o(dt) = 0$, that is, per $dt \to 0$, $o(dt)$ goes to zero much faster than dt.

The first and second conditions in *b*) mean that in any infinitesimal interval dt either no event or only one event occurs. The probability of occurrence of such a unique event is given by the frequency multiplied by the interval length. The third condition in *b*) means that the probability of occurrence of more than one event in any infinitesimal interval dt is negligible with respect $\mathsf{P}\{X(dt) = 1\}$.

The random variable $X(\Delta t)$, the number of arrivals in an interval Δt of *finite* length, can be demonstrated to follow the Poisson distribution by showing that it is an approximation to the binomial distribution:

$$Bin(n, N, p) = \frac{N!}{n!(N-n)!}\, p^n (1-p)^{N-n}$$

in the limits $N \to \infty$, $p \to 0$, and if $\mathsf{E}\,[Bin]$ tends to a finite constant Np. In the above equation, n is the number of successes n in N independent trials, each of which has the same success probability p. Imagine dividing the interval Δt in N infinitesimal subintervals of length dt, such that $N = \Delta t/dt \gg 1$, with Δt a finite quantity (see Fig. 4.2). Furthermore, by virtue of the above conditions in a tiny interval dt at most

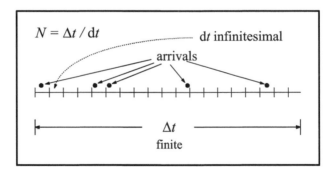

Fig. 4.2 Poisson distribution derived from the binomial distribution. The finite interval Δt is divided in N infinitesimal subintervals of length dt.

one arrival may occur. That is like saying that in the time interval Δt, N trials are executed (arrivals observation) and in each subinterval dt the event either happens (arrival) or it does not (no arrival). Let $p = \lambda dt$ be the success probability, the average number of successes is given by:

$$Np = N\lambda dt, \quad \text{but} \quad N = \frac{\Delta t}{dt},$$

$$\text{then} \quad Np = \frac{\Delta t \lambda dt}{dt} = \lambda \Delta t \ \ \textit{finite quantity}$$

Since dt is an infinitesimal interval, if λ is a finite quantity p is very small. Therefore very few successes may occur in Δt, that is $n \ll N$. In such conditions, we can write:

$$\frac{N!}{(N-n)!} = N(N-1)(N-2)\ldots,(N-n+1) \overset{n\ll N}{\approx} N^n$$

Consider, for example, $N = 9$ and $n = 4$:

$$\frac{9!}{(9-4)!} = \frac{1\cdot2\cdot3\cdot4\cdot5\cdot6\cdot7\cdot8\cdot9}{1\cdot2\cdot3\cdot4\cdot5} = 9\cdot8\cdot7\cdot6 = 3024$$

which is not close to $9^4 = 6561$, because $n = 4$ in not much smaller than $N = 9$. But if we take $N = 90$ and still $n = 4$, we have:

$$\frac{90!}{(90-4)!} = 61\,324\,560\ldots, \quad \text{while} \quad 90^4 = 65\,610\,000$$

the two quantities differing by about 6%. Increasing N by a factor of 10 the discrepancy becomes as small as 0.6%.

If we put:

$$y = (1-p)^{N-n}$$

we have:

$$\log y = (N-n)\log(1-p) \xrightarrow{p\ll1} -p(N-n) \xrightarrow{n\ll N} -Np$$

for instance, if $p = 0.01$, $\log(1 - 0.01) = -0.01005$. With these approximations:

$$y = (1-p)^{N-n} \to e^{-Np}$$

then:

$$Bin(n,N,p) \xrightarrow[\substack{N\to\infty\\p\to0}]{} \frac{N^n}{n!}\,p^n\,e^{-Np} = \frac{(Np)^n}{n!}\,e^{-Np}$$

and being $Np = \lambda\Delta t$:

$$= \frac{(\lambda\Delta t)^n}{n!}\,e^{-\lambda\Delta t}$$

Notice the substantial difference between a Bernoulli process and a Poisson process: in the former, events are bound to occur in fixed time intervals, while in the latter they occur in continuous time.

Up to now we have been faced with the number of events occurring in a finite time interval Δt. Let us now consider two more random variables: the waiting time T_n of the n-th arrival eqn (4.2), and $A_n = T_n - T_{n-1}$, the time between the $(n-1)$-th and n-th events.

Let us focus on the latter. Take a certain instant t_0 as the origin of the time scale, by putting $t_0 = 0$ and consider $A_1 = T_1 - 0$, elapsed time up to the first arrival. We ask ourselves: for every $t > 0$, what is the probability that the random variable T_1 takes a value $t_1 > t$, that is:

$$P\{T_1 > t\} \equiv P\{A_1 > t\}$$

Such a probability is the same as:

$$P\{0 \text{ arrivals in } [0, t]\}$$

Then, for every $t > 0$ the event $T_1 > t$ is realized if and only if there are no events in the interval $[0, t]$ (including t). By putting $n = 0$ in eqn (4.4) and since $\Delta t = [0, t] = t$, it results in:

$$P\{X(t) = 0\} = P\{A_1 > t\} = e^{-\lambda t} \qquad (\text{since } A_1 = T_1)$$

The functional form of A_1 is therefore a negative exponential. The repartition function of A_1 is given by:

$$F_{A_1} = P\{A_1 \le t\} = 1 - P\{A_1 > t\} = 1 - e^{-\lambda t}$$

which is the repartition function of the exponential random variable with parameter λ, denoted as $Exp(\lambda)$.

Remark 4.1 Recall that the continuous random variable X has exponential distribution $Exp(\lambda)$ if its probability density function is:

$$f(x) = \begin{cases} \lambda e^{-\lambda x}, & 0 \leqslant x < \infty \\ 0, & \text{else} \end{cases} \qquad (4.5)$$

Its repartition function is:

$$F = P\{Exp \le t\} = \int_0^t \lambda e^{-\lambda y}\, dy = -\lambda \frac{1}{\lambda} e^{-\lambda t} \Big|_0^t = 1 - e^{-\lambda t}$$

To derive the expected value $E[Exp(\lambda)] = 1/\lambda$ and the variance $\text{Var}[Exp(\lambda)] = 1/\lambda^2$, we shall make use of the 'moment-generating function' of the random variable X given by:

$$M_X(y) = E\left[e^{yX}\right] = \begin{cases} \displaystyle\sum_{x \in S_X} e^{yx} p(x) & X \text{ discrete} \\ \displaystyle\int_{-\infty}^{\infty} e^{yx} f(x)\, dx & X \text{ continuous} \end{cases}$$

Note that $M_X(y)$ is a function of the numerical variable $y \in \mathbb{R}$, since the random variable X is fixed and it is defined for the values of the y's for which the expected value of $\exp(yX)$ exists. If $M_X(y)$ is differentiated n times in the origin, the n-th moment of X is found, which is:

$$M_X^{(n)}(0) = E[X^n], \qquad n = 1, 2, \ldots$$

Then:

$$M_X'(y) = \frac{d}{dy} E\left[e^{yX}\right] = E\left[\frac{d}{dy} e^{yX}\right] = E\left[X e^{yX}\right]$$

and for $y = 0$ it is $M_X'(0) = E[X]$.

Similarly:

$$M_X''(0) = \frac{d^2}{dy^2} \, \mathrm{E}\left[e^{yX}\right] = \mathrm{E}\left[\frac{d^2}{dy^2} \, e^{yX}\right] = \mathrm{E}\left[X^2 \, e^{yX}\right]$$

and for $y = 0$ it is $M_X''(y) = \mathrm{E}\left[X^2\right]$. Turning to the random variable *Exp*, it results in:

$$M_{Exp}(y) = \mathrm{E}\left[e^{y\,Exp(\lambda)}\right] = \int_0^\infty e^{yx}\,\lambda\,e^{-\lambda x}\,dx$$

$$= \int_0^\infty e^{-(\lambda-y)x}\,dx = \frac{\lambda}{\lambda-y}, \qquad y < \lambda$$

from which:

$$\mathrm{E}\left[Exp(\lambda)\right] = M_{Exp}'(0) = \frac{d}{dy}\,\frac{\lambda}{\lambda-y}\bigg|_{y=0}$$

$$= \frac{\lambda}{(\lambda-y)^2}\bigg|_{y=0} = \frac{1}{\lambda}$$

and:

$$\mathrm{Var}\left[Exp(\lambda)\right] = \mathrm{E}\left[Exp(\lambda)^2\right] - \left(\mathrm{E}\left[Exp(\lambda)\right]\right)^2 = \frac{1}{\lambda^2}$$

since it is the second moment:

$$\mathrm{E}\left[Exp(\lambda)^2\right] = M_{Exp}''(0) = \frac{2\lambda\,(\lambda-y)}{(\lambda-y)^4}\bigg|_{y=0} = \frac{2}{\lambda^2}.$$

We ask now the probability that the random variable A takes a certain value $t_2 > t_1$, given that $A_1 = t_1$, that is:

$$\mathrm{P}\left\{0 \text{ arrivals in } (t_1, t_1 + t] \middle| A_1 = t_1\right\}$$

To get the functional form of $A_2 = T_2 - T_1$, the elapsed time between the event 1 and the event 2, the conditional probability given $A_1 = t_1$ is required:

$$\mathrm{P}\left\{A_2 > t \middle| A_1 = t_1\right\} = \mathrm{P}\left\{0 \text{ arrivals in } (t_1, t_1 + t] \middle| A_1 = t_1\right\}$$

Now the events $\{A_1 = t_1\}$ and $\{0 \text{ arrivals in } (t_1, t_1 + t]\}$ are independent for the independence property of increments, hence:

$$\mathrm{P}\left\{A_2 > t \middle| A_1 = t_1\right\} = \mathrm{P}\left\{0 \text{ arrivals in } (t_1, t_1 + t]\right\}$$

$$= e^{-\lambda t}, \text{ for the stationary property of increments}$$

By induction, we can conclude that the random variables $A_1, A_2, \ldots$ are independent and identically distributed, according to an exponential function, with mean

$$\frac{1}{\lambda}$$

average time between the $(n-1)$-th and the n-th event. The independence property of increments ensures that what takes place from some time onwards is independent of what happened before. Moreover, due to the stationarity property of increments, any time instant $t = t^*$ behaves as $t = 0$, i.e. the initial time does not affect the statistical properties of the process. In other words, the process is *memoryless*, as we shall see in detail in the following.

Let us turn to the random variable T_n, waiting time of the n-th arrival, that can be also written as:

$$T_n = \sum_{i=1}^{n} A_i, \qquad n \geqslant 1 \tag{4.6}$$

In Fig. 4.1, the realizations $t_1, t_2, t_3, \ldots$ of the random variables $T_1, T_2, T_3, \ldots$ are given by $6\,(a_1 = 6), 14\,(a_1 + a_2 = 6 + 8), 24\,(a_1 + a_2 + a_3 = 6 + 8 + 10), \ldots$. So the random variable T_n appears as the sum of the independent exponential function, all with mean $1/\lambda$, and it has a gamma distribution with parameters (n, λ), $n, \lambda > 0$, denoted as $Gam(n, \lambda)$. Its probability density function is given by:

$$f_{Gam}(t) = \lambda\, e^{-\lambda t}\, \frac{(\lambda t)^{n-1}}{(n-1)!}, \qquad t \geqslant 1 \tag{4.7}$$

Remark 4.2 The distribution (4.7) is actually a special case of the gamma distribution, attributed to A.K. Erlang and consequently often denoted as an 'Erlang' distribution. The generalized gamma distribution involves a real parameter α, instead of the integer n, and its probability density is:

$$f_{Gamma}(t) = \frac{\lambda^\alpha}{\Gamma(\alpha)} t^{\alpha-1} e^{-\lambda t}$$

where $\Gamma(x)$ is the Euler gamma function, generalizing the factorial to real values:

$$\Gamma(x) = \int_0^\infty \xi^{x-1} e^{-\xi} d\xi$$

If x = k, integer: $\Gamma(k) = (k-1)!$

To derive such a result, note that if $n = 1$, the first arrival takes place at the time t if and only if there are no events in $[0, t)$ (excluding t). Since for $n = 1$ is $T_1 = A_1$ (elapsed time up to the first arrival), for $t \geqslant 0$ we have:

$$P\{T_1 \leqslant t\} = 1 - P\{A_1 > t\} = 1 - e^{-\lambda t}$$
$$= F(t) = \int_0^t \lambda\, e^{-\lambda y}\, dy \tag{4.8}$$

Therefore the random variable T_1 is the exponential random variable $Exp(\lambda)$. We already know it, since $T_1 = A_1$, which here represents the waiting time before the occurrence of the first arrival, where λ, it was said, is the average frequency of occurrence and $1/\lambda$ the average time between one event and the next one.

To recover eqn (4.7), the result obtained for the waiting time of the first arrival is extended to the waiting time of the n-th arrival. The n-th arrival occurs at an instant $\leqslant t$, if and only if the number of arrivals in the interval $[0, t]$ is $\geqslant n$.

As previously described, see eqn (4.3), the events: $\{n$-th arrival before of, or at, time $t\}$ and $\{$number of arrivals up to $t \geqslant n\}$ are equivalent. Therefore the repartition function of the random variable T_n is given by:

$$F_{T_n}(t) = \mathsf{P}\{T_n \leqslant t\} = \mathsf{P}\{Poiss(t) \geqslant n\} = \sum_{i=n}^{\infty} \frac{(\lambda t)^i}{i!} e^{-\lambda t} \tag{4.9}$$

where *Poiss* denotes the random variable with Poisson distribution.

Given the repartition function of the random variable T_n, the probability density function $f_{T_n}(t)$ is obtained by differentiating with respect to t:

$$f_{T_n}(t) = -\sum_{i=n}^{\infty} \lambda \frac{(\lambda t)^i}{i!} e^{-\lambda t} + \sum_{i=n}^{\infty} \lambda \frac{(\lambda t)^{i-1}}{(i-1)!} e^{-\lambda t}$$

$$= \lambda e^{-\lambda t} \left[-\sum_{i=n}^{\infty} \frac{(\lambda t)^i}{i!} + \sum_{i=n}^{\infty} \frac{(\lambda t)^{i-1}}{(i-1)!} \right] = \lambda e^{-\lambda t} \left[\sum_{i=n}^{\infty} \frac{(\lambda t)^{i-1}(i - \lambda t)}{i!} \right]$$

$$= \lambda e^{-\lambda t} \frac{(\lambda t)^{n-1}}{(n-1)!} = e^{-\lambda t} \frac{\lambda^n}{(n-1)!} t^{n-1}$$

that is the density f_{Gam} of $Gam(n, \lambda)$ in eqn (4.7).

Remark 4.3 A small clarification. We said that the random variable T_n, waiting time, has a gamma distribution with probability density function given by eqn (4.7). This equation is a particular case, of a more general gamma distribution function:

$$f(x) = \frac{\lambda^r x^{r-1} e^{-\lambda x}}{\Gamma(x)}, \qquad x > 1, \quad \lambda, r > 0 \tag{4.10}$$

where λ is the scale parameter and r the shape parameter, and:

$$\Gamma(x) = \int_0^\infty x^{r-1} e^{-x}\, dx, \qquad n > 0$$

The following inequalities are notable:

$$\Gamma(1) = 0! = 1$$
$$\Gamma(r+1) = r\Gamma(r)$$
$$\Gamma(r+1) = r!$$
$$\Gamma(1/2) = \sqrt{\pi}$$

The mean is $\mu = r/\lambda$ and the variance is $\sigma^2 = r/\lambda^2$. If r is a positive integer $= 1, 2, \ldots$, then eqn (4.10) becomes eqn (4.7) that, in the case, is also named the *Erlang distribution*.

We claimed before that the Poisson process is memoryless. Let us clarify such a concept. First of all, let us also see how, and in what sense, the exponential random variable $Exp(\lambda)$ is memoryless. Consider a random variable X and let $\{X > t\}$ be the event {an arrival event occurs at a time greater than t}, equivalent to {no arrivals occurred up to the time t}, and let $\{X > t + \Delta t\}$ be the event {an arrival occurs at a time greater than Δt, starting from t}, equivalent to {no arrivals occurred up to the time $t + \Delta t$}. We say that the random variable X is memoryless if:

$$\mathsf{P}\{X > t + \Delta t | X > t\} = \mathsf{P}\{X > \Delta t\} \qquad (4.11)$$

Described differently, the probability of an arrival at a time $t + \Delta t$ is always the same, regardless of t, i.e. regardless of the time considered as origin. Let us consider the negative exponential random variable. Recall the definition of conditional probability: given events A and B, it holds $\mathsf{P}\{A \cap B\} = \mathsf{P}\{A|B\}\mathsf{P}\{B\}$. Hence, it is:

$$\mathsf{P}\{X > t + \Delta t | X > t\} = \frac{\mathsf{P}\{\{X > t + \Delta t\} \cap \{X > t\}\}}{\mathsf{P}\{X > t\}}$$

$$= \frac{\mathsf{P}\{X > t + \Delta t\}}{\mathsf{P}\{X > t\}}$$

since $\{X > t + \Delta t\} \subset \{X > t\}$. If $X \sim Exp(\lambda)$ it results in:

$$= \frac{\exp[-\lambda(t + \Delta t)]}{\exp[-\lambda t]} = \exp[-\lambda \Delta t] = \mathsf{P}\{X > \Delta t\}$$

It can be proved that the negative exponential random variable is the only continuous random variable with such lack-of-memory property.

Let us look at the condition $\{X > t + \Delta t\} \subset \{X > t\}$ by means of an example. Let E_1 be the event {a customer enters after 12 o'clock}, which means that the customer entered from 12 to 19 o'clock (closing time). Let E_2 be the event {a customer enters after 12.30}. It is clear that if the event E_2 occurred, also the event E_1 previously occurred. Obviously the converse is not true: just think of a customer entered at 12.10. Then, since $E_2 \subset E_1$, it is $\mathsf{P}\{E_1 \cap E_2\} = \mathsf{P}\{E_2\}$.

A very recurrent example concerns the life time of a device before failing. The event 'failure' is what we named 'arrival', so eqn (4.11) says that the probability that the device fails after a certain time is the same as it is new, just out of the factory, even though it has worked for some time t. The device, indeed, 'does not remember' that it was working for some time.

This does not mean that the exponential function is always a good choice for modelling systems, electronic, mechanical, biological, etc., affected by aging. In such situations the Weibull (*Weib*) random variable is exploited. Its probability density function is:

$$f_{Weib}(t) = at^{b-1}\exp(-at^b/b), \quad t \geqslant 0, \quad a \text{ and } b \text{ parameters} > 0$$

and its cumulative distribution function is:

$$F_{Weib}(t) = \int_0^t f_{Weib}(t')\,dt' = 1 - e^{-(a/b)t^b}$$

If $a = b = 1$, we come back to the exponential random function, while if, for instance, $a = b = 2$, it is:

$$f_{Weib}(t) = 2t\,e^{-t^2}$$

hence:

$$\mathsf{P}\{X > t + \Delta t | X > t\} = \exp[-(\Delta t)^2 - 2t\Delta t]$$

that is the probability of no events (failures) up to $t + \Delta t$ (life expectancy) decreases with increasing t.

Resorting to the Poisson process, a further important property concerns the precise instant at which an event occurs. We know that in the interval $(0, t]$ an arrival took place. We ask ourselves: what is the probability that the arrival occurred at the instant $s \in (0, t]$? The probability of the first arrival, given an arrival in $(0, t]$, is:

$$\mathsf{P}\{A_1 \leqslant s | X(t) = 1\} = \frac{\mathsf{P}\{\{A_1 \leqslant s\} \bigcap \{X(t) = 1\}\}}{\mathsf{P}\{X(t) = 1\}}$$

$$= \frac{\mathsf{P}\{\{X(t) - X(s) = 0\} \bigcap \{X(s) = 1\}\}}{\mathsf{P}\{X(t) = 1\}}$$

Because of the independence property of increments, the events in the intersection are independent. But if A and B are stochastically independent, we have $\mathsf{P}\{A \bigcap B\} = \mathsf{P}\{A\}\,\mathsf{P}\{B\}$. Now the probability of the first event in the intersection is given by:

$$\mathsf{P}\{X(t) - X(s) = 0\} = \frac{[\lambda(t - s)]^0}{0!}\,e^{-[\lambda(t-s)]} = \exp[-\lambda(t - s)]$$

while for the second event in the intersection, we have:

$$\mathsf{P}\{X(s) = 1\} = \frac{(\lambda s)^1}{1!}\,e^{-\lambda s} = \lambda s\,\exp(-\lambda s)$$

and similarly for $\mathsf{P}\{X(t) = 1\}$ at the denominator. Therefore:

$$\mathsf{P}\{A_1 \leqslant s | X(t) = 1\} = \frac{\exp[-\lambda(t - s)]\,\lambda s\,\exp(-\lambda s)}{\lambda t\,\exp(-\lambda t)} = \frac{s}{t}$$

that is the repartition function of the random variable $U(0, t)$, uniformly distributed in the interval $[0, t]$.

Indeed, if $U(a, b)$ is the uniform random variable defined in the interval $[a, b]$, the probability that a point x is inside the interval $[a_1, b_1] \subset [a, b]$, is proportional to its width:

$$\mathsf{P}\{a_1 \leqslant X \leqslant b_1\} = \int_{b_1}^{a_1} p(x)dx = \int_{b_1}^{a_1} \frac{1}{b - a}\,dx = \frac{b_1 - a_1}{b - a}$$

in our case, $[a_1, b_1] = [0, s]$ and $[a, b] = [0, t]$.

So, we have seen that, given an arrival in the $[0, t]$, this arrival may occur with uniform probability at any point of the interval. It follows, if we know that n arrivals occurred in a certain interval $[0, t]$, the arrival times $T_1, \ldots, T_n$ form an ordered n size sample, extracted from the random variable $U(0, t)$.

We said that, if the random variables A_n, describing the time between two consecutive arrivals, have exponential distribution, then there may not be memory between such events. However, stationarity and independence properties entail the memorylessness of the process and, therefore, random variables modelling inter-arrival times can only be of the exponential type. These considerations allow for definition of the Poisson process in a further way, through the random variables A_n.

4.4 Poisson process through the inter-arrival time

The counting process $\{X(t), t \geqslant 0\}$ is a Poisson process with intensity λ if:

a) $X(0) = 0$.

b) The random variables A_n describing the time between two consecutive arrivals are independent and identically distributed exponential random variables with mean $1/\lambda$.

The reasoning is the following. Since A_n $i.i.d \sim Exp(\lambda)$, we have $T_n = \sum_{i=1}^{n} A_i \sim Gam(n, \lambda)$ and its probability density function is given by eqn (4.7). The repartition function of the random variable T_n, that is $\mathsf{P}\{T_n \leqslant t\}$ which can be written as:

$$F_{T_n}(t) = \mathsf{P}\{T_n \leqslant t\} = 1 - \sum_{i=0}^{n-1} \frac{(\lambda t)^i}{i!} e^{-\lambda t}$$

Now, for $n = 1, 2, \ldots$, the event $\{X(t) = n\}$ occurs if the process at time t is in state n, that is the following events are equivalent (counting from $n = 1$):

$$\{X(t) = n\} = \{A_1 + \cdots + A_n \leqslant t < A_1 + \cdots + A_n + A_{n+1}\}$$
$$= \{T_n \leqslant t < T_{n+1}\}$$

For instance, with reference to Fig. 4.1, the number of arrivals at time 30 is equal to 4, if the fourth arrival occurs before or at time 30, and the fifth arrival occurs after time 30. Indeed, $T_4 = 29$ and $T_5 = 33$. Similarly, by summing the A_i: $6 + 8 + 10 + 5 \leqslant 30 < 6 + 8 + 10 + 5 + 4$. Therefore:

$$\mathsf{P}\{X(t) = n\} = \mathsf{P}\{T_n \leqslant t\} - \mathsf{P}\{T_{n+1} \leqslant t\} = \frac{(\lambda t)^n}{n!} e^{-\lambda t}$$

which is a Poisson distribution with intensity λt. For $n = 0$:

$$\mathsf{P}\{X(t) = 0\} = \mathsf{P}\{T_1 > t\} = e^{-\lambda t}$$

i.e. the probability of not having any arrival exponentially decreases with time. To summarize, we have seen three random variables playing a major role in the Poisson process:

- Poisson random variable: it describes the number of events in a certain interval.
- exponential random variable: it describes the time between two consecutive arrivals.
- random variable gamma (Erlang): it describes the waiting time of the n-th event.

4.5 Poisson processes simulations

Let us begin with a simple example. Suppose that, in the morning, usually 10 customers per hour arrive on average in a pub. The average time between two arrivals is given by

$$\frac{1}{\lambda} = \frac{3600}{10} = 360\,\text{s} = 6 \text{ minutes}$$

and the average frequency is $\lambda = 2.8 \times 10^{-3}\,\text{s}^{-1}$. This does not mean that a customer arrives every about 6 minutes (i.e. the first customer after 6 minutes, the following after 12 minutes, etc.) We could think that customers arrive at times 'centered' on multiples of 6 minutes, like $6 \pm 1, 12 \pm 1, \ldots$, or $6.0 \pm 1.5, 12.0 \pm 1.5, \ldots$ or, in other words, we could think to Gaussian functions centred at 6, 12, 18, ... But this is not the correct picture: the shape of the density function eqn (4.5) of the exponential random variable $Exp(\lambda)$ says that events occurring closer in time are more likely than relatively long waits between successive events. Don't forget also that the random variable $Exp(\lambda)$ is memoryless, therefore there is no correlation between arrivals. Indeed, we derived it under this hypothesis of no correlation. Perhaps the barman believes that those few customers agreed with each other to have a drink together, but we know he is wrong. It is just the absence of correlation between arrivals that could lead to the existence of temporal correlation.

Someone might recognize from the above the 'scientific demonstration' of the popular saying (*vox populi, vox Dei*) 'bad luck never comes alone'. Misfortunes, as strokes of luck, are rare events and, often, are not related to each other. We expect from theory that occurrences of rare events in close times are more likely than relative long waits.

Remark 4.4 The demonstration of the last statement is not simple, but an elementary simulation of a Poisson process should convince the reader about its truth. Consider the following R code, simulating a Poisson process with $\lambda = 1/10$.

```
t <- 300
lambda<-1/10
set.seed(2)
N<-rpois(1,lambda*t)
set.seed(4)
unifs<-runif(N,0,t)
arrivals<-sort(unifs)
plot(arrivals,rep(1,length(arrivals)),yaxt="n",ylab="",xlab="arrival times")
```

The result is shown in Figure 4.3. We observe that inter-arrival times have an exponential distribution $e^{-\lambda t}$, whose maximum is at $t = 0$. Therefore, short inter-arrival times are rather frequent and, as a consequence, arrival times tend to cluster. This phenomenon is also known as Poisson clumping or Poisson burst.

The quantification of the Poisson burst phenomenon requires us to compute the probability that a large time interval of length T contains a smaller interval (a 'window') W_n of length w in which n or more arrivals occur. This probability can be computed by means of the so-called Alm's approximation (Glaz and Balakrishnan, 1999)

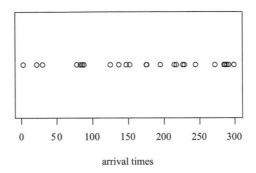

arrival times

Fig. 4.3 Arrival times in a Poisson process.

$$P(W_n \subset T) = 1 - P_{n-1}(\lambda w)\exp\left[-(1 - \frac{\lambda w}{n})\lambda(T - w)p_{n-1}(\lambda w)\right] \qquad (4.12)$$

where

$$p_k(\lambda w) = \frac{e^{-\lambda w}(\lambda w)^k}{k!}$$

is the Poisson probability of having k arrivals in time w, if λ is the arrival rate, and

$$P_n(\lambda w) = \sum_{k=0}^{n} p_k(\lambda w)$$

An example, borrowed from (Tijms, 2003), will help in verifying the above formula. The average annual number of people hit and killed by a tram in Amsterdam is about 3.7. If you read in a newspaper that seven people have been killed by a tram in the first five months this year, you may ask yourself if such a "cluster" of events is exceptional and in contrast with the known average. While it is clearly exceptional that seven people will be killed *exactly* in those five months, it is not exceptional at all to find a cluster of seven (or even more) fatal accidents in a window of five months in a long period. Indeed, the probability of having seven or more accidents in t = 5 months, given a monthly rate $\lambda = 3.7/12$ is:

$$Pr(k \geq 7) = 1 - Pr(k < 7) = 1 - \sum_{k=0}^{6} \frac{e^{-\lambda t}(\lambda t)^k}{k!}$$

which gives $Pr(k \geq 7) = 0.001$. Instead, the probability of seven or more accidents in **any** period of five months in ten years, computed by means of the Alm's formula (4.12) is 0.105, i.e. 100 times greater! A rare event, like seven accidents in five months, becomes increasingly more probable when increasing the total observation time. With the above numbers, that event would become almost certain in a period of 300 years (that is of course unrealistic in a rapidly changing world). The following code implements the above computations.

```
# rate
lam <- 3.7/12
# window
w <- 5
# number of accidents
```

```
n <- 7

# Probability of k<=6
P <- 0
for (i in 0:(n-1)){
P <- P + exp(-lam*w)*(lam*w)^i/factorial(i)
}
# Probability of 7 or more accidents
1-P

# total time 10 years
T <- 120
p <- exp(-lam*w)*(lam*w)^(n-1)/factorial(n-1)
prob <- 1 - P*exp(-(1-lam*w/n)*lam*(T-w)*p)
prob
```

Some examples follow of problems concerning the Poisson process.

Example 4.1 What is the probability of an arrival in the first minute, that is the probability of 1 event in $[0, 60]$ s?

It is given by:

$$P\left\{0 \leqslant T_1 \leqslant 60\right\} = \int_0^{60} \lambda e^{-\lambda t}\, dt = 1 - e^{-\lambda t}$$
$$= 1 - e^{-0.002860} = 0.1546 \approx 15\%$$

Example 4.2 What is the probability of an arrival after 20 minutes?

With time in seconds, it is:

$$P\left\{T_1 > 1200\right\} = 1 - P\left\{0 \leqslant T_1 \leqslant 1200\right\} = 1 - (1 - e^{-\lambda t})$$
$$= \exp[-2.8 \times 10^{-3} \times 1200] = 0.0347 \approx 3.5\%$$

Example 4.3 How much time has the barman to wait on average for the arrival of the third customer?

Compute the value t_3 assumed by the random variable T_3 distributed as $Gam(3, 2.8 \times 10^{-3})$. Hence $E[T_3] = 3/2.8 \times 10^{-3} = 1071$ s, which is about 18 minutes.

Not always can an analytical answer be given as in the examples above, for instance if the Poisson process is non-homogeneous (see Section 4.6). The way forwards is computer simulation.

The Monte Carlo method lends itself to computer experiments on stochastic processes. Turning to the example of customers going to a pub, a code modelling their random arrivals is discussed below.

```
## Code_4_1.R
# distribution of: number of arrivals, time of arrivals (waiting time),
# inter arrival time
# the average time between two arrivals is 1/lambda = Deltat/events
# lambda is the events average frequency.

# The process is discretized. The time unit is the second (infinitesimal dt).

Deltat<-3600        # the process is followed for 1 hour (3600 sec)
nevents<-10         # average number of arrivals in 1 hour
nhistories<-1000    # number of simulated histories (sample size)
lambda<-nevents/Deltat
lambda
ev<-numeric()               # vector of the number of arrivals in the l-th history
at<- numeric()              # vector of the inter arrival times in the l-th history
intert<- numeric()          # vector of the all inter-arrival times
t.arr<- numeric()           # vector of the arrival times in the l-th history
# to save each history (arrival times):
matr.history.t<-matrix(,nhistories,Deltat)
# to save each history (inter arrival times):
matr.history<-matrix(,nhistories,Deltat)
lst<-    numeric() # vector of the number arrival times in the l-th history
l_int<-  numeric() # vector of the number inter arrival times in the l-th history
###  2 loops: most external loop on histories, inner loop on steps
set.seed(2)     # reset random numbers
for(l in 1:nhistories)      {          # starting loop on histories
# at the start of each history:
t.arr<- 0
at<- 0
t1<- 0
i<-  0    # 'i' counts the seconds
ev[l]<-0
while(i <= Deltat)    {                # starting loop on steps
i<-i+1 # the following step
u<- runif(1)      # random number in [0, 1]
# if u<=lambda, an arrival occurs
if(u<=lambda) {  # starting if
# when an event occurs, the event and the occurrence time are saved
ev[l]<-ev[l]+1      # events in the l-th history are counted
t2<- i              # the occurrence time is saved in t2
t.arr <- c(t.arr,t2) # all the arrivals time in the l-th history are saved
# in the vector inter-arrival time the time interval between
#     the two occurrences is saved
at[ev[l]]<- t2-t1
t1<- t2    # t2 becomes t1
} # # ending 'if(u<=lambda)'
}               # ending loop on steps
lst[l]<-length(t.arr)    # No. arrival times in this l-th history
# save the number of arrivals in this l-th history:
matr.history.t[l,1:lst[l]] <- t.arr
l_int[l]<-length(at)     # No. of inter-arrival times in this l-th history
# save the number of inter arrival times in this l-th history:
matr.history[l,1:l_int[l]] <- at
}    ###############       # ending loop on histories

# ============================== analysis of results ======================

#................ number of arrivals ............

# ev               # uncomment to print all the arrivals
```

```
lbin<- 1
hist(ev,br=seq(0.5,24.5,by=lbin),main=" ",
freq=T,xlab="No. arrivals",ylab="counts",cex.lab=1.1,font.lab=3)
quant<-quantile(ev, c(0.025, 0.1587, 0.5, 0.8413, 0.975))
quant
abline(v=quant[2],lty=2,col="black",lwd=1)
abline(v=quant[4],lty=2,col="black",lwd=1)
mev<- mean(ev)
sdev<-  sd(ev)
mev
sdev
minev<- min(ev)
maxev<- max(ev)
minev
maxev


#..................... arrival time .........................

# matr.history.t[1,1:lst[l]]  to print the arrival times of l-th history.
# examples: histories 1, 2, 3 and last history
matr.history.t[1,1:lst[1]]
matr.history.t[2,1:lst[2]]
matr.history.t[3,1:lst[3]]
matr.history.t[nhistories,1:lst[nhistories]]
# in the plots: lst[l]-1, to ignore the initial 0
# type="s" for stair step look
par(mfrow=c(2,2))
plot(matr.history.t[1,1:lst[1]],0:(lst[1]-1),type="s",xlab="time (s)",
ylab="No. arrivals",cex.lab=1.,font.lab=3)
plot(matr.history.t[2,1:lst[2]],0:(lst[2]-1),type="s",xlab="time (s)",
ylab="No. arrivals",cex.lab=1.,font.lab=3)
plot(matr.history.t[3,1:lst[3]],0:(lst[3]-1),type="s",xlab="time (s)",
ylab="No. arrivals",cex.lab=1.,font.lab=3)
plot(matr.history.t[nhistories,1:lst[nhistories]],0:(lst[nhistories]-1),
type="s",xlab="time (s)",ylab="No. arrivals",cex.lab=1.,font.lab=3)


# the following lines for further analysis of arrival times
# all the arrival times in all the histories are saved in the vector t.arrT
t.arrT<- numeric()
# first history, after summing on the remaining histories:
t.arrT<- matr.history.t[1,1:(lst[1]-1) ]
for (l in 2:nhistories)         {
t.arrT<- c(t.arrT,matr.history.t[l,1:(lst[l]-1) ])
}
# t.arrT              # uncomment to print all the arrival times
t.arrT_c<- numeric()
t.arr_c<- t.arrT[t.arrT>1200]        # arrival times > 20 minutes ...
# t.arr_c                            # uncomment to print
nt.arr_c<- length(t.arr_c)
nt.arr_c
t.arr_c<- t.arr_c[t.arr_c<1500]    # ... but < 30 minutes
# t.arr_c                            # uncomment to print
nt.arr_c<- length(t.arr_c)
nt.arr_c


#..................... inter arrival time ..............................

# matr.history[l,1:l_int[l]] to print the inter arrival times of l-th history.
# for instance l=3 it is:
matr.history[3,1:l_int[3]]
```

```
# summing the inter arrival times of the l-th history (example l=3):
sum_int_sl<- sum(matr.history[3,1:l_int[3]])
sum_int_sl
# examples: histories 1, 2, 3 and last history
xmax<- 2000
ymax<- 6
lbin<-120
nclass<- seq(0,xmax,by=lbin)  #vector of breakpoints
par(mfrow=c(2,2))
hist(matr.history[1,],breaks=nclass,ylim=c(0,ymax),xlab="int_arr (s)",
ylab="counts",main=" ",font.lab=3,cex.lab=1.)
hist(matr.history[2,],breaks=nclass,ylim=c(0,ymax),xlab="int_arr (s)",
ylab="counts",main=" ",font.lab=3,cex.lab=1.)
hist(matr.history[3,],breaks=nclass,ylim=c(0,ymax),xlab="int_arr (s)",
ylab="counts",main=" ",font.lab=3,cex.lab=1.)
hist(matr.history[nhistories,],breaks=nclass,ylim=c(0,ymax),xlab="int_arr (s)",
ylab="counts",main=" ",font.lab=3,cex.lab=1.)
# plot=F to have a list of breaks and counts (example l=3):
hist(matr.history[3,],breaks=nclass,plot=F)
# all the inter arrival times in all the histories are saved
# first history, after summing on the remaining histories
int_arr<- matr.history[1,1:l_int[1] ]
for (l in 2:nhistories) {
int_arr<- c(int_arr,matr.history[l,1:l_int[1] ])
}
min.int<- min(int_arr)
min.int
max.int<- max(int_arr)
max.int
max.int.minutes<- max.int/60
max.int.minutes
mean.int = Deltat/nevents  # (=1/lambda)
mean.int
n.int_arr<- length(int_arr)
n.int_arr
lbin.int<-120
hist(int_arr,freq = T,main=" ",xlab="int_arr (s)")
# to compare with analytical density
hist(int_arr,freq=F,main=" ",xlab="int_arr(s)",ylab="density",
cex.lab=1.2,br=seq(0,2640,by=lbin.int),font.lab=3,axes=F)
axis(2)
x.minutes<- seq(0,2400,by=480)
axis(1,int_arr,at=x.minutes,lab=x.minutes,srt=0)
x<-rexp(n.int_arr,lambda)
curve(dexp(x,lambda),add=T,col="black",0,2640,lty=2,lwd=2)
histog<-hist(int_arr,plot=F,br=seq(0,2640,by=lbin.int))
histog
lb<-lbin.int
y.rescale <- pretty(range(histog$density*n.int_arr*lb))
axis(4, at = y.rescale / n.int_arr/lb, lab = y.rescale, srt = 90)
mtext("No. arrivals", side=4, line=-2,cex=1.2,font=3)
```

The most intuitive idea is to choose a tiny interval dt as time unit. According to the theory of Poisson stochastic processes, in dt only one event may occur with probability $\lambda dt \ll 1$. In our case, we choose $\Delta t = 3600$ s, that is 60 minutes, $dt = 1$ s, and suppose we know the average number of arrivals to be equal to 10. Therefore, as seen above, the events average frequency is $\lambda = 2.8 \times 10^{-3}$ s^{-1}. It should be obvious that we speak of 'second', 'minutes', etc., but they are not real physical time, they are – as they are called – 'computer steps' or, more specifically, 'Monte Carlo steps'.

A loop from 1 to 3600 (Δt) is executed; at each step (in each dt) an arrival may or may not occur; such an event is decided by a random number u uniformly distributed in $[0, 1]$. If $u < \lambda$, a customer has arrived: the time of occurrence and the elapsed time from the previous arrival are recorded. In any case, the loop goes one step further.

At the end of the loop, a computed time series (or history) is obtained. A certain number, for instance 1000, of such histories, all different from each other, are constructed by means of different sequences of random numbers. In this manner, the parameters of interest can be estimated. The code is divided into two parts. In the first part, the two loops, on the time steps and on the histories, respectively, are performed. In the second part, results produced by simulations are statistically analysed.

Figure 4.4 shows the distribution of the number of arrivals in an hour computed on all the 1000 histories. Note `freq=T` in the command `hist(ev, ...`, to plot the bin counts, or frequencies, of resulted values. With `freq=F`, probability densities are plotted, so that the total area of the histogram is equal to 1. As we shall see in the following, the density scale is used for comparison with a mathematical density model. The dashed lines give the limits of the confidence level with coverage probability = 68%. The command `quant<-quantile(ev, c(0.025, 0.1587, 0.5, 0.8413, 0.975),type=2)` estimates sample quantiles corresponding to chosen percentiles. In the R help (`?quantile`) page, we can find nine methods for computing sample quantiles from Hyndman and Fan (1996). Types 1, 2 and 3 refer to sample quantiles of discrete distributions. The computed quantiles are: $4, 7, 10, 13, 17$. The mean, `mean(ev)`, and the standard deviation, `sd(ev)`, result in being, as expected, 10 and 3.25, respectively.

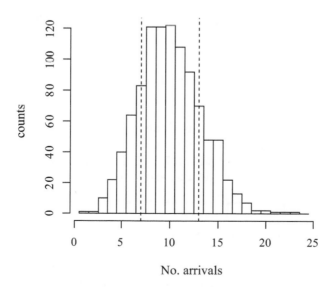

Fig. 4.4 Number of arrivals computed on 1000 histories. The dashed lines give the limits of the confidence level with coverage probability = 68%.

The printed counts and the figure describe, for instance, that in 122 out of 1000 histories, 10 arrivals occurred, but also that there are 5 histories in which only one event took place, while the maximum number of events in a history is 23.

Figure 4.5 shows the number of arrivals as a function of time for four different histories, the first, second, third histories and the last one.

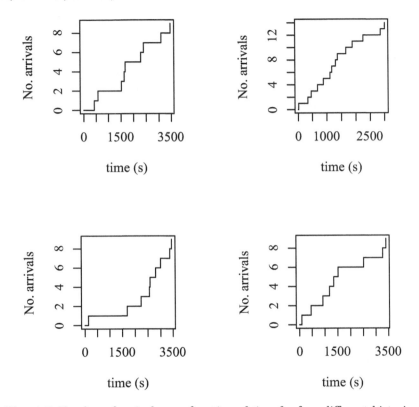

Fig. 4.5 Number of arrivals as a function of time for four different histories.

The histograms report that the events up to the last one occurred just before the end time (1 hour). The histograms are of the type seen before in Fig. 4.1, where the realizations of the random variable $X(t)$ are the 'arrivals'.

The command `matr.history.t[1,1:lst[1]]` prints the arrival times of the l-th history. For instance, with $l = 3$, the nine arrival times are:

```
matr.history.t[3,1:lst[3]]
```

```
0 136 1687 2239 2568 2589 2812 3000 3374 3447
```

The last arrival is at 3447 s, just before $\Delta t = 3600$ s, and the next arrival would be at 3704 s, if Δt were increased to 3800 s. The following lines can be executed if the number of arrivals within a certain time interval is required. The total number of arrival times is saved in `t.arrT`, and, for instance, those in $(1200, 1500)$ are selected.

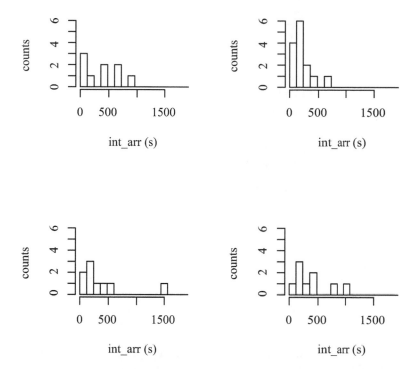

Fig. 4.6 Number of the time intervals between two consecutive arrivals as a function of their length, for four different histories.

Figure 4.6 shows the number of the time intervals between two consecutive arrivals as a function of their length. Four different histories, first, second, third and last one, are plotted. The command `matr.history[l,1:l_int[l]]` prints the inter-arrival times of the *l*-th history. For instance, with $l = 3$, the inter-arrival times are:

```
matr.history.t[3,1:l_int[3]]
136 1551 552 329 21 223 188 374 73
```

As an example, in the first bin between 0 and 120 (see the list `$'breaks'` by `hist(matr.history[3,],breaks=nclass,plot=F)` there are two inter-arrival times, 21 and 73. The inter-arrival time equal to 1551 is the time between the two arrivals at 136 and 1687 (see above `matr.history.t[3,1:lst[3]]`).

In the vector `int_arr`, all the inter-arrival times are saved. The shortest time (`min.int`) results to be 1 s, while the longest (`max.int`) is 2539 s, a little bit more than 42 minutes. Further information is found in Fig. 4.7, comparing the normalized histogram with an exponential fitting. The printed counts and the figure show, for instance, that in the first two minutes (120 s) there are 3140 inter arrival times, between 2 and 4 minutes they are 2079, etc.

As we discussed, simulations involve discretized events, while the exponential random variable, describing the waiting times, is continuous. Figure 4.7 shows how well an exponential model fits data derived by a discrete approximation. The reason for such an excellent agreement is that in simulations the arrival time is ruled by the

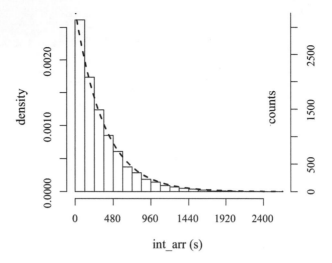

Fig. 4.7 Number of inter-arrival times computed on 1000 histories. The y-axis on the left refers to the normalized histogram for comparison with the probability density function of the exponential random variable (dashed line); the y-axis on the right refers to the number of inter-arrival times in each interval of 120 s, that is two minutes.

geometric discrete random variable (*Geom*), whose probability distribution function (better: probability mass function) is analogous, both in its form and properties, to the exponential random variable.

The probability mass function of the geometric random variable is given by:

$$p(k) = p\,(1-p)^{k-1}, \qquad k = 1, 2, \ldots \tag{4.13}$$

sometimes written as:

$$p(k) = p\,(1-p)^{k}, \qquad k = 0, 1, 2, \ldots$$

which is essentially the same. In fact, in eqn (4.13), the random variable *Geom* describes the 'number k of independent trials needed to get one success', each with success probability p. In the second definition, the random variable, which we call *Geom'*, describes the number of independent trials *before* the first success, that is the 'number of failures until the first occurrence of success', therefore $Geom' = Geom - 1$.

In the code `Code_4_1.R` above, the process evolves second after second and, at each instant, an arrival may or may not take place. In the following code `Code_4_2.R`, the step length is decided by the exponential distribution, and at each step an arrival occurs.

```
## Code_4_2.R
# Poisson process simulation by exponential generation of arrival times
# lambda is known

Deltat<-3600       # the process is followed up to 1 hour (3600 sec)
lambda<-0.003
set.seed(2)        # reset random numbers
```

```
nhistories<-4       # number of simulated histories
# to save arrival times in each history:
matr.history.t<-matrix(,nhistories,Deltat)
lst<- numeric()
###  2 loops: most external loop on histories,
#               inner loop generates arrival times
set.seed(2)     # reset random numbers
for(l in 1:nhistories)      {        # starting loop on histories
# at the start of each history:
t.arr<- 0    # arrival times
t.exp<- 0    # exponential generated times
while(t.exp<Deltat) {                 # arrival times generation
# 1 random deviate from an exponential distribution is generated:
t.exp<- t.exp + rexp(1,lambda)
# the arrivals time in the l-th history are saved:
t.arr<- c(t.arr,t.exp)
# print(t.arr)     # uncomment to see how events occur over time
}                # ending while loop
lst[l]<-length(t.arr)            # No. arrivals in this l-th history
# save the number of arrivals in this l-th history:
matr.history.t[l,1:lst[l]] <- t.arr
}              # ending loop on histories
#  matr.history.t[l,1:lst[l]] to print the arrival times of l-th history.
#  the four histories
matr.history.t[1,1:lst[1]]
matr.history.t[2,1:lst[2]]
matr.history.t[3,1:lst[3]]
matr.history.t[nhistories,1:lst[nhistories]]
# in the plots: lst[l]-1, to ignore the initial 0
# type="s" for stair step look
ymax<- 12
par(mfrow=c(2,2))
plot(matr.history.t[1,1:lst[1]],0:(lst[1]-1),type="s",font.lab=3,
xlab="time (s)",ylab="No. arrivals",ylim=c(0,ymax),cex.lab=1.)
abline(v=Deltat,lty=3)
plot(matr.history.t[2,1:lst[2]],0:(lst[2]-1),type="s",font.lab=3,
xlab="time (s)",ylab="No. arrivals",ylim=c(0,ymax),cex.lab=1.)
abline(v=Deltat,lty=3)
plot(matr.history.t[3,1:lst[3]],0:(lst[3]-1),type="s",font.lab=3,
xlab="time (s)",ylab="No. arrivals",ylim=c(0,ymax),cex.lab=1.)
abline(v=Deltat,lty=3)
plot(matr.history.t[nhistories,1:lst[nhistories]],0:(lst[nhistories]-1),
type="s",xlab="time (s)",ylab="No. arrivals",ylim=c(0,ymax),
font.lab=3,cex.lab=1.)
abline(v=Deltat,lty=3)
```

The code simulates only four histories to stress the difference in their construction with respect to the code `Code_4_1.R`. The R function `rexp(n,lambda)` generates n random deviates from an exponential distribution with rate parameter λ. As in the code `Code_4_1.R`, the command `matr.history.t[l,1:lst[l]]` prints the arrival times of history l-th. Arrival times are also called above waiting times. For instance, with $l = 2$, the arrival times are (with two decimal places):

```
matr.history.t[2,1:lst[2]]
 0.0000 224.59 754.69 1115.44 1380.97 1862.73 3360.04 3927.85
```

Note that now the last arrival is after $\Delta t = 3600$, since the loop stops when the condition `t.exp<Deltat` results FALSE.

Figure 4.8 shows the number of arrivals as a function of time for the four simulated histories, as in Fig. 4.5. Once the trajectories generation is executed, the number of histories can be increased and results produced by simulations can be statistically analysed with the instructions of code `Code_4_1.R`.

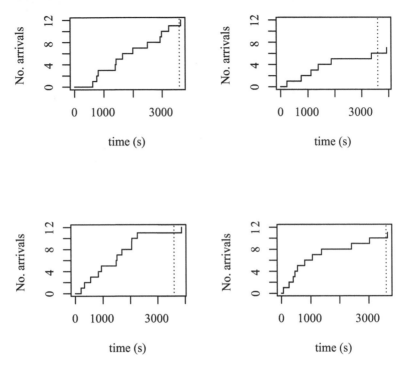

Fig. 4.8 Number of arrivals as a function of time for four histories. The dotted line is at $\Delta t = 3600\,\text{s}$

We recall that random deviates can also be generated from an exponential distribution by means of a well-known algorithm that exploits random deviates from a uniform distribution. The R function `runif(n,a,b)` generates n random deviates from a uniform distribution in $[a, b]$, if a and b are omitted the default is $[0, 1]$. So, the command `t.exp<- t.exp + rexp(1,lambda)` may be replaced by the command `t.exp<- t.exp-1/lambda*log(runif(1))`.

There exists a more elegant (efficient) way to generate the arrival times, by using vectorized operations of R. We have said that the random variable T_n, waiting time of the n-th arrival, may be written as the sum of independent exponential function, all with mean $1/\lambda$: $T_n = A_1, A_2, \ldots, A_n$. The code `Code_4_3.R` exploits this property.

```
## Code_4_3.R
# Poisson process simulation by exponential vectorized generation of arrival times
# lambda is known

Deltat<-3600        # the process is followed up to 1 hour (3600 sec)
lambda<-0.003
set.seed(2)         # reset random numbers
nhistories<-4       # number of simulated histories
# to save arrival times in each history
matr.history.t<-matrix(,nhistories,Deltat)
t.max<- 50 # alleged max number of arrivals
lst<- numeric()
t.exp<- numeric()
t.arr<- numeric()
###  1 loop on histories
set.seed(2)     #  reset random numbers
for(l in 1:nhistories)        {           # starting loop on histories
# at the start of each history:
t.exp<- rexp(t.max,lambda) # generation of t.max arrival times
# print(t.exp) # uncomment to print the generated arrival times
t.arr<-c(0,cumsum(t.exp)) # since T1=A1, T2=A1+A2,... Tk=A1+A2+...+Ak
length(t.arr)<- max(which(t.arr<Deltat)) + 1
# +1 if desired the arrival immediately after Deltat
lst[l]<- length(t.arr) # No. arrivals in this history l-th
# save the number of arrivals in this l-th history
matr.history.t[l,1:lst[l]] <- t.arr
}               # ending loop on histories
#  matr.history.t[l,1:lst[l]]  to print the arrival times of l-th history.
#  examples: histories 1, 2, 3 and last history
matr.history.t[1,1:lst[1]]
matr.history.t[2,1:lst[2]]
matr.history.t[3,1:lst[3]]
matr.history.t[nhistories,1:lst[nhistories]]
# in the plots: lst[l]-1, to ignore the initial 0
# type="s" for stair step look
ymax<- 12
xmax<- 4000
par(mfrow=c(2,2))
plot(matr.history.t[1,1:lst[1]],0:(lst[1]-1),type="s",font.lab=3,
xlab="time (s)",ylab="No. arrivals",ylim=c(0,ymax),
xlim=c(0,xmax),cex.lab=1.)
abline(v=Deltat,lty=3)
plot(matr.history.t[2,1:lst[2]],0:(lst[2]-1),type="s",font.lab=3,
xlab="time (s)",ylab="No. arrivals",ylim=c(0,ymax),
xlim=c(0,xmax),cex.lab=1.)
abline(v=Deltat,lty=3)
plot(matr.history.t[3,1:lst[3]],0:(lst[3]-1),type="s",font.lab=3,
xlab="time (s)",ylab="No. arrivals",ylim=c(0,ymax),
xlim=c(0,xmax),cex.lab=1.)
abline(v=Deltat,lty=3)
plot(matr.history.t[nhistories,1:lst[nhistories]],0:(lst[nhistories]-1),
type="s",xlab="time (s)",ylab="No. arrivals",ylim=c(0,ymax),font.lab=3,
cex.lab=1.,xlim=c(0,xmax))
abline(v=Deltat,lty=3)
```

The loop `while(t.exp<Deltat)` is replaced by the unique command `t.exp<-rexp(t.max,lambda)` to generate `t.max` deviates from an exponential distribution with parameter λ. The number of deviates `t.max` is quite arbitrary, but it must be not less than the presumed number of events in the considered Δt interval. It is advisable to choose it with a bit of generosity, but taking into account that the average number of arrivals in $[0, \Delta t]$ is $\lambda \Delta t$. The cumsum function returns the cumulative sum, for instance if x<- c(1,2,3,4), the command cumsum(x) returns 1 3 6 10. So in the code, the first t.exp generated values are (with two decimal places): 621.78 134.91 48.88 576.90 ..., the first waiting times are: 0.00 621.78 756.70 805.58 1382.48 .... Further on the code is the same as Code_4_1.R and Code_4_2.R.

We have seen that, given an arrival in the $[0, t]$, this arrival may occur with uniform probability in any point of the interval. If I know that n arrivals occurred in a certain interval $[0, t]$, it follows that the arrival times $T_1, \ldots, T_n$ form an ordered n size sample, extracted from the random variable $U(0, t)$. The code Code_4_4.R (only the initial part is reported below) generates arrival times by implementing the above Poisson process property.

```
## Code_4_4.R
# If I know that N arrivals occurred in a certain interval
# [0,Deltat], the arrival times T_1, ..., T_N
#              form an ordered N size sample, extract from
# the random variable U[0,Deltat].
# rpois(n,lambda) returns n random numbers from the Poisson distribution
# the number N of points on an interval [0,Deltat] is Poisson distributed
# with mean lambda*Deltat

Deltat<-3600        # the process is followed for 1 hour (3600 sec)
nevents<-10         # average number of arrivals in 1 hour
nhistories<-1000    # number of simulated histories (sample size)
lambda<-nevents/Deltat
lambda
ev<- numeric()      # vector of the number of arrivals in the l-th history
lst<-numeric()      # vector of the number of arrival times in the l-th history
# to save arrival times in each history:
matr.history.t<-matrix(,nhistories,Deltat)

### 1 loop on histories
set.seed(2)      # reset random numbers
for(l in 1:nhistories)      {        # starting loop on histories
# at the start of each history:
# N: a random number generated from the Poisson distribution with mean nevents
N<- rpois(1,nevents)
# print(N)    # uncomment to print N
Un<- runif(N,0,Deltat) # NOT ORDERED arrival times from U[0,3600]
# print(Un)   # uncomment to see N not ordered arrivals
t.arr<- sort(Un)     # arrival times (Un ordered) in the l-th history
# print(t.arr)    # uncomment to see how events occur over time
ev[l]<- max(which(t.arr<= Deltat)) # to check if arrivals are in [0,Deltat]
lst[l]<-length(t.arr)           # No. arrival times in this l-th history
# save the number of arrival times in this l-th history:
matr.history.t[l,1:lst[l]] <- t.arr
}        # ending loop on histories

#................ number of arrivals ............
############  to follow as  Code_4_1:R ############
```

The R function rpois(n,lambda) is used to generate n numbers from the Poisson distribution with mean λ. In the code Code_4_4.R, the command N<- rpois(1,nevents) gives how many Poisson observations, with mean nevents equal to 10, are generated in the $[0, \Delta t]$ interval, for instance, for the first history, N = 7. The command Un<-

`runif(N,0,Deltat)` generates the following seven uniform random numbers in $[0, \Delta t]$ (with two decimal places):

604.98 3397.82 3396.50 464.97 3000.41 1684.86 1979.94

and the command `sort(Un)` orders them, giving the `t_arr` arrival times in the first history:

464.97 604.98 1684.86 1979.94 3000.41 3396.50 3397.82

Once the trajectories generation is executed, the code follows the instructions of the code `Code_4_1.R`.

4.5.1 Merging of independent Poisson processes

Let $\{N_1(t), t \geq 0\}$ and $\{N_2(t), t \geq 0\}$ be two independent Poisson processes having rates λ_1 and λ_2, respectively. Suppose $N_1 = (5, 12, 25, ...)$ and $N_2 - (7, 26, 30, ...)$ are "arrivals". The process defined by

$$N(t) = N_1(t) + N_2(t) = (5, 7, 12, 25, 26, 30, ...)$$

obtained combining the arrival times in N_1 and N_2 is a Poisson process too, with a rate $\lambda = \lambda_1 + \lambda_2$. The demonstration is left as an exercise.

Passing from the arrival times $N_1(t)$ and $N_2(t)$ to the inter-arrival times $X(t)$ of the merged process (that is exponentially distributed), the inter-arrival time X_k between the $(k-1)$-th and k-th arrival is such that the probabilities of occurrence of an event of type "1" (i.e. coming from N_1) or "2" are:

$$P(\text{``1''}|X_k = t) = \frac{\lambda_1}{\lambda_1 + \lambda_2}$$

$$P(\text{``2''}|X_k = t) = \frac{\lambda_1}{\lambda_1 + \lambda_2}$$

and they are independent of the value t.

This result can be generalized to n independent processes. In particular, for the minimum M of a set of independent exponential processes:

$$M = min(X_1, X_2, ...)$$

having rates $\lambda_1, \lambda_2, ...$, the following relations can be demonstrated (Dobrow, 2016) for any time $t > 0$:

$$P(M > t) = e^{-(\lambda_1 + \lambda_2 + ...)t}$$

i.e., the combined process has an exponential distribution, and for any k > 1:

$$P(M = X_k) = \frac{\lambda_k}{\sum_i \lambda_i}$$

4.6 Nonhomogeneous Poisson process

We said that a counting process has *stationary*, or *homogeneous* increments, if the number of arrivals in any interval $[t, t + \Delta t]$ depends only on the interval length Δt, but not on t. In the *non-homogeneous* Poisson process, increments are not stationary, without prejudice to the validity of their independence. This means that the arrivals frequency varies along the time axis, so that λ is no more constant, but a function of time $\lambda(t)$.

The definition of a Poisson process from a counting process is rewritten as follows. The counting process $\{X(t)\}$ is a *non-homogeneous* Poisson process if:

a) $X(0) = 0$. The counting process begins at time $t = 0$.

b) The process has independent increments.

c) The probability of n arrivals in any interval of finite length $[s, t]$ is given by:

$$P\{X(t) - X(s) = n\} = \frac{1}{n!} \left[\int_s^t \lambda(y)dy \right]^n \exp\left[-\int_s^t \lambda(y)dy \right]$$

Therefore, we see that the distribution of the number of arrivals explicitly depends on both s and t, and not simply on the length of the interval $(s, t]$. Define:

$$m(s, t) = \int_s^t \lambda(y)dy$$

If $s = 0$, $m(t)$ is called the *mean value function* of the process. Then the random variable $X(t) - X(s)$ has Poisson distribution with mean $m(s, t)$, $\forall 0 \leqslant s < t$.

The definition of the Poisson process from the Bernoulli process is modified in a quite analogous way. The counting process $\{X(t), t \geqslant 0\}$ is a non-homogeneous Poisson process with intensity $\lambda(t)$ if:

a) $X(0) = 0$.

b) The following conditions hold:

$$P\{X(t + dt) - X(t) = 0\} = 1 - \lambda(t)dt + o(dt)$$
$$P\{X(t + dt) - X(t) = 1\} = \lambda(t)dt + o(dt)$$
$$P\{X(t + dt) - X(t) \geqslant 2\} = o(dt)$$

c) The process has independent increments.

We have seen that, for a homogeneous process, the waiting time of the first arrival $T_1 \equiv A_1$ has distribution $\exp(-\lambda t)$, that is:

$$P\{X(t) = 0\} = P\{T_1 > t\} = e^{-\lambda t}$$

If the process is non-homogeneous the above equation becomes:

$$P\{X(t) = 0\} = P\{T_1 > t\} = \exp\left[-\int_0^t \lambda(y)dy \right]$$

The derivative with respect to t is the probability density function as a generalization of eqn (4.7):

$$f_{T_1}(t) = \lambda(t) \exp\left[-\int_0^t \lambda(y)dy \right] = \lambda(t)\, e^{-m(t)}$$

It should be clear that the waiting times T_n are not independent random variables.

To simulate a non-homogeneous Poisson process, a technique often adopted is what is called *thinning*. The idea is to start from a homogeneous Poisson process with constant frequency λ. Suppose that an event occurs at time s. Such an event is not always counted, but only with a certain probability $p(s)$. Counted events form a Poisson process with not constant frequency, but with frequency $\lambda(s) = \lambda p(s)$, where $s = 1, 2, \ldots$ are the counted events.

Basically, to simulate a non-homogeneous process, a command has to be added, which allows to count or not the event. Let us suppose that events occur according to a function $\lambda(t)$ of t, such that $\lambda(t) \leq \lambda$, $\forall t \geq 0$. We want to deal with a homogeneous process with intensity λ, and to generate the arrival times. If s is one of these instants, we ask whether it has to be counted, even for the non-homogeneous process. The answer is given by the usual random number u between 0 and 1: the event occurred at time s for the Poisson homogeneous process with constant frequency λ is also an event for the Poisson process with time dependent $\lambda(t)$, if:

$$u \leq \frac{\lambda(s)}{\lambda}$$

Naturally, it is necessary to know (or to conjecture) the functional form of $\lambda(t)$.

As an example, let us consider the non-homogeneous Poisson process with $\lambda(t) = \sin(t)$ (code **Code 4_5**, below). We choose $\lambda = 1$, so that the condition $\lambda(t) \leq \lambda$ always holds. The process is followed for, say, 60 minutes, now the time unit is the minute.

```
## Code_4_5.R
# Non homogeneous Poisson process is a process with intensity lambda not constant,
# but depending on time: lambda(t).
# It is a not-stationary increments process,
# not depending only on the length interval, but also on its time position.
# The simulation is performed by sampling from a homogeneous process,
# arrivals occur according the time function. Here: lambda(t) = sin(t)
# lambda1 (homogeneous process) is known and lambda(t) <= lambda1, all t

Deltat<- 60  # minutes
lambda1<- 1.
set.seed(1)    # reset random numbers
# vector of the number of accepted arrivals:
ev<-  0
# vector of the number of arrivals if it were a homogeneous process:
evt<- 0
# vector of values lambda(t)/lambda1:
evs<- 0
t<-   0
while (t<Deltat) {       # arrival times generation
t<- t + rexp(1,lambda1)
evt<- c(evt,t)   # arrival times for a homogeneous process
fl=sin(t)/lambda1   # lambda(t)/lambda1
evs<- c(evs,fl)  # fl values, but not all are accepted
if (runif(1) < fl) ev<- c(ev,t)  #  <-----  crucial point
}        # ending loop while
length(ev)
length(evt)
plot(evt,0:(length(evt)-1),type="s",ylim=c(-4,length(evt)),xlab="time (min)",
ylab="No. arrivals",lty=3,col="black",cex.lab=1.1,font.lab=3,lwd=2)
curve(sin(x),0,Deltat,add=T,lty=2,lwd=1.6)
lines(ev,0:(length(ev)-1),type="s",lwd=2.2,col="black")
```

The result is displayed in Fig. 4.9, showing the histories of both non-homogeneous (continuous line) and homogeneous (dotted line) process. At the bottom, the function $\sin(t)$ (dashed line) is also reported.

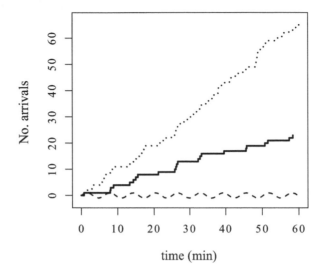

Fig. 4.9 Number of arrivals as a function of time for a non-homogeneous Poisson process (continuous line), with intensity $\lambda(t) = \sin(t)$ (dashed line, at the bottom). The dotted line refers to the homogeneous Poisson process, with $\lambda = 1$.

It is clear how the 'thinning' operation works. In the homogeneous process, there are 67 arrivals (`length(evt)`) in 60 minutes, while they decrease to 24 (`length(ev)`) in the non-homogeneous process. Notice that the arrivals occur in correspondence of $\sin(t)$ values close to near 1, for which the inequality $u \leq \lambda(t)/\lambda$ holds almost always. On the contrary, the process languishes for small or negative values of $\sin(t)$.

Some comments are in order. In experimental practice, not always are all the *occurred* arrivals *counted* arrivals. Suppose we have a source emitting charged particles and a counter to record the events. There exists for the instrument a threshold under which it is not able to register all the events anymore: such a threshold is the *temporal resolution* of the instrument. For instance, if the temporal resolution is equal to a tenth of a second, and two events occur at a distance of a hundredth of a second from one another, only one event is registered by the instrument.

There is a further factor which might alter experimental observations, the so-called *dead time*. It is the time interval during which the instrument remains inactive after the registration of an event. If we wish to count by eye the flashes of light arriving from a source and these flashes are extremely intense, the eye may remain dazzled for a moment after a flash, so that further flashes in rapid succession cannot be seen. The same might happen with an instrument. We wish to stress, in short, that there might be events following a Poisson distribution, but that their registration might not look like it, unless the temporal resolution and the dead time are negligible with respect to the arrival time.

4.7 Exercises

Exercise 4.1 The following is a fun example extracted from the R. McElreath book *Statistical Rethinking* (CRC Press, 2020). You are in the business with a monastery that you own. The monastery employs 1000 monks, independently working from one another. They copy by hand ancient manuscripts of varying length, and produce every day a variable number of manuscripts, some day more than three, other day none. You know that on the average one out of 1000 monks finishes a manuscript.

That is clearly a binomial process, so you can compute the mean μ and variance v by the well-known formulae:

$$\mu = Np$$
$$v = Np(1 - p)$$

where $p = 1/1000$ and $N = 1000$.

Simulate a true manuscript production in 10 years (3650 days) by an R code, to verify how close the 'true' average production and variance are to the theoretical quantities. Why are you allowed to affirm that the number of finished manuscripts follows a Poisson distribution? *Hint: use the binomial random number generator* `rbinom`

Exercise 4.2 With reference to exercise 4.1, suppose that you want to monopolize the market for manuscript production. You are offered to buy a new monastery, but you don't know how many monks work there. The only information you have is that it produces, on the average, two manuscripts per day. Use this information to infer the distribution of the number of manuscripts completed each day.

Exercise 4.3 Your sister constantly watches her smartphone, waiting for messages from her friends. Starting from 8 a.m. she gets messages at the rate of five per hour. Assuming that the messages arrival is a Poisson process, find:
(1) The probability that she receives exactly 20 messages before eleven o'clock.
(2) The probability of receiving at least 20 messages by the same hour.
(3) The probability that she receives exactly 20 messages by 11 a.m. and exactly 40 by 4 p.m.

Exercise 4.4 Use the R function `dpois` to compute the probabilities of exercise 4.3.

Exercise 4.5 A motorway rescue service receives calls according to a Poisson process at the rate of six calls per hour.
(1) Find the probability that they do not receive calls over a period of two hours.
(2) Find the probability of having exactly six calls in the next hour and a half.

Exercise 4.6 Write an R code to simulate the Poisson process of exercise 4.5 and numerically compute what is required in that exercise.
Hint: remember that inter-arrival times are exponentially distributed.

5
Random Walk

A man starts from a point O and walks L yards in a straight line; he then
turns through any angle whatever and walks another L yards in a straight line.
He repeats this process N times. I require the probability that after these N
stretches he is at a distance between R and R+δR from his starting point O.

Karl Pearson,   The Problem of the Random Walk

5.1   Definitions and examples

This epigraph is well known, since the term 'random walk' was introduced for the
first time by K. Pearson in a letter to *Nature* (Pearson, 1905). He asked if there
was any reader able to solve the above problem. The answer came from Lord Rayleigh
(1842−1919) only a week later published in the same volume of *Nature*, (Lord Rayleigh,
1905, 318). He wrote that he had faced similar problems 25 years earlier, in his studies
concerning the superposition of sound waves of equal frequency and amplitude but
with random phases (Lord Rayleigh, 1880; Lord Rayleigh, 1899); the complete citation
should be John William Strutt, third Baron Rayleigh, but we simply cite him as Lord
Rayleigh.

Pearson, a real gentleman, in a subsequent page (*Nature* p. 342), recognized that:
'I ought to have known [Lord Rayleigh's solution], but my reading of late years has
drifted into other channels, and one does not expect to find the first stage in a biometric
problem provided in a memoir on sound.'

Let us start with the *simple random walk*. This random walk is described through
a discrete-time Markov chain in the discrete state space. The walker flips a coin (even
not fair). If it lands on heads he moves one unit step to the left, and if the outcome
is tails the step is to the right. If we call 0 the starting point, after, say, 8 steps the
walker could be on $0, -1, -2, -1, 0, +1, +2, +3, +2$. The metaphor of a coin toss
will follow us for the whole chapter.

More formally, let $X_1, X_2, X_3, \ldots, X_n, \ldots$ be *i.i.d.* random variables, where X_i is
the walk distance on the ith step. Each step i has a probability $\mathsf{P}\{X_i\}$ such that:

$$\begin{cases} \mathsf{P}(X_i = +1) = p \\ \mathsf{P}(X_i = -1) = q = 1 - p \end{cases}$$

with $0 < p < 1$. If $p = q = 1/2$, the walk is called *symmetric*. The motion is schematized
in Fig. 5.1. Let us see if the walk is irreducible, that is whether all states are recurrent.
In that case all states communicate with each other, then we have to estimate the
probability p_{00}^m of a path of the type: $0, -1, -2, -3, -2, -1, 0$. But starting at 0, it

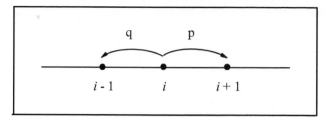

Fig. 5.1 Random walk on a line. Transitions from state i to state $i+1$ with probability p or to state $i-1$ with probability $q=1-p$.

is not possible to return to 0 in an odd number of steps, therefore m must be equal to $2n$. We can think again of the walker flipping a coin, winning or losing £1 at each flip. Starting at 0, he returns to 0 after $2n$ steps, if he won £n and lost £n. The probability of such an event is that of n successes in $2n$ independent Bernoulli trials. Recall that the probability of x successes in n independent trials is:

$$p(x) = \binom{N}{x} p^x (1-p)^{N-x} = \frac{N!}{x!(N-x)!} p^x (1-p)^{N-x}$$

in the present case $x = n$, $N = 2n$ and $(N-x) = n$. So, we write:

$$p_{00}^{(2n)} = \frac{(2n!)}{(n!n!)} p^n (1-p)^n = \frac{(2n!)}{(n!)^2} p^n (1-p)^n$$

For large n, we can exploit the Stirling's approximation:

$$n! \approx \sqrt{2\pi n}\,(n/e)^n,\ n \to \infty$$

then:

$$\frac{(2n!)}{(n!)^2} \approx \frac{\sqrt{2\pi \cdot 2n}\left(\dfrac{2n}{e}\right)^{2n}}{\left(\sqrt{2\pi n}\left(\dfrac{n}{e}\right)^n\right)^2} = \frac{2^{(2n)}}{\sqrt{\pi n}}$$

therefore, we have:

$$p_{00}^{(2n)} \approx \frac{(4p(p-1))^n}{\sqrt{\pi n}},\ n \to \infty$$

If $p = q = 1/2$, $4pq = 1$, and:

$$p_{00}^{(2n)} = \frac{1}{\sqrt{n\pi}}$$

so it is:

$$p_{00}^{(2n)} \geq \frac{1}{2\sqrt{n\pi}}$$

then:

$$\sum_{n\geqslant 1} p_{00}^{(2n)} \geq \frac{1}{2\sqrt{\pi}} \sum_{n\geqslant 1} \frac{1}{\sqrt{n}} = \infty$$

which means that the walk is recurrent. On the contrary, if $p \neq q$, $4pq = r < 1$, so now we have:

$$\sum_{n \geqslant 1} p_{00}^{(2n)} \leq \frac{1}{\pi} \sum_{n \geqslant 1} r^n < \infty$$

then the walk is transient. Also in two dimensions, all states are recurrent, but in three dimensions no state is recurrent; they are all transient.

Let $S_n = X_1 + X_2 + X_3 \cdots + X_n$ be the position of the walker at time n, or after n steps, since the time unit is the step of size one. For convenience, the starting point is $S_0 = 0$. If $x_1, x_2, \ldots$ are the realization of $X_1, X_2, \ldots$, we have, for instance: $S_8 = 0 + (-1) + (-2) + (-1) + 0 + (+1) + (+2) + (+3) = 2$. Note that the steps $x_1, x_2, x_3, \ldots, x_n$ are realizations of independent random variables, either $+1$ or -1, while the positions at S_n is correlated with the position at S_{n-1}.

In Chapter 3, we introduced the number of visits V_i to the state i and its expected value:

$$E\left[V_i | X_0 = i\right] = \sum_{n=0}^{\infty} p_{ii}^n$$

We saw that if i is recurrent, the number of visits to i is infinite, so the expected value of V_i is also infinite, while if i is transient V_i is finite and the expected value of V_i is finite too. We can *estimate* the position of the particle after n steps, by evaluating the expectation of the walk given by:

$$E\left[S_n\right] = E\left[\sum_{i=1}^{n} X_i\right] = \sum_{i=1}^{n} E\left[X_i\right] = \sum_{i=1}^{n} p - (1-p) = n(2p-1)$$

equal to 0, if the walk is symmetric. The uncertainty in this estimate is given by the variance:

$$\text{Var}\left[S_n\right] = \sum_{i=1}^{n} \left[E\left[X_i^2\right] - \left[E(X_i)\right]^2\right] = p + (1-p) - (2p-1)^2 = 1 - 4p^2 + 4p - 1 = 4np(1-p)$$

equal to n, if $p = 1 - p$. Then, if the walk is symmetric, the standard deviation after n steps is $\sqrt{n}$, perhaps a bit of a counter intuitive result.

A viewable approach, which we will also follow later on for two-dimensional walks, consists in considering again the total displacement after n steps, each of length l_i (not necessarily of unit length):

$$S_n = \sum_{i=1}^{n} l_i$$

whose mean is:

$$\langle S_n \rangle = \sum_{i=1}^{n} \langle l_i \rangle = 0$$

since, if n is large enough, there are as many positive as negative l_i values. Consider now S_n^2, which is:

$$S_n^2 = (l_1, l_2, \ldots, l_n)(l_1, l_2, \ldots, l_n) = \sum_{i=1}^{n} l_i^2 + \sum_{i \neq j}^{n} l_i l_j$$

Suppose now the steps are all equal to l, the first sum is nl^2, while the second sum is zero, since $\langle l_i \, l_j \rangle = 0$, there being no correlation between steps. Then we have the previous derived result for *unit step*:

$$\langle S_n^2 \rangle = n$$

The random walk can be graphically represented as an XY plot with the number of steps on the x-axis there are the number of steps and the value of S_n on the y-axis. The points (n, S_n) and $(n + 1, S_{n+1})$ are joined with a straight line segment. An R code with somewhat naïve instructions could be of the type:

```
x <- numeric()
x[1] <- 0
n <- 20
for(i in 2:n)
{
  u<-runif(1,0,1)
  if (u<=p) z<- +1
  if (u>p) z<- -1
  x[i]<-x[i-1]+z
}
```

so the S_n path is constructed. However, this way is not the most efficient one. A better version is the `Code_5_1.R` below, in which one path for a few steps is shown. The R code is based on the Bernoulli distribution given by `rbinom(n,size,prob)` where `n` is the number of independent random variables to be generated, `size` is the number of trials, and `prob` the probability of success.

```
## Code_5_1.R
# simple symmetric random walk, one path
set.seed(1)
nsteps<- 20          # total number of steps
p<-0.5               # transition probability
x<-numeric()
n<- numeric()        # time after n steps
x0<- 0               # starting point
x<-rbinom(nsteps,1,p)
x[which(x==0)]<- -1  # also  x<- 2*x-1
#  Sn: position of the walker after n steps
Sn<-c(x0,cumsum(x))  # adding the starting point
n<- 0:nsteps              # time axis
par(mai=c(1.02,1.,0.82,0.42)+0.1)
plot (n,Sn,type="o",xlab="number of steps",ylim=c(-4,4),lty=1,
ylab=expression(italic("S"[n] )~~ italic((position)) ),
cex.lab=1.2,font.lab=3,pch=19,xaxt="none"   )
axis(1, at=seq(0,nsteps,1))
abline(h=0,lty=2)
```

Ten realizations from the Bernoulli random variable are (added $x_1 = 0$):

```
 0 -1 -1  1  1 -1  1  1  1  1 -1
```

from which the S_n values are:

```
0 -1 -2 -1  0 -1  0  1  2  3  2
```

Note the 'time axis' command `n <- 0:nsteps` creates a timescale from 0 to `nsteps`, one step at a time, with $S_0 = 0$.

Figure 5.2 shows the position of the walker S_n as a function of the number of steps. We see that the *first return* to the origin occurs at time $n = 4$, that is to say after four

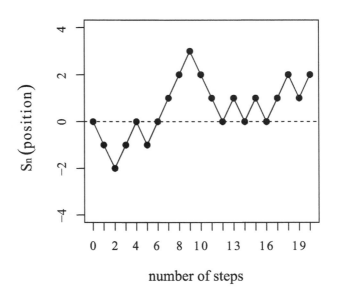

Fig. 5.2 Symmetric random walk on a line. Position S_n of the walker after n steps. The dots are the reached points.

steps. The following `Code_5_2.R` extends the previous `Code_5_1.R` to simulate several different random paths. The *estimate* $\hat{S}_n$ of the expectation of the walk $\mathrm{E}\,[S_n]$ and the *estimate* $\hat{\sigma}_n^2$ of the variance $\mathrm{Var}\,[S_n]$ are also computed.

```
## Code_5_2.R
# simple symmetric random walk, nhists paths
set.seed(1)
p<- 0.5        # transition probability
nhists<- 100   # number of  histories
nsteps<- 200   # total numero of steps
x0<-0          # starting point
splot<- 30     # plotted only splot histories out of all nhists
#    starting vector: each history starts from x0
#    column vector with all x0 repeated nhists times
start_matr<-matrix(x0,nhists,1)
Sn<-matrix(,nhists,nsteps+1)
y<-matrix(,nhists,nsteps)
for (i in 1:nhists)            { # starting loop on histories
x<-rbinom(nsteps,1,p)
x[which(x==0)]<- -1    # also  x<- 2*x-1
y[i,]<-cumsum(x)   # Sn without the starting point
                          } # ending loop on histories
```

```
Sn <- matrix(c(start_matr,y+x0),nhists,nsteps+1)  # adding the starting vector
n<- seq(0, nsteps)          # time axis
par(mai=c(1.02,1.,0.82,0.42)+0.1)
plot (n,Sn[1,],xlab="number of steps",ylim=c(-40,40),type="l",lty=1,
ylab=expression(italic("S"[n] )~~ italic((position)), ),
cex.lab=1.2,font.lab=3,lwd=0.5,xaxt="none"    )
axis(1, at=seq(0,nsteps,20))
# Sn[1,]      # uncomment to print the first path
for (i in 2:splot)
{
    lines(Sn[i,],type="l",lty=1,lwd=0.5)
}
par(mai=c(1.02,1.,0.82,0.42)+0.1)
hist(Sn[,nsteps+1],freq=F,main="",
xlim=c(-50,50),font.lab=3,
xlab=expression(italic("S"[n] )~~ italic((position)) ),
cex.lab=1.2,font.lab=3)
# freq=F to compare with analytical density by 'curve(.)'
mSn<-mean(Sn[,nsteps+1])
mSn
VarSn<- var(Sn[,nsteps+1])
VarSn
se.Sn<- sqrt(var(Sn[,nsteps+1]))
se.Sn
curve (dnorm(x, mean=mSn, sd=se.Sn),lty=2,lwd=2,add=T)
shapiro.test(Sn[,nsteps+1])
mSn.an<- nsteps*(2*p-1)        # .an stays for 'analytical'
mSn.an
VarSn.an<- 4*nsteps*p*(1-p)
VarSn.an
seSn.an<- sqrt(VarSn.an)
seSn.an
```

Figure 5.3 shows the time evolution of 30 paths out of 100 simple symmetric random walks. The number of different symmetric walks of n steps is 2^n, and each of these walks is equally likely. From the simulation, we can estimate where on average the walker is after n steps. It results in $\hat{S}_n = -0.82$ and $\hat{\sigma}_n^2 = 211.40$, in agreement with the theory $E[S_n] = 0$ and $\text{Var}[S_n] = 200$, respectively.

In the space (n, S_n) of possible paths, the ensemble of paths covers a bounded 'triangular' area. If we plot a histogram of the endpoints, Fig. 5.4 is obtained. The y-axis refers to the normalized histogram for comparison with the probability density function of the normal random variable (dashed line). One sees that the histogram is approximately normal in shape, suggesting that the distribution of the end points might tend, in probability, to a normal one with mean $\hat{S}_n$ and standard deviation $\sqrt{\hat{\sigma}_n^2}$, as the central limit theorem states. This claim is evidenced by superimposing a normal probability density function (dashed lines in Fig. 5.4) with mean $= -0.82$ and standard deviation $= \sqrt{211.40}$: the agreement appears to be very good.

To further assess the normality of the histogram data, the Shapiro-Wilk normality test is exploited with the R command **shapiro.test(Sn[,nsteps+1])**. For this test, the null hypothesis is that the data come from a normal distribution, and commonly it is accepted (better: it cannot be rejected) if the p-value is greater or equal than 0.05. In our case, the p-value is equal to 0.1487, so we cannot reject the hypothesis that the distribution is normal.

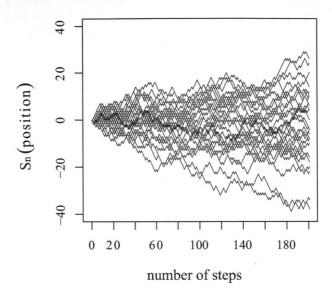

Fig. 5.3 Symmetric random walk on a line. Position S_n of the the walker after n steps of 30 paths out of 100.

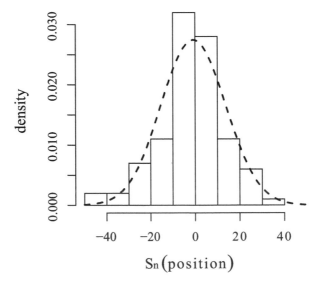

Fig. 5.4 Symmetric random walk on a line. Histogram of the endpoints of 100 simulated paths. Comparison with the probability density function of the normal random variable $\mathcal{N}\left(\hat{S}_n, \sqrt{\hat{\sigma}_n^2}\right)$ (dashed line) is also shown.

Let us compare the above results with those of an asymmetric simple random walk. In the `Code_5_2.R`, we set the transition probability to `p <- 0.7`. The result is shown in Fig. 5.5. The shape of the 'triangular' area of possible paths does not change, it

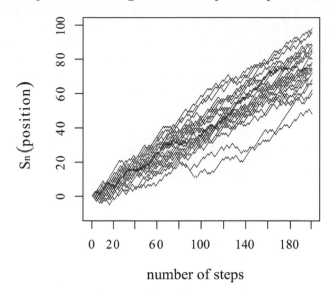

Fig. 5.5 As in Fig. 5.3, but with $p = 0.7$.

looks a bit narrower and points upwards. Moreover, the estimates result $\hat{S}_n = 79.58$ and $\hat{\sigma}_n^2 = 156.95$, in agreement with the theory $\mathrm{E}[S_n] = 80$ and $\mathrm{Var}[S_n] = 168$, respectively.

5.1.1 Barriers

Barriers are states with characteristic properties. If a state is absorbing, we say that it is an absorbing barrier. If the walk reaches an absorbing barrier, the walk ends and the walker stays there forever. The matrix of transition probabilities is:

$$
\begin{pmatrix}
1 & 0 & 0 & 0 & \cdots & 0 & 0 & 0 \\
q & 0 & p & 0 & \cdots & 0 & 0 & 0 \\
0 & q & 0 & p & \cdots & 0 & 0 & 0 \\
& & & \cdots\cdots\cdots & & & \\
0 & 0 & 0 & 0 & \cdots & q & 0 & p \\
0 & 0 & 0 & 0 & \cdots & 0 & 0 & 1
\end{pmatrix}
$$

If the states are numbered $0, 1, 2, \ldots, n$, the state space is partitioned into three classes:

$$\underset{\text{absorbing state}}{\{0\}} \qquad \underset{\text{transient states}}{\{1, 2, \ldots, n-1\}} \qquad \underset{\text{absorbing state}}{\{n\}}$$

The walk is restricted to the internal states $\{1, 2, \ldots, n-1\}$ where transitions are allowed either to the right $p_{i,i+1} = p$ or to the left $p_{i,i-1} = q = 1 - p$. If the walker ends up in 0 or n, then $p_{00} = p_{nn} = 1$. If there is no barrier, the walk is called *unrestricted*.

The following code `Code_5_3.R` simulates several *restricted* random paths with two absorbing barriers. The code can plot on the same graph, trajectories of unrestricted and restricted walks. The first part of the code is a copy of `Code_5_2.R`, but the plot is limited to only one unrestricted trajectory (`plot (n,Sn[1,] ...)`), for clarity of plots. The barriers `b1` and `b2` are in the positions S_{10} and S_{-10}, respectively. If the walker at a certain step reaches one of the barriers, he gets absorbed and the position of such event is recorded. The code gives the estimates of the expected duration of the walk and other quantities.

```
## Code_5_3.R
# restricted random walk with absorbing barriers

# beginning with walk without barriers
set.seed(3)
p<-0.5
x0<-0
nsteps<- 200
nhists<- 100
#   starting vector: each history starts from x0
#   column vector with all x0 repeated nhists times
start_matr<-matrix(x0,nhists,1)
y<-matrix(,nhists,nsteps)
for (i in 1:nhists)          {  # starting loop on histories
x<-rbinom(nsteps,1,p)
x[which(x==0)]<- -1     # also  x<- 2*x-1
y[i,]<-cumsum(x)    # Sn without the starting point
} # ending loop on histories
Sn <- matrix(c(start_matr,y+x0),nhists,nsteps+1)  # adding the starting vector
n<- seq(0, nsteps)          # time axis
par(mai=c(1.02,1.,0.82,0.42)+0.1)
plot (n,Sn[1,],xlab="number of steps",ylim=c(-15,15),type="l",lty=2,
ylab=expression(italic("S"[n] )~~ italic((position)), ),
cex.lab=1.2,font.lab=3,lwd=1,xaxt="none"   )
axis(1, at=seq(0,nsteps,20))

#-----------------------------------------
# continue to plot the walks on a single graph

set.seed(3)     # reset random numbers if wanted
t1<-numeric()
t2<-numeric()
t1b1<-numeric()
t2b2<-numeric()
no.abs<-numeric()
# the following parameters can be redefined
x0<- 0
nsteps<- 200
nhists<- 100
x<-numeric()
p<-0.5
y<-matrix(,nhists,nsteps)
b1<-   10
b2<- -10
for(i in 1:nhists)          {   # starting loop on histories
x[1]<-x0
for(j in 2:nsteps)     {  # starting loop on steps
u<-runif(1,0,1)
if(u<p)z<- +1
```

```
else z<- -1
x[j]<-x[j-1]+z
if(x[j-1] == b1)  x[j]<- b1
if(x[j-1] == b2)  x[j]<- b2
}   # ending loop on steps
y[i, ]<- x
t1[i]<- min(which(x==b1))
t2[i]<- min(which(x==b2))
}   # ending loop on histories
Sn <- matrix(c(start_matr,y+x0),nhists,nsteps+1)  # adding the starting vector
#y[i, ]  # if wanted to print entire i-th selected history
t1[!is.finite(t1)] <- 0
t1
t1b1 <- t1[t1!=0]
t1b1
t2[!is.finite(t2)] <- 0
t2
t2b2 <- t2[t2!=0]
t2b2
t12<- 0
for(i in 1:nhists)
if(t1[i]==0 & t2[i]==0) {t12<-t12+1
no.abs[t12]<- i }
no.abs
par(lwd=1.7,font=3)
lines(y[17, ], type="l",lty=1)
text(180,5,"c")
lines(y[60, ], type="l",lty=1)
text(115,12,"a")
lines(Sn[2, ], type="l",lty=1)
text(135,-12,"b")
lines(y[60, ], type="l",lty=1)
abline(h=b1,lty=4,col="black",lwd=2)
abline(h=b2,lty=4,col="black",lwd=2)
l_t1b1<- length(t1b1)
l_t1b1
l_t2b2<- length(t2b2)
l_t2b2
l_no.abs<-length(no.abs)
l_no.abs
m.t1b1<- mean(t1b1)
m.t1b1
quant<-quantile(t1b1)
quant
se.t1b1<-  sqrt(var(t1b1))
se.t1b1
m.t2b2<- mean(t2b2)
m.t2b2
quant<-quantile(t2b2)
quant
se.t2b2<-  sqrt(var(t2b2))
se.t2b2

hist(t1b1,freq=F,main=" ",xlab="number of steps",
col=rgb(0.8275,    0.8275,    0.8275,1),
cex.lab=1.2,font.lab=3,ylab="absorbed states",xaxt="none",
ylim=c(0,0.015),right = TRUE    )
axis(1, at=seq(0,nsteps,20))
lines(density(t1b1),lwd=2,lty=1)
op <- par(cex = 1.3,font=3)
```

```
text(160,0.010,"b1")
histb1<- hist(t1b1,plot=F)

hist(t2b2,freq=F,main=" ",xlab="number of steps",
col=rgb(0.8275,    0.8275,    0.8275,1),
cex.lab=1.2,font.lab=3,ylab="absorbed states",xaxt="none",
ylim=c(0,0.015),right = TRUE   )
axis(1, at=seq(0,nsteps,20))
lines(density(t2b2),lwd=2,lty=1)
op <- par(cex = 1.3,font=3)
text(160,0.010,"b2")
histb2<- hist(t2b2,plot=F)
histb1
histb2
```

The command y[i,] prints the entire *i*-th path, for instance y[60,] prints:

```
[1]  0 1 0 1 2 3 2 3 2 1
..............................
[76]  9 8 7 6 5 6 5 6 7 8 7 6 7 6 7 6 7 8 9 8 9 8 9 10 10
..............................
```

It shows that at step 99, the walker encounters the higher barrier and thereafter the path does not change anymore.

In the vector **t1** the absorbing time for each history is recorded, where 0 means that for this history, the walker does not reach the barrier **b1**. For example, for the first five histories no absorbtion in **b1** occurs. The 6-th path is absorbed at the 193-th step. Note that the history 60 is absorbed at the step 99.

```
[1]  0  0  0  0  0 193 0 53 ......
..............................
[55]  0  37  51   0  91  99  63   0   0 ......
..............................
```

In the vector **t1b1**, the 0 values are removed to print only the absorbing time on the higher barrier **b1** for each absorbed path.

```
[1] 193  53  95 103  61 125  59 101 109  29  27 109  75 123  33  25 123  27  87
[20]  51  65  37  51  91  99  63 117  77 193 137  63  23  33  43  93  89  61  33
[39] 113  51  17  85 151  33
```

The same procedure is applied for the lower barrier **b2=-10**.

There are paths that do not end on one barrier, but continue up to the given number of steps. The vector **no.abs** reports such not absorbed histories:

```
[1]   1 17 18 24 31 38 54 62 77 94
```

Figure 5.6 shows one unrestricted path (dashed line) and three restricted paths (continuous line), out of 100. The walk (*a*) is absorbed on **b1** barrier at step 99 (history 60). The walk (*b*) reaches the **b2** barrier at the step 113 (history 2). The walk (*c*) (history 17) is never absorbed during 200 steps.

The number of absorbing paths on the **b1** barrier is **1_t1b1** = 44, and on the **b2** barrier is **1_t2b2** = 46, then **no.abs** = 10 paths do not fall on a barrier. The mean of the number of steps at which the path is absorbed is **m.t1b1** = 77.86, and **m.t2b2** =

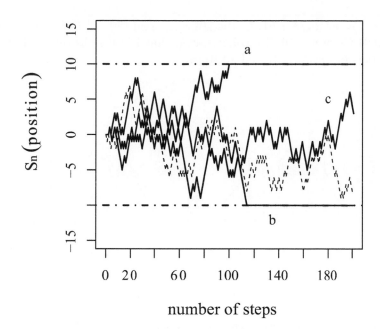

Fig. 5.6 One unrestricted path (dashed line) and three restricted paths (continuous line), out of 100, of a random walk with barriers. Dot-dashed lines: barriers positions S_{10} and S_{-10}, respectively.

84.74, for the higher and lower barrier, respectively. The command `quantile(t1b1)` gives the quantile estimates corresponding to chosen percentiles, for the **b1** barrier:

```
> quant
   0%   25%   50%    75%   100%
 17.0  41.5  70.0  104.5  193.0
```

and for the **b2** barrier:

```
> quant
  2.5%  15.87%      50%  84.13%    97.5%
22.000  47.283  72.000 140.151  181.000
```

Figure 5.7 shows the number of absorbing states as a function of the number of steps. The figure refers to the barriers **b1** and **b2** respectively. Note that with `freq=F`, probability densities are plotted instead of frequencies, so the histogram can be overlaid with a kernel density plot. Use the command `plot=F` to have a list of breaks and counts, as reported below:

```
breaks
histogram
 [1]   0  20  40  60  80 100 120 140 160 180 200
counts
 [1]  1 10  6  7  7  6  4  1  0  2    b1 barrier
counts
 [1]  1  5 12 10  4  3  3  4  1  3    b2 barrier
```

For instance, in histogram `b1`, in the steps interval $(20, 40]$ (left-hand endpoints are not included) there are 10 absorbing states, that is 10 paths out of 100 end on the higher `b1` barrier between 21 and 40 steps.

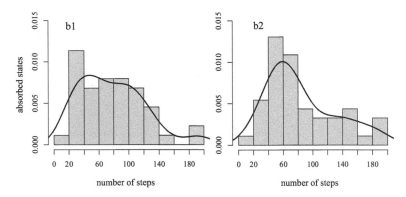

Fig. 5.7 Histograms of the absorbing states on `b1` and `b2` barriers of 100 simulated paths. A kernel density curve is also overlaid.

Since the walk is symmetric, the two barriers are equiprobable, indeed the values of the means `m.t1b1` $= 77.86$, and `m.t2b2` $= 84.74$ are close to each other. On the contrary, for instance, with $p = 0.7$, no history reaches the `b2` barrier, but all end on `b1` (for the $p \neq q$ case, see later).

We now run the `Code_5_3.R`, with `nsteps` $= 400$, and barriers at 0 and 40. The starting point is `x0` $= 20$. Before discussing the results, let us imagine the walker not as a pilgrim carrying out *el camino* step by step. He (or she) is actually a gambler, tossing a coin at every step. We ask ourselves whether he is winning or he has lost everything. The answer is dealt with in the next section by facing the well-known gambler's ruin problem.

5.1.2   Gambler's ruin

Figure 5.8 is obtained with `Code_5_3.R`. We see again three different behaviours: absorption on `b1` $= 40$, absorption on `b2` $= 0$, and no absorption. However, Fig. 5.8 can now be interpreted as the result of 400 tosses of a *fair* coin, that is the probability of heads and the probability of tails are $p = 1/2$. The gambler plays against the house (a casino) and he wins £1, if heads occurs, or loses £1, in the case of tails. His initial amount of money is £20, that is the initial value (`x0`, in the code) and it has to be imagined as the coins in his pockets before starting the game, rather than the beginning of the path. Similarly, the quantity S_n is no more the *position* of the walker after n steps, but the *capital* of the gambler after n tosses. The game ends when:

i) either the gambler has finished all the money; he is ruined,

ii) or the gambler has decided to retire after a certain winning, for instance when his initial capital is doubled.

In real life, we know that losing is more likely than winning, and the ruin is the ultimate fate of all gamblers, as, for instance, William Makepeace Thackeray writes in

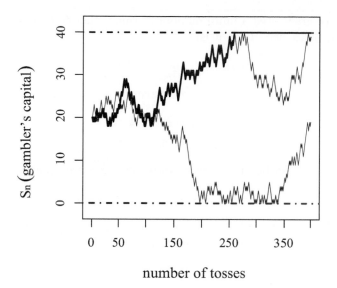

number of tosses

Fig. 5.8 Three restricted paths, out of 100, of a random walk with barriers at 0 and 40.

The Luck of Barry Lyndon (first original edition 1844):

Barry was born clever enough at gaining a fortune, but incapable of keeping one. For the qualities and energies which lead a man to triumph in the former case are often the very cause of his undoing in the latter.

Later, Stanley Kubrick, in 1975, directed and produced the famous and Oscar-winning film *Barry Lyndon*, inspired by the novel by Thackeray.

In our simulation, the walker (sorry, the gambler) returns home, winning 31 (`1_t1b1`), evenings out of 100, he is ruined 27 (`1_t1b1`) times, but there are also 42 (`1_no.abs`) games in which the gambler is neither winner nor loser, and the number of tosses is not enough to go to one end. Let us look into the matter by applying the method of the difference equation.

The Gambler's ruin problem has a long history tracing it back to Pascal who proposed this problem to Fermat in 1656. Many mathematicians dealt with this subject: Huygens, James Bernoulli, De Moivre, up to Ampère in 1802. How the problem was formulated over time and how solutions were proposed are discussed in Song and Song (2013). Here we follow the exposition which is found in the 'immortal' *An Introduction to Probability Theory and Its Applications* by Feller (1970), adopting also the same symbology.

Let z be the *initial* capital of the gambler and let q_z be probability of his ruin and p_z of his winning, that is when his capital becomes a, for instance $2z$. We are at the first toss. After that, the capital becomes either $z+1$ or $z-1$. We have seen that the state space of the process is partitioned into three classes: recurrent states $\{0\}$ and $\{n\}$, transient states: $\{1, 2, \ldots, n-1\}$. In the gambler's case the transient states are:

$$z = 1, 2, \ldots, a-1$$

Transient states are visited a finite number of times, therefore after a certain number of steps 0 or a is reached. Then it is:

$$q_z = p\,q_{z+1} + q\,q_{z-1}, \quad 1 < z < a - 1$$

Considering also $z = 1$ and $z = a - 1$, we have the additional equations:

$$q_1 = pq_2 + q, \quad z = 1$$
$$q_{a-1} = qq_{a-2}, \quad z = a - 1$$

All these equations can be incorporated into the *difference equation*:

$$q_z = p\,q_{z+1} + q\,q_{z-1}, \quad 1 \leqslant z \leqslant a - 1 \tag{5.1}$$

Equation (5.1) does not hold for $z = 0$ or $z = a$, since the game terminates for these values of z. We have the *boundary conditions*:

$$q_0 = 1 \qquad q_a = 0$$

clearly, if $q_0 = 1$, the gambler has no money to begin with (he is certain to lose), i.e. the game is already over; similarly if $q_a = 0$ (he already has £a) the probability of ruin is zero, so there is no reason to play again. Equation (5.1) is *linear*, so if q_z and q'_z are solutions, also $Aq_z + Bq'_z$ $(A, B \in \mathbb{R})$ is a solution. The above difference equation can be derived by the *partition theorem*. This says that the probability that an event A occurs, considering the fact that another event B may or may not have occurred, is given by:

$$\mathsf{P}\{A\} = \mathsf{P}\{A|B\}\,\mathsf{P}\{B\} + \mathsf{P}\{A|B^c\}\,\mathsf{P}\{B^c\}$$

Here A is the event 'gambler's ruin' and B the event 'gambler wins'. It is $\mathsf{P}\{B\} = p$, $\mathsf{P}\{B^c\} = 1 - p = q$. If the first game is won, the capital becomes $z + 1$ and the game goes on:

$$\mathsf{P}\{A|B\} = q_{z+1}$$

and if the first game is lost:

$$\mathsf{P}\{A|B^c\} = q_{z-1}$$

Therefore, for the partition theorem:

$$q_z = q_{z+1}p + q_{z-1}q, \quad 1 \leqslant z \leqslant a - 1$$

with the above boundary conditions.

Equation 5.1 can be solved by introducing an auxiliary equation of the type:

$$p\alpha^2 - \alpha + q = 0$$

where $\alpha^z = q_z$. The above equation has the roots $\alpha_1 = 1$ and $\alpha_2 = q/p$. If $p \neq 1/2$, α_1 and α_2 are different, the general solution of eqn (5.1) can be written as:

$$q_z = c_1(1)^z + c_2 \left(\frac{q}{p}\right)^z$$

with the condition $p \neq q$. The boundary conditions are satisfied if c_1 and c_2 are the solutions of the equations:

$$1 = c_1 + c_2 \qquad 0 = c_1 + c_2 \left(\frac{q}{p}\right)^z$$

The solutions are:

$$c_1 = \frac{\left(\frac{q}{p}\right)^a}{\left(\frac{q}{p}\right)^a - 1} \qquad c_2 = -\frac{1}{\left(\frac{q}{p}\right)^a - 1}$$

Finally, the solution for q_z is given by:

$$q_z = \frac{\left(\frac{q}{p}\right)^a - \left(\frac{q}{p}\right)^z}{\left(\frac{q}{p}\right)^a - 1}, \qquad p \neq q \tag{5.2}$$

Then we have found the probability of the gambler's ruin, when he begins to play with an initial capital z. We have presented a scenario in which a gambler plays against the house; historically and in modern literature, the problem is seen also as a game between two players G_1 ad G_2. The initial capital of G_1 is z, the initial capital of G_2 is $a - z$. If G_1 wins, his capital becomes $z + 1$ and G_2's capital $a - z - 1$. Similarly, if G_1 loses, his capital decreases to $z - 1$, while the G_2's capital increases to $a - z + 1$. In any case the total capital a stays constant. The game continues as we have already seen, with £1 going from G_1 to G_2 or vice versa. The game finishes when either G_1 or G_2 has no more money. In conclusion: either G_1 has won £a and G_2 is ruined, or G_1 is eventually ruined and G_2 has gained £a.

Turning to eqn (5.2), we have found the probability of the gambler's ruin, when he began to play with £a, we can ask ourselves what is the probability of winning the desired amount of money a. It is the probability that his adversary is ruined, given by interchanging in eqn (5.2) p and q, and putting $a - z$ in the place of z, obtaining:

$$p_z = \frac{\left(\frac{q}{p}\right)^z - 1}{\left(\frac{q}{p}\right)^a - 1} \tag{5.3}$$

From eqn (5.2) and eqn (5.3) it follows that:

$$q_z + p_z = 1$$

which means that formally the game cannot continue indefinitely, but ends with probability 1. We have seen in Fig. 5.8 the results of fair games, that is the toss of a fair coin, for which the probabilities of making heads or tails are equal to $1/2$. In this case, eqn (5.2) is no longer valid, since the particular solutions $q_z = 1$ and $q_z = (q/p)^z$ are identical. However, also $q_z = z$ is a solution of eqn (5.2), therefore also $q_z = c_1 + c_2 z$ is a solution of eqn (5.2). The boundary conditions are satisfied if:

$$1 = c_1 \qquad\qquad 0 = c_1 + c_2\, a$$

Then:

$$q_z = 1 - \left(\frac{z}{a}\right), \quad p = q \tag{5.4}$$

Another simple way suggested by Feller (1970, p. 345), to obtain the above result, consists of computing the limit $p \to \frac{1}{2}$ through the L'Hospital rule. Putting $u = q/p$ in eqn (5.2), gives:

$$q_z = \lim_{u \to 1} \frac{u^a - u^z}{u^a - 1} = \lim_{u \to 1} \frac{au^{a-1} - zu^{z-1}}{au^{a-1}} = 1 - \frac{zu^{z-1}}{au^{a-1}} = 1 - \left(\frac{z}{a}\right)$$

Figure 5.8 shows three trajectories of fair games in which, with the present symbology, $z = 20$ and $a = 40$. Then $z = \frac{1}{2}a$, and $q_z = \frac{1}{2}$. This is sensible, since, in the random walk language, the walker starts half-way between the boundaries and he has an equal chance of reaching either the boundary 0 or the boundary a. In the `Code_5_3.R`, the number of steps is now 1000 and, to increase the sample size, the number of histories is put equal to 1000. As expected, the number of ruins (`l_t1b2`= 481) is very close to the number of wins (`l_t1b1`= 451), also in agreement with eqn (5.4). The number of unending paths, up to 1000 steps, is 68. Unending paths are present also in a run up to 5000 steps. We will return later to this topic.

Feller (1970, p. 346) notes that from eqn (5.4), when $p = q$, if the gambler has an initial capital $z = £999$, has a probability 0.999 to win £1 before losing all. If $q = 0.6$ and $p = 0.4$ the game is unfair, nevertheless the gambler has a probability about 2/3 to win £1 before being ruined. If the initial capital z is conspicuous, the gambler can hope for a small quantity $a - z$ before losing his wealth.

Suppose now that there is no limiting value a, since the gambler wants to win an 'infinite' fortune and continues to play stubbornly until he is ruined. This case is known as a game against an *infinitely rich adversary*, for instance a casino. Such a fanciful probability of success can be obtained by $a \to \infty$ in eqn (5.2) and in eqn (5.4). In eqn (5.2), it gives:

$$\lim_{a \to \infty} \frac{\left(\dfrac{q}{p}\right)^a - \left(\dfrac{q}{p}\right)^z}{\left(\dfrac{q}{p}\right)^a - 1} = \lim_{a \to \infty} \frac{1 - \left(\dfrac{p}{q}\right)^{a-z}}{1 - \left(\dfrac{p}{q}\right)^a} = 1$$

In conclusion:

$$\begin{cases} q_z = 1 & \text{if } p \leqslant q \\ (q/p)^z & \text{if } p > q \end{cases}$$

Therefore:

$p \leqslant q$. The ruin has a probability equal to one to occur, even though p is slightly less than $1/2$. Clearly, the probability of ruin decreases as the gambler's initial capital increases, that is to say that the starting point of the walk is nearer to the barrier b1 $= 40$.

$p > q$. There exists a non-zero probability that the gambler becomes 'indefinitely' rich. However, even though the bias is in favour of the gambler, there still exists the possibility of ruin.

In the random walk language, a walker starting at the origin $z = 0$ has probability 1 of reaching the position $z > 0$, if $p \geqslant q$, and equal to $(p/q)^z$ if $p > q$.

Let us address the problem by simulating with the Code_5_3.R, in which now $p = 0.45$ and the number of histories is 1000. The results show that the ruin happens 976 (1_t1b2) times out of 1000, while 22 (1_t1b1) histories reach the barrier b1 — 40, and two do not terminate on one of the barriers. By the way, note that the above number can change slightly, with different sequences of random numbers.

Suppose now that the game is biased in favour of the gambler, with $p = 0.55$. In this case, the gambler wins 987 (1_t1b1) times and loses 11 games, two histories have not ended on one of the barriers. The peculiar implementation of the computer simulations leads to a symmetry of the set-ups: the number of games ending on the barrier at 0 when $p = 0.45$ is about the number of games reaching the barrier at $a = 40$ when $p = 0.55$.

A further important quantity to characterize the gambler's ruin problem is the *expected duration* of the game, which is the number of tosses until one of the barriers is reached. Let d_z be the *final expectation* of the game starting at z. The first toss can be either a success (occurring with probability p) or a failure (occurring with probability q). In the first case the conditional expectation of the duration is $d_{z+1}+1$, analogously in the second case it is $d_{z-1}+1$. Therefore the expected duration is given by the following difference equation:

$$
\begin{aligned}
d_z &= p(d_{z+1}+1) + q(d_{z-1}+1) \\
&= 1 + pd_{z+1} + qd_{z-1}, \quad 1 \leqslant z \leqslant a-1
\end{aligned}
\tag{5.5}
$$

with the boundary conditions:

$$
d_0 = 0 \quad \text{and} \quad d_a = 0
$$

Clearly, no further trials are to be expected if a barrier has been reached. Equation (5.5) is non-homogeneous, for the presence of the term $+1$. The general solution is:

$$
d_z = A + B \left(\frac{q}{p}\right)^z + \frac{z}{q-p}, \quad p \neq q
$$

and from the boundary conditions:

$$
A + B = 0 \quad \text{and} \quad A + B \left(\frac{q}{p}\right)^a + \frac{a}{q-p} = 0
$$

Then the expected duration of the game, when $p \neq q$ is:

$$d_z = \frac{z}{q-p} - \frac{a}{q-p} \times \frac{1 - \left(\dfrac{q}{p}\right)^z}{1 - \left(\dfrac{q}{p}\right)^a}$$

When $p = q = \frac{1}{2}$, the particular solution $d_z = z/(q-p)$ is invalid, so it is replaced by $d_z = -z^2$, another particular solution of eqn (5.5). Then:

$$A + Bz - z^2$$

and using again the boundary conditions, we obtain:

$$d_z = z(a - z)$$

As previously, consider the game against an infinitely rich adversary, that is when $a \to \infty$. When $p > q$, the game may continue forever and the quantity 'expected duration' has no more sense. If $p < q$, the expected duration is:

$$d_z = z \frac{1}{q-p}, \quad p < q \tag{5.6}$$

. Lastly, if $p = q$, the expected duration becomes infinite.

By simulation, it is possible to estimate the *distribution* of the expected duration, as it has already appeared in Fig. 5.7 showing results concerning rather short path lengths and a reduced number of histories. With reference to `Code_5_3.R` with increasing numbers of steps and histories, for $p = q$ we have found that a considerable number of games do not end on a barrier, even after 5000 tosses.

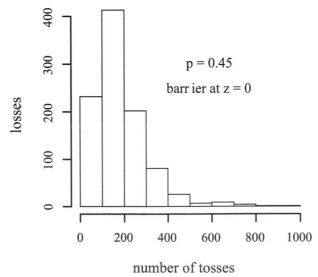

Fig. 5.9 Distribution of the expected duration of 976 games out of 1000 ending on the barrier at $z = 0$.

The histogram in Figure 5.9 is obtained by `Code_5_3.R` with $p = 0.4$. We see, for instance, that 413 games out of 1000 reach the barrier $z = 0$ after an expected duration in the interval $(101, 200]$. The duration of the games is between 29 and 925 tosses, and the average duration is 184.37 in agreement with the value 200 from eqn (5.6).

The gambler might be so lucky to generously receive £1 from the casino when he has lost all his money. Similarly, the gambler could decide to bet £1 more, even if he has already reached the prefixed winning. Clearly, with such terms the game could continue indefinitely: we are in the presence of reflecting barriers situations.

5.1.3 Reflecting barriers

Consider a *perfectly* reflecting barrier. If the random process is in the state 0, it goes to state 1 with probability 1: $p_{01} = 1$. Just like that, $p_{n,n-1} - 1$, if n is the number of states of the process. Note that $p_{00} = 0$, i.e. the walker cannot stay in 0. The matrix of transition probabilities is:

$$\begin{pmatrix} 0 & 1 & 0 & 0 & \cdots & 0 & 0 & 0 \\ q & 0 & p & 0 & \cdots & 0 & 0 & 0 \\ 0 & q & 0 & p & \cdots & 0 & 0 & 0 \\ & & & \cdots\cdots\cdots\cdots & & & \\ 0 & 0 & 0 & 0 & \cdots & q & 0 & p \\ 0 & 0 & 0 & 0 & \cdots & 0 & 1 & 0 \end{pmatrix}$$

At a non-perfectly reflecting barrier the walker can stay at 0, not with certainty, but with probability $p_{00} = q$. So it can go to 1 with probability $p_{01} = p$. Similarly, $p_{nn} = p$ and $p_{n,-1} = q$. The code to simulate reflecting barriers derives from the `Code_5_3.R`. The `Code_5_4.R` is adapted to deal with the gambler's ruin scenario, when the game might go even further. As before, the barriers are at 0 and £40, respectively, with an initial capital of £20. If desired, we can obtain estimates of quantities about reflecting barriers, as was done by `Code_5_3.R` about absorbing barriers.

```
## Code_5_4.R
# restricted random walk with reflecting barriers

# beginning with walk without barriers
set.seed(3)
p<-0.5
x0<-20
nsteps<- 400
nhists<- 100
#    starting vector: each history starts from x0
#    column vector with all x0 repeated nhists times
start_matr<-matrix(x0,nhists,1)
y<-matrix(,nhists,nsteps)
for (i in 1:nhists)          { # starting loop on histories
 x<-rbinom(nsteps,1,p)
 x[which(x==0)]<- -1     # also   x<- 2*x-1
 y[i,]<-cumsum(x)     # Sn without the starting point
                          } # ending loop on histories
Sn <- matrix(c(start_matr,y+x0),nhists,nsteps+1)  # adding the starting vector
# Sn: position of the walker after n steps
n<- seq(0, nsteps)          # time axis
par(mai=c(1.02,1.,0.82,0.42)+0.1)
plot (n,Sn[1,],xlab="number of tosses",ylim=c(-2,42),type="n",lty=2,
```

```
ylab=expression(italic("S"[n] )~~
italic(("gambler's"~~capital)), ),
cex.lab=1.2,font.lab=3,lwd=1,xaxt="none"   )
axis(1, at=seq(0,nsteps,50))

#----------------------------------------
# continue to plot the walks on a single graph
set.seed(3)     # reset random numbers if wanted
r1<-numeric()
r2<-numeric()
r1b1<-numeric()
r2b2<-numeric()
no.refl<-numeric()
# the following parameters can be redefined
x0<- 20
nsteps<- 400
nhists<- 100
x<-numeric()
p<-0.5
y<-matrix(,nhists,nsteps)
b1<- 40
b2<- 0
for(i in 1:nhists)           {   # starting loop on histories
x[1]<- x0
for(j in 2:nsteps)     {   # starting loop on steps
u<-runif(1,0,1)
if(u<p)z<- +1
else z<- -1
x[j]<-x[j-1]+z
if(x[j-1] == b1)  x[j]<- b1-1
if(x[j-1] == b2)  x[j]<- b2+1
                         }   # ending loop on steps
y[i, ]<- x
r1[i]<- min(which(x==b1))
r2[i]<- min(which(x==b2))
                         }   # ending loop on histories
#y[i, ]  # if wanted to print entire i-th selected history
r1[!is.finite(r1)] <- 0
r1
r1b1 <- r1[r1!=0]
r1b1
r2[!is.finite(r2)] <- 0
r2
r2b2 <- r2[r2!=0]
r2b2
r12<- 0
for(i in 1:nhists)
if(r1[i]==0 & r2[i]==0) {r12<-r12+1
no.refl[r12]<- i }
no.refl
Sn <- matrix(c(start_matr,y+x0),nhists,nsteps+1)  # adding the starting vector
par(lwd=1.7,font=3)
lines(y[1, ], type="l",lty=1)
lines(y[60, ], type="l",lty=1)
lines(y[5, ], type="l",lty=1)
abline(h=b1,lty=4,col="black",lwd=2)
abline(h=b2,lty=4,col="black",lwd=2)
l_r1b1<- length(r1b1)
l_r1b1
l_r2b2<- length(r2b2)
```

```
l_r2b2
l_no.refl<-length(no.refl)
l_no.refl
m.r1b1<- mean(r1b1)
m.r1b1
se.r1b1<-  sqrt(var(r1b1))
se.r1b1
m.r2b2<- mean(r2b2)
m.r2b2
se.r2b2<-  sqrt(var(r2b2))
se.r2b2
```

Figure 5.10 shows two trajectories hitting the barriers. In the lower one, we see that after 197 tosses the gambler should be ruined, but thanks to of gift of £1 he goes on, beginning a series of bounces up to 339 tosses. Let us see the higher trajectory: the absorbed trajectory in Fig. 5.8 is reported in Fig. 5.10, to compare it with the reflected trajectory. Both trajectories are the results of the same history, i.e. the same initial values and the same random numbers are employed. The absorbed trajectory is overlapped on the reflected one for the first 259 tosses, after that it changes no more, while the reflected trajectory continues its race, reaching the barrier again after 273 tosses and one more time after 395 tosses. Barriers may also be partially reflecting, that

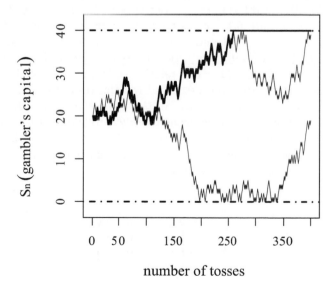

Fig. 5.10 Three restricted paths out of 100, of a random walk with barriers at 0 and 40. Reflected trajectories: thinner lines; absorbed trajectory: thicker line.

is the reflection is not certain, but only probable. The matrix of transition probabilities is:

$$\begin{pmatrix} q\ p\ 0\ 0 \cdots 0\ 0\ 0 \\ q\ 0\ p\ 0 \cdots 0\ 0\ 0 \\ 0\ q\ 0\ p \cdots 0\ 0\ 0 \\ \cdots\cdots\cdots\cdots\cdots \\ 0\ 0\ 0\ 0 \cdots q\ 0\ p \\ 0\ 0\ 0\ 0 \cdots 0\ q\ p \end{pmatrix}$$

In gambling language this situation may be interpreted as mentioned at the end of the previous section. The house, or the adversary, gives or does not give £1 to the ruined gambler allowing him to play again. If the gambler receives no money, he continues nonetheless to toss a coin, waiting for £1. Similar observations can be made when the gambler has won £a.

In the `Code_5_4.R`, reflection is implemented by the lines:

```
if(x[j-1] == b1)  x[j]<- b1-1
if(x[j-1] == b2)  x[j]<- b2+1
```

become:

```
if(x[j-1] == b1) {u<-runif(1,0,1)
if(u>c_refl){x[j]<- b1-1}
else{x[j]<- b1}        }
if(x[j-1] == b2)  {u<-runif(1,0,1)
if(u>c_refl){x[j]<- b2+1}
else{x[j]<- b2}        }
```

where we have introduced the *coefficient of reflection* `c_refl`, that is the probability that the trajectory reaching the barrier will be reflected (the game goes on). Clearly, it is $0 \leqslant$ `c_refl` $\leqslant 1$. If `c_refl` $= 0$, the barrier is perfectly reflecting, if `c_refl` $= 1$, the barrier is absorbing. Figure 5.11 shows four examples of histories with `c_refl` $= 0.2$ (dotted line) and `c_refl` $= 0.8$ (solid line). At the barrier 0, we see that the trajectory with `c_refl` $= 0.8$ (solid line) has a series of bounces from about 131 to 258 tosses, while the trajectory with `c_refl` $= 0.2$, relative small probability of reflection (dotted line), reaches the barrier after 355 tosses and remains on the barrier up to 400 tosses, except some isolated small reflections. At the barrier $a = 40$, the trajectory with `c_refl` $= 0.8$ (solid line) has bounces from about 227 to 360 tosses, while the trajectory with `c_refl` $= 0.2$ (dotted line) stays on the barrier from about 303 to 326 tosses.

Let us find now the invariant distribution of a Markov chain with partially reflecting barrier. Consider, for example, a chain with four states:

$$\mathbf{P} = \begin{array}{c} \\ 0 \\ 1 \\ 2 \\ 3 \end{array} \begin{array}{cccc} 0 & 1 & 2 & 3 \\ \begin{pmatrix} q & p & 0 & 0 \\ q & 0 & p & 0 \\ 0 & q & 0 & p \\ 0 & 0 & q & p \end{pmatrix} \end{array}$$

where:

$$p_{01} = p_{12} = p_{23} = p_{33} = p$$
$$p_{00} = p_{10} = p_{21} = p_{32} = 1 - p = q$$

then $\mathbf{P}$ can be written:

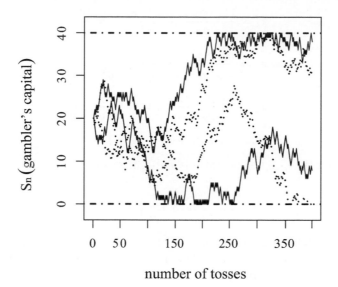

number of tosses

Fig. 5.11 Four restricted paths out of 100, of a random walk with partially reflected barriers at 0 and 40. Dotted line: `c_refl` $= 0.2$; solid line: `c_refl` $= 0.8$

$$\mathbf{P} = \begin{array}{c} \\ 0 \\ 1 \\ 2 \\ 3 \end{array} \begin{array}{cccc} 0 & 1 & 2 & 3 \\ \begin{pmatrix} p_{00} & p_{01} & p_{02} & p_{03} \\ p_{10} & p_{11} & p_{12} & p_{13} \\ p_{20} & p_{21} & p_{22} & p_{23} \\ p_{30} & p_{31} & p_{32} & p_{33} \end{pmatrix} \end{array}$$

There are entries equal to 0, for instance:

$$\pi_2 \, p_{20} = 0, \pi_3 \, p_{30} = 0, \quad \text{etc.}$$

So we have:

$$\pi_0 = \pi_0 \, p_{00} + \pi_1 \, p_{10}$$
$$\pi_1 = \pi_0 \, p_{01} + \pi_2 \, p_{21}$$
$$\pi_2 = \pi_1 \, p_{12} + \pi_3 \, p_{32}$$
$$\pi_3 = \pi_2 \, p_{23} + \pi_3 \, p_{33}$$

With some simplification, we have:

$$\pi_1 = \frac{(1 - p_{00}) \pi_0}{p_{10}} = \frac{p_{01}}{p_{10}} \pi_0 = \frac{p}{q}$$

$$\pi_2 = \frac{p_{01} p_{12}}{p_{10} p_{21}} \pi_0 = \frac{p \cdot p}{q \cdot q} = \frac{p^2}{q^2}$$

$$\pi_3 = \frac{p_{01} p_{12} p_{23}}{p_{10} p_{21} p_{32}} \pi_0 = \frac{p \cdot p \cdot p}{q \cdot q \cdot q} = \frac{p^3}{q^3}$$

with: $\pi_0 + \pi_1 + \pi_2 + \pi_3 = 1$

In general, when the number of states is finite, it gives:

$$\pi_k \propto \left(\frac{p}{q}\right)^k \tag{5.7}$$

If we have a countably infinite number of states, we may speak again of a 'transition matrix', but it is an infinite matrix. For instance, the transition matrix of the chain with partially reflecting barrier, if the states are numbered $0, 1, 2, \ldots$, is:

$$\mathbf{P} = \begin{pmatrix} q\, p\, 0\, 0\, 0\, \ldots 0\, 0\, 0\, \ldots \\ q\, 0\, p\, 0\, 0\, \ldots 0\, 0\, 0\, \ldots \\ 0\, q\, 0\, p\, 0\, \ldots 0\, 0\, 0\, \ldots \\ 0\, 0\, q\, 0\, p\, \ldots 0\, 0\, 0\, \ldots \\ \cdots\cdots\cdots\cdots\cdots\cdots \\ 0\, 0\, 0\, 0\, 0\, \ldots q\, 0\, p\, \ldots \\ \cdots\cdots\cdots\cdots\cdots\cdots \end{pmatrix}$$

If the number of states is infinite, the above equation eqn (5.7) holds again, but for $k \to \infty$, there are three possible cases.

- $p > 1/2$. In that case:

$$\pi_k \propto \left(\frac{p}{1-p}\right)^k$$

and since $p/(1-p) > 1$:

$$\pi_k \propto \left(\frac{p}{1-p}\right)^k \xrightarrow[n \to \infty]{} \infty$$

then there is no invariant distribution. All the states are transient, at that situation is possible since the chain is not finite.

- $p < 1/2$. In this case, the states are *positive recurrent*. Let c be the constant of proportionality, which is:

$$\pi_k = c \left(\frac{p}{1-p}\right)^k$$

Let:

$$\theta = \left(\frac{p}{1-p}\right)$$

it becomes (geometric series):

$$\sum_{k=0}^{\infty} \theta^k = \frac{1}{1-\theta}, \quad \text{since } \theta < 1$$

If it has to be $\sum_{k=0}^{\infty} \pi_k = 1$, c has to be:

$$\sum_{k=0}^{\infty} c\,\theta^k = 1, \quad \text{that is} \quad c\,\frac{1}{1-\theta} = 1$$

from which:

$$c = 1 - \theta = \frac{1-2p}{1-p}$$

Then, in conclusion:

$$\pi_k = (1-\theta)\,\theta^k = \frac{1-2p}{1-p}\left(\frac{p}{1-p}\right)^k$$

- $p = 1/2$. In this case, the states are recurrent, but *null recurrent*. It gives:

$$\pi_k = c\left(\frac{1/2}{1/2}\right)^k = c$$

But it must be:

$$\sum_{k=0}^{\infty} \pi_k = 1, \quad \text{that is} \quad \sum_{k=0}^{\infty} c = 1$$

which is not possible.

5.1.4 Two-dimensional random walk

The walker, whatever direction of origin is, can go towards north (N), south (S), east (E) or west (W). Assuming a symmetric random walk, the four directions have the same probability, to be chosen equal to $1/4$, as sketched in Fig. 5.12. The transition probabilities are:

$$\begin{cases} p_{ij} = 1/4 & \text{if } |i - j| = 1 \\ 0 & \text{otherwise} \end{cases}$$

Each position has four neighbours: $(+1, +1), (+1, -1), (-1, +1), (-1, -1)$. At each time point, the walker takes a step in one of the above positions, with the same probability equal to $1/4$. We can ask ourselves firstly whether the process is irreducible, that is if all states are recurrent, or more simply if the walk is recurrent. The answer goes back to a theorem proved by G. Pólya (1921) which says that a simple random walk in one and two dimensions returns to its starting position with probability 1, but if the dimensions are greater than or equal to three, then with positive probability, the walk will never return where it started from. We can say equivalently that a walker may start from any point, but certainly he will reach the origin, or pass through some

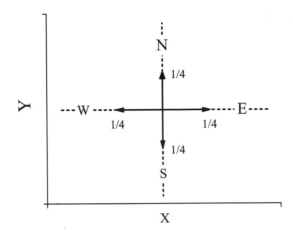

Fig. 5.12 Symmetric random walk in the plane. All the four directions have the same probability 1/4.

other possible points infinitely often. Feller (1970) cites the old saying 'all roads lead to Rome'. A further equivalent formulation of the theorem is that two walkers (in one and two dimensions) are certain to meet infinitely often.

To show that in two dimension the walk is recurrent, it is useful to look at the process as two one-dimensional processes, as sketched in Fig. 5.13. The trajectory of

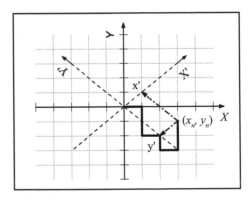

Fig. 5.13 Orthogonal projection on the X' and Y' axes of some steps of a trajectory taking place on the (X, Y) plane.

the walk described in the (X, Y) plane is projected onto orthogonal axes X' and Y' rotated by $\pi/4$. The unit steps in the new coordinates become:

$$
\begin{array}{ll}
(X, Y) \text{ axes,} & (X', Y') \text{ axes} \\
\pm (1, 0) & \pm (1/\sqrt{2}) \, (1, 1) \\
\pm (0, 1) & \pm (1/\sqrt{2}) \, (-1, 1)
\end{array}
$$

For instance, after six steps the trajectory is in the point ($x_6 = 3, y_6 = -1$), which is projected in the points x' and y'. In this way, the two-dimensional random walk is reconstructed in a pair of one-dimensional independent symmetric simple random walks. Now if the walk returns to the origin $(0,0)$ in the (X, Y) plane, this means that it returns to 0 in the X' and Y' axes. Then:

$$p_{00}^{(2n)} = \left(\frac{(2n)!}{n!n!} \frac{1}{2^{(2n)}}\right)^2 \approx \frac{1}{\pi n}$$

then:

$$\sum_{n \geqslant 1} p_{00}^{(2n)} = \infty$$

and the walk is recurrent. With a similar approach it is proved that in three dimensions (or more) the walk is transient.

A further result obtained by considering the projection of the walk along the X- and Y-axes concerns the estimate of the final distance reached by the walker (see Peterson and Noble (1972) and also for extended applications Bertozzi (2008)). Let us look at Fig. 5.14. The walker starts at the origin and takes a step length l_1 up to the point 1.

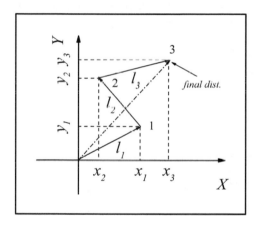

Fig. 5.14 Three step trajectory with the arrival points projected on the X- and Y-axes, the final distance d_f is also indicated.

The projections of this point on the X- and Y-axes are x_1 and y_1. The walker takes a second random step length l_2 (not necessary equal to l_1) and the new coordinates are $X = x_1 + x_2$, where x_2 is the projection of l_2 on the X-axis. Similarly, $Y = y_1 + y_2$. The point 3 with coordinates (x_3, y_3) determines the *final distance* $d_f^2 = x_3^2 + y_3^2$ of this three step trajectory, given by:

$$d_f^2 = (x_1^2 + 2x_1x_2 + 2x_1x_3 + \cdots + x_2^2 + 2x_2x_3 + \cdots + x_3^2) + \\ (y_1^2 + 2y_1y_2 + 2y_1x_3 + \cdots + y_2^2 + 2y_2y_3 + \cdots + y_3^2)$$

The final distance after n steps is then $d_f^2 = X^2 + Y^2$, where:

$$X^2 = (x_1 + x_2 + x_3 \cdots + x_n) \quad \text{and} \quad Y^2 = (y_1 + y_2 + y_3 \cdots + y_n)$$

By performing the calculations and considering that for n large positive x and y values balance the negative ones, it gives:

$$d_f^2 = x_1^2 + y_1^2 + x_2^2 + y_2^2 + x_3^2 + y_3^2 + \cdots + x_n^2 + y_n^2$$

so that for each l_i, it is $l_i^2 = x_i^2 + y_i^2$. In this way $d_f^2 = l_1^2 + l_2^2 + l_3^2 + \cdots + l_n^2$ or:

$$d_f = \sqrt{\sum_{i=1}^{n} l_i^2}$$

If the step length is the same, it is:

$$d_f = \sqrt{\sum_{i=1}^{n} l^2} = l\sqrt{n} \quad \text{and for unit step} \quad d_f = \sqrt{n}$$

the same as for the one-dimensional case.

The following `Code_5_5.R` simulate 5000 histories with 1000 steps each, for a two-dimensional random walk. Note firstly that each neighbour `xdir` and `ydir`, in the four direction, is chosen by a random number u.

```
## Code_5_5.R
# Two-dimensional random walk

set.seed(1)
nsteps<- 1000
nhists<- 5000
distf<-matrix(,nhists,nsteps)
dist<-matrix(,nhists,nsteps)
dist_max<-matrix(,nhists,nsteps)
dist_min<-matrix(,nhists,nsteps)
dist0<-matrix(,nhists,nsteps)
rmax<- 10                    # max dist about 100
xf<-matrix(,nhists,nsteps)
yf<-matrix(,nhists,nsteps)
for (i in 1:nhists)          { # starting loop on histories
xdir<-0
ydir<-0
x<-vector()
x[1]<-xdir
y<-vector()
y[1]<-ydir
khist<- vector()
for (j in 1:(nsteps-1))      {    # starting loop on steps
u<-runif(1,0,1)
if(u<=0.25) {xdir<-xdir+1}
if(u>0.25 & u<=0.5) {xdir<-xdir-1}
if(u>0.5 & u<=0.75) {ydir<-ydir +1}
if(u>0.75) {ydir<-ydir-1}
x[j+1]<-xdir
y[j+1]<-ydir
}    # ending loop on steps
if(i==3){         # to plot this history as example
xmin<- -45
xmax<-  45
```

```
ymin<- -45
ymax<-  45
plot(x,y,type="l",xlab="x",ylab="y",main=" ",lwd=1,
col="black",xlim=range(xmin:xmax),ylim=range(ymin:ymax),
font.lab=3,cex.lab=1.4)
}
if(i==15){lines(x,y,col="black",lty=1,lwd=2)} # to plot this history as example
dist[i, ]<- sqrt(x**2 + y**2)
#print(dist[i, ])        # uncomment to print dist
for (j in 1:nsteps)          {     # instructions  for the minimum distances
if(dist[i,j]>rmax)  {
break                }
khist[i]<- j
}
dist0[i,khist[i]:nsteps] <- dist[i,khist[i]:nsteps]
dist_min[i,  ]<- min(dist0[i,khist[i]:nsteps])
dist_max[i, ]<- max(dist[i, ])
distf[i, ]<- sqrt(x[nsteps]**2 + y[nsteps]**2)
xf[i, ]<- x[nsteps]
yf[i, ]<- y[nsteps]
}  # ending loop on histories

xfmin<- -80
xfmax<-  80
yfmin<- -80
yfmax<-  80
nfpoint<- 500
plot(xf[1:nfpoint],yf[1:nfpoint],type="p",xlab="x",ylab="y",
main=" ",lwd=1, col="black",xlim=range(xfmin:xfmax),
ylim=range(yfmin:yfmax), font.lab=3,cex.lab=1.4,pch=19,cex = .7)
abline(v=0,lwd=2,lty=4)
abline(h=0,lwd=2,lty=4)
text(70,-70,"IV",cex=1.1,font=2)
text(-70,-70,"III",cex=1.1,font=2)
text(-70,70,"II",cex=1.1,font=2)
text(70,70,"I",cex=1.1,font=2)
dist_f<-distf[1:nhists]
#dist_f                          # uncomment to print dist_f
dist_max<- dist_max[1:nhists]
#dist_max                        # uncomment to print dist_max
dist_min<- dist_min[1:nhists]
#dist_min                        # uncomment to print dist_min
xf_f<-xf[1:nhists]
#xf_f                            # uncomment to print x_f
yf_f<-yf[1:nhists]
#yf_f                            # uncomment to print y_f
xf_fm<- max(abs(xf_f))
xf_fm
yf_fm<- max(abs(yf_f))
yf_fm
nq1<- sum(xf_f >= 0 & yf_f >= 0)
nq1
nq2<- sum(xf_f >= 0 & yf_f <  0)
nq2
nq3<- sum(xf_f <  0 & yf_f <  0)
nq3
nq4<- sum(xf_f <  0 & yf_f >= 0)
nq4

opar <- par(lwd=2)
```

```
hist(dist_f,freq=F,main="",xaxt="none",
xlim=range(0:100), ylim=c(0,0.035),right = TRUE,
xlab="distance",ylab="density",font.lab=3,cex=1.1,
cex.lab=1.2,lwd=1.5 )
axis(1, at=seq(-10,90,20))
m_distf<-mean(dist_f)
v_distf<-var(dist_f)
m_distf
v_distf
sd_distf=sqrt(v_distf)
sd_distf
curve (dnorm(x, mean=m_distf, sd=sd_distf),
lwd=1, add=TRUE, yaxt="n")
opar <- par(lwd=2,lty=2,font.lab=3)
hist(dist_max,freq=F,xaxt="none",
right = TRUE,add=TRUE)
text(63,0.03,"max distance")
text(10,0.03,"final dist.")
m_distm<-mean(dist_max)
v_distm<-var(dist_max)
m_distm
v_distm
sd_distm=sqrt(v_distm)
sd_distm

opar <- par(lwd=2)
hist(dist_min,freq=T,main="",xaxt="none",
xlim=range(0:10), ylim=c(0,1500),right = TRUE,
xlab="min distance",ylab="counts",font.lab=3,cex=1.1,
cex.lab=1.2,lwd=1.5 )
axis(1, at=seq(0,10,2))
# plot=F to print a list of breaks and counts
hist_f<- hist(dist_f,plot=F)
hist_f
hist_max<- hist(dist_max,plot=F)
hist_max
hist_min<- hist(dist_min,plot=F)
hist_min
```

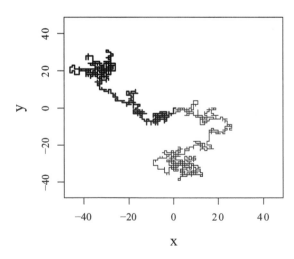

Fig. 5.15 Two trajectories 1000 steps long of a two-dimensional random walk.

Figure 5.15 shows, as examples, two trajectories in the (X, Y) plane, both starting at $(0, 0)$. The plane is divided in four quadrants and in each of them the final points of the trajectories are recorded. The results confirm that the trajectories arrive on the plane with a constant probability equal to $1/4$. The code gives the following number of the final points: $1287, 1280, 1201, 1232$. Figure 5.16 shows the final points of 500 out of 5000 trajectories.

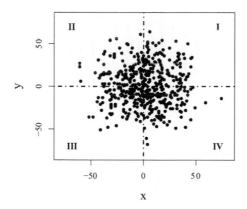

Fig. 5.16 Final points of 500 out of 5000 trajectories 1000 steps long of a two-dimensional random walk.

For each walk, the maximum distance from the origin (`dist_max`) and the final distance (`distf`) are computed. By 'final distance' we mean the distance from the origin to the last step n (`nsteps`) of the trajectory. Also the minimum distance (`dist_min`) is computed, but requires a small discussion later. The estimated mean of the 5000 final distances is `m_distf`= 28.00, in agreement with $\sqrt{n}$ ($n = 1000$), and the standard error `sd_distf` = 14.75. For the maximum distance, it is `m_distm` = 36.80, and `sd_distf` = 12.39. Figure 5.17 shows the distributions of the final and the maximum distances. The histograms are normalized for comparison with the probability density function of the normal random variable, (`dnorm(x, mean=m_distf, sd=sd_distf)`).

To estimate the minimum distance (`dist_min`) consider Fig. 5.18 where a trajectory 1110 steps long is plotted.

The trajectory starts at the point $(0, 0)$ and it remains very close to the origin for a few steps. At step $= 36$, for instance, the trajectory returns to $(0, 0)$, so that `dist_min` $= 0$. But can we really say that the walk returns where it started from? We ask the walker to move far enough and only after we may say that he returned actually to the origin. So we have introduced a circle of radius `rmax` equal to 10 and consider as 'minimum distance' only the distances of the trajectories that have gone outside that circle. Of course a trajectory, after moving outside the circle, may return inside. The value 10 is somewhat arbitrary, but checks performed with reasonably different values do not invalidate such an approach. From the output of the code, it results that at step 36 the final distance is zero, but only after 230 steps do the distances become stably

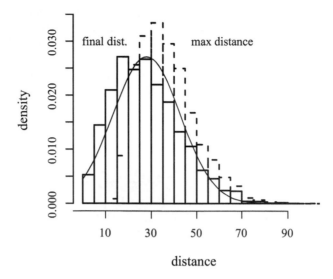

Fig. 5.17 Distributions of the final (continuous line) and the maximum (dotted line) distances of 5000 trajectories 1000 steps long of a two-dimensional random walk. Comparison with the probability density function of the normal random variable is also shown.

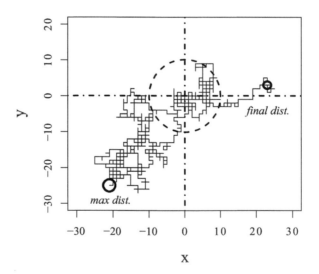

Fig. 5.18 Trajectory 1110 steps long of a two-dimensional random walk. The final distance = 23.19 and the maximum distance = 32.65 are marked. For the circle see text.

greater than `rmax`. At step = 988 an actual first return to the origin takes place. The distribution of the minimum distances are reported in Fig. 5.19. Note that the 30% of `dist_min` is in the interval $(0, 10]$, and by increasing the number of steps to 5000, the percentage becomes 43%, as the walker has more possibilities to find the road to

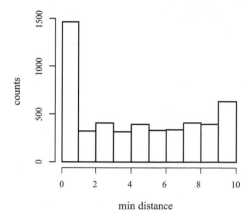

Fig. 5.19 Distributions of the minimum distances of 5000 trajectories 1000 steps long of a two-dimensional random walk.

home.

5.2 Some topics on Brownian motion

The term 'Brownian motion' has physical origins: it refers to the disordered movement of small particles as a consequence of collisions. Let $X(t)$ be the position of a particle at time t. Now $t \in [0, \infty)$ and $X(t)$ takes values on the real line. It is assumed that:

1. $X(0) = 0$, i.e. at time $t = 0$ the particle is conventionally in 0. Otherwise, if $X(0) = x_0 \neq 0$, x_0 is taken as initial point.
2. The process is 'completely random'.

The last item is quoted from Lawler (1995). Let us take two times s and t, with $s < t$. Here 'random' means that the positions x_s and x_t are independent of each other and, furthermore, that the motion of the particle after time s, $X(t) - X(s)$, is independent of the 'past' $x_i, i < s$, and also of the 'present' x_s. The particle is at $X(s) = x_s$, it moves to $X(t) = x_t$, then the step $x_t - x_s$ is independent of trajectories inside the interval $[0, s]$, s included, provided that $X(s) = x_s$. This assumption must hold for every s_i and t_i, $i = 1, \ldots, n$. If:

$$0 \leqslant s_1 < t_1 \leqslant s_2 < t_2 \leqslant \cdots \leqslant s_n < t_n < \infty$$

then the random variables:

$$X(t_1) - X(s_1), X(t_2) - X(s_2), \ldots, X(t_n) - X(s_n)$$

are jointly independent. These independent increments are also stationary increments. This means that, for any $0 < s, t < \infty$, the distribution of the increment $X(t+s) - X(s)$ has the same distribution as $X(t) - X(0) = X(t)$.

Recall also that the Poisson process has independent and stationary increments, but their distribution is quite different. For the Poisson process it is required that the sample path is a stepped motion. For the Brownian motion, we require a continuous path, that is $X(t)$ must be a continuous function of t.

Formally, a stochastic process $X(t)$ is a Brownian motion or a *Wiener process* with variance σ^2 if:

1. $X(0) = 0$
2. The process $X(t)$ has stationary increments. The distribution of any increment $X(t) - X(s)$ only depends on the length of the time interval $(t - s)$.
3. The process $X(t)$ has independent increments. For any non-overlapping time intervals $(t_1, t_2]$ and $(t_3, t_4]$, the random variables $X(t_2) - X(t_1)$ and $X(t_4) - X(t_3)$ are independent.
4. $\forall s < t$, the increment $X(t) - X(s)$ has a normal distribution with mean 0 and variance $(t - s)\sigma^2$.
5. The trajectories of a Brownian motion are continuous.

Some comments are in order. If $\sigma^2 = 1$ the motion is called *standard Brownian motion*.

Point 1 is conventional. $X(0)$ can assume any other value.

Point 2 is a consequence of *time homogeneity*. The dynamics of a particle in the time interval $(t - s)$ depends only on the length of the interval, not where it is on the time axis.

Point 3 represents a physical assumption. It means that the time intervals are much larger than the 'infinitesimal' interval between the collisions of the particles.

Point 4 is a redundant condition. It can be proved that a stochastic process with stationary and independent increments, having also a continuous sample path is a Brownian motion with normally distributed increments. Similarly in discrete state space, the Poisson process is the unique counting process that has stationary and independent increments.

Point 5 is essential, since we have to describe the trajectory of a physical particle as a function of time.

Remark 5.1 To say 'Brownian motion' is to say 'history of Physics', 'philosophy of probability', 'mathematics achievements', and more. Galavotti (2005) writes:

The process by which probability gradually entered physical science, not only in connection with errors of measurement, but more penetratingly as a component of physical theory, can be traced back to the work of the Scottish traveller and botanist Robert Brown (1773-1858).

Our botanist (Brown, 1828) in 1827 demonstrated that any small particle suspended in a fluid, even an inorganic grain, is subject to an incessant and irregular motion (Brown, 1828). Actually, explanations of such motion did not enter the field of biology, rather that of physics, or even economics. In physics, theories were advanced by Einstein in 1905 and Marian von Smoluchowski in 1906. These theories were confirmed by a series of experiments on fluctuation phenomena. It is interesting from an historical and philosophical perspective that these theories are founded on different bases (see the comprehensive von Plato (1994), also Brush (1968)). Einstein writes at the beginning of his (Einstein, 1905):

In this paper it will be shown that according to the molecular-kinetic theory of heat, bodies of microscopically-visible size suspended in a liquid will perform movements of such magnitude that they can be easily observed in a microscope, on account of the molecular motions of heat. It is possible that the movements to be discussed here are identical with the so-called 'Brownian molecular motion'

and says: 'an exact determination of actual atomic dimensions is then possible'.

In his doctoral dissertation written before the paper quoted above, he determines the diffusion coefficient D as:

$$D = \frac{k_B T}{6\pi k r}$$

where k_B is the Boltzmann's constant, T the absolute temperature, k the viscosity coefficient, and r the radius of Brownian particles (that is the diffusing molecules).

Reasoning on the basis of statistical mechanics, with some probabilistic assumptions, (for instance, independency of the movements of particles), he gives the probability density function of a Brownian particle satisfying the following diffusion equation:

$$\frac{\partial f(x,t)}{\partial t} = D\,\frac{\partial^2 f(x,t)}{\partial x^2}$$

whose solution is:

$$f(x,t) \;=\; \frac{1}{\sqrt{4\pi Dt}}\,\exp(-x^2/4Dt)$$

Einstein suggests that the mean square displacements of the particles, $\langle \Delta x^2(t)\rangle = 2Dt$ being related to their diffusion coefficient D, rather their velocities, are observable and measurable quantities.

A probabilistic account of the Brownian motion was put forward by Smoluchowski (1906) based on the following assumptions, which are the same as for random walk, using the current terminology. The particle velocity remains constant, and very small changes in direction occur at each collision. Successive collisions are independent of each other, the length of the steps are all the same. von Plato (1994, 129-130) writes:

The individual motions are described as consisting of a succession of nearly linear parts with randomly changing direction at points of collision. The random changes occur at discrete intervals. Being based on assumptions about collisions, it is an approach in the style of kinetic theory, whereas Einstein's theory has the abstract character of statistical mechanics.

In 1900 Louis Bachelier, independently of Brown's work, developed a mathematical theory of Brownian motion, anticipating the theories by Einstein (1905) and Smoluchowski (1906). Bachelier submitted his work in his PhD dissertation, with H. Poincaré as examiner (Bachelier, 1900). One of his achievement was to grasp the Markovian character of the Browning motion: to give an idea, the future prices of assets are independent from information of past prices. His analysis of markets allowed him to advance a theory for continuous time random process with independent increment.

5.2.1 Brownian motion as limit of random walks

Consider the symmetric random walk:

$$S_n = X_1 + X_2 + \cdots + X_n$$

The X_i's are independent random variables such that:

$$\mathsf{P}\{X_i\} = 1 = \mathsf{P}\{X_i\} = -1 = 1/2$$

so S_n is the position of the particle after n steps. Recall that:

$$\mathrm{E}\,[S_n] = \sum_{k=1}^{n}\mathrm{E}\,[X_k] = 0 \qquad \mathrm{Var}\,[S_n] = \sum_{k=1}^{n}\mathrm{E}\,[X_k^2] = n$$

Divide the half-line into intervals of length δ (see Figure 5.20):

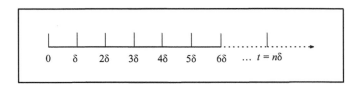

Fig. 5.20 The half-line $[0, \infty)$ is divided into intervals of length δ.

The time unit for each step is δt, that is in each interval $[(j-1)\delta, j\delta]$, the particle moves forwards or backwards. Suppose δx to be the length of the step to the left or to the right. After $n \times \delta t$ time, S_t will be the position of the particle at time t. Then:

$$S_t = \delta x \left(X_{\delta t} + X_{2\delta t} + \cdots + X_{n\delta t} \right)$$

To sum up: t is the time elapsed, δt is the time unit, then $t/\delta t$ is the number of the taken steps. It gives:

$$\mathrm{E}\left[S_t\right] = 0 \qquad \mathrm{Var}\left[S_t\right] = \frac{t}{\delta t}\left(\delta x\right)^2$$

But if $\delta x = \delta t \to 0$, $S_t \to 0$ and the motion of the particle will become meaningless. The idea is to choose δx proportional to $\sqrt{\delta t}$, that is $\delta x = c\sqrt{\delta t}$. Define X_i as (see Figure 5.21):

$$X_i = \begin{cases} \sqrt{\delta t} & \text{with} \quad \mathsf{P} = 1/2 \quad \text{to the right} \\ -\sqrt{\delta t} & \text{with} \quad \mathsf{P} = 1/2 \quad \text{to the left} \end{cases}$$

The motion begins at $(0,0)$, and after one time step δt, the particle is in the point $-\sqrt{\delta t}$; after two steps $2\delta t$, it is at 0, and so on. With this choice, the size of an increment is:

$$|X_{t+\delta t} - X_t| \approx \sqrt{\delta t}$$

and, if $\delta t \to 0$, $\sqrt{\delta t} \to 0$, thus fulfilling the request of the continuity of the trajectories.
If $c = 1$ (standard motion), we have:

$$\mathrm{E}\left[S_t\right] = 0 \qquad \mathrm{Var}\left[S_t\right] = t$$

Applying the central limit theorem to a sequence of i.i.d. random variables

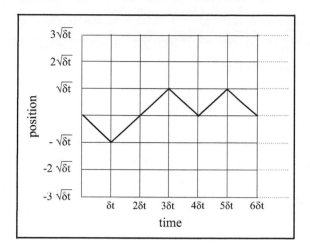

Fig. 5.21 Sketch of the construction of a Brownian motion from a limiting case of a random walk.

$$X_1, X_2, \ldots, X_n$$

with finite mean μ, finite non-zero variance σ^2, and $S_n = \sum_{i=1}^{n} X_i$, we have:

$$Z_n = \frac{S_n - \mathrm{E}\,[S_n]}{\sqrt{\mathrm{Var}\,[S_n]}} = \frac{S_n - n\mu}{\sigma\,\sqrt{n}}$$

For $n \to \infty$, this converges in distribution to a standard normal random variable $\mathcal{N}(0,1)$. In our case, $\mu = \mathrm{E}\,[S_n] = 0$ and $\sigma^2 = \mathrm{Var}\,[S_n] = 1$ and we can write:

$$S_t - \mathrm{E}\,[S_t] \xrightarrow[n\to\infty]{d} \mathcal{N}(0,t)$$

Then $\forall s \geqslant 0$ and $\forall t \geqslant 0$:

$$X_{s+t} - X_s \sim \mathcal{N}(0,t)$$

with $\forall t \geqslant 0$, $X_t \sim \mathcal{N}(0,t)$.

The following short code, `Code_5_6.R`, illustrates how to think intuitively of the Brownian motion as a limiting case of a random walk as its time increment approaches zero. The first part of the code is the as same already presented in `Code_5_1.R`.

```
## Code_5_6.R

set.seed(3)
x0<- 0
nsteps<-100
p<-0.5
x<-rbinom(nsteps,1,p)
x<- 2*x-1
y<- cumsum(x)
Sn<- c(x0,y+x0)     # adding the starting point x0
```

```
n<- 0:nsteps              # time axis
par(mai=c(1.02,1.,0.82,0.42)+0.1)      # to control the margin size
plot (n,Sn,type="l",xlab="time",ylim=c(-15,7),lty=3,
ylab=expression(italic("S"[t] )~~ italic((position)) ),lwd=4,
cex.lab=1.2,font.lab=3,xaxt="none"   )
axis(1, at=seq(0,nsteps,20))
set.seed(3)
dt<- 0.01
x0<- 0
y<-numeric()
y[1]<- x0
n<-  10000
t<- 100
for(i in 2:(n+1)) {
y[i]<- y[i-1] + rnorm(1,mean=0,sd=sqrt(dt))
                }
lines(seq(0,t,dt),y,type="l",lty=1,lwd=1)
```

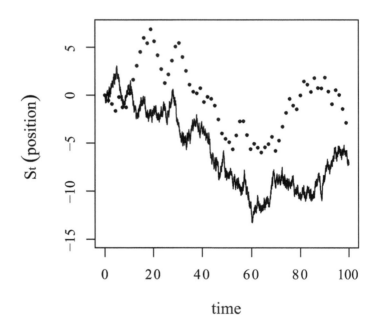

Fig. 5.22 Brownian motion from limiting case of random walk. Dotted lines: sample path of a random walk. Continuous line: sample path of a Brownian motion.

With $\delta t \to 0$ the segmented line approximates to a continuous curve, that is; the process becomes a continuous-time Markov process in a continuous state space. In Fig. 5.22 the x-axis label is denoted as 'time', not 'number of steps', in view of the limit continuous approximation.

The trajectories of a Brownian motion $\{X(t)\}$ possess some peculiarities:

i) They have infinite length in any finite time interval.

ii) They are not differentiable anywhere. We imagine that the particle moves more and more in different directions in each infinitesimal interval.

The proof of the above statements requires advanced mathematics, but an intuitive idea might be the following. For item *i*), we have seen that in the interval $[0, t]$, $t/\delta t$ is the number of the steps taken, each of length $\sqrt{\delta}$, so the total path length λ is $\lambda = t/\sqrt{\delta}$, which is:

$$\lim_{\delta \to 0} \frac{t}{\sqrt{\delta}} = +\infty$$

For item *ii*), We should compute the derivative dX_t/dt, that is:

$$\frac{dX_t}{dt} = \lim_{\Delta t \to 0} \frac{dX_{t+\Delta t} - X_t}{\Delta t}$$

This limit does not exist, since we have already seen that the numerator $|X_{t+\delta t} - X_t|$ is on the order of $\sqrt{\Delta t}, \gg \Delta t$.

The standard Brownian motion is a time-homogeneous Markov process, since, as said above, for the independent increments property, the increment $X(s+t) - X(s)$ does not depend on the past time before time s. So, the future time $X(s+t)$, given the present state $X(s)$ only depends on $X(s+t) - X(s)$. Moreover, since the increments are also stationary, the process is a time-homogeneous Markov process. Being a Markov process, the Brownian process also has some properties that hold in discrete time.

The random variable T with values in $[0, \infty)$ is a *stopping time* for Brownian motion $X(t)$ if the event $\{T \leqslant t\}$ only depends on $\{X(s), 0 \leqslant s \leqslant t\}$, but not on $\{X(s), s > t\}$.

Let $Y(t)$ be the process after the stopping time T, that is: $Y(t) = X(t+T) - X(T)$, then $Y(t)$ is also a standard Brownian motion. The *strong Markov property* says that $Y(t)$ is a Brownian motion that does not depend on T, that is:

$$\forall t > 0, \ \{Y(s), 0 \leqslant s \leqslant t\} \quad \text{is independent of} \quad \{X(u), 0 \leqslant u \leqslant T\}$$

Now we are interested in the first passage times. Following Ross (2019), let T_a be the first time the standard Brownian motion hits $a, a > 0$. We compute the probability of the event $\{T_a \leqslant t\}$, that is $\mathsf{P}\{T_a \leqslant t\}$, and the probability $\mathsf{P}\{X(t) \geqslant a\}$, conditioned on the occurrence or not of the event $\{T_a \leqslant t\}$. We have:

$$\mathsf{P}\{X(t) \geqslant a\} = \mathsf{P}\{X(t) \geqslant a | T_a \leqslant t\}\, \mathsf{P}\{T_a \leqslant t\}$$
$$+ \underbrace{\mathsf{P}\{X(t) \geqslant a | T_a > t\}}_{=0}\, \mathsf{P}\{T_a > t\}$$

The second right-hand term is equal to 0, since the term between curly braces is zero. Indeed, the path of the motion is continuous, so the process cannot go beyond a without having yet hit a. For the first addendum, if $T_a \leqslant t$, the process has hit a at some instant in $[0, t]$ and, by symmetry, it is just as likely to be either above or below a at time t, that is:

$$\mathsf{P}\{X(t) \geqslant a | T_a \leqslant t\} = \frac{1}{2}$$

Then the partition function of the random variable T_a is given by:

$$P\{T_a \leqslant t\} = 2 \int_a^\infty \frac{1}{2\pi t} \exp -\frac{x^2}{2t} \, dx \quad \text{and putting } x/\sqrt{t} = u:$$

$$= \frac{2}{\sqrt{2\pi}} \int_{a/\sqrt{t}}^\infty \exp -\frac{u^2}{2} \, du$$

It follows:

$$P\{T_a < \infty\} = \lim_{t\to\infty} P\{T_a \leqslant t\} = 2 \int_0^\infty \mathcal{N}(0,1) \, du = 1$$

therefore the path hits the point a with probability 1. From the above equation, we can derive the expected value of the first passage time $E[T_a]$ which results to be infinite. Then, it is true that the process will hit any point a with certainty, but it is equally true it needs an infinite time to arrive to a, however close it is.

From the notion above, we introduce the *reflection principle*, as follows. Let T_a be the first passage time to the value a of the Brownian process $X(t), t \geqslant 0$. Let us consider the new process $Y(t)$:

$$\begin{cases} Y(t) = X(t), & t < T_a \\ Y(t) = 2a - X(t), & t \geqslant T_a \end{cases}$$

From the strong Markov property, it can be proved that $Y(t)$ is also a standard Brownian motion. In other words, the reflection principle states that if the path of a Brownian process $X(t)$ reaches the value a at time $t = T_a$, then the path after T_a has the same distribution as the path reflected from the horizontal line $X(t) = a$. Figure 5.23 shows the underlying idea of the reflection principle. Note that the paths are segmented as for a discrete time stochastic process.

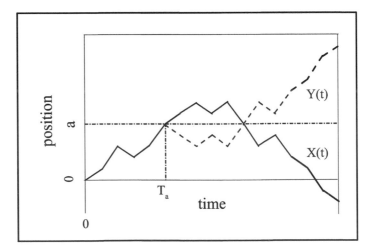

Fig. 5.23 Sketch showing the basic of the reflection principle. Continuous line: sample path of a Brownian motion. Dashed line: portion of the reflected Brownian motion after T_a.

From the reflection principle, the probability that the process hits the origin, that is it crosses the x-axis at some time s, with $1 \leqslant s \leqslant t$, is (see Lawler 1995):

$$\mathsf{P}\left\{X(s) = 0, \text{ for some } 1 \leqslant s \leqslant t\right\} = 1 - \frac{2}{\pi} \arctan \frac{1}{\sqrt{t-1}} \qquad (5.8)$$

For $t \to \infty$ such quantity $\to 1$, so that the process returns to the origin infinite times (strong Markov property), and the motion assumes positive and negative values for large t, Let us see what happens near $t = 0$. Consider the process $Y(t) = t\,X(1/t)$. Also, $Y(t)$ is a standard Brownian motion, therefore when $t \to \infty$ in the process $X(t)$, we have $t \to 0$ in the process $Y(t)$. Since the process $X(t)$, for arbitrarily large t, assumes positive and negative values, $Y(t)$ assumes arbitrarily small positive and negative values. Then in any interval near the origin, the Brownian motion assumes positive and negative values and, for continuity also 0.

Consider the '*random set*' Z defined as:

$$Z = \{t : X(t) = 0\}$$

and also $Z_1 = Z \cap [0,1]$. We wish to cover Z_1 by intervals of length $\epsilon = 1/n$ of the type:

$$\left[\frac{k-1}{n}, \frac{k}{n}\right], \; k = 1, 2, \ldots, n$$

How many n intervals do we need to cover Z_1? Before answering, a brief digression on the concept of '*fractal dimension*' seems appropriate. For a more extended discussion of such an argument, see Huffaker *et al.* (2017) and references therein.

A line segment of length L can be covered with another segment of the same length, or with two segments of length $L/2$, or with four segments of length $L/4$, etc. If $N(\epsilon)$ is the number of segments of length ϵ covering the segment of length L, it is:

$$N(\epsilon) = L\,\frac{1}{\epsilon}$$

Analogously when we wish to cover a square of side L, it needs:

$$N(\epsilon) = L^2\,\frac{1}{\epsilon^2}$$

squares of side ϵ. In d dimension we have:

$$N(\epsilon) = L^d\,\frac{1}{\epsilon^d}$$

where ϵ is the side of the d-dimensional hypercube. Taking the logarithms, we have:

$$d = \frac{\log N(\epsilon)}{\log L + \log(1/\epsilon)}$$

If $\epsilon \to 0$, we can formulate the definition of the `Hausdorff-Besicovitch dimension` (H–B) as:

$$D_{HB} = \lim_{\epsilon \to 0} \frac{\log N(\epsilon)}{\log(1/\epsilon)} \qquad (5.9)$$

For a line segment, for a surface, and analogously for a cube and a hypercube, the H–B dimension coincides with the usual Euclidean dimension, but let us see when it is applied to the fractal subset of $[0,1]$, that is to the Cantor set, or more exactly *the middle third Cantor set*.

The rules of the game to construct the Cantor set are the following. At the beginning the unit interval, say I_0, is divided into three equal parts and the middle part is deleted, that is the open interval $(1/3, 2/3)$ is cancelled, so remains the set of points $I_1 = [0, 1/3] \bigcup [2/3, 1]$. The same procedure is executed in the successive iterations. Then, I_2 consists of four segments, each of length $1/9$, and in general I_n consists of 2^n segments, each of length $(1/3)^n$. As the process moves forwards the number of small segments increases, but their length decreases. The geometrical figure so created, the middle third Cantor set, is formally defined as:

$$C = \lim_{n \to \infty} I_n = \bigcap_{n=0}^{\infty} I_n$$

The set C is the set of points that belong to all I_n, $\forall n$. Without going into more detail, let us compute the H–B dimension on C. Clearly, we can cover the initial interval $I_0 = [0, 1]$ by $N(\epsilon) = 1$ segment of length $\epsilon = 1$ and the interval $I_1 = [0, 1/3] \cup [2/3, 1]$ by $N(\epsilon) = 2$ segments of length $\epsilon = 1/3$. In this manner, we arrive at:

$$D_{HB} = \lim_{\epsilon \to 0} \frac{\log N(\epsilon)}{\log(1/\epsilon)} = \lim_{\epsilon \to 0} \frac{\log 2^n}{\log 3^n} = \frac{\log 2}{\log 3} \approx \frac{0.6931}{1.0986} \approx 0.63$$

which is not an integer number, that is to say D_{HB} is a fractional dimension, and we say that the Cantor set is a fractal. It is a geometric figure consisting of infinite points, but its dimension is less than 1, so to speak, 'a little lesser' than a segment (dimension 1), and 'a little greater' than a point (dimension 0).

Turning to the set Z_1 defined before, to cover it we must have:

$$Z_1 \bigcap \left[\frac{k-1}{n}, \frac{k}{n} \right] \neq \emptyset$$

or, in other words, searching for:

$$\mathsf{P}(k, n) = \mathsf{P}\left[\frac{k-1}{n}, \frac{k}{n} \right] \neq 0$$

If $k = 0$, the probability is 1, since $0 \in Z$, then we assume $k \geqslant 1$. If $X(t)$ is a standard Browning motion, then:

$$Y(t) = \left((k-1)/n \right)^{-1/2} X\left(nt/(k-1) \right)$$

is a standard Brownian motion. We have:

$$\mathsf{P}(k, n) = \mathsf{P}\{ X(s) = 0, \text{ for some } 1 \leqslant s \leqslant t \}$$

Now it is:

$$\mathsf{P}\left\{ Y(t) = 0, \text{ for some } 1 \leqslant t \leqslant \frac{k}{k-1} \right\}$$

therefore from eqn (5.8):

$$P(k,n) = 1 - \frac{2}{\pi} \arctan \frac{1}{\sqrt{\frac{k-(k-1)}{k-1}}} = 1 - \frac{2}{\pi} \arctan \sqrt{k-1}$$

Then the number of intervals to cover Z_1 is of the type:

$$\sum_{k=1}^{n} P(k,n) = \sum_{k=1}^{n} \left(1 - \frac{2}{\pi} \arctan \sqrt{k-1}\right)$$

By developing arctan(.) into Taylor series, the final result is:

$$\sum_{k=1}^{n} P(k,n) = \frac{4}{\pi}\sqrt{n}$$

therefore, $\sqrt{n}$ intervals of length $1/n$ are necessary to cover Z_1, and

$$D_{BH} = \frac{\log \sqrt{n}}{\log n} = \frac{1}{2}\frac{\log n}{\log n}$$

Conclusion: the dimension of the set Z_1 is $1/2$.

Until now we have dealt with the *standard Brownian motion*, characterized by zero mean μ and variance $\sigma^2 = 1$, that is $\forall s < t$, and the random variable $X(t) - X(s)$ has distribution $\mathcal{N}(0,1)$. Consider now a non-zero mean process: $W(t) = X(t) + \mu t$, where $X(t)$ a Brownian motion and μ is the *drift* parameter (a stochastic process is said to be 'with drift', if its mean value is not zero). With respect to a standard motion, the difference is that the distribution of the increments is:

$$W(t) - W(s) \sim \mathcal{N}\big(\mu(t-s), \sigma^2(t-s)\big)$$

To account for stochastic processes we have used a mathematical formulation describing the *random state* of the system in different times of type $\mathsf{P}\{X(t) = i\}$. Now we write the probabilistic law describing the *motion* of the system, that is the steps between states. The process is given in differential formulation and to obtain $W(t)$ we have to integrate:

$$dW(t) = \mu\big(W(t),t\big)dt + \sigma\big(W(t),t\big)dX(t)$$

with $X(t)$ a standard Brownian motion. The first term on the right is a deterministic term that stand for the expected move of the system, while the second is a stochastic term that represents the variability of this expected value. The above equation is a stochastic differential equation, whose solution is the 'path':

$$W(T) = \int_0^T \mu\big(W(t)\big)dt + \int_0^T \sigma\big(W(t)\big)dX(t)$$

For a rigorous introduction to stochastic differential equation, see Øksendal (2003).

We simulate the process on the basis of random walk. We choose a small dt and simulate a random walk with time increment dt and spatial increment $\sqrt{dt}$. Then we can write:

$$dW(t) = Z(t)\sqrt{dt}$$

If $Z(t) \sim \mathcal{N}(0,1)$, we have $Z(t)\sqrt{dt} \sim \mathcal{N}(0, dt)$. The following `Code_5_7.R` is an example of simulation of a stochastic differential equation (see also Iacus and Masarotto 2003).

```
## Code_5_7.R

set.seed(2)
n<- 1000
T<- 1
dt<- T/n
x0<- 1                        # starting point
mu     <-function(x,t) {-x*t}
sigma <-function(x,t) {+x*t}
w<-numeric()
w[1]<- x0
for(i in 2:(n+1)) {
t<- dt*(i-1)
w[i]<- w[i-1] + mu(w[i-1],t) * dt + sigma(w[i-1],t)*rnorm(1,sd=sqrt(dt))
plot(seq(0,T,dt),w,type="l",ylim=c(0.,2),font.lab=3,lty=1,lwd=2,
ylab="W(t)",cex.lab=1.2)
# second trajectory
set.seed(4)
w[1]<- x0
for(i in 2:(n+1)) {
t<- dt*(i-1)
w[i]<- w[i-1] + mu(w[i-1],t)*dt + sigma(w[i-1],t)*rnorm(1,sd=sqrt(dt))
lines(seq(0,T,dt),w,type="l",lty=2,lwd=2 )
```

We have chosen $\mu(x,t) = -xt$ and $\sigma(x,t) = xt$. Figure 5.24 shows two trajectories of a Brownian motion with drift, obtained as a random walk with time increment dt and spatial increment $\sqrt{dt}$. Note that the walk begins at x0 = 1.

Finally, the motion they call Brownian has never stopped over the years. In its journey it visited a variety of lands, statistical physics, interpretation of probability, stochastic differential equations, entropic forces, and more. There have been several attempts to involve Brownian motion in alternative conceptions of quantum mechanics. Also the Feynman path integral formulation has been compared to stochastic approaches. Robert Brown wrote (1828): 'These motions were such as to satisfy me ...that they arose neither from currents in the fluid, nor from its gradual evaporation, but belonged to the particle itself', and Richard Feynman (1948): 'The paths involved are, therefore, continuous but possess no derivative. They are of a type familiar from study of Brownian motion.

5.3   Exercises

Exercise 5.1 A flea lives on a finite segment of a line, of length D. It is able to do jumps towards left or right, except in the starting position $x = 0$ and in the ending position $x = D$, i.e. the two barriers are not surmountable with a jump. The insect jumps to the left with probability p_L, to the right with probability p_R or it decides to rest where it is, with probability p_S. The jump length is such that D is covered with 10 jumps in the same direction (starting from 0). A rather limited world, poor little bugs!

(1) How can you write the transition matrix of this random walk? And, (2) What kind of random walk is it?

Exercise 5.2 If, in the previous exercise, you take $p_R = 1 - p_L$ and, consequently, $p_S = 0$, what does this new random walk represent? Write its transition matrix.

Exercise 5.3 What kind of random walk is represented by the following transition matrix $(q = 1 - p)$?

$$\mathbf{P} = \begin{pmatrix} q\ p\ 0\ \ldots\ 0\ 0\ 0 \\ q\ 0\ p\ \ldots\ 0\ 0\ 0 \\ \vdots \qquad\qquad \vdots \\ 0\ 0\ 0\ \ldots\ q\ 0\ p \\ 0\ 0\ 0\ \ldots\ 0\ q\ p \end{pmatrix}$$

Exercise 5.4 Draw a graph like that in Fig. 5.1 to represent a transition matrix of exercise 5.3 for five states.

Exercise 5.5 We borrow from H. Tijms (*Understanding Probability*, Cambridge University Press, 2007), the depiction of the random walk as the walk of a drunk man. A drunkard moves on a line, at each step going to the left or to the right with equal probability. Call S_n the distance covered by a drunkard on a line, starting from the origin (the pub exit) after n

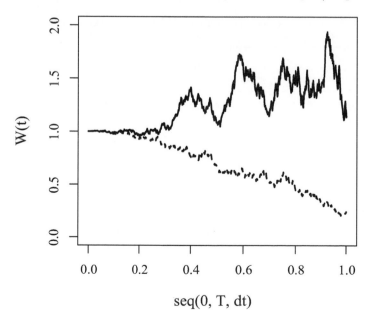

Fig. 5.24 Two trajectories of Brownian motion with drift, obtained as a finite time approximation of a random walk.

steps. We have seen in Section 5.1 that for n large the average of S_n is zero, because there are as many positive as negative steps. But nevertheless, with the passing of time, the drunkard gets tired of walking because he actually 'travelled' for a distance that is

$$D_n = |X_1 + \ldots + C_n|$$

where $X_1 \ldots X_n$ are the n steps taken by the drunkard.

(1) Numerically compute $\mathrm{E}\,[D_n]$ and $\mathrm{E}\,[S_n]$ with increasing n, for a random walk simulated with R. (2) For n large: demonstrate that $\mathrm{E}\,[D_n] \approx \sqrt{\frac{2n}{\pi}}$

Hint: apply the central limit theorem to $X_1 \ldots X_n$

Exercise 5.6 A common approach to simulating spatial dispersal of insects with time is to use the random walk. A two-dimensional random walk is used to simulate the degree of insect diffusion over an area with time. Write a code to simulate a two-dimensional movement of a female insect that starts with 100 eggs and deposits one egg at each step. Assume an equiprobability for the movement in four directions: left, right, top, bottom. The program ends when the female has finished the eggs. Then assume there are N females and modify the program by setting a variable *steps to oviposit* that is the number of steps the insect must take before depositing one egg. Play with the variables: *number of females*, *number of eggs per female* and *steps to ovideposit*, to see how the distribution of eggs changes.

Exercise 5.7 With reference to exercise 5.6, suppose there are two species of ants: black and red. Suppose the queens of both species have 100 eggs, and assign a different number N_s of *steps to oviposit* to the species, for example $N_s = 2$ for black ants, $N_s = 5$ for red ants. On a graph, represent black and red eggs with a coloured dot and see the behaviour of the two distributions of eggs.

Exercise 5.8 Simulate a trajectory of a three-dimensional random walk.

6
ARMA Processes

Statistics are the only tools by which an opening can be cut
through the formidable thicket of difficulties that bars
the path of those who pursue the Science of man.

Sir Francis Galton, Natural Inheritance

In this chapter we will introduce a class of stochastic processes which are of great importance in the study of random phenomena, e.g. for time-series forecasting. It is the subject of auto-regressive processes, moving-average processes and, more generally, of auto-regressive moving-average (ARMA) ones. ARMA models essentially describe a time series in terms of a combination of polynomials. The first thorough description of such models was probably given in 1970 in the first edition of the book (Box *et al.*, 1994), although they are a generalization of auto-regressive models, described more than forty years earlier by Yule (1927). As we shall see, ARMA processes are based on *white noise*, which therefore deserves a short introduction.

6.1   White noise and other useful definitions

Some of the concepts introduced in this chapter will be clearer after reading Chapter 7, where the spectral analysis of stochastic processes is treated in more detail. We anticipate here, without any proof, that a time series (as a time-dependent signal) can also be analysed in the frequency domain, exactly as an electrical circuit or a mechanical system can be studied in physics in terms of its time response or of its frequency behaviour.

White noise is a simple stationary process $\{\epsilon_t\}$, defined as follows

$$
\begin{aligned}
E[\epsilon_t] &= 0, && \forall t \in \mathbb{N} \\
\text{var}[\epsilon_t] &= \sigma^2, && \forall t \in \mathbb{N} \\
\text{corr}[\epsilon_t, \epsilon_s] &= 0, && \forall t, s \in \mathbb{N}
\end{aligned}
\tag{6.1}
$$

where the stochastic process has been assumed discrete. In other words, ϵ_t is an *i.i.d.* sequence of zero-mean, constant variance, uncorrelated random variables. For example, if every ϵ_t is normally distributed $\epsilon_t \sim \mathcal{N}(0, \sigma^2)$, the sequence $\{\epsilon_t\}$ is white noise. Anyway, ϵ_t is not necessarily Gaussian.

The following R example code shows how to simulate white noise.

```
## Code_6_1.R
#White noise
```

```
# for reproducibility: change seed to change noise
set.seed(3)
# create noise time series
w <- rnorm(1000)
plot(as.ts(w),ylab="White noise")
# verify w is zero mean
mean(w)
var(w)
hist(w,main="",col="white")
# verify w is uncorrelated
acf(w,main="",lag=20,ylim=c(-0.2,1))
# verify w is really white
w.spec <- spectrum(w,log="no",span=200,plot=FALSE)
plot(w.spec$freq,w.spec$spec,t="l",ylim = c(0,1.2),
xlab="frequency (cycles/sample interval)",ylab="spectral density")
```

Figure 6.1 shows the time series generated by the R code above.

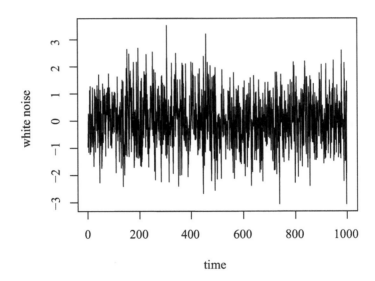

Fig. 6.1 White noise time series.

The mean and variance of the generated series are very close to 0 and 1 respectively, as the histogram in Fig. 6.2 also shows. The **acf** R function allows us to verify that data are uncorrelated (see Figure 6.3).

The reason for the 'white' attribute comes from the analogy with light signals. A white light is composed of all colours or, which is the same, its frequency spectrum contains all optical frequencies with approximately the same intensities. Indeed, by performing a spectral analysis on the time series (see Chapter 7) we obtain what figure 6.4 shows.

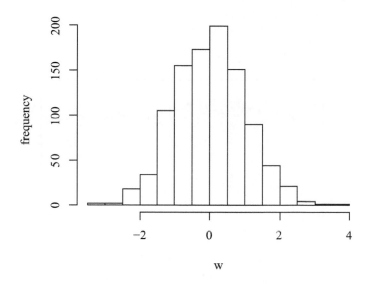

Fig. 6.2 White noise histogram.

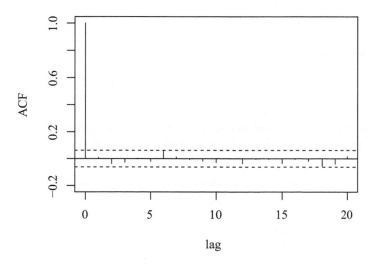

Fig. 6.3 White noise auto-correlation function.

All frequencies (note, by default the frequency axis of the **spectrum** function is expressed in cycles per sample interval) of the spectrum are present with comparable intensities in the *periodogram*, or spectral density plot.

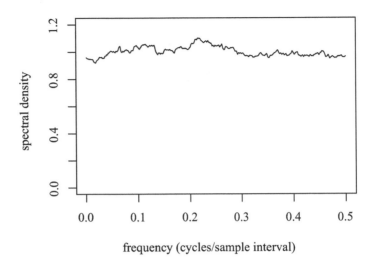

Fig. 6.4 White noise spectral analysis.

White noise, as we will see shortly, plays a fundamental role in auto-regressive moving-average processes.

6.1.1 The lag operator

Given a discrete stochastic process $\{X_t\}$ we formally define a 'lag' operator L establishing a correspondence between the random variable X_t at time t and the variable X_s at time s, a step backwards or forwards with respect to t, i.e. $s = t-1$ or $s = t+1$. The lag operator is often named 'backwards shift' or 'backshift' operator[1], usually denoted by B and defined by $BX_t = X_{t-1}$. Choosing the backwards time, for the moment, L is defined by:

$$LX_t = X_{t-1} \tag{6.2}$$

where the 'standard' notation LX is not a product but actually means L(X), i.e. L *operates* on X. L is linear, because:

---

[1] Lag is more general than backshift. The lag operator allows relating X_t with X values at previous or successive times, while the backshift only considers previous times.

$$L(\beta X_t) = \beta X_{t-1}$$
$$L(X_t + X_s) = X_{t-1} + X_{s-1}$$

As a consequence, L can be recursively applied:

$$L^2 X_t = L(LX_t) = X_{t-2}$$
$$...$$
$$L^n X_t = X_{t-n}$$

and it also admit an *inverse* operator L^{-1} such that $L^{-1}X_t = X_{t+1}$. The above relationships allows us to formally treat expressions involving L as algebraic ones. L also allows us to express more complex relations between the terms of time series. For example, the difference operator, defined as the difference between random variables at adjacent times $\Delta X_t = X_t - X_{t-1}$, is simply given by:

$$\Delta X_t = X_t - X_{t-1} = (1 - L)X_t \tag{6.3}$$

Complex relationships can be easily written by symbolic manipulation, for example one involving X_t, X_{t-1} and X_{t-2} in terms of two constants ϕ_1 and ϕ_2:

$$(\phi_1 + \phi_2 L)\, LX_t = \phi_1 X_{t-1} + \phi_2 X_{t-2}$$

The algebraic representation also allows us to define a polynomial of degree n in L, in terms of $n+1$ coefficients $\theta_0, \theta_1, ...\theta_n$ as in (6.4). We will make extensive use of the lag operator and lag polynomials in the succeeding sections.

$$\Theta(L) = \sum_{k=0}^{n} \theta_k L^k \tag{6.4}$$

6.2 Moving-average processes

By using the lag polynomial (6.4) it is possible to construct a weighted moving average of white noise processes $\{\epsilon_t\}$, as follows:

$$X_t = \Theta(L)\epsilon_t = \epsilon_t + \theta_1 \epsilon_{t-1} + \theta_2 \epsilon_{t-2} + ... + \theta_q \epsilon_{t-q} \tag{6.5}$$

where θ_k, for $k = 1..q$ are the weights of the average. Equation (6.5) formally defines a moving-average process of q-th order, or shortly an MA(q) process. To understand what MA(q) processes are, let us start with q=1.

An MA(1) process is defined by:

$$X_t = \epsilon_t + \theta\epsilon_{t-1} = (1 + \theta L)\epsilon_t \tag{6.6}$$

which is a zero mean process, remembering that ϵ_t is white noise for any t. The definition (6.6) can be made more general by including the possibility of a non-zero mean: $X_t = \mu + \epsilon_t + \theta\epsilon_{t-1}$, but this does not add anything as we are allowed to consider $X_t - \mu$ as a random variable (μ does not depend on t if the process is stationary in the weak sense).

Before briefly analysing the peculiar characteristics of MA(1) and, subsequently, of MA(q) processes, a simple fictional example can help in understanding what an MA(1) process is, and how a process at time t can depend on noise at the same time and at a previous time $t-1$. The MA(1) process has three ingredients: a random variable X_t at time t and two white noise variables ϵ_t and ϵ_{t-1}.

Suppose X_t is the difference with respect to the average value, say in one year and in a great number of cities in a given country, of the number of umbrellas bought by the country's inhabitants. Qualitatively, the number of sold umbrellas in a locality increases when it rains or when the weather forecasts predict rain with a high probability. Now, forcing somewhat, suppose the rain 'expectation' to be a white noise process. That is not actually true, but what is certainly true is that the daily amount of precipitation in a place at a given time is described by a probability distribution (Ye *et al.*, 2018), and that the so-called 'probability of precipitation' (PoP) we find in the 'meteo', is based on it (and, of course, on climatological models!). Therefore, although it is not scientifically based, we can give to ϵ_t a meaning related to the PoP in a certain area at a certain time. Essentially, suppose that a high value of PoP pushes people to get an umbrella, or to buy it if you do not possess one. Continuing with the artificial example, we can suppose that the expected number of bought umbrellas Y_t is given by an expression like:

$$Y_t = \mu + \beta(\epsilon_t + \theta\epsilon_{t-1})$$

telling us that the number of sold umbrellas is the average number plus (or minus; it depends on the sign of β) something *proportional* to the PoP. By normalizing, to the β coefficient, the difference with respect to the annual average value, $X_t = (Y_t - \mu)/\beta$:

$$X_t = \epsilon_t + \theta\epsilon_{t-1}$$

Let us arbitrarily suppose that $\theta = -0.5$. If the rain probability/amount at time $t-1$ was 0 (dry weather), at time t we have $X_t = \epsilon_t$. In other words, the increment of the number of bought umbrellas with respect to the mean depends on ϵ_t, i.e. on the weather forecast at a given place on day t, (remember: X_t and ϵ_t are random variables, not deterministic!). If the PoP is very high somewhere, e.g. $\epsilon \simeq 1$, we have $X_t \simeq 1$, a high number of sold umbrellas.

Now, if the bad weather persists at time $t+1$, at that time we have:

$$X_{t+1} = \epsilon_{t+1} - 0.5\epsilon_t = 0.5$$

that is, the increment in the number of sold umbrellas decreases because most people bought it the day before.

The example is, of course, fictitious and cannot be pushed too far. What we want to point out from this is that random phenomena exist that can be described by models accounting for a relationship between a random variable and noise at different times. A moving-average process has some kind of 'memory', but at time t it does not remember where it was at $t-1$; it only remembers what the random noise component was at that previous time. And, because of the limited memory (limited to two time steps in the MA(1) case) an MA process can produce 'pairs' of observations that are likely to be both high or low. After two time steps (or $q+1$ in the MA(q) case) the process is again random.

Let us briefly look at the main characteristics of an MA(1) process, defined by equation (6.6). Taking into account the characteristics of white noise (eqn 6.1), the autocovariances of different orders are easy to compute:

$$\gamma(0) = E[X_t X_t] = E[(\epsilon_t + \theta \epsilon_{t-1})^2] = \sigma^2(1+\theta^2)$$
$$\gamma(1) = E[X_t X_{t-1}] = E[X_t L X_t] - E[(\epsilon_t + \theta \epsilon_{t-1})(\epsilon_{t-1} + \theta \epsilon_{t-2})] - \sigma^2 \theta \qquad (6.7)$$
$$\gamma(k) = E[X_t X_{t-k}] = 0, \text{ for } k > 1$$

As a consequence, the autocorrelation coefficient $\rho(1)$ is given by:

$$\rho(1) = \frac{\gamma(1)}{\gamma(0)} = \frac{\theta}{1+\theta^2}$$

while $\rho(k) = 0$ for any k greater than one. This means that MA(1) processes are 1-correlated, while white noise is 0-correlated. Note that by substituting the θ coefficient with its reciprocal $1/\theta$, the first-order autocorrelation coefficient does not change.

The following code shows how to simulate an MA(1) process.

```
## Code_6_2.R
# Moving-average process
# MA(1) positive coefficient
# for reproducibility
set.seed(13)
# generate white (Gaussian) noise
w <- rnorm(100)
x <- rep(0,100)
# generate MA(1)
x[1] <- w[1]
for (t in 2:100) x[t] <- w[t] + 0.5*w[t-1]
# plot time series
plot(as.ts(x),ylab="MA(1) process")
# plot autocorrelation function
acf(x,main="")
# plot spectral density
x.spec <- spectrum(x,log="no",span=30,plot=FALSE)
plot(x.spec$freq,x.spec$spec,t="l",ylim = c(0,3),
  xlab="frequency (cycles/sample interval)", ylab="spectral density")
```

Figure 6.5 shows the MA(1) process generated in the example above. Exactly the same data can be generated with a single R function (**arima.sim**), present in the standard package *stats*, which we will find useful in the following:

```
set.seed(13)
x <- arima.sim(model = list(ma = 0.5), n = 100)
```

At first sight, the process appears similar to white noise. But, looking more thoroughly, some kind of correlation seems to exist between data at consecutive times; in

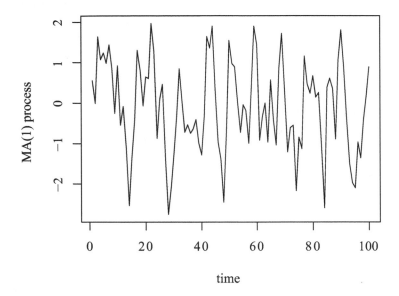

Fig. 6.5 MA(1) process

particular it appears that every 'up' in the plot is followed by a 'down'. Indeed, that is the case: Fig. 6.6 shows that the first autocorrelation is not zero, as it comes from eqn 6.7, and also that the spectrum of the process is not flat.

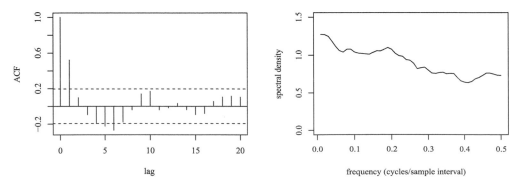

Fig. 6.6 ACF (left) and spectral density (right) of an MA(1) process with $\theta = 0.5$.

Changing the sign of the θ coefficient a very similar process is obtained, the main difference being the reversed sign of the first-order autocorrelation coefficient $\rho(1)$ (see Fig. 6.7). Moreover, MA(1) processes with negative coefficient tend to be less 'smooth' than the positive counterpart, by virtue of the negative correlation between consecutive times. The spectrum is not flat, as before, and the sign of the θ coefficient affects the frequency distribution. We will come back to this example in the next chapter.

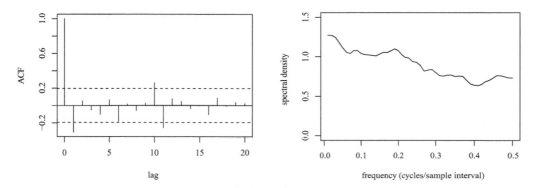

Fig. 6.7 ACF (left) and spectral density (right) of an MA(1) process with $\theta = -0.5$.

6.2.1   Moving-average processes of higher order

Equation (6.5) generalizes the MA(1) case to an order q. By computing the autocovariance and autocorrelation coefficients, it is easy to verify that an MA(q) process is q-correlated:

$$\gamma(0) = \sigma^2(1 + \theta_1^2 + ... + \theta_q^2)$$

$$\gamma(k) = \sigma^2 \sum_{j=0}^{q-k} \theta_j \theta_{j+k}, \text{ for } k \leq q \tag{6.8}$$

$$\gamma(k) = 0, \text{ for } k > q$$

The k-order autocorrelation for $k \leq q$ is expressed by:

$$\rho(k) = \frac{\sum_{j=0}^{q-k} \theta_j \theta_{j+k}}{1 + \sum_{j=1}^{q} \theta_j^2}$$

and it is zero for $k > q$. Without going too deep inside the connection between MA(q) processes and general stationary processes, we mention the fact that the opposite is also true: any stationary q-correlated process can be represented by an MA(q) model.

Remark 6.1 The observed quantity in a time series is X_t, and the value at time t, while the noise ϵ_t is unknown. It appears from (6.5) that if the lag polynomial $\Theta(L)$ is invertible, we can formally obtain the noise:

$$\epsilon_t = \Theta(L)^{-1} X_t$$

An in-depth discussion of the invertibility property of an MA process is beyond the purpose of this book. The reader can find it in any specific textbook dealing with time series, e.g. (Box *et al.*,

1994), (Shumway and Stoffer, 2006). Inverting the lag polynomial is in principle rather easy. It means finding a series:

$$\Psi(L) \equiv \Theta(L)^{-1} = \psi_0 + \psi_1 L + \psi_2 L^2 + \dots.$$

such that $\Psi(L)\Theta(L) = 1$ which, by matching the coefficients of the product, brings:

$$\psi_0 = 1$$
$$\psi_1 + \psi_0 \theta_1 = 0$$
$$\psi_2 + \psi_1 \theta_1 + \psi_0 \theta_2 = 0$$
$$\dots$$

Aa MA(q) process is invertible if the roots of the lag polynomial, i.e. the solutions of a q-order equation $1 + \theta_1 z + \dots + \theta_q z^q = 0$ lie *all* outside the unit circle. In the MA(1) case, for example, this means $|\theta| < 1$. We have seen that θ and $1/\theta$ give the same autocorrelation coefficient, e.g. $\theta = 0.5$ and $\theta = 2$ give $\rho(1) = 0.4$. But the former is invertible, while the latter is not. A true comprehension of what invertibility implies requires a knowledge of autoregressive processes, and their relationship with moving-average ones.

6.3 Autoregressive processes

The time series $\{X_k\}$ is an autoregressive process of order p, denoted by AR(p), if X at time $k = t$ depends **only** on the *previous* p times:

$$X_t = \phi_1 X_{t-1} + \phi_2 X_{t-2} + \dots + \phi_p X_{t-p} + \epsilon_t = \sum_{k=1}^{p} \phi_k X_{t-k} + \epsilon_t \qquad (6.9)$$

where $\{\epsilon_t\}$ is white noise as defined in Section 6.1. For example, $\{\epsilon_t\}$ can be a Gaussian error with zero average and variance σ^2, with zero autocovariance ($\text{cov}(\epsilon_k, \epsilon_j) = 0$ for any k and j) and uncorrelated with the signal X: $\text{cov}(X_t, \epsilon_k) = 0$, for any $k < t$. The process defined by (6.9) is defined 'autoregressive' because it represents X at time t as a linear regression on past terms of the same series.

Equation (6.9) can be expressed in terms of the lag operator L introduced earlier. Recursively applying it to the terms of the time series (6.9) we obtain:

$$\left(1 - \sum_{k=1}^{p} \phi_k L^k\right) X_t = \epsilon_t \qquad (6.10)$$

In terms of a lag polynomial $\Phi(L)$, equation (6.9) can be re-written as follows:

$$\Phi(L) X_t = \epsilon_t$$

from which we note the formal analogy between $\Phi(L)$ and $\Theta(L)^{-1}$ introduced for MA(q) processes.

Let us start with the simplest AR process, that is for $p = 1$:

$$(1 - \phi L)X_t = \epsilon_t$$

and invert the lag polynomial:

$$X_t = (1 - \phi L)^{-1}\epsilon_t = \sum_{i=0}^{\infty}(\phi L)^i\epsilon_t = \sum_{i=0}^{\infty}\phi^i\epsilon_{t-i}$$

which formally identifies the AR(1) as an MA(∞) process. We can easily check that in the AR(1) case the condition for stationarity is $|\phi| < 1$, which implies that the root of the characteristic equation $\Phi(z) = 1 - \phi z = 0$, associated to the lag polynomial $\Phi(L)$, must be greater than 1.

In the general case of p order the stationarity of the process depends on the roots of the polynomial $\Phi(L)$. Indeed, the AR process is stationary if and only if all the solutions of $\Phi(L) = 0$, in general complex, are outside the unit circle in the complex plane (Box *et al.*, 1994). $\Phi(L)$ is the lag polynomial of the process, and $\Phi(z) = 1 - \phi_1 z - \phi_2 z^2 - ....\phi_p z^p = 0$ is the characteristic equation.

6.3.1 Low-order autoregressive processes

We will analyse the main properties of AR(p) process in the $p = 1$ and $p = 2$ cases.

If $p = 1$ the process is autoregressive of order 1, AR(1), i.e.:

$$X_t = \phi X_{t-1} + \epsilon_t \tag{6.11}$$

In this case, the lag polynomial is simply $\Phi(L) = 1 - \phi L$ and, explicitly:

$$(1 - \phi L)X_t = \epsilon_t \tag{6.12}$$

The AR(1) process is stationary if $|\phi| < 1$. This condition implies, as discussed above, that the absolute value of the root of the characteristic equation must be greater than 1.

Proof Given (6.11), the characteristic equation is simply:[2]

$$1 - \phi z = 0$$

whose root is $z_1 = 1/\phi$. $|z_1| > 1$ only if $|\phi| < 1$.

$\square$

In the above, the (wide-sense) stationary AR(1) process was supposed to have zero mean or the X_t are differences with respect to the mean value. The expected value μ_X and the variance σ_X^2 are readily computed:

---

[2]We use z instead of L to avoid confusion between the operator and the polynomial variable

$$\mu_X \equiv E[X_t] = \phi E[X_{t-1}] + E[\epsilon_t] = 0$$

$$\sigma_X^2 \equiv var[X_t] = E[\phi^2 X_{t-1}^2 + \epsilon_t^2 + 2\phi X_{t-1}\epsilon_t] = \phi^2 E[X_{t-1}^2] + \sigma^2$$

therefore:

$$\sigma_X^2 = \frac{\sigma^2}{1 - \phi^2}$$

In a similar way, it is rather simple to compute the k-th autocovariance γ_k:

$$\gamma_k = E[X_t, X_{t-k}] = \phi^k \frac{\sigma^2}{1 - \phi^2}$$

If $p = 2$, we have an AR(2) process:

$$X_t = \phi_1 X_{t-1} + \phi_2 X_{t-2} + \epsilon_t \tag{6.13}$$

In this case the lag polynomial is $\Phi(L) = 1 - \phi_1 L - \phi_2 L^2$. The characteristic equation can be written in terms of its two (possibly complex or coincident) roots $z_1 = \lambda_1^{-1}$ and $z_2 = \lambda_2^{-1}$:

$$(1 - \lambda_1 z)(1 - \lambda_2 z) = 0 \tag{6.14}$$

The AR(2) process is stationary if both roots are outside the unit circle: $|z_1| > 1$ and $|z_2| > 1$, which is true if and only if the parameters ϕ_1 and ϕ_2 satisfy the following conditions:

$$\phi_1 + \phi_2 < 1 \qquad \phi_2 - \phi_1 < 1 \qquad |\phi_2| < 1 \tag{6.15}$$

Expected value μ_X and variance σ_X^2 of AR(2) process are listed below. The demonstration is left to the reader as an exercise.

$$\mu_X \equiv E[X_t] = 0$$

$$\sigma_X^2 \equiv E[X_t^2] = \frac{(1 - \phi_2)\sigma^2}{(1 + \phi_2)(1 - \phi_1 - \phi_2)(1 + \phi_1 - \phi_2)}$$

The autocovariance γ_k can be computed by means a recursive relationship:

$$\gamma_k \equiv E[X_t X_{t-k}] = \phi_1 \gamma_{k-1} + \phi_2 \gamma_{k-2}$$

which, knowing $\gamma_0 = \sigma^2$ allows to compute any order.

Let us look at a couple of simple R examples, to gain familiarity with the above concepts. The first shows the dependence of an AR(1) process (6.11) on the parameter ϕ. The following code shows how a time series with $\phi = 0.3$ (stationary) and one with $\phi = 1.01$ (not stationary) will differ. Here, the white noise is Gaussian with unitary variance.

```
## Code_6_3.R
#Autoregressive model (1), stationary and not stationary
# Set seed for random number generation, for reproducibility of results
set.seed(45)
# Common parameters
N <- 101
t <- 0:(N-1)
# white (gaussian) noise with variance = 1
epsilon <- rnorm(N)

# AR(1) stationary
phi <- 0.3
Xs <- rep(0,N)
for (i in 2:N) {
Xs[i] <- phi*Xs[i-1] + epsilon[i]
}

# AR(1) not stationary
phi <- 1.01
X <- rep(0,N)
for (i in 2:N) {
X[i] <- phi*X[i-1] + epsilon[i]
}

# Comparison
plot(t,X,t="l")
lines(t,Xs,col="red")
```

Figure 6.8 compares the time series with $\phi = 0.3$ (dashed line) and $\phi = 1.01$ (solid line).

The second example shows, through computation in R, how the parameters of a stationary AR(2) process are related to the roots of the characteristic equation. Suppose that $\phi_1 = 0.5$ and $\phi_2 = 0.2$ in (6.13), which fulfils the stationarity conditions (6.15). The following code computes such a time series.

```
## Code_6_4.R
#Autoregressive model (2) parameters
# AR(2)
phi1 <- 0.5
phi2 <- 0.2
X <- rep(0,N)
for (i in 3:N) {
X[i] <- phi1*X[i-1] + phi2*X[i-2] + epsilon[i]
}
plot(t,X,t="l")
```

The series is stationary (evident also by eye), as the computation of (6.15) confirms. But, what about the roots of the lag polynomial?

In the present case, the second-order characteristic equation $1 - 0.5z - 0.2z^2 = 0$ is (easily) analytically solvable, giving the following roots z_1 and z_2:

$$z_1 = \frac{-2.5 + \sqrt{26.25}}{2} = 1.31$$

$$z_2 = \frac{-2.5 - \sqrt{26.25}}{2} = -3.81$$

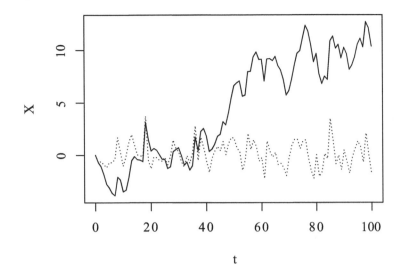

Fig. 6.8 Comparison of AR(1) time series with $\phi = 0.3$ (dashed line) and $\phi = 1.01$ (solid line).

Therefore, as it should be, both $|z_1|$ and $|z_1|$ are greater than one. Figure 6.9 shows the stationarity of the time series.

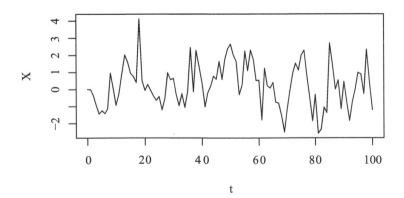

Fig. 6.9 AR(2) time series with $\phi_1 = 0.5$ and $\phi_2 = 0.2$

6.3.2   Autocorrelation structure and model analysis

When faced with a time series, suppose a stationary one, we might be interested in determining which model best represents the data. Is the process MA(q)? Or is it better described by an AR(p)? With which order? A first useful step in this direction is the analysis of the autocorrelation and partial autocorrelation at different time lags. The latter, a_k, is defined in terms of the autocorrelation between two times t and $t-k$, conditional on all other times:

$$a_k \equiv \text{Corr}(X_t, X_{t-k}|X_{t-1}...X_{t-k+1})$$

with this definition, $a_1 = \text{Corr}(X_t, X_{t-1}) = \rho(1)$

Intuitively, while the autocorrelation among X_t and X_{t-k}, i.e. the correlation of the random variable X at time t with the same variable at a time distance k, depends on X at all other intermediate times $t-1, t-2, ... t-k+1$, the partial autocorrelation only depends on the "direct" effect of time $t - k$ on time t.

The meaning of partial autocorrelation can be better understood considering the partial correlation among three correlated random variables, say X, Y and Z. The partial correlation between X and Y given Z is obtained regressing X on Z (obtaining $\hat{X}$), regressing Y on Z (obtaining $\hat{Y}$) and eventually computing the correlation among $X - \hat{X}$ and $Y - \hat{Y}$:

$$\text{Corr}(X, Y|Z) = \text{Corr}(X - \hat{X}, Y - \hat{Y})$$

In other words, partial correlation is the correlation between the prediction errors of the regressions of X and Y on Z. Now, if X, Y, Z are respectively substituted by X_t, X_{t-k} and $X_{t-1} \ldots X_{t-k+1}$, we obtain an operative definition of the partial autocorrelation for any k is readily:

$$a_k = \text{Corr}(X_t - \hat{X}_t, X_{t-k} - \hat{X}_{t-k})$$

i.e. a_k is the autocorrelation between X_t and X_{t-k} with the linear dependence on $X_{t-1} \ldots X_{t-k+1}$ removed. In the above formula, $\hat{X}_t$ is the linear regression estimate of X_t over $X_{t-1} \ldots X_{t-k+1}$, and $\hat{X}_{t-k}$ is the best prediction at time $t - k$ based on the same $X_{t-1} \ldots X_{t-k+1}$. To be explicit, thanks to the assumed stationarity, if $\hat{X}_t = \alpha_1 X_{t-1} + \alpha_2 X_{t-2} + \ldots + \alpha_{k-1} X_{t-k+1}$, reverting the time direction: $\hat{X}_{t-k} = \alpha_1 X_{t-k+1} + \alpha_2 X_{t-k+2} + \ldots + \alpha_{k-1} X_{t-1}$.

Remark 6.2 The particular case of AR(1) can help to clarify the issue. At any time t the regression of X_{t+2} on X_{t+1} is:

$$\hat{X}_{t+2} = \beta X_{t+1}$$

with no constant term, because $E[X_t] = 0$. We choose β so as to minimize the variance of $X_{t+2} - \hat{X}_{t+2}$:

$$E[X_{t+2} - \beta X_{t+1}]^2 = \gamma(0) - 2\beta\gamma(1) - \beta^2\gamma(0)$$

Differencing with respect to β and equating to zero, we obtain:

$$\beta = \frac{\gamma(1)}{\gamma(0)} \equiv \rho(1) = \phi$$

We obtain the same result for the regression of X_t on X_{t+1}, hence:

$$Corr(X_{t+2} - \hat{X}_{t+2}, X_t - \hat{X}_t) = Corr(X_{t+2} - \phi X_{t+1}, X_t - \phi X_{t+1}) = Corr(\epsilon_{t+2}, X_t - \phi X_{t+1})$$

the last is equal to zero because $X_t - \phi X_{t+1}$ is, by definition, not correlated with ϵ_{t+2}. The same is true for any k, so we conclude that for an AR(1) process the partial autocorrelations are $a_1 = \phi$ and $a_k = 0$ for any $k > 1$.

The computation of partial autocorrelations for an AR(p) process is similar to that developed above but, of course, a bit more complex. Referring to classical texts on time series analysis, e.g. (Box *et al.*, 1994), the partial autocorrelations a_k are all zeros for $k > p$, while they can be non-zero for $k \leq p$. Remember that, instead, the autocorrelations are exponentially decreasing with k: $\rho(k) = \phi^k$. The duality between AR and MA processes suggests that, because a finite-order stationary AR corresponds to an infinite-order invertible MA process, and a finite-order invertible MA process can be expressed as an infinite-order AR, an invertible MA(q) process has exponentially decreasing partial autocorrelation, while we have seen (6.8) that its autocorrelations are zero for lags $k > q$.

In conclusion, the autocorrelation function (ACF) and the partial autocorrelation function (PACF) play dual roles in AR and MA processes. The analysis of the PACF of an AR(p) process helps in determining the order p, as well as the ACF of an MA(q) allows to estimate q. Furthermore, looking at the autocorrelation and partial autocorrelation structure of a process can shed light on its nature. Simplifying things (a process can contain both an AR and an MA structure, as we will see in Section 6.4), if the ACF dies out slowly and PACF is non zero only for a few lags, the process is probably described by an MA model, while in the opposite case the model is probably an AR. A few lines of code follow to show the different autocorrelation structure of AR(2) and MA(2) example processes.

```
## Code_6_5.R
#Autocorrelation structure of AR(2) and MA(2)
# stationary AR(2)
set.seed(4321)
x.AR2 <- arima.sim(model = list(order=c(2,0,0), ar = c(0.5,-0.7)),
n = 200)
plot(x.AR2,main="")
acf(x.AR2,main="")
pacf(x.AR2,main="")

# invertible MA(2)
set.seed(4321)
x.MA2 <- arima.sim(model = list(order=c(0,0,2), ma = c(0.2,-0.5)),
n = 200)
plot(x.MA2,main="")
acf(x.MA2,main="")
pacf(x.MA2,main="")
```

Figure 6.10 compares the plot (a) of a time series generated from an AR(2) process with coefficients 0.5 and -0.7, therefore stationary, with an invertible MA(2) time series (b) with coefficients 0.2 and −0.5. Plots (d) and (e), respectively showing the ACF of the MA process and the PACF of the AR process, inform us about the order of the two processes. Plots (c) and (f) show an oscillatory exponential decay, confirming that generated data are AR and MA, respectively.

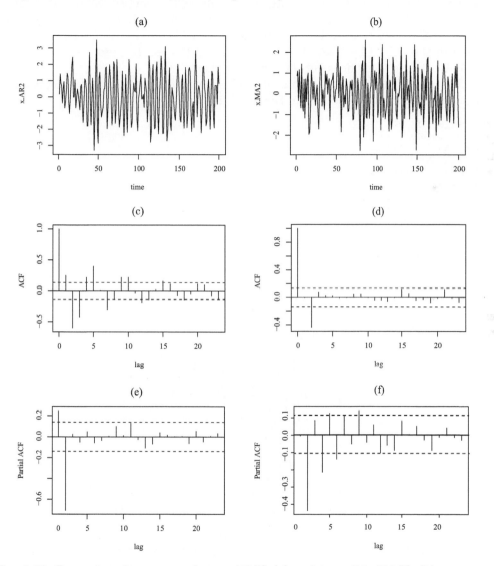

Fig. 6.10 Comparison between stationary AR(2) (a) and invertible MA(2) (b) processes; ACF of AR(2) (c) and MA(2) (d); PACF of AR(2) (e) and MA(2) (f)

6.4 Autoregressive moving-average processes (ARMA)

The autocorrelation structures we have learned in the previous sections are of two types: (1) slowly decaying ACF and sharp cut-off of PACF (autoregressive process), (2) slowly decaying PACF and sharp cut-off of ACF (moving-average process). Of course, not all stationary processes can be represented by AR(p) or MA(q) models.

For example, a time series like that shown in Fig. 6.11 has the autocorrelation structure shown in Fig. 6.12, clearly pointing out that such a process is neither AR nor MA because no evident cut-off is present there.

The data in Fig. 6.11 has been generated by the following chunk of code, using the function **arima.sim** introduced in Section 6.2.

```
## Code_6_6.R
# Mixed models
# for reproducibility
set.seed(5678)
# ARMA(2,2) process with
# AR parameters: phi_1 = 0.5, phi_2 = -0.7
# MA parameters: theta_1 = 0.2, theta_2 = -0.5
x.ARMA <- arima.sim(model = list(order=c(2,0,2), ar = c(0.5,-0.7),
ma = c(0.2,-0.5)), n = 200)
plot(x.ARMA,main="",ylab="time series")
acf(x.ARMA,main="")
pacf(x.ARMA,main="")
```

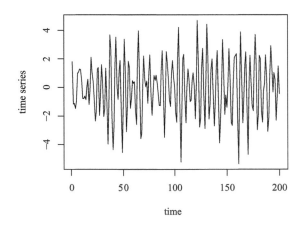

Fig. 6.11 Time series: ARMA(2,2) process.

A generalization of pure autoregressive or pure moving-average processes is a combination of them. Writing AR(p) and MA(q) in terms of the relevant lag polynomials $\Phi(L)$ and $\Theta(L)$, and solving for $X_t - \epsilon_t$ we obtain:

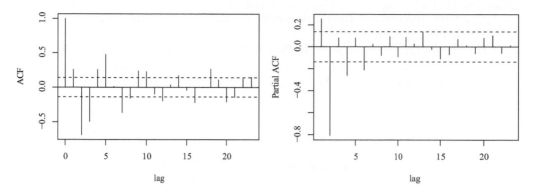

Fig. 6.12 Autocorrelation structure of the time series of Fig. 6.11.

$$\Phi(L)X_t = \Theta(L)\epsilon_t \qquad (6.16)$$

Such a process is named auto-regressive-moving-average (ARMA). Explicitly, an ARMA(p, q) process is:

$$X_t = \phi_1 X_{t-1} + ... + \phi_p X_{t-p} + \epsilon_t - \theta_1 \epsilon_{t-1} - ... - \theta_q \epsilon_{t-q} \qquad (6.17)$$

where the choice of the sign of θ_k is of course arbitrary.

From the previous discussion it should not be a surprise that a stationary and invertible ARMA process has both the infinite moving-average representation and the infinite autoregressive one. We will give the main properties of ARMA models without mathematical proof (the interested reader can refer to books like (Box *et al.*, 1994)).

Essentially, the properties of ARMA(p, q) models derive from those of the AR(p) and MA(q) of which ARMA is a mixture. In particular:

- The condition of stationarity, i.e. of stability, is that of the AR(p) process. Given the characteristic equation associated with the lag polynomial

$$1 - \phi_1 z - \phi_2 z^2 - ... - \phi_p z^p = 0$$

the process is stationary if all roots lie outside the circle $|z| = 1$, i.e. when all roots exceed unity in absolute value.
- The condition of invertibility is 'inherited' from the MA(q) process. Given the characteristic equation associated with the lag polynomial:

$$1 + \theta_1 z + \theta_2 z^2 + ... + \theta_q z^q = 0$$

the process is stationary if all the roots lie outside the circle $|z| = 1$.
- An ARMA(p, q) model possessing both the above properties is called *causal*.
- The mean of an ARMA process also derives from those of the constituting AR and MA. Having subtracted the possible non-zero mean μ: $E[X_t] = 0$.
- The ACF of an ARMA(p, q) process dies out as that of its AR(p) component.
- The PACF of an ARMA(p, q) process dies out as that of its MA(q) component.

Note that, from (6.16), the properties of an ARMA process are related to the roots of the ratio $\Theta(z)/\Phi(z)$.

The computation of autocorrelations and partial autocorrelations for a generic-order ARMA model is rather cumbersome. The R functions **acf** and **pacf** do the job for us. In the first-order case, ARMA(1,1) the autocovariances are given by:

$$
\begin{aligned}
\gamma_0 &\equiv \text{Var}[X_t] = \frac{1 + 2\phi\theta + \theta^2}{1 - \phi^2}\sigma^2 \\
\gamma_1 &\equiv \text{Cov}[X_t, X_{t-1}] = \frac{(\phi + \theta)(1 + \phi\theta)}{1 - \phi^2}\sigma^2 \\
\gamma_k &\equiv \text{Cov}[X_t, X_{t-k}] = \phi\gamma_{k-1}, \text{ for } k > 1
\end{aligned}
\tag{6.18}
$$

where σ^2 is the white noise variance.

A few words are in order about *redundancy*. If you try to fit data coming from an ARMA(p, q) model, e.g. those in Fig. 6.11, by means of an ARMA(P,Q) model, with $P \geq p$ and $Q \geq q$ you can always find several couples (P,Q) fitting well the data. A nice example from (Shumway and Stoffer, 2006) clears the concept of parameter redundancy.

Suppose you have a purely white noise process, i.e. $X_t = \epsilon_t$. Multiplying both members at a previous time by the same coefficient (e.g. 0.5) and subtracting them from the left- and right-hand side we obtain a 'new' process:

$$
X_t - 0.5X_{t-1} = \epsilon_t - 0.5\epsilon_{t-1}
$$

which can be re-written:

$$
X_t = 0.5X_{t-1} + \epsilon_t - 0.5\epsilon_{t-1}
$$

Now, X_t is still white noise, but it appears as an ARMA(1,1) process. Writing it in terms of the lag polynomials:

$$
(1 - 0.5L)X_t = (1 - 0.5L)\epsilon_t
$$

i.e. we observe that the ratio between the lag polynomials in (6.16) is 1 or, said differently, $\Theta(z)$ and $\Phi(z)$ have a common factor $1 - 0.5z$. The common factor identifies the redundancy of the parameters: once removed, both lag polynomials reduce to $\Theta(z) = 1$ and $\Phi(z) = 1$. We conclude that, when trying to fit an ARMA(p, q) model to data, the common roots of the AR(p) and MA(q) polynomials constituting it must be ruled out, decreasing the order or both the AR and MA components of the ARMA model.

The R package **stats** includes a function for fitting an ARMA model to an univariate time series: **arima**. As we will briefly introduce in the next section, that function actually deals with 'integrated' ARMA models, or ARIMA, but it can also be used for ARMA by taking as zero the *integration* order. As a conclusive example of this section we will show how to fit an ARMA(p, q) model to data, taking as an example the ARMA(2,2) data of Fig. 6.11, having parameters $\phi_1 = 0.5, \phi_2 = -0.7, \theta_1 = 0.2, \theta_2 = -0.5$. The following code tests an ARMA(2,2) model with the data.

```
fit <- arima(x.ARMA,order = c(2,0,2))
fit
```

The output shows that the fitted parameters are rather close to the actual ones.

```
Call:
arima(x = x.ARMA, order = c(2, 0, 2))

Coefficients:
          ar1      ar2     ma1      ma2   intercept
       0.4548  -0.6676  0.3076  -0.5485    -0.0380
s.e.   0.0625   0.0557  0.0732   0.0736     0.0433

sigma^2 estimated as 0.9385:  log likelihood = -279.38,  aic = 570.76
```

Determining the 'true' values of the order p and q is not a straightforward task. One possibility is that of fitting ARMA models of increasing AR and MA orders, and to look at some 'performance' parameter of the fitting. The most intuitive approach could be to choose the fitting giving the maximum likelihood value, i.e. that minimizing the residual error. But the error has been proved to decrease with increasing order of the ARMA model. A solution to that problem has been proposed by Akaike (Akaike, 1974) whose idea was to penalize the error variance by a term proportional to the number of parameters. The so-called AIC (Akaike's information criterion) assumes that the best model is that having an order $k = p + q$ such as to minimize:

$$AIC = 2(k + 1) - 2\log(L) \tag{6.19}$$

where L is the likelihood of the model, and $k + 1$ is the number of parameters in the model (AR and MA orders, plus the residual variance). Just to show an application of the AIC, we can try to compute a sequence of ARMA(p, q) fittings with p and q both in the range $[1, 4]$. The following listing shows how to do it in R.

```
aic.fit <- c()
for (p in 1:4){
  for (q in 1:4) {
    fit <- arima(x.ARMA,order = c(p,0,q))
    aic.fit <- c(aic.fit,fit$aic)
  }
}
AIC.fit <- matrix(data=aic.fit,nrow=4,ncol=4)
AIC.fit
```

The result fit of an **arima** fitting is a list, one of whose components is the computed AIC. The code above collects the AIC values in a list and then converts it to a $p \times q$ matrix. The result:

```
          [,1]      [,2]      [,3]      [,4]
[1,] 700.7636 600.6213 578.7266 575.0947
[2,] 658.2359 570.7618 572.6152 574.3528
[3,] 613.6939 572.6444 574.5500 576.3364
[4,] 605.5990 574.2344 576.1547 574.4908
```

indicates that the minimum AIC value is 570.76, and that it occurs for $p = 2$ and $q = 2$ as it does.

6.5 An introduction to non stationary and seasonal time series

All the processes considered so far were stationary. Consider random walk (see Chapter 5 for a description of random walk), which is an AR(1) process with unitary parameter, thus not fulfilling the stationarity requirements:

$$X_t = X_{t-1} + \epsilon_t \tag{6.20}$$

By differencing (6.20) we obtain a stationary time series:

$$\nabla X_t \equiv X_t - X_{t-1} = \epsilon_t \tag{6.21}$$

i.e. a stationary white noise. The following code generates a random walk X_t and computes its first difference ∇X_t.

```
## Code_6_7.R
# Random walk as autoregressive process
# for repeatability
set.seed(123456)
# gaussian white noise
w <- rnorm(200)
x <- rep(0,200)
# Random walk
x[1] <- w[1]
for (t in 2:200) x[t] <- w[t] + x[t-1]

# Plot time series
plot(as.ts(x),ylab="Random walk")
# Compute and plot the ACF
acf(x,main="")
# Compute and plot the spectrum
x.spec <- spectrum(x,log="no",span=30,plot=FALSE)
plot(x.spec$freq,x.spec$spec,t="l",
  xlab="frequency (cycles/sample interval)", ylab="spectral density")

# difference
d <- rep(0,199)
d[1] <- x[2]-x[1]
for (t in 3:200) d[t-1] <- x[t] - x[t-1]

# Plot time series
plot(as.ts(d),ylab="Difference(Random walk)")
# Compute and plot the spectrum
d.spec <- spectrum(d,log="no",span=30,plot=FALSE)
plot(d.spec$freq,d.spec$spec,t="l",
  xlab="frequency (cycles/sample interval)", ylab="spectral density",
  ylim = c(0,1.2))
acf(d,main="")
```

The results of the above computation are shown in Fig. 6.13. Figures (a) and (b) compare the random walk X_t to ∇X_t. Figures (c) and (d) compare the relative ACFs and, eventually, figures (d) and (e) compare the spectra, clearly showing that ∇X_t is white noise.

The ACF behaviour of Fig. 6.13 (c), very slowly going to zero, is typical of non-stationary processes. Actually, the analysis of the autocorrelation structure of a time series is very helpful in determining if it is stationary or not (of course, specific statistical tests also exist for detecting non-stationarity).

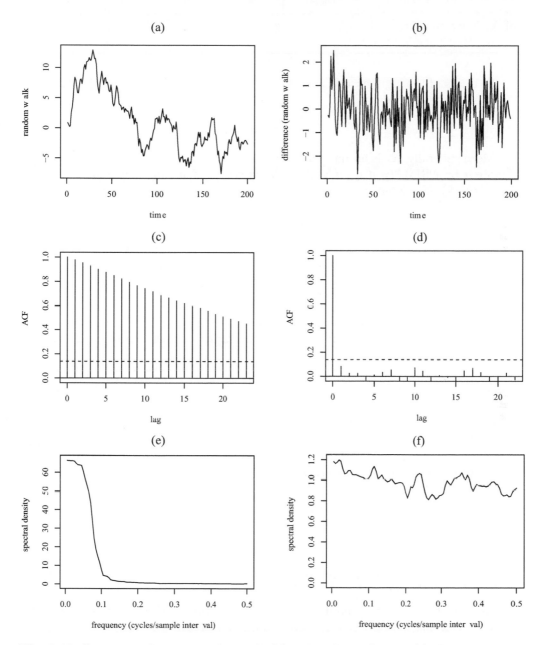

Fig. 6.13 Comparison between random walk (a) and its first difference (b); ACF of random walk (c) and first difference (d); spectrum of random walk (e) and first difference (f).

6.5.1 Integrated ARMA models

The difference operator introduced in eqn 6.21 can be recursively applied to any time series. Given a time series X_t, we say it is 'integrated of order d' if the d-th difference of X_t is white noise:

$$\nabla^d X_t = \epsilon_t$$

In terms of the lag operator L we have: $\nabla \equiv 1 - L$ and $\nabla^d \equiv (1 - L)^d$. Therefore, X_t is integrated of order d if:

$$(1 - L)^d X_t = \epsilon_t$$

More generally, X_t is an integrated autoregressive moving-average process, and we denote it by ARIMA(p, d, q), if $\nabla^d X_t$ is ARMA(p, q). Using lag polynomials, if we set: $Y_t = (1 - L)^d X_t$ we have:

$$\Phi_p(L)Y_t = \Theta_q \epsilon_t$$

where we have explicitly declared the order of the AR and MA polynomials. Or, which is the same, the lag-polynomial representation of an ARIMA(p, d, q) model is:

$$\Phi_p(L)(1 - L)^d X_t = \Theta_q(L)\epsilon_t$$

The following simple example shows how the difference operator, simply implemented in R by means of the **diff** function, transforms a non stationary ARIMA(2,1,1) series into a stationary ARMA(2,1).

```
## Code_6_8.R
# Difference operator
# for repeatability
set.seed(1234)
# generate data from an ARIMA(2,1,1) model
# AR parameters: 0.5, -0.7
# MA parameter: 0.6
x.ARIMA <- arima.sim(model = list(order=c(2,1,1), ar = c(0.5,-0.7), ma = 0.6), n = 200)
plot(x.ARIMA,main="",ylab="time series")
acf(x.ARIMA,main="")
```

The non-stationary character is evident from the plot, Fig. 6.14 (left), and confirmed from the behaviour of the ACF (right). Applying the difference operator we obtain the time series shown in Fig. 6.15.

```
x.ARMA = diff(x.ARIMA)
plot(x.ARMA,main="",ylab="time series")
acf(x.ARMA,main="")
```

In order to verify that the difference series is actually ARMA(2,1) we proceed as before, by trying to apply ARMA(p, q) models of increasing orders.

```
aic.fit <- c()
for (p in 1:3){
  for (q in 1:3) {
    fit <- arima(x.ARMA,order = c(p,0,q))
    aic.fit <- c(aic.fit,fit$aic)
  }
}
AIC.fit <- matrix(data=aic.fit,nrow=3,ncol=3)
AIC.fit
```

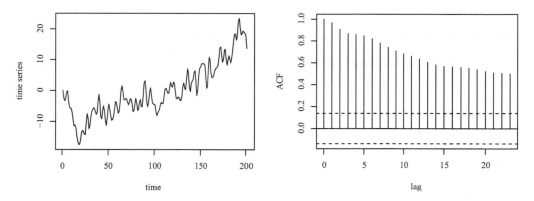

Fig. 6.14 Plot of an ARIMA(2,1,1) time series (left), and of its ACF (right).

The matrix *AIC.fit* actually shows that the (2,1) case is that with the lowest AIC.

```
          [,1]      [,2]      [,3]
[1,] 681.1871 590.3703 592.3307
[2,] 686.1645 592.3207 594.2920
[3,] 629.1372 594.1055 595.9257
```

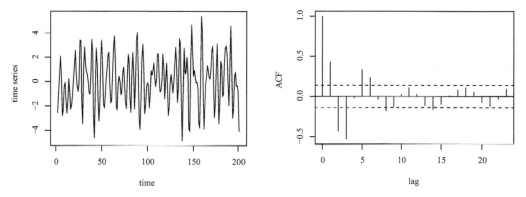

Fig. 6.15 Plot of the ARMA(2,1,1) time series (left) obtained from ARIMA(2,1,1), and of its ACF (right).

The **arima** function fits an ARIMA(p, d, q) model to the data. The resulting AR and MA parameters, and the error variance, are rather close to those used for the simulation.

```
arima(x.ARMA,order=c(2,0,1))

# RESULT:
Call:
arima(x = x.ARMA, order = c(2, 0, 1))

Coefficients:
         ar1      ar2     ma1  intercept
      0.5653  -0.6772  0.6669     0.0627
```

s.e.   0.0585   0.0565   0.0622   0.1088

sigma^2 estimated as 1.05:   log likelihood = -290.19,   aic = 590.37

6.5.2   Seasonal ARIMA models

We conclude this chapter with a brief introduction to seasonal ARIMA (or SARIMA) models.

A seasonal model is something of a generalization of the models we encountered in the preceding sections, where in a time series $\{X_t\}$ the random variable X_t was related to X_s and/or to white noise ϵ_s at previous times $s = t - 1, t - 2, ...$ For example, in an MA(1) process, as discussed in Section 6.2:

$$X_t = \epsilon_t + \theta\epsilon_{t-1}$$

which has the property that the autocorrelation ρ_k is zero for $k > 1$. It could happen that the non-zero autocorrelation occurs between X_t and X_{t-k} with $k \neq 1$. That is the case of a seasonal MA(1) where, for example, we observe in the data a 'periodicity' of seven time intervals (e.g. if the time scale is the day, and data has a weekly periodicity):

$$X_t = \epsilon_t + \theta\epsilon_{t-7}$$

The same can be done for an AR(1) process. The seasonality can be extended to any order MA(q) processes, AR(p), ARMA(p, q) and also to non-stationary ARIMA(p, d, q) processes. Using the lag operator notation (we substitute ϕ_k with ψ_k and θ_k with ω_k to avoid confusion between non-seasonal and seasonal lag polynomials), we define:

Definition 6.1 *Seasonal MA(q) with period s*

$$X_t = (1 + \omega_1 L^s + \omega_2 L^{2s} + ... + \omega_q L^{qs})\epsilon_t$$

Definition 6.2 *Seasonal AR(p) with period s*

$$(1 - \psi_1 L^s - \psi_2 L^{2s} - ... - \psi_p L^{ps})X_t = \epsilon_t$$

With these definitions, a seasonal ARMA(P,Q) model with period s is defined in terms of lag polynomials in L^s instead of L. Generalizing (6.16) we have:

$$\Psi_P(L^s)X_t = \Omega_Q(L^s)\epsilon_t \tag{6.22}$$

where the subscripts explicitly declare the order of the lag polynomials.

Finally, we observe that 'conventional' ARMA models contain autocorrelation at time lags $1, 2, ...k$, while seasonal ARMA models incorporate autocorrelations only at seasonal lags $s, 2s, ...ks$. A more general model should include both correlations, taking into account that a seasonal effect can be superimposed on another autocorrelation effect acting on a lower time lag. One of the most used seasonal models doing that is the *multiplicative* seasonal ARMA (Box, Jenkins and Reinsel, 1994), which can be concisely defined by:

$$\Psi_P(L^s)\Phi_p(L)X_t = \Omega_Q(L^s)\Theta_q(L)\epsilon_t \qquad (6.23)$$

where all orders have been explicitly given[3].

An example should help to better understand what (6.23) means. Suppose you have monthly data showing an annual seasonality and that would be AR(1) if the seasonality could be removed. Such data could be represented, for example, by:

$$(1 - \phi L)X_t = (1 - \omega L^{12})\epsilon_t$$

i.e. by a multiplicative AR(1) × seasonal MA(1)$_{12}$ model, or: ARMA(1,0) × ARMA(0,1)$_{12}$. In the notation of (6.23):

$$\Psi_0(L^{12})\Phi_1(L)X_t = \Omega_1(L^{12})\Theta_0(L)\epsilon_t$$

with:

$$\Psi_0(L) = 1$$
$$\Phi_1(L) = 1 - \phi L$$
$$\Omega_1(L^{12}) = 1 + \omega L^{12}$$
$$\Theta_0(L) = 1$$

A last consideration about the power of the lag operator formalism: suppose the above mixed model consisting of a non-seasonal AR(1) and a seasonal MA(1)$_{12}$ also has a seasonal AR(1)$_{12}$, i.e. it is represented by the equation:

$$(1 - \phi L)(1 - \psi L^{12})X_t = (1 + \omega L^{12})\epsilon_t$$

Expanding the above we obtain:

$$(1 - \phi L - \psi L^{12} + \phi\psi L^{13})X_t = (1 + \omega L^{12})\epsilon_t$$

or, solving for X_t:

$$X_t = \phi X_{t-1} + \psi X_{t-12} - \phi\psi X_{t-13} + \epsilon_t + \omega\epsilon_{t-12}$$

which states the dependence of X_t on the variable at lags 1, 12 and 12 + 1, and on white noise at the current time and at lag 12. The multiplicative nature of the mixed model manifests in the presence of the product $\phi\psi$ as a coefficient for X_{t-13}, instead of having a free parameter there.

---

[3]If X_t is replaced by $Y_t = (1 - L)^d(1 - L^s)^D$, with d and D respectively integration orders of non-seasonal and seasonal components, (6.23) transforms into a multiplicative $ARIMA(p, d, q) \times (P, D, Q)_s$

Remark 6.3 The multiplicative form (6.23) is actually a consequence of the conceptual construction of the seasonal ARMA model. For example, if we are analysing monthly observations of some meteorological quantity (temperature, pressure, wind speed, precipitation amount) we often find that a relation exists between observations for the same month in different years, i.e. 12 months apart. Therefore, observations of X_t one year apart can be related by a seasonal ARMA model like (6.22):

$$\Psi_P(L^{12})X_t = \Omega_Q(L^{12})\alpha_t \tag{6.24}$$

where, this time, α_t is not white noise, but a residual 'error' component that is not explained by the model. The errors α_t will usually be correlated, as a consequence of a correlation between observations in successive months. A non-seasonal ARMA model can be suitable to explain that relationship:

$$\Phi(L)\alpha_t = \Theta(L)\epsilon_t \tag{6.25}$$

with ϵ_t white noise. By 'multiplying' the left- and right-hand sides of (6.24) by $\Phi(L)$ and using (6.25), we obtain (6.23) with s=12.

6.5.3   An example

We will apply a SARIMA model to the analysis of the maximum temperatures registered in Bologna, Italy, from 1961 to 1991. Figure 6.16 (left) clearly suggests the presence of a seasonal behaviour, as confirmed by the ACF of the time series, shown in Fig. 6.16 (right).

The R code for obtaining Fig. 6.16 and 6.17 is the following.

```
## Code_6_9.R
# Seasonal ARIMA
setwd("C:/RPA/code/ARMA")
# Maximum temperature in Bologna (Period: 1961-1991)
Tmax.data <- read.table("data/Bologna_Tmax_1961_1991.txt",h=FALSE)
Tmax <- Tmax.data$V1
# Compute lag-12 differences
Tmax.diff12 <- diff(Tmax,12)
# Plots
ts.plot(Tmax,xlab="Months from January 1961")
ts.plot(Tmax[1:60],xlab="Years 1961-1991",ylab="Tmax")
# ACF
acf(Tmax,lag.max = 36,main="")
acf(Tmax.diff12,lag.max = 36,main="")
pacf(Tmax.diff12,lag.max = 36,main="")
```

The correlation structure shown in Fig. 6.16 (right) clearly suggests a seasonal ARIMA model, which means that data must be seasonally-differenced, i.e. transformed by taking differences at lag 12: $\nabla_{12}X_t \equiv (1 - L^{12})X_t$.

Figure 6.17 shows the ACF and PACF of the lag-12 differences. Note that the lagged-difference ∇_s can be combined with the difference operator ∇^D with integration order D defined in 6.5.1, to give:

$$\nabla_s^D \equiv (1 - L^s)(1 - L)^D$$

which, when applied to a time series X_t gives $(1 - L)^D(X_t - X_{t-s})$.

The ACF of $\nabla_{12}X_t$ has a large autocorrelation at lag 12, suggesting the presence of a moving-average MA(1) term in the seasonal component, while the shape of the PACF

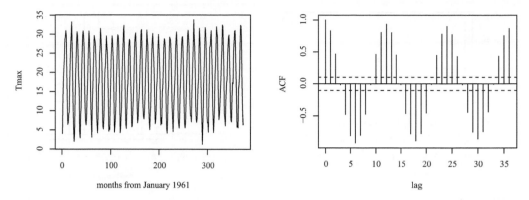

Fig. 6.16 Temperature in Bologna from 1961 to 1991 (left) and ACF of the time series (right).

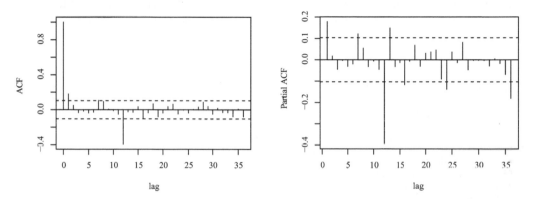

Fig. 6.17 ACF (left) and PACF (right) of the yearly differences of the time series of figure 6.16.

suggests the presence of both AR and MA terms. By a procedure like that implemented in Subsection 6.5.1, it is simple to find the set of numbers (p, q, P, Q) giving rise to the minimum AIC value for a mixed (SARIMA) model $(p.0, q) \times (P, 1, Q)_{12}$.

The R code that follows computes the parameters of an ARIMA $(1, 0, 0) \times (1, 1, 1)_{12}$ model.

```
## Code_6_10.R
# best fit SARIMA orders
p = 1
q = 0
P = 1
Q = 1
# SARIMA fit
(arima(Tmax,order=c(p,0,q),list(order=c(P,1,Q),period=12)) -> fit)
acf(residuals(fit),48,main="")
hist(residuals(fit),main="")
```

The residuals of the SARIMA fitting appear to have the characteristics of white noise, as Fig. 6.18 shows, confirming the goodness of the model.

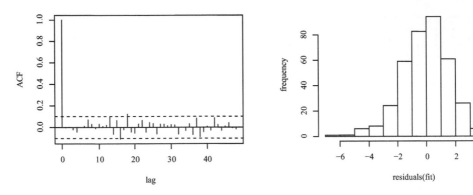

Fig. 6.18 ACF (left) and histogram (right) of the ARIMA $(1, 0, 0) \times (1, 1, 1)_{12}$ residuals.

The fitted model can be 'tested' by using the generic R function **predict**, which is an overloaded one, i.e. depending on the context. When the fitted model is ARIMA, R actually invokes **predict.Arima**. Suppose we are at the beginning of the year 1990, and the maximum temperatures from January 1961 to December 1989 are available to us. The following code fits a multiplicative ARIMA $(1, 0, 0) \times (1, 1, 1)_{12}$ to the above time series.

```
## Code_6_11.R
# convert Tmax into a time series
# X.true is the full time series
X.true <- ts(Tmax)
# X is the time series of the period 1961-1989
X <- ts(Tmax[1:(29*12)],start=1,end=(1989-1961+1)*12)
# seasonal ARIMA model
fit <- arima(X,order=c(1,0,0),list(order=c(1,1,1),period=12))
# make prediction for next 2 years
X.pred <- predict(fit, 24)
# Plot predicted values
plot(X[313:348],t="l",xlim=c(1,60),xlab="January 1987 to December 1991",ylab="Tmax")
lines(37:60,X.pred$pred,col="red")
lines(36:60,X.true[(313+35):372],lty=2)
```

Figure 6.19 shows the temperatures in the period 1987:1989 (solid line), and the known values between 1990 and 1991 (dashed line). The prediction based on the seasonal ARIMA model (red line) is rather good.

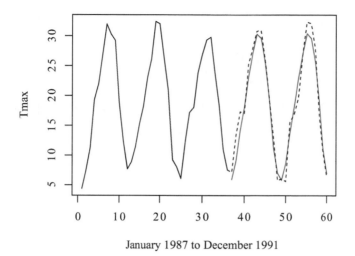

January 1987 to December 1991

Fig. 6.19 True data (solid line) vs fitting results (dashed line).

6.6 A physical application

Physical quantities related to non-stationary phenomena, when repeatedly measured over time, often appear as random. Of course, randomness usually is not a true physical characteristics but, rather, it reflects our incomplete knowledge of the phenomenon at hand. For example, referring to the same data to which Fig. 6.16 refers, the maximum daily temperatures in Bologna in December 2018 (see Fig. 6.20) do not appear to have any evident relation with the day number or, in other words, they look random.

River streamflow time series exhibit a stochastic nature too. The flow is certainly deterministic, but the multiplicity and complexity of factors influencing it - rainfall, evaporation, ice melting and human activities - are such as to make it a random quantity or better as employing random variables to describe it. As an example, Fig. 6.21 (left) plots the time series of the average monthly water flow of the Tiber river, in Italy, from January 1921 to December 1979 (data are extracted from the database Oak Ridge National Laboratory Distributed Active Archive Center (ORNL DAAC) for Biogeochemical Dynamics, NASA (Vorosmarty *et al.*, 1998)).

Figure 6.21 (right) plots a time series relative to the Dow Jones index in a 25-year period. Do you see any real difference among the properties of the two series? Both appear as random, in spite of their totally different origin. By looking at the two time series could you guess which one refers to a river flow and which to a stock exchange index?

Without entering into the realms of philosophy (what is actually random?), we have no doubt about the apparent randomness of time series like those in Fig. 6.21.

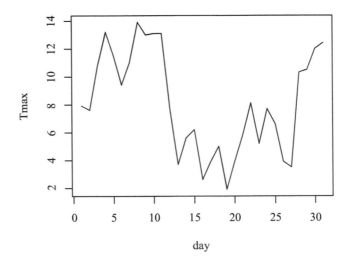

Fig. 6.20 Maximum daily temperatures measured in Bologna (Italy) in December 2018

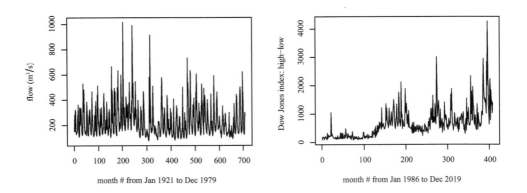

Fig. 6.21 Average monthly water flow of Tiber (Italy) (left), and High-Low Dow Jones index (right).

The question is: can we establish any kind of relationship among physical laws and the flow behaviour of Fig. 6.21(left) (or, in perfect analogy between economic laws and Fig. 6.21(right))?

Concerning the river flow, and its relation to rainfall, the answer is yes, actually that is the subject of stochastic hydrology (Hipel and McLeod, 1994). Indeed, stochastic hydrology, a statistical branch of hydrology dealing with probabilistic models of hydrological processes, has been shown to be physically based several years ago (Salas and Smith, 1981). What follows is known to be an over-simplification of a very complex problem, but it is nevertheless instructive.

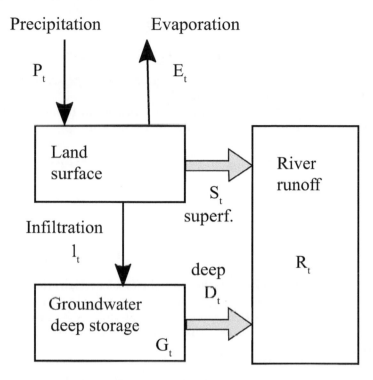

Fig. 6.22 Hydrological cycles in a watershed.

Figure 6.22 shows a simplified schematic of the hydrological cycle in a watershed. The precipitation P_t is the rainfall quantity affecting a river watershed during the year t. Rain hitting the land surface is partially absorbed, partially evaporates and a fraction of water flows into the river (this is called *runoff*). Let $E_t = e \, P_t$ represent the fraction of water that evaporates, with e being an evaporation coefficient. $I_t = iP_t$ represents the water amount that infiltrates through the soil.

The surface runoff S_t is given by $P_t - E_t - I_t$ or, in other words, it consists of the water amount coming from rain precipitation that is neither absorbed by the land, nor evaporated. We assume that $G_{t-1} = I_t$ represents the ground water storage at the beginning of year t. A fraction r of that storage contributes to the total river runoff in the year t, as a deep runoff term D_t. As a consequence of the above mechanisms, the total runoff R_t during year t is:

$$R_t = S_t + D_t = (1 - e - i)P_t + rG_{t-1} \tag{6.26}$$

The mass balance requires:

$$G_t = (1 - r)G_{t-1} + iP_t \tag{6.27}$$

which we can read as 'water stored at the end of year t is given by the initial storage at $t - 1$, minus the deep runoff rG_{t-1}, plus the i fraction of precipitation during year t.'

Combining eqns (6.26) and (6.27), we obtain:

$$R_t = (1 - r)R_{t-1} + sP_t - (s(1 - r) - ir) \tag{6.28}$$

where $s = 1 - e - i$ is the superficial water fraction coefficient, i.e. such that sP_t is the surface contribution to the total river runoff.

Equations (6.26), (6.27) and (6.28) can be rewritten in terms of differences with respect to the mean values of R_t, G_t an P_t, respectively μ_R, μ_G, μ_P:

$$\hat{R}_t \equiv R_t - \mu_R = r(G_{t-1} - \mu_G) + s(P_t - \mu_P) \tag{6.29}$$

$$\hat{G}_t \equiv G_t - \mu_G = (1 - r)(G_{t-1} - \mu_G) + i(P_t - \mu_P) \tag{6.30}$$

$$\hat{R}_t \equiv R_t - \mu_R = (1 - r)(R_{t-1} - \mu_R) + s(P_t - \mu_P) - (s(1 - r) - ir)(P_{t-1} - \mu_P) \tag{6.31}$$

Suppose, as in Section 6.2, that the probability of precipitation can be approximated by a normal distribution or, in other words, that rainfall is a Gaussian white noise. In practice, this means that rain quantity at time τ_1 is assumed independent from rain quantity at time $\tau_2 \neq \tau_1$. Under such hypothesis $\hat{P}_t$ is i.i.d. $(0, \sigma_P^2)$ and (6.30), rewritten as follows:

$$\hat{G}_t = (1 - r)\hat{G}_{t-1} + i\hat{P}_t \tag{6.32}$$

very closely resembles an AR(1) model, because if $\hat{P}_t$ is white noise and its fraction $i\hat{P}_t$ is.

With a pretty similar reasoning, eqn (6.31) can be written:

$$\hat{R}_t = (1 - r)\hat{R}_{t-1} + s\hat{P}_t - (s(1 - r) - ir)\hat{P}_{t-1} \tag{6.33}$$

which looks like an ARMA(1,1).

To summarize, a normally distributed rainfall gives rise to a groundwater storage that is described by an AR(1) model. In the same conditions, the river runoff is described by an ARMA(1,1) model. More generally, it can be expected that the ARMA model for runoff-rainfall, i.e. relating the river flow $\hat{R}_t$ to the rainfall $\hat{P}_t$ (both difference with respect to their average values), is something like:

$$\hat{R}_t = \alpha_1 \hat{R}_{t-1} + \alpha_2 \hat{R}_{t-2} + \dots + \beta_0 \hat{P}_t + \beta_1 P_{t-1} + \dots + v_t$$

where v_t is a 'residual' term at time t and the α and β play the role of AR an MA coefficients, respectively, allowing us to interpret the runoff-rainfall stochastic behaviour of a river basin in terms of an ARMA(p, q) model.

What happens if rainfall is not independent? A plausible assumption, particularly on short timescales (days, weeks) is that precipitations are autocorrelated. Therefore, suppose that P_t is given by:

$$(P_t - \mu_P) = \phi_1(P - t - 1 - \mu_P) + \epsilon_t \tag{6.34}$$

with ϵ_t white noise. By substituting (6.34) in eqn (6.30) we get the model for groundwater storage:

$$(G_t - \mu_G) = (1 - r + \phi_1)(G_{t-1} - \mu_G) + (1 - r)\phi_1(G_{t-2} - \mu_G) + i\epsilon_t$$

which appears to be an AR(2). Similarly, by substituting (6.34) in eqn (6.31), the river runoff coming from an AR(1) rainfall is:

$$R_t - \mu_R = (1 - r + \phi_1)(R_{t-1} - \mu_R) - (1 - r)\phi_1(R_{t-2} - \mu_R) + s\epsilon_t - (s(1 - r) - ir)\epsilon_{t-1}$$

The above equations say that an AR(1) precipitation, causing the groundwater storage to be AR(2), makes the river flow ARMA(2,1).

The procedure can be repeated for any kind of rainfall statistics. For example, assuming an ARMA(1,1) model for precipitation P_t, the groundwater storage can be demonstrated (Hipel and McLeod, 1994) to be described by an ARMA(2,1), while the river runoff is ARMA(2,2). The demonstration is left as an exercise.

6.6.1 Runoff-rainfall relationship in a real case: the Loire river

Loire is a very long river (about 1000 km), crossing France from south-east to north-west, with a basin extending for about one fifth of the country area. We will make use of flow and precipitation data coming from the Oak Ridge National Laboratory Distributed Active Archive Center (ORNL DAAC) for Biogeochemical Dynamics *https://daac.ornl.gov/*, a NASA Earth Observing System Data and Information System (EOSDIS) data centre. This open archive contains a huge number of geophysical data, including the Global Monthly River Discharge Data Set (Vorosmarty, Fekete and Tucker, 1998) and the Global Monthly Precipitation (Hulme, 1999). The former contains monthly averaged discharge measurements (years 1807:1991) for 1018 stations located throughout the world, and the latter is a historical monthly precipitation data set for global land areas (years 1900:1999).

We choose a river station located at Blois (47°34'48.0"N latitude, 1°19'48.0"E longitude). Due to the length of the river, the flow at Blois clearly depends on the precipitation in the whole upstream area. Nevertheless, it appears reasonable to relate the annual flow to the annual precipitation amount at a couple of locations close to the discharge station, located upstream. The closest grid-points for rainfall data appear to be at coordinates (47°30'00.0"N, 2°30'00.0"E) and (47°30'00.0"N 3°45'00.0"E). The precipitation stations are distant from Blois about 90 km and 180 km, respectively.

Figure 6.23 shows the time series of the annual rainfall at the chosen locations, that appear to be strongly correlated, as expected.

It is easy to show that both the time series shown in Fig. 6.23 have neither AR nor MA behaviour Also the annual rainfall values averaged between the two locations are normally distributed, as Fig. 6.24 shows.

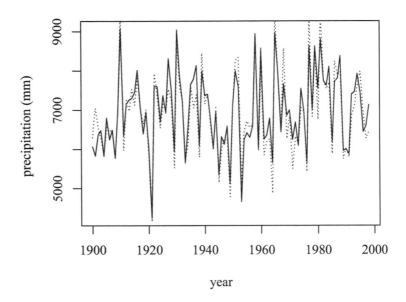

Fig. 6.23 Annual rainfall recorded at 47°30′00.0″N latitude, 2°30′00.0″E longitude (solid line) and at 47°30′00.0″N 3°45′00.0″E (dashed line).

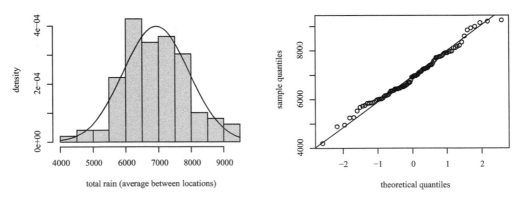

Fig. 6.24 Histogram of total precipitation (left) averaged between the two locations, and normality test (right).

The above analysis is conducted by the following R code (see Code_6_12.R). Data files *Loire_rainfall1.txt* and *Loire_rainfall2.txt* have been extracted from the Global Monthly Precipitation data cited above (ID numbers 1981 and 5234), and they can be downloaded from the book website.

```
## Code_6_12.R
#  Loire river analysis
setwd("C:/RPA/code/ARMA")
read.table("data/Loire_rainfall1.txt",h=T) -> rain1
read.table("data/Loire_rainfall2.txt",h=T) -> rain2
# annual data
plot(rain1$Y,rain1$TOT,t="l",xlab="Year",ylab="Precipitation (mm)")
lines(rain2$Y,rain2$TOT,col="red")
# average between locations
rain.AVG <- (rain1$TOT+rain2$TOT)/2
hist(rain.AVG,freq=FALSE,main="",xlab="Total rain (average between locations)")
x <- rain.AVG
curve(dnorm(x, mean=mean(rain.AVG), sd=sd(rain.AVG)), add=TRUE, col="red")
qqnorm(x,main="")
qqline(x,col="red")
```

A normally-distributed precipitation should imply an ARMA(1,1) river runoff, based on the stochastic hydrological model introduced before. The monthly discharge data for the Loire river at the Blois station are extracted from the cited Global Monthly River Discharge Data Set (ID number 179), and are reported in the file *Loire_runoff.txt*. Before analysing annual data, let us take a look at the monthly flow time series, shown in Fig. 6.25.

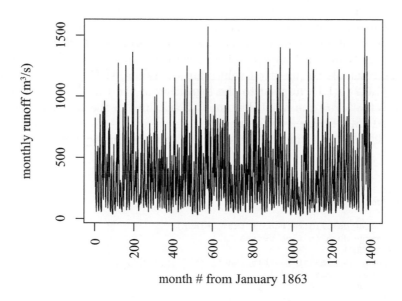

Fig. 6.25 Monthly runoff recorded at Blois (47°34′48.0″N latitude, 1°19′48.0″E longitude)

Figures 6.26 (left) and (right) respectively show the sample ACF and PACF of the flow data. The oscillating patterns clearly suggest a seasonal component in a multiplicative ARIMA, with period 12.

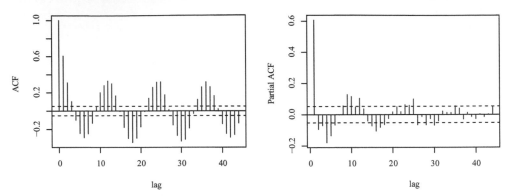

Fig. 6.26 ACF (left) and PACF (right) of the time series of Figure 6.25.

A cut-and-try procedure allows us to establish a reasonable ARIMA model:

$$(1,0,2) \times (2,1,1)_{12}$$

i.e. an ARIMA(2,1,1) (AR2, MA1, integration order 1) for the seasonal part, and an ARMA(1,2) for the residual stationary part. Once the ARIMA model is computed we observe from a plot of the residuals versus fitted values that the residual variance is rather uneven (see Fig. 6.27). An increasing variance is never a good thing in parametric statistical analysis, and it usually calls for some kind of transformation.

Indeed, by analysing the logarithm of the river flow the situation gets better. Figure 6.28 plots the residuals versus fitted values for a seasonal ARIMA model, having the same orders (1,0,2)(2,1,1), applied to the logarithm of the runoff time series. Figure 6.29 compares the log of the monthly runoff (black line) to the values expected from the seasonal ARIMA fitting (red line).

The code for analysing monthly runoff data, and producing Figs 6.25 to 6.29, is the following.

```
## Code_6_13.R
# Analyse monthly runoff
setwd("C:/RPA/code/ARMA")
# read data
dati <- read.table("data/loire_runoff.txt",h=TRUE)
# convert discharge data to time series
X <- ts(dati$DISCHRG)
plot(X,ylab=expression(paste("Monthly runoff (m"^"3","/s)"))),
xlab="Month # from January 1863",las=3)
# seasonal ARIMA analysis on monthly data
arima(X,order=c(1,0,2),seasonal=list(order=c(2,1,1),period=12)) -> fit
X.fitted <- X-fit$residuals
acf(X,na.action=na.pass,44,main="")
pacf(X,na.action=na.pass,44,main="")
plot(X.fitted,fit$residuals,xlab="Fitted",ylab="Residuals")
# seasonal ARIMA analysis on log of monthly data
plot(log(X),ylab=expression(paste("log(Monthly runoff) (m"^"3","/s)"))),
xlab="Month # from January 1863",las=3)
```

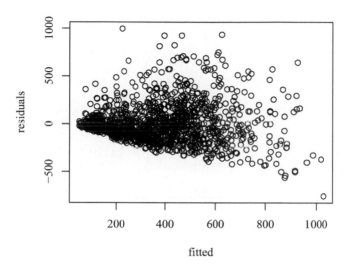

Fig. 6.27 Residuals of the ARIMA $(1,0,2) \times (2,1,1)_{12}$ model versus fitted values.

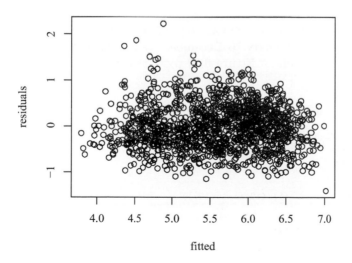

Fig. 6.28 Residuals of the ARIMA $(1,0,2) \times (2,1,1)_{12}$ model versus fitted values, applied to log(runoff).

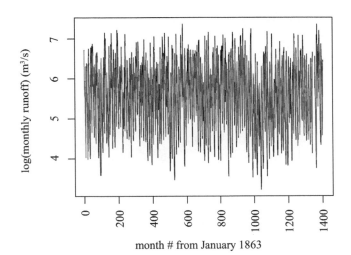

Fig. 6.29 Logarithm of monthly runoff recorded at Blois (black line) and fitted values from seasonal ARIMA (red line).

```
arima(log(X),order=c(1,0,2),seasonal=list(order=c(2,1,1),period=12)) -> fit.log
lines(fitted(fit.log),col="red")
plot(fitted(fit.log),fit.log$residuals,xlab="Fitted",ylab="Residuals")
```

We can now focus on annual runoff, to verify if a relation like (6.33) holds. Annual runoff, computed from January to December[4] can be straightforwardly obtained by summing up monthly values. Figure 6.30 plots the standardized annual runoff time series.

```
## Code_6_14.R
# Compute annual average
Y <- dati$YEAR
N <- length(dati$YEAR)
years <- Y[1]:Y[N]
n <- Y[N]-Y[1]+1
x <- rep(0,n)
for (i in 1:n){
  x[i] <- sum(X[(12*(i-1)+1):(12*(i-1)+12)])
}
# recover NA value
x[113] <- 0.5*(x[112]+x[114])
# Standardized runoff x
x.m <- mean(x)
x.sd <- sd(x)
x <- ts((x-x.m)/x.sd,start=1863)
```

[4]That is an arbitrary choice, of course. Annual runoff can be computed from any day, the only constraint being the total period of 365 days, and annual values can slightly depend on such a choice.

```
plot(x,xlab="Year",ylab=expression(paste("Annual runoff (m"^"3","/s)")),mgp=c(2,0.7,0))
```

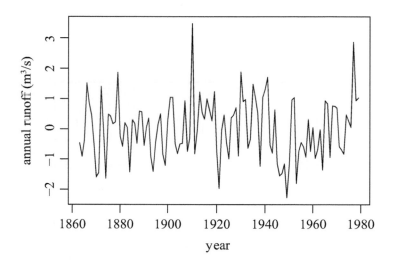

Fig. 6.30 Standardized annual runoff at Blois computed from monthly time series.

In order to obtain the proper ARMA model parameters, we can adopt the approach introduced in Section 6.5.1, computing the AIC (6.5.1) for various AR and MA orders, as follows.

```
## Code_6_15.R
# Manual search for best fitting:
# AR1, MA1, ARMA11, AR2, MA2, ARMA12, ARMA21, ARMA22

fit_AR1 <- arima(x,order = c(1,0,0))
fit_MA1 <- arima(x,order = c(0,0,1))
fit_ARMA11 <- arima(x,order = c(1,0,1))
fit_AR2 <- arima(x,order = c(2,0,0))
fit_MA2 <- arima(x,order = c(0,0,2))
fit_ARMA12 <- arima(x,order = c(1,0,2))
fit_ARMA21 <- arima(x,order = c(1,0,2))
fit_ARMA22 <- arima(x,order = c(2,0,2))
fits <- c(fit_AR1$aic,fit_MA1$aic,fit_ARMA11$aic,fit_AR2$aic,fit_MA2$aic,
  fit_ARMA12$aic,fit_ARMA21$aic,fit_ARMA22$aic)
plot(fits, xlab="Model" ,ylab="aic",xaxt="n")
xtick = 1:8
# grids
for (ii in xtick){
  abline(v=ii,lty=1,lwd=0.5)
}
axTicks(2) -> ygrid
for (jj in ygrid){
  abline(h=jj,lty=1,lwd=0.5)
}
```

```
#
axis(side=1, at=xtick, labels = FALSE)
text(x=xtick, par("usr")[3], labels =c("AR1","MA1","ARMA11","AR2","MA2","ARMA12","ARMA21",
  "ARMA22"),srt = 90, pos = 2, xpd = TRUE,cex=0.85)
```

The result of the computation, graphically shown in Fig. 6.31, confirms that the most suitable model is ARMA(1,1).

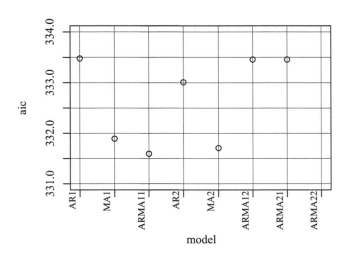

Fig. 6.31 ARMA model choice for annual runoff time series.

6.7 Exercises

Exercise 6.1 A non-stationary process, consisting in a monotonic trend and a white noise, exhibits a slowly decreasing linear ACF. For example, the process defined by:

$$X_t = \beta t + \epsilon_t$$

with β constant, has an ACF like that in Fig. 6.32. Remembering the definition of autocovariance at lag k:

$$\gamma_k = \mathrm{E}\left[(X_t - \mu)(X_{t+k} - \mu)\right]$$

justify the linear behaviour in that particular case.

Hint: take $\beta = 1$ and consider the difference between subsequent lags, $\gamma_k - \gamma_{k+1}$

Exercise 6.2 With reference to Exercise 6.1, write an R code for computing the autocorrelation function by implementing:

$$\gamma_k = \frac{1}{N-k} \sum_{i=1}^{N-k} (X_i - \mu)(X_{i+k} - \mu)$$

and remembering that $\rho_k = \gamma_k/\gamma_0$. Use the code to show that a process involving a quadratic trend:

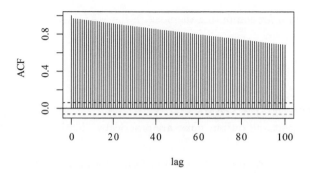

Fig. 6.32 ACF of a linear trend

$$X_t = \beta t^2 + \epsilon_t$$

also has a linear slowly decreasing ACF. Using the following time series data:

```
ti <- 0:1000
beta <- 1e-4
err <- rnorm(1001,sd=2)
x <- beta*ti^2+err
```

show that the 100 lag ACF is very similar to that of the linear trend process.

Exercise 6.3 The following code:

```
n <- 10
x_n <- runif(n)
Nr <- 100
X <- rep(x_n,Nr)
acf(X)
# series length
N <- Nr*n
```

shows that the ACF of a repeated pattern of period n has almost unitary values at lags n and multiples of n. How can you explain it?

Hint: use the definition of autocorrelation, considering that $N - n \approx N$.

Exercise 6.4 An invertible MA(1) process with parameter θ can be expressed as an infinite AR. In fact, expressing the former in terms of the lag polynomial $\Theta(L) = 1 + \theta L$, if $\Psi(L)$ denotes the inverse of Θ, we have:

$$\Psi(L)X_t = \Psi(L)\left(\Theta(L)w_t\right) = w_t$$

because $\Psi\Theta = 1$, with Ψ polynomial of infinite order:

$$\Psi(L) = 1 + \psi_1 L + \psi_2 L^2 + \psi_3 L^3 + \ldots$$

(i) Obtain the expression of the coefficients ψ_k.

(ii) How many of them are necessary to regain the noise w_t?

(iii) Write an R program for computing the noise, if the MA(1) is given by the following code:

```
N <- 200
theta <- 0.5
w <- rnorm(N)
X <- rep(0,N)
# generate MA(1)
X[1] <- w[1]
for (t in 2:N) X[t] <- w[t] + theta*w[t-1]
```

and verify that the input and output noise are practically identical with just 5-6 terms.

Exercise 6.5 Repeat Exercise 6.4, items (i) and (iii), for an invertible MA(2) process with MA coefficients $\theta_1 = 0.8$ and $\theta_2 = 0.5$. In this case, the process can be generated by:

```
N <- 200
theta1 <- 0.8
theta2 <- 0.5
w <- rnorm(N)
X <- rep(0,N)
# generate MA(2)
X[1] <- w[1]
X[2] <- w[2]
for (t in 3:N) X[t] <- w[t] + theta1*w[t-1] + theta2*w[t-2]
```

Exercise 6.6 A theorem coming from general linear process theory asserts that an AR(p) process is stationary if and only if the modulus of all the roots of the characteristic polynomial (associated to the lag polynomial) is greater than one. Demonstrate that, for an AR(2) process with coefficients ϕ_1 and ϕ_2 and roots z_1 and z_2:

$$|z_1| > 1 \text{ and } |z_2| > 1$$

implies the following relations:

$$\phi_1 + \phi_2 < 1$$

$$\phi_2 - \phi_1 < 1$$

$$|\phi_2| < 1$$

Hint: consider the roots λ_1 and λ_2 of the 'reciprocal' characteristic equation $\lambda^2 - \phi_1 \lambda - \phi_2 = 0$ instead of those of $-\phi_1 z - \phi_2 z^2$, and determine conditions on the solutions satisfying $|\lambda_1| < 1$ and $|\lambda_2| < 1$.

Exercise 6.7 Write an R code graphically showing that an AR(1) process $X_t = \phi X_{t-1} + \epsilon_t$ gradually changes from random noise ($\phi = 0$) to a random walk ($\phi = 1$), passing through an infinite number of stationary processes (for $0 < \phi < 1$). Which parameter can you compute to distinguish between different AR(1) processes?

Exercise 6.8 Which model does the following expression represent?

$$\Phi_1(L)(1-L)X_t = \Theta_2(L)\epsilon_t$$

with:

$$\Phi_1(L) = 1 - \phi L$$

and:

$$\Theta_2(L) = 1 + \theta_1 L + \theta_2^2 L$$

Explicitly write down the model represented by it. Is that a stationary process? If not, how can you get a stationary process from it?

Exercise 6.9 Simulate in R a series of 100 values extracted from an ARMA(1,2) model with autoregressive coefficient $\phi = 0.5$ and moving average coefficients $\theta_1 = 0.4$ and $\theta_2 = -0.3$, using the function `arima.sim`. Refer to *help(arima.sim)* for the proper syntax.

Analyse its plot, and investigate its autocorrelation structure by computing and plotting the ACF and PACF.

Exercise 6.10 With reference to Exercise 6.9, compare the ACF and PACF plots of the ARMA(1,2) model with those of AR(1) and MA(2) processes with the same parameters. Try to generate data several times and see how the autocorrelation functions change.

Hint: generate AR(1) and MA(2) data with arima.sim, and superimpose ACF and PACF of the three models.

7
Spectrum Analysis

Fourier's Theorem ...
is not only one of the most beautiful results of modern analysis,
but it may be said to furnish an indispensable instrument
in the treatment of nearly every recondite question in modern physics.

Lord William Thomson Kelvin, Treatise on Natural Philosophy

Fourier's Theorem, as Lord Kelvin wrote, was actually a turning point in the history of physics and applied mathematics. Essentially, that theorem asserts that any periodic function $f(x)$ of a variable x (not necessarily time), continuous and limited on a given domain, can be represented as a sum of harmonic functions (sines and cosines or complex exponentials) of suitable frequencies and amplitudes, the so-called *Fourier series*. Spectrum analysis concerns the detection and computation of such frequencies.

The Fourier Theorem has a mathematical realization in the Fourier transform, which essentially extends the Fourier series representation to aperiodic functions. Given a time signal $s(t)$ describing some physical process, the same information about the process is contained in a function $S(f)$ of the frequency, referred to as the Fourier transform of $s(t)$ and defined by:

$$S(f) = \int_{-\infty}^{\infty} s(t)e^{-2\pi i f t}dt \qquad (7.1)$$

A similar relation exists between $s(t)$ and $S(f)$:

$$s(t) = \int_{-\infty}^{\infty} S(f)e^{2\pi i f t}df \qquad (7.2)$$

which defines the 'inverse' Fourier transform. If $s(t)$ is a discrete time sequence: $\{s(t); t = 0, \pm 1, \pm 2, ...\}$ the integral (7.1) becomes a summation:

$$S(f) = \sum_{t=-\infty}^{\infty} s(t)e^{-2\pi i f t} \qquad (7.3)$$

which is the definition of the discrete Fourier transform (DFT). Note also that if the angular frequency $\omega = 2\pi f$ is used in place of f, and if ω is measured in radians per sampling interval, i.e. it is assumed to belong to the interval $[-\pi, \pi]$, the inverse transform (7.2) becomes:

$$s(t) = \frac{1}{2\pi} \int_{-\pi}^{\pi} S(\omega)e^{i\omega t} d\omega \tag{7.4}$$

Among the mathematical requisites for a function to posses a Fourier transform, the most stringent is that of being square-summable:

$$\sum_{t=-\infty}^{\infty} |s(t)|^2 < \infty$$

a characteristic that, for an actual physical signal, means it has finite energy (Stoica and Moses, 2005).

Before tackling the spectral analysis of stochastic processes, consider a simple deterministic signal $s(t)$ consisting of a sinusoidal function of time:

$$s(t) = A\sin(2\pi f t)$$

A is the signal amplitude and f its frequency, i.e. the number of times the signal repeats itself in one second or, in other terms, the inverse of the *period* of the sinusoid. Figure 7.1 plots such a signal for $A = 1$ and $f = 0.05$ Hz, corresponding to a period $T = 20$ s. In this simple case, f is the only frequency, and we do not need to invoke Fourier to see that. But what can we say of the signal in Figure 7.2?

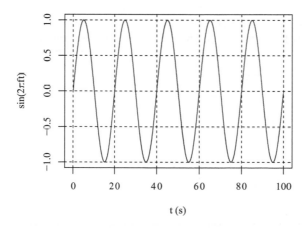

Fig. 7.1 Time plot of a sinusoidal signal.

Looking at the time behaviour of that signal, it appears that more than one repetition frequency is present there. Indeed, we can count five repetitions in 100 s of the whole signal $s(t)$ (having several positive and negative peaks of different amplitudes) but $s(t)$ is anything but sinusoidal. In terms of the *angular* frequency $\omega = 2\pi f$, the

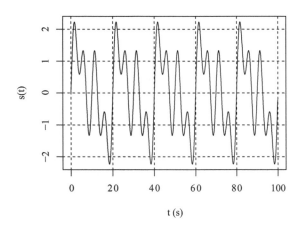

Fig. 7.2 Time plot of a complex deterministic signal.

spectrum $S(\omega)$ of $s(t)$ is that shown in Figure 7.3(right), compared to the spectrum of a single sinusoid (left). Indeed, the signal in Fig. 7.2 is:

$$s(t) = A_1\sin(2\pi f_1 t) + A_2\sin(2\pi f_2 t) + A_3\sin(2\pi f_3 t)$$

with $A_1 = A_2 = A_3 = 1$, $f_1 = 0.05$ Hz, $f_2 = 0.1$ Hz, $f_3 = 0.2$ Hz.

The spectrum is computed through the FFT (*fast Fourier transform*) procedure, a fast numerical implementation of the *discrete Fourier transform*, or DFT, which is the Fourier transform (7.1) applied to discrete data. [1] The R code generating Figures 7.1, 7.2 and 7.3 is shown below.

```
## Code_7_1.R
# Single sinusoid
deltaT <- 1/10
T <- 1000
N <- T/deltaT
t <- seq(0,T,deltaT)
A1 <- 1
f1 <- 0.05

s <- A1*sin(2*pi*f1*t)
# plot sample of 100 seconds
plot(t[1:1000],s[1:1000],t="l",xlab="t (s)",ylab=expression(paste("sin(2",pi,"f t)")))
abline(v=c(0,T/50,2*T/50,3*T/50,4*T/50,T/10),lty=2,col="red")
abline(h=c(-1,-0.5,0,0.5,1),lty=2,col="red")

fs <- 1/deltaT
deltaf <- fs/N
```

[1] We will not go into the depths of FFT, which is exhaustively explained in hundreds of texts. We only warn the reader that the FFT algorithm requires the sample length to be a power of 2. Such a limitation can be circumvented by the zero-padding procedure (Stoica and Moses, 2005).

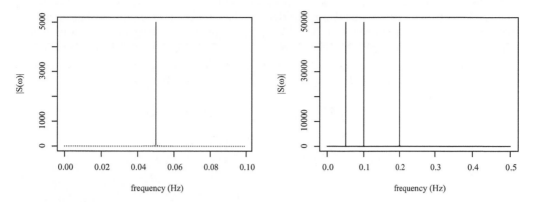

Fig. 7.3 (Left) Spectrum of a sinusoid (with reference to Fig. 7.1) and (right) spectrum of a composition of sinusoids (with reference to Fig. 7.2).

```
S <- fft(s);
Sm <- abs(S)
fr <- deltaf*(0:(N/2))
# plot frequencies up to 0.1 Hz
plot(fr[1:100],Sm[1:100], type="h",xlab="frequency (Hz)",
    ylab=expression(paste("|S(",omega,")|")))

## Code_7_2.R
# Sum of three sinusoidal signals
deltaT <- 1/10
T <- 10000
N <- T/deltaT
t <- seq(0,T,deltaT)
A1 <- 1
f1 <- 0.05
A2 <- 1
f2 <- 0.1
A3 <- 1
f3 <- 0.2

s <- A1*sin(2*pi*f1*t)+A2*sin(2*pi*f2*t)+A3*sin(2*pi*f3*t)
# plot sample of 100 seconds
plot(t[1:1000],s[1:1000],t="l",xlab="t (s)",ylab="s(t)")
abline(v=c(0,T/500,2*T/500,3*T/500,4*T/500,T/100),lty=2,col="red")
abline(h=c(-2,-1,0,1,2),lty=2,col="red")

fs <- 1/deltaT
deltaf <- fs/N

S <- fft(s);
Sm <- abs(S)
fr <- deltaf*(0:(N/2))
# plot frequencies up to 0.5 Hz
plot(fr[1:5000],Sm[1:5000], type="h",xlab="frequency (Hz)",
    ylab=expression(paste("|S(",omega,")|")))
```

Figure 7.4 summarizes the transformation between time and frequency domains. Note that N time samples (input) produce N frequency samples (output), but only half of them are useful for the spectrum analysis because the Fourier transform produces a symmetric frequency range between $-f_s/2$ and $f_s/2$. Given the sampling time Δt, which is the total time duration divided by the number of samples, the sampling frequency f_s is the reciprocal of Δt. $f_c \equiv f_s/2$ is the critical frequency, also known as the *Nyquist* frequency, the highest frequency that can be detected from data having a time spacing Δt. Any frequency component outside the range $(-f_c, f_c)$ is wrongly translated into that interval, the phenomenon known as 'aliasing'.

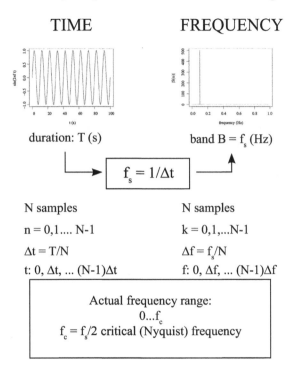

TIME   FREQUENCY

duration: T (s)   band B = f_s (Hz)

$$f_s = 1/\Delta t$$

N samples   N samples

n = 0,1.... N-1   k = 0,1,...N-1

Δt = T/N Δf = f_s/N

t: 0, Δt, ... (N-1)Δt f: 0, Δf, ... (N-1)Δf

> Actual frequency range:
> 0...f_c
> $f_c = f_s/2$ critical (Nyquist) frequency

Fig. 7.4 Transformation between time and frequency domain.

We briefly come back to the definitions (7.3) and (7.4). If $S(\omega)$ is a physical signal, its squared absolute value:

$$E_{SD}(\omega) = |S(\omega)|^2$$

represents its energy spectral density.

Remark 7.1 The above definition has a physical origin. If $v(t)$ is the voltage across a resistor of unitary resistance

$$\int_{-\infty}^{\infty} v(t)^2 \, dt$$

represents the total energy dissipated in the resistor ($v(t)$ is a real function). By the Parseval's theorem, if the complex function $V(\omega)$ denotes the Fourier transform of $v(t)$, we have:

$$\int_{-\infty}^{\infty} v(t)^2 \, dt = \frac{1}{2\pi} \int_{-\infty}^{\infty} |V(\omega)|^2 \, d\omega$$

and $|V(\omega)|^2$ is an 'energy spectral density'.

If we compute the total 'energy', substituting (7.3) in the above definition we obtain:

$$\frac{1}{2\pi} \int_{-\pi}^{\pi} E_{SD}(\omega)d\omega = \sum_{t=-\infty}^{\infty} |s(t)|^2 \tag{7.5}$$

where the angular frequency is in radians per cycle, as in (7.4).

The energy spectral density is related to the autocorrelation function, which is analogous to the statistical autocovariance defined in eqn (3.4). We are dealing with a deterministic signal, so that definition is modified to express the autocorrelation $R(k)$ of a deterministic, finite-energy signal:

$$R(k) = \sum_{t=-\infty}^{\infty} s(t)s(t-k)$$

As for random time series, the autocorrelation of a deterministic signal measures the similarity of it with delayed versions of the signal itself, the first-order $R(0)$ expressing the signal energy. It is easy to verify that the following relationship holds:

$$\sum_{k=-\infty}^{\infty} R(k)e^{-i\omega k} = E_{SD}(\omega)$$

i.e. the energy spectral density is the DFT of the autocorrelation function. In other words, the energy spectral density and the autocorrelation function are Fourier transform pairs.

7.1 Spectrum of stochastic signals

The application of the Fourier analysis to a stochastic process is pretty similar to that seen for deterministic signals. Indeed, a time series – a particular realization of a stochastic process (see Chapter 3) – is nothing different from a function $s(t)$. Therefore, the spectrum (Fourier transform) $S(\omega)$ would appear to be a well-defined quantity. Actually it is not, for several reasons.

First of all, there is a conceptual problem. A finite time series is only a particular 'realization' of a stochastic process, therefore its Fourier transform does not necessarily shed light on the underlying process itself. In other words, only probabilistic statements can be made on a random phenomenon. A key assumption in the spectral analysis of stochastic processes is ergodicity: if the equivalence between ensemble averages and time averages would not hold, we should not be allowed to infer process characteristics from a single realization of it.

Moreover, a discrete-time sequence $\{s(t); t = 0, \pm1, \pm2, ...\}$ does not have finite energy, if it is supposed to be extracted from a 'virtually' infinite time series including all possible realizations of the process. Thus, the finite-energy requirement cannot be assumed with certainty, and $s(t)$ does not rigorously[2] possess a DFT.

---

[2]Making the DFT of a finite discrete time series is of course possible, but it corresponds to take $s(t) = 0$ outside the sampling time interval.

In the following, the process $s(t)$ will be assumed to be wide-sense stationary. We can furthermore assume the time series to have zero average, $E[s(t)] = 0$, without any undesired counter-effect. By analogy with the definition of the energy spectral density of a deterministic signal, we can define the power spectral density (Jenkins and Watts, 1968) P_{SD} of a stochastic process as the (discrete) Fourier transform of the autocovariance function γ_k defined in (3.4):

$$P_{SD}(\omega) = \sum_{k=-\infty}^{\infty} \gamma_k e^{-i\omega k} \tag{7.6}$$

whose inverse transform gives the autocovariance function:

$$\gamma(k) \equiv \gamma_k = \frac{1}{2\pi} \int_{-\pi}^{\pi} P_{SD}(\omega) e^{i\omega k} d\omega$$

If $s(t)$ is real-valued, the power spectral density (7.6) can be expressed in terms of the Fourier cosine transform:

$$P_{SD}(\omega) = \gamma_0 + 2 \sum_{k=1}^{\infty} \gamma_k \cos(\omega k) \tag{7.7}$$

7.1.1 Periodogram and power spectral density (PSD) estimation

Both previous definitions, eqns (7.6) and (7.7), involve an infinite sample. We might be tempted to modify the P_{SD} definition in terms of a finite length N of the time series:

$$\phi_P(\omega) = \sum_{k=-N}^{N} \gamma_k e^{-i\omega k} \tag{7.8}$$

or

$$\phi_P(\omega) = \gamma_0 + 2 \sum_{k=1}^{N} \gamma_k \cos(\omega k) \tag{7.9}$$

for a real-valued series.

Equations (7.8) and (7.9) define the *periodogram* for $\omega \in [-\pi, \pi]$. In terms of the time series $s(t)$ it is obviously defined as:

$$\phi_P(\omega) = \frac{1}{N} \left| \sum_{k=1}^{N} s(k) e^{-i\omega k} \right|^2 \tag{7.10}$$

The periodogram has a very simple interpretation when $s(t)$ is periodic, and thus can be represented on a basis of orthogonal harmonic functions:

$$s(t) = a_0 + \sum_{j=1}^{N} [a_j \cos(2\pi f_j t) + b_j \sin(2\pi f_j t)]$$

and the $f_j = j/N$ are Fourier frequencies, i.e. integer multiples of $f_1 = 1/N$. The above relation, the Fourier series, appears as a regression with coefficients a_j, b_j, which can be estimated, for example by means of a least-squares procedure (Box, Jenkins and Reinsel, 1994), as:

$$a_0 = \bar{s}$$

$$a_j = \frac{2}{N} \sum_{t=1}^{N} s_t \cos\left(2\pi \frac{j}{N} t\right)$$

$$b_j = \frac{2}{N} \sum_{t=1}^{N} s_t \sin\left(2\pi \frac{j}{N} t\right)$$

for $j = 1...n$ if $N = 2n + 1$ is odd, and as:

$$a_j = \frac{1}{N} \sum_{t=1}^{N} (-1)^t s_t$$

$$b_j = 0$$

for $j = 1...n - 1$, if $N = 2n$ is even.

In those cases the periodogram consists of n values:

$$\phi_P(f_j) - \frac{N}{2}(a_j^2 + b_j^2) \qquad\qquad j = 1..(n\text{-}1)$$

$$\phi_P(f_n) = \frac{N}{2}(a_n^2 + b_n^2) \qquad\qquad \text{if N is odd} \qquad (7.11)$$

$$\phi_P(f_n) = N a_n^2 \qquad\qquad\qquad \text{if N is even}$$

Note that with the adopted units the highest frequency is $1/2$ cycles per sampling interval. Limiting ourselves to the case N odd, there are $n = (N-1)/2$ pairs of coefficients. The j-th component of the periodogram defined by (7.11), $\phi_P(j) = N/2(a_j^2 + b_j^2)$, is associated with the variance explained by the $j-$th couple of coefficients in the regression, because:

$$\sum_{t=1}^{N}(s_t - \bar{s})^2 = \sum_{k=1}^{n} \phi_P(f_k)$$

In the most general case, $s(t)$ not periodic and frequencies not Fourier harmonics (i.e. not integer multiples of $1/N$), the periodogram still maintains a relationship with the 'variance' of the process, being related to its autocovariance function by (7.9).

As mentioned before, a single finite time succession is not sufficient for inferring the properties of the underlying stochastic process. If we could repeat the observations we would obtain a population of pairs of coefficients (a_f, b_f) related to any frequency value f, due to the randomness of the process. Therefore, we could think that a better estimate of the periodogram is its expected value. For example, from (7.8), we obtain:

$$E[\phi_P(\omega)] = \sum_{k=-N}^{N} E[\hat{\gamma}_k] e^{(-i\omega k)} \tag{7.12}$$

where we have used the ˆ symbol to specify that $\hat{\gamma}_j$ are sample quantities.

In the limit $N \to \infty$, the expected value of $\hat{\gamma}_j$ of the sample autocovariances tends to the theoretical γ_j, so the sample autocovariances can be substituted for the theoretical, unknown values. But in general the sample spectrum fluctuates about the theoretical spectrum, and its variance does not vanish with increasing length of the time series (Jenkins and Watts, 1968) or, in other words, $E[\phi_P(\omega)]$ is an inconsistent estimator of the theoretical periodogram.[3] If you find that result strange, you should reflect about this: the fact that a quantity y is a consistent estimate of a statistics Y does not imply at all that the Fourier transform $\mathcal{F}(y)$ consistently estimates $\mathcal{F}(Y)$.

Let us summarize what we learnt and what we still need to define.

- The knowledge of the power spectrum density and of the autocovariance structure are perfectly equivalent, by virtue of (7.7).
- The periodogram $\phi_P(\omega)$ is an estimate of the power spectral density $P_{SD}(\omega)$ which consists of 'truncating' the infinite time series to the available sample, which has a finite length.
- The expected value $E[\phi_P]$ is not consistent (we will see some numerical demonstrations in the following).

The final item needs an in-depth analysis. First of all, suppose you have an estimator $\hat{\theta}$ of a quantity θ. The 'correctness' of the estimator is usually determined in terms of unbiasedness and consistency. Consider how close $\hat{\theta}$ is to θ, for example by computing the mean squared error of the estimator:

$$\text{MSE}(\hat{\theta}) \equiv E[(\hat{\theta} - \theta)^2]$$

which, once developed, gives:

$$\text{MSE}(\hat{\theta}) = E[(\hat{\theta} - E[\hat{\theta}])^2] + (E[\hat{\theta}] - \theta)^2$$

The first term is the variance, the second is the square of the bias:

$$\text{var}(\hat{\theta}) = E[(\hat{\theta} - E[\hat{\theta}])^2] \tag{7.13}$$

$$\text{bias}(\hat{\theta}) = (E[\hat{\theta}] - \theta) \tag{7.14}$$

---

[3] Anyway, the limit operation cannot be actually performed: in the end, we have the data we have!

The computation of the bias and variance of the periodogram as an estimator of the power spectral density is not easy; in particular the latter is not possible in the general case while it is affordable for a Gaussian white noise process (Stoica and Moses, 2005). The main result is that the periodogram $\phi_P(\omega)$ is an unbiased estimator for the power spectral density $P_{SD}(\omega)$ of a Gaussian noise, while it is asymptotically unbiased for other stationary processes. But, $\mathrm{var}(\phi_P(\omega))$ does not vanish in any case with increasing sample size.

In order to compute the bias of the periodogram:

$$\mathrm{bias}(\phi_P(\omega)) = E[\phi_P(\omega)] - P_{SD}(\omega)$$

of course, we need to know the theoretical P_{SD} but, anyway, the quantity depending on the characteristics of the sample (in particular its length N) is $E[\phi_P(\omega)]$. From (7.12) we need to compute the expected values of the autocovariance functions. The sample autocovariances, computed from a sample $\{s(t) : t = 1...N\}$, are defined by:

$$\hat{\gamma}_k = \frac{1}{N} \sum_{t=0}^{N-1-|k|} (s(t) - \bar{s})\,(s(t-k) - \bar{s})$$

for $k = -(N-1)...(N-1)$. As a consequence:

$$E[\hat{\gamma}_k] = \frac{N - |k|}{N}\gamma_k \tag{7.15}$$

and we are able to compute the expected value of the periodogram:

$$E[\phi_P(\omega)] = \sum_{k=-(N-1)}^{N-1} \left(1 - \frac{|k|}{N}\right)\gamma_k e^{-i\omega k} \tag{7.16}$$

The quantity in parenthesis inside the summation is the so-called 'Bartlett' or triangular window:

$$w_B(k) = \begin{cases} 1 - |k|/N \text{ , if } k \in (-N, N) \\ 0 \qquad\qquad \text{ , otherwise} \end{cases}$$

which allows us to rewrite (7.16) between infinite limits:

$$E[\phi_P(\omega)] = \sum_{k=-\infty}^{\infty} w_B(k)\gamma(k)e^{-i\omega k} \tag{7.17}$$

Note that $w_B(k)$ has the shape shown in Fig. 7.5.

Equation (7.17) is the DFT of a product, therefore it is the convolution between the DFT of the autocovariance, i.e. the P_{SD} (7.6), and that of the Bartlett window, $W_B(\omega)$:

$$E[\phi_P(\omega)] = \frac{1}{2\pi} \int_{-\pi}^{\pi} P_{SD}(\xi)W_B(\omega - \xi)d\xi \tag{7.18}$$

The Fourier transform $W_B(\omega)$ can be easily demonstrated to be:

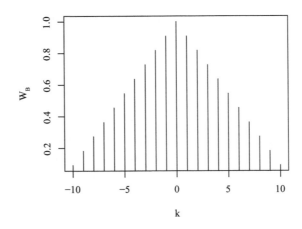

Fig. 7.5 Bartlett window.

$$W_B(\omega) = \frac{1}{N}\left(\frac{\sin(\omega N/2)}{\sin(\omega/2)}\right)^2 \tag{7.19}$$

which exhibits an oscillatory behaviour depending on N as shown in Fig. 7.6 for $N = 11$ (black line) and $N = 110$ (red line):

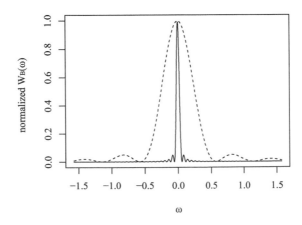

Fig. 7.6 DFT of the Bartlett window.

Figure 7.6 shows that increasing N by an order of magnitude, the width of the main 'lobe' of (7.18) rapidly decreases, W_B approaches a *Dirac delta* function and, finally, $\phi_P(\omega)$ tends to $P_{SD}(\omega)$. In other words, the periodogram is an asymptotically unbiased estimator of the power spectral density.

Computing the variance of the periodogram, in order to complete the evaluation of its goodness as an estimator of the spectral power, is not a simple task, as we mentioned before. Instead of replicating a result that the interested reader can find in the literature (Stoica and Moses, 2005), simply we will show the behaviour of the variance of the periodogram in a specific case. Suppose the process, whose the series $s_t : t = 1..N$ is one realization, is a Gaussian noise with unitary standard deviation. The variance of the periodogram, from (7.13), is:

$$\text{var}(\phi_P(\omega)) = E\left[(\phi_P(\omega) - E[\phi_P(\omega)])^2\right] \tag{7.20}$$

in particular, we are interested to the asymptotic value of variance. We have seen that ϕ_P asymptotically tends to P_{SD} and we also asserted (the demonstration is left to the reader) that it is unbiased for a white noise process. White noise was introduced in Section 6.1 and we will see just ahead that a Gaussian noise is characterized by constant variance (as a function of frequency), equal to the noise variance, unitary in the case at hand. For a Gaussian process $\sim \mathcal{N}(0,1)$, therefore, (7.20) reduces to:

$$\text{var}(\phi_P(\omega)) = E[\phi_P(\omega)^2] - E[P_{SD}(\omega)^2] = E[\phi_P(\omega)^2] - 1 \tag{7.21}$$

Thus, $E[\phi_P(\omega)^2]$ is the only quantity we need to compute. The code below computes that quantity by running 100 simulations for $N = 10$ (thin line), $N = 100$ (dashed line) and $N = 10000$ (thick line) to investigate the asymptotic behaviour of the periodogram variance. Figure 7.7 shows that $\text{var}(\phi_P(\omega))$ tends to a constant value (one, as the process variance) with increasing N. Thus, the periodogram is an inconsistent estimator of power spectral density.

```
## Code_7_3.R
# Gaussian noise
set.seed(246)
phi2.10 <- rep(0,100)
phi2.100 <- rep(0,100)
phi2.10000 <- rep(0,100)
for (i in 1:100)
{
  w <- rnorm(10)
  w.spec <- spectrum(w,log="no",plot=FALSE)
  phi <- w.spec$spec
  phi2.10[i] <- mean(phi^2)-1
  w <- rnorm(100)
  w.spec <- spectrum(w,log="no",plot=FALSE)
  phi <- w.spec$spec
  phi2.100[i] <- mean(phi^2)-1
  w <- rnorm(10000)
  w.spec <- spectrum(w,log="no",plot=FALSE)
  phi <- w.spec$spec
  phi2.10000[i] <- mean(phi^2)-1
}
plot(phi2.10,t="l",xlab="run",ylab=expression(paste("E[",phi[P]^2,"]")))
lines(phi2.100,col="red")
```

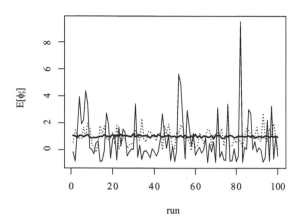

Fig. 7.7 Variance of the periodogram of Gaussian white noise with $\sigma^2 = 1$: $N = 10$ (thin line), $N = 100$ (dashed line), $N = 10000$ (thick line).

```
lines(phi2.10000,col="blue")
```

7.1.2   Consistent estimation of power spectral density

It should be clear from the previous discussion that the 'raw' periodogram is not very suitable for representing the power spectrum of a time series. As a numerical confirmation, Fig. 7.8 shows the raw periodogram of a time series extracted from a Gaussian white noise process $\sim \mathcal{N}(0, 4)$ with variance $\sigma^2 = 4$ (dashed line). Could you infer that we are dealing with a frequency-independent noise from that figure?

It appears reasonable to apply some kind of smoothing procedure to the periodogram in order to obtain a less variable spectrum. The shape of equation (7.17) suggests how to do it: the Bartlett window w_B can be substituted with a more suitable window function w, whose Fourier transform - or, practically, whose DFT − has lower side lobes compared to W_B. In other words, the smoothed periodogram is computed by:

$$\hat{\phi}_P(\omega) = \sum_{k=-M}^{M} w(k)\hat{\gamma}(k)e^{-i\omega k} \tag{7.22}$$

w is sometimes called 'lag' window, because it operates on the *lag k*.

The raw periodogram corresponds to $M = N$ and $w(k) = 1 - |k|/N$, the Bartlett window. Windowing is a well-established technique, commonly used in signal processing (Poularikas, 1999). A variety of smoothing windows can be applied for our purpose, for example a 'cosine-taper' such as the Hann window which, for a sample length $N = 2M + 1$, is defined by:

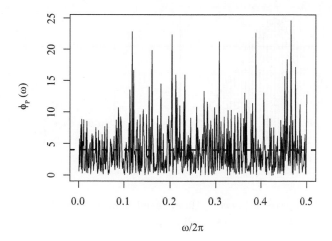

Fig. 7.8 Periodogram of Gaussian white noise with $\sigma^2 = 4$ (solid line), compared to the theoretical PSD (dashed line).

$$w(k) = \frac{1}{2}\left[1 + \cos\left(\frac{\pi k}{M}\right)\right]$$

Its DFT is:

$$W(\omega) = \frac{1}{4}\left[2D(\omega) + D(\omega - \frac{\pi}{M}) + D(\omega + \frac{\pi}{M})\right]$$

where D is the Dirichlet kernel:

$$D(\omega) = \frac{\sin(M\omega)}{\sin(\omega/2)}$$

The effect of that smoothing window on the spectrum estimation should be clear from Fig. 7.9(left) and (right), respectively comparing the Hann (solid line) and Bartlett (dots) window and their DFTs, with $N = 51$, $M = 25$.

The purpose of the window is to provide a smooth decay to zero for the extreme values of k. That reflects in lower side lobes in the Fourier transform of the window, producing a smoothing in the power spectral density estimation. Several other windows give lower side lobes than Hann's, for example the Blackman-Tukey, which is a slight modification of the former:

$$w(k) = 0.42 + 0.5\cos\left(\frac{\pi k}{M}\right) + 0.08\cos\left(\frac{2\pi k}{M}\right)$$

The choice of the window is related to the amount of smoothing we can tolerate (in order to reduce variance) without reducing too much the resolution of the spectral analysis. The R function **spectrum**, in the base *stats* package allowing statistical calculations, invokes the **spec.pgram** which allows various kinds of smoothing. By defining the 'span' argument, the Daniell moving-average window is used. This last

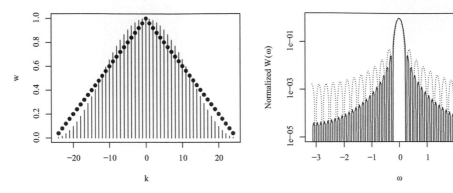

Fig. 7.9 (Left) Comparison between Hann (solid line) and Bartlett window (dots); (right) Comparison between their DFTs (same line specifications).

kind of smoothing is a 'local' one where, based on the *spans* value m, the value of $\hat{\phi}_P(\omega)$ at $\omega = \omega_k$ is computed by averaging on the values at $\omega_{k-m}...\omega_{k+m}$:

$$\hat{\phi}_P(\omega_k) = \frac{1}{2m+1} \sum_{i=-m}^{m} \phi_P\left(\omega_k + \frac{i}{m}\right)$$

We will see application examples and operative details of the **spectrum** function in the next sections.

7.2 Noise spectrum

We have encountered white noise several times throughout the book. It's time to give it a more deep look.

A seemingly good definition of noise is 'spurious' or 'unwanted' signal. When the meaning of signal is well-defined, think for example to a piece of music or to a photographic image, such a definition works: noise is everything confusing the signal containing information, i.e. acoustic or visual disturbances. Anyway, the physical nature of noise and signal is the same: acoustics waves in music, electromagnetic waves in image visualization.

It therefore appears that what is noise and what is signal depends on the 'receiver'. The distinction between *wanted* and *unwanted* signals is a matter of information content. Thus, a better definition is that noise is the part of data we choose not to explain.

In Chapter 6 we saw that white noise, i.e. noise having a flat frequency spectrum, plays a fundamental role in ARMA models. We go back to the definition (6.1) of white noise of variance σ^2. Substituting in (7.7) the zero-th order autocovariance $\gamma_0 = \sigma^2$, all other γ_k being zero for $k > 0$, we obtain the constant power spectral density that motivates the 'white' attribute. In order to estimate the power spectrum, we apply some smoothing procedure to the time series giving rise to Fig. 7.8.

In function **spectrum** the argument *kernel* can be used in place of *spans* to use a different Daniell kernel. the R function **kernel** (also belonging to the *stats* package) can be used inside *plot* to show the kernel shape. Figure 7.10 compares a simple

Daniell moving average window, based on five points, with a more complex window with different weights. The two figures are generated by the following code:

```
# Daniell kernels
plot(kernel("daniell",5),main="")
plot(kernel("daniell",c(3,3)),main="")
```

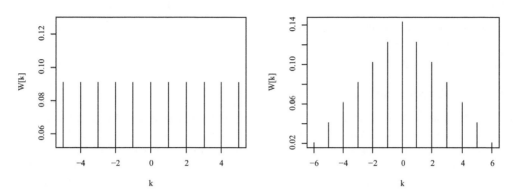

Fig. 7.10 Daniell kernel with moving average over five points (left) and modified Daniell kernel with smooth behaviour (right).

Kernels such those shown in Fig. 7.10 (a,b) are employed inside the *spectrum* command as shown below. Figure 7.11 shows the spectrum of a Gaussian white noise series of length 1000 and variance 4, highlighting the effect of the smoothing window on the power spectral density estimation.

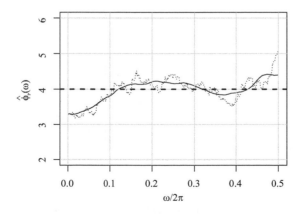

Fig. 7.11 Smoothed periodogram of Gaussian white noise with $\sigma^2 = 4$ (Daniell kernel: dotted line, modified Daniell kernel: solid line), compared to the theoretical PSD (thick dashed line). See the R code producing this figures.

```
## Code_7_4.R
# Gaussian white noise smoothed periodogram
set.seed(123)
w <- rnorm(1000,mean=0,sd=2)

w.spec <- spectrum(w,log="no",kernel=kernel("daniell",51),plot=FALSE)
plot(w.spec$freq,w.spec$spec,t="l",ylab=expression(paste(hat(phi)[P],"(",omega,")")),
xlab=expression(paste(omega,"/2",pi)),ylim=c(2,6),mgp=c(2.2,1,0),col="blue",lty=3)

w.spec <- spectrum(w,log="no",kernel=kernel("daniell",c(51,51)),plot=FALSE)
lines(w.spec$freq,w.spec$spec,col="black")
grid(lty=1)
abline(h=4,col="red",lwd=2)
```

7.2.1 Red and blue spectrum

The definition of white noise as a constant-spectrum signal has been extended to coloured noise, meaning with that term 'noise with a non constant spectral density'. We look at two examples, treated in Section 6.2, dealing with two MA(1) processes with $\theta = 0.5$ and $\theta = -0.5$ respectively, and $\sigma^2 = 1$. Equation (6.7) allows computing the two non-zero autocovariances for such processes, and the theoretical power spectral density by means of (7.7):

$$P_{SD}(\omega) = \sigma^2 \left(1 + \theta^2 + 2\theta\cos(\omega)\right), \ \omega \in [0, \pi]$$

The spectrum relative to an MA(1) process with positive coefficient concentrates its power density at low frequency, thus resembling a red noise (remember that the visible spectrum of electromagnetic energy goes from red to blue, with increasing frequency). Conversely, a negative coefficient brings a power density increasing with frequency, i.e. shifted towards blue.[4] The above two spectra are shown in Fig. 7.12, comparing the theoretical spectra with the smoothed periodograms, as detailed in the R code that follows.

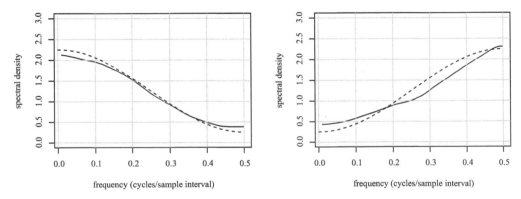

Fig. 7.12 Estimated (solid line) and theoretical (dashed line) spectrum for MA(1) processes representing red noise (left) and blue noise (right).

---

[4]The above depends on the choice of the sign in the definition (6.6) of the MA(1) process. If the process were defined by $X_t = \epsilon_t - \theta\epsilon_{t-1}$, as some authors do, red and blue spectra would correspond to negative and positive θ coefficients, respectively.

```
## Code_7_5.R
## red and blue noise
##MA(1) positive coefficient
set.seed(246)
# generate white (Gaussian) noise
w <- rnorm(100)
x1 <- rep(0,100)
x2 <- rep(0,100)
theta1 <- 0.5
theta2 <- -0.5
# generate MA(1)
x1[1] <- w[1]
x2[1] <- w[1]
for (t in 2:100) {
  x1[t] <- w[t] + theta1*w[t-1]
  x2[t] <- w[t] + theta2*w[t-1]
}
# plot spectral densities
omega <- seq(0,pi,pi/100)

x1.spec <- spectrum(x1,log="no",kernel=kernel("daniell",c(11,11)),plot=FALSE)
plot(x1.spec$freq,x1.spec$spec,t="l",ylim = c(0,3),
xlab="frequency (cycles/sample interval)",ylab="spectral density")
grid(lty=1)
lines(omega/(2*pi),1+theta1^2+2*theta1*cos(omega),col="red")

x2.spec <- spectrum(x2,log="no",kernel=kernel("daniell",c(11,11)),plot=FALSE)
plot(x2.spec$freq,x2.spec$spec,t="l",ylim = c(0,3),
xlab="frequency (cycles/sample interval)",ylab="spectral density")
grid(lty=1)
lines(omega/(2*pi),1+theta2^2+2*theta2*cos(omega),col="red")
```

7.3 Applications of spectrum analysis

As a simple example of application we will conduct an exploratory analysis on a time series containing the maximum temperature values measured monthly in Bologna (Italy) from 1961 to 1991 (see the example in Section 6.5.3). Figure 6.16 shows the time series and the ACF. Note that this time the frequency peak (see Fig. 7.13) is so sharp and well defined that a smoothing window would introduce an artefact The code below shows the R code for computing the spectrum; the last two lines find the index of the peak of the spectral density and use this index to compute the period, i.e. the reciprocal of the frequency, which is nearly 12 months as expected.

```
## Code_7_6.R
## Temperature analysis
# Maximum (monthly averages) temperatures in Bologna: years 1961-1991
# change directory to that containing the data
setwd("C:\RPA\code\spectrum_analysis\data")
maxTempData <- read.table("Bologna_Tmax_1961_1991.txt")
Tmax <-maxTempData$V1
T.spec <- spectrum(Tmax,log="no",plot=FALSE)
plot(T.spec$freq,T.spec$spec,t="l",ylab=expression(paste(phi[P],"(",omega,")")),
xlab=expression(paste(omega,"/2",pi)),mgp=c(2.2,1,0))
grid(lty=1)
# Find index of the spectrum peak
max_index <- which(T.spec$spec==max(T.spec$spec))
period <- 1/T.spec$freq[max_index]
```

That's exactly the same result we found from analysing the autocorrelation structure (Fig. 6.16). Indeed, spectral density and autocorrelation exactly convey the same
information, as the relationships (7.6)-(7.7) indicate.

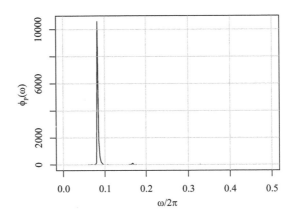

Fig. 7.13 Raw periodogram of the maximum monthly temperatures in Bologna (Italy) from
1961 to 1991.

7.3.1 Searching for hidden periodicity

The search of hidden periodicity is one of the objectives of spectral analysis. In this last
example we will analyse a time series of rainfall data, extracted from the UK National
River Flow Archive (NRFA).[5] NRFA is the primary archive of daily and peak river
flows for the UK: it incorporates daily, monthly and flood peak data from over 1500
gauging stations, and it also provides catchment rainfall time series consisting of the
total rainfall averaged over the catchment in millimetres per day. We will consider a
catchment area of 122 km^2 in the south of UK (Hampshire) where the river Dever
flows. The time series (Fig. 7.14) covers 57 years (684 months) from January 1961 to
December 2017.

Monthly and annual time series are computed from the above data, and a spectral
analysis is conducted on both series. Figure 7.15 shows the smoothed periodogram of
the annual time series. If we compute the frequency position of the first two maxima,
with reference to the R code shown below, we obtain periods of 6.67 and 2.30 years.

Figure 7.16 shows the power spectral density estimated for the monthly time series.
The solid line confirms the existence of an expected periodicity with annual frequency.
The dashed lines correspond to the first two maxima found in the annual series (see

[5]Website: *https://nrfa.ceh.ac.uk/*

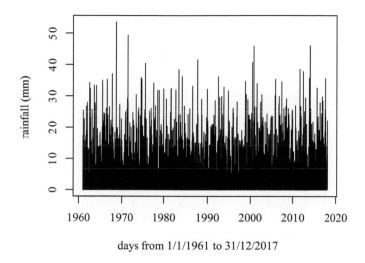

Fig. 7.14 Daily rainfall at Bransbury (UK) from 1961 to 2017.

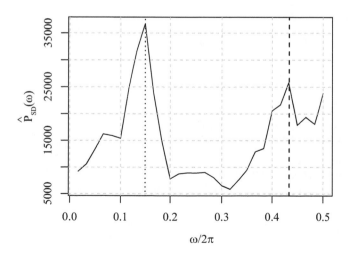

Fig. 7.15 Smoothed periodogram of annual rainfall at Bransbury (UK) from 1961 to 2017.

Fig. 7.15). The dotted lines suggest the presence of other characteristic frequencies corresponding to periods of about 6 and about 4 months.

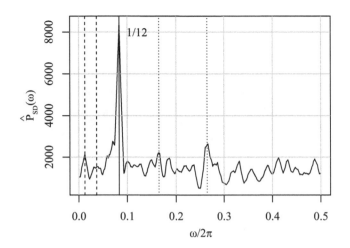

Fig. 7.16 Smoothed periodogram of monthly rainfall at Bransbury (UK) from 1961 to 2017.

```
## Code_7_7.R
# Searching hidden periodicity
library(lubridate)
# change directory to that containing the data
setwd("C:\RPA\code\spectrum_analysis\data")
# read data
cdr <- read.table("42027_cdr.txt")
as.Date(cdr$V1) -> cdr.date
rain.daily <- cdr$V2
# plot daily time series
plot(cdr.date,rain.daily,t="l",ylab="Rainfall (mm)",
    xlab="Days from 1/1/1961 to 31/12/2017")

# Compute monthly series
rain.monthly <- rep(0,(2017-1961)*12)
i <- 1
for (Y in 1961:2017) {
  for (j in 1:12){
    tmp <- rain.daily[year(cdr.date)==Y & month(cdr.date)==j]
    rain.monthly[i] <- sum(as.numeric(as.character(tmp)))
    i <- i+1
  }
}

rain.annual <- rep(0,57)
for (i in 1:57){
```

```
  rain.annual[i] <- sum(rain.monthly[(1+(i-1)*12):(12+(i-1)*12)])
}

# spectrum of annual series
spectrum(rain.annual,log="no",kernel("daniell",c(1,1)),plot=FALSE) -> annual
max_index <- which(annual$spec==max(annual$spec))
annual.period <- 1/annual$freq[max_index]
plot(annual$freq,annual$spec,xlab=expression(paste(omega,"/2",pi)),
     ylab=expression(paste(hat(P)[SD],"(",omega,")")),t="l",mgp=c(2.2,1,0))
grid(lty=2)
abline(v=annual$freq[max_index],col="red")
abline(v=0.434,col="red",lty=2)

# spectrum of monthly series
spectrum(rain.monthly,log="no",kernel("daniell",c(3,3)),plot=FALSE) -> monthly
plot(monthly$freq,monthly$spec,xlab=expression(paste(omega,"/2",pi)),
     ylab=expression(paste(hat(P)[SD],"(",omega,")")),t="l",mgp=c(2.2,1,0))
grid(lty=1)
abline(v=1/12,col="green")
text(0.12,8000,"1/12")
abline(v=1/(12*annual.period),col="red",lty=2)
abline(v=0.434/12,col="red",lty=2)
abline(v=0.165,col="blue",lty=3)
abline(v=0.265,col="blue",lty=3)
```

7.4 Singular Spectrum Analysis

In several processes wide-sense stationarity cannot be assumed valid. In those cases the Wiener–Khinchin theorem, expressing the power spectral density (PSD) as the Fourier transform (FT) of the autocorrelation function, does not hold. Actually, if the FT of the ACF exists, a version of that theorem can be demonstrated also in the non-WSS case, but that is not true in general.

As a consequence, when the process at hand is non-stationary (in particular when it has a trend) the PSD estimation does not give meaningful results. Consider, for example, a random walk process (see the example in Section 6.5). Figure 6.13 shows that the spectral density tends to become virtually infinite at zero frequency and shows a steep decrease to zero with increasing frequency.

A similar situation occurs with random signals having a trend. The following artificial example shows the spectral analysis of a quasi-periodic time series having a trend (see Fig. 7.17).

```
## Code_7_8.R
# Signal with trend
set.seed(2)
ti <- seq(0:0.1:100)
A <- 5+rnorm(length(ti))
omega <- 2*pi*(0.05+0.01*rnorm(length(ti)))
X <- A*sin(omega*ti)+10*ti/100
X <- ts(X)
plot(X,mgp=c(2.2,1,0))
PSD <- spectrum(X,spans=5,log="no",plot=FALSE)
plot(PSD$freq,PSD$spec,xlab=expression(paste(omega,"/2",pi)),
     ylab=expression(paste(hat(P)[SD],"(",omega,")")),t="l",mgp=c(2.2,1,0))
```

Figure 7.18 'correctly' shows the main frequency, which is a random value because of the definition (see R listing). We would like to separate the clear trend from the

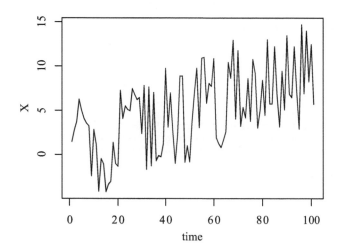

Fig. 7.17 Time series.

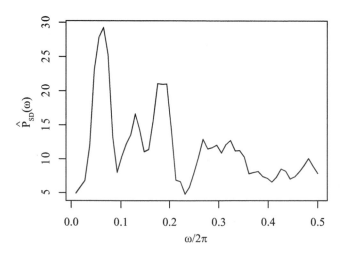

Fig. 7.18 Smoothed spectrum.

quasi-periodic oscillation. Here singular spectrum analysis (SSA) comes into play. We can think of SSA as a sort of principal component analysis: it operates by decomposing a matrix identifying the process (the *trajectory* matrix) by means of the singular value decomposition (SVD) (see Huffaker *et al.* (2017) and Golyandina *et al.* (2018) for more details about the method). The trajectory matrix M is built from the original time series $\{x_t; t : 1..T\}$ by choosing a window length L and rewriting x_t in a matrix format (by column) as follows:

$$\mathbf{M} = \begin{bmatrix} x_1 & x_2 & \cdots & x_N \\ x_2 & x_3 & \cdots & x_{N+1} \\ \vdots & \vdots & \ddots & \vdots \\ x_L & x_{L+1} & \cdots & x_T \end{bmatrix}$$

where $N = T - L + 1$. $\mathbf{M}$ is written as a 'sequence' of N column vectors $\mathbf{X_1} = (x_1, x_2, .., x_L)'$, $\mathbf{X_2} = (x_2, x_3, .., x_{L+1})'$, ..., $\mathbf{X_N} = (x_N, x_{N+1}, .., x_T)'$, where the single prime denotes transposition.

$$\mathbf{M} = (\mathbf{X_1 X_2 .. X_N}) \tag{7.23}$$

The SVD decomposition consists of expressing $\mathbf{M}$ as a product:

$$\mathbf{M} = \mathbf{U}\mathbf{D}^{1/2}\mathbf{V}'$$

between the matrices:

- $\mathbf{U}$: consisting of the eigenvectors of $\mathbf{MM}'$
- $\mathbf{D}^{1/2}$: a diagonal matrix containing the square root of the eigenvalues of $\mathbf{MM}'$
- $\mathbf{V}'$: consisting in the eigenvectors of $\mathbf{M}'\mathbf{M}$ associated with non-zero eigenvalues

In practice the above matrices are the result of the command **svd** in R. An example of a step-by-step application of the construction of the trajectory matrix and its decomposition is reported in Huffaker *et al.* (2017). The explorative part of SSA is conducted in R by the function **ssa**, included in the *Rssa* package. With reference to the artificial data example before, we can do it with very few lines of code:

```
library(Rssa)
s <- ssa(X)
plot(s,numvalues = 10,main="",ylab="Eigenvalue norms")
plot(s,type="vectors",idx=1:10,main="")
```

Figures 7.19 and 7.20 respectively show the first ten eigenvalues and the relative eigenvectors.

The SSA allows us to partition the trajectory matrix by grouping the $\mathbf{X_j}$ vectors in (7.23) in groups of vectors corresponding to eigenvalues λ_j of similar value, for example (λ_1, $\lambda_2 - \lambda_3$, $\lambda_4 - \lambda_{10}$) with reference to Fig. 7.19. We see from Fig. 7.20 that the first two groups (principal components...), corresponding to λ_1 and (λ_2, λ_3), explain 71% of the variance. Therefore, the subsequent step (grouping plus series reconstruction) is accomplished as follows:

```
r <- reconstruct(s, groups = list(1, c(2,3)))
plot(r,main="")
```

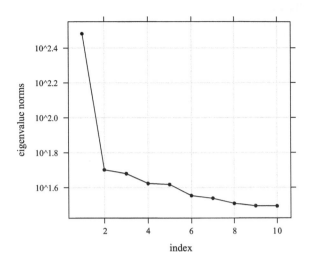

Fig. 7.19 Eigenvalues of the trajectory matrix SVD decomposition.

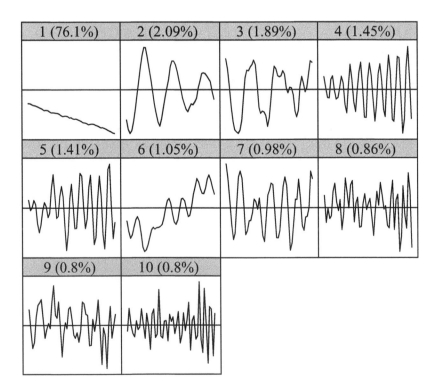

Fig. 7.20 Eigenvectors of the trajectory matrix SVD decomposition.

with the result shown in Fig. 7.21. From top to bottom: the original time series, the first two 'principal components' and the residual term (bottom line).

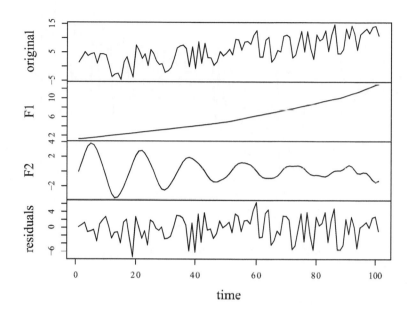

Fig. 7.21 Reconstruction phase of SSA.

The series F_1 and F_2 correspond to the trend and oscillatory part of the original one. Figure 7.22 superimposes the trend on the original time series.

The oscillatory part is shown in Fig. 7.23 (left) together with its smoothed spectrum (right).

Compare Fig. 7.23 (b) with Fig. 7.18 to understand the power of SSA: we have obtained the trend and a more reliable estimation of the spectrum of the quasi-periodic part of the series in one shot. The code for producing the above plots is the following:

```
# trend part
plot(X); lines(r$F1,col="red")
# oscillatory part
plot(r$F2,t="l")
PSD <- spectrum(r$F2,spans=5,log="no",plot=FALSE)
plot(PSD$freq,PSD$spec,xlab=expression(paste(omega,"/2",pi)),
    ylab=expression(paste(hat(P)[SD],"(",omega,")")),t="l",mgp=c(2.2,1,0))
```

In many cases the time series can contain gaps and missing data. The analysis performed with the *Rssa* package supports a solution to perform the analysis when the original time series has missing data. Golyandina *et al.* (2018) presented two ap-

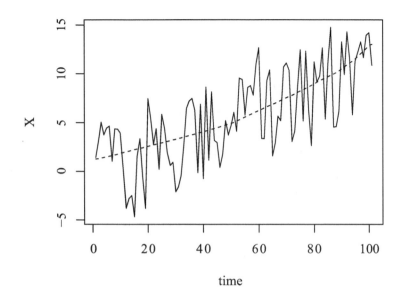

Fig. 7.22 Trend computed by means of the SSA.

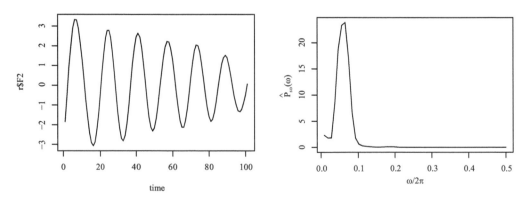

Fig. 7.23 Time plot (left) and spectrum (right) of the oscillatory part of the SSA reconstruction.

proaches, the subspace-based approach and the iterative one. For details and application with R, see Golyandina *et al.* (2018).

7.4.1 Application to real data: the average temperatures in Switzerland in one century

The dataset concerns the historical monthly temperatures in Switzerland from 1901 to 2015, downloaded from *climateknowledgeportal.worldbank.org*. The SSA analysis is performed as follows:

```
## Code_7_9.R
# Application to real data (temperature)
# Read data
setwd("C:\RPA\code\spectrum_analysis\data")
read.csv("Temp_Switzerland.csv",h=TRUE) -> dati
X <- ts(dati$Temperature)
# First stage of SSA: Embedding and Decomposition
s <- ssa(X)
plot(s,numvalues = 10,ylab="Eigenvalue norms",main="")
plot(s,type="vectors",idx=1:10,main="")
# Second stage of SSA: Grouping and Reconstruction
r <- reconstruct(s, groups = list(1, c(2,3)))
plot(r,main="")
signal <- r$F1+r$F2
noise <- X-signal
plot(X); lines(r$F1,col="red")
PSD <- spectrum(r$F2,spans=5,log="no",plot=FALSE)
plot(PSD$freq,PSD$spec,xlab=expression(paste(omega,"/2",pi)),
    ylab=expression(paste(hat(P)[SD],"(",omega,")")),t="l",mgp=c(2.2,1,0))
plot(noise)
PSD <- spectrum(noise,spans=20,log="no")
plot(PSD$freq,PSD$spec,xlab=expression(paste(omega,"/2",pi)),
    ylab=expression(paste("noise ",hat(P)[SD],"(",omega,")")),t="l",mgp=c(2.2,1,0))
```

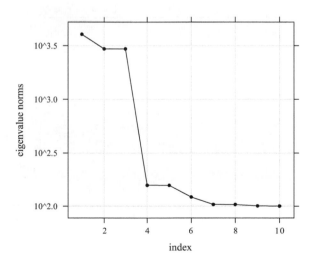

Fig. 7.24 Eigenvalues.

Figures 7.24 and 7.25 show the results of the first stage of the SSA analysis.

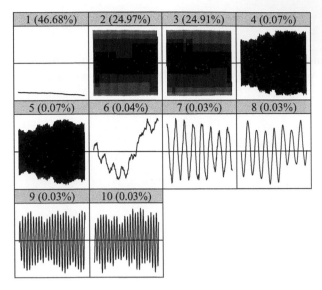

Fig. 7.25 Eigenvectors.

From Figs 7.24 and 7.25 we can decide how to partition the trajectory matrix. Figure 7.26 shows the reconstruction based on the first three eigenvalues, grouped as λ_1 and (λ_2, λ_3).

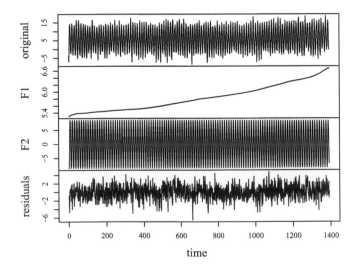

Fig. 7.26 SSA reconstruction of the original time series.

Figure 7.27 show how well the trend fits the original time series. The following figures show the spectrum of the periodic component and, finally, time series and

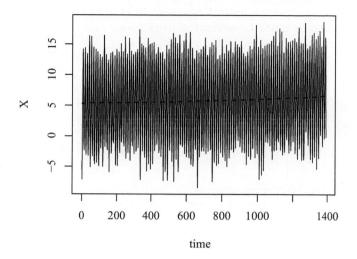

Fig. 7.27 Original time series with SSA trend.

smoothed spectrum of the residual noise.

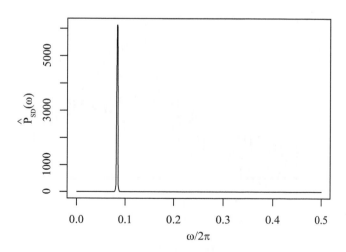

Fig. 7.28 Spectrum of the periodic component.

SSA analysis showed a clear trend in the time series that was successfully identified and removed from the original time series. The pairs of eigenvectors indicated a clear quasi-periodicity (F2) explaining a large part of the signal. The residuals (noise) did not display clear frequency spectra indicating either blue or red coloured noise, therefore additional information cannot be extracted from noise.

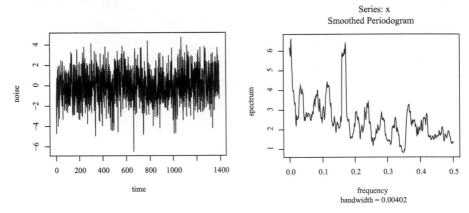

Fig. 7.29 Time plot (left) and spectrum (right) of the residual noise.

7.4.2　An SSA application for beer lovers

This final dataset concerns the monthly beer production in Australia from 1956 to 1995. Data come from the Australian Bureau of Statistics. Figure 7.30 shows the time series.

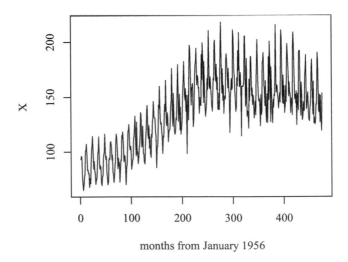

Fig. 7.30 Monthly beer production in Australia in millions of litres, from 1956 to 1995.

Note, in particular, that here we do not use the 'rule of thumb' used before for grouping: although the first eigenvalue is higher that the second, third and fourth (that are comparable to one another), we group as (λ_1, λ_2) and (λ_3, λ_4) because that grouping scheme appears to give a better trend line. Moreover, looking at the eigenvectors (Fig. 7.32) this choice appears reasonable, as the 'oscillating' eigenvectors are the third and fourth. Figure 7.33 superimposes the trend on the beer time series.

The SSA analysis is performed in two stages, as in the preceding section.

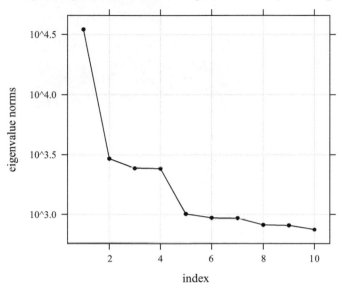

Fig. 7.31 Eigenvalues.

```
## Code_7_10.R
# Application to real data (beer)
setwd("C:\RPA\code\spectrum_analysis\data")
# Read data
dati <- read.csv("monthly-beer-production-in-austr.csv",h=TRUE)
X <- ts(dati$Monthly.beer.production)
plot(X,xlab="Months from January 1956")
# First stage of SSA: Embedding and Decomposition
s <- ssa(X)
plot(s,numvalues = 10,ylab="Eigenvalue norms",main="")
plot(s,type="vectors",idx=1:10,main="")
# Second stage of SSA: Grouping and Reconstruction
r <- reconstruct(s, groups = list(c(1,2), c(3,4)))
plot(r,main="")
signal <- r$F1+r$F2
noise <- X-signal
# time series and trend
plot(X,xlab="Months from January 1956"); lines(r$F1,col="red")
PSD <- spectrum(r$F2,log="no",spans=5,plot=FALSE)
plot(PSD$freq,PSD$spec,xlab=expression(paste(omega,"/2",pi)),
    ylab=expression(paste(hat(P)[SD],"(",omega,")")),t="l",
    mgp=c(2.2,1,0))
abline(v=1/12,col="red")
# residual noise
plot(noise)
```

```
PSD <- spectrum(noise,spans=5,log="no",plot=FALSE)
plot(PSD$freq,PSD$spec,xlab=expression(paste(omega,"/2",pi)),
    ylab=expression(paste("noise ",hat(P)[SD],"(",omega,")")),
    t="l",mgp=c(2.2,1,0))
```

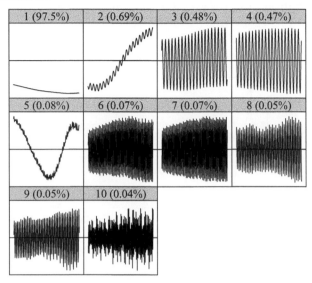

Fig. 7.32 Eigenvectors.

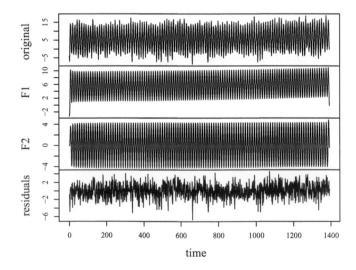

Fig. 7.33 Original time series with SSA trend.

Finally, the spectrum of the periodic component is shown in Fig. 7.34, with the dotted line denoting the annual frequency, and the remaining figures concern the residual noise. Note that Fig. 7.35 clearly shows that the residual is far from being white, or also coloured, noise. Going back to the definition of 'noise', here it represents the part of the signal that we are not interested - or capable - to explain.

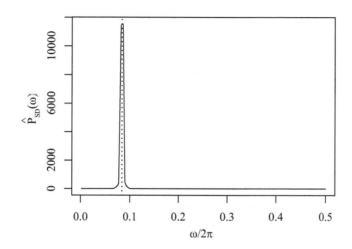

Fig. 7.34 Spectrum of the periodic component.

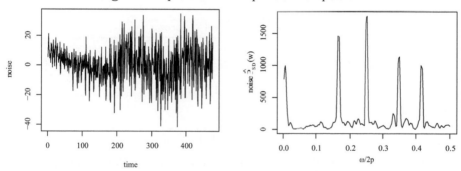

Fig. 7.35 Time plot (left) and spectrum (right) of the residual noise.

7.5 Exercises

Exercise 7.1 Given a zero-mean, stationary real-valued time series $X_t = \{x_t\}$, $t = 1..N$, the power spectral density is defined as:

$$P_{SD}(\omega) = \frac{1}{N} \left| \sum_{k=0}^{N} x_k e^{-i\omega k} \right|$$

Prove that:

$$P_{SD} = \frac{1}{N} \sum_{k=-N}^{N} \hat{\gamma}(k) e^{-i\omega k}$$

where $\hat{\gamma}(k)$ is the sample autocovariance:

$$\hat{\gamma}(k) = \frac{1}{N} \sum_{j=0}^{N-k} x_j x_{j+k}$$

for $k \geq 0$, and:

$$\hat{\gamma}(k) = \hat{\gamma}(-k)$$

for $k < 0$.

Hint: use the notable relation:

$$\sum_{t=1}^{N} \sum_{s=1}^{N} f(t-s) = \sum_{\tau=-N+1}^{N-1} (N - |\tau|) f(\tau)$$

Exercise 7.2 The autocovariance function is defined in terms of the periodogram $\phi(\omega)$, as:

$$\gamma(k) = \frac{1}{2\pi} \int_{-\pi}^{\pi} \phi(\omega) e^{i\omega k} d\omega$$

Prove that $|\gamma(k)| \leq \gamma(0)$ for any $k > 0$.

Exercise 7.3 Prove that for a causal AR(1) process $X_t = \phi X_{t-1} + \epsilon_t$, with noise variance σ^2:

$$P_{SD}(\omega) = \frac{\sigma^2}{1 + \phi^2 - 2\phi\cos\omega}$$

Exercise 7.4 If X_t an MA(1) process $X_t = \epsilon_t + \theta\epsilon_{t-1}$, its lag-polynomial is defined by:

$$\Theta(L) = 1 + \theta L$$

Prove that, given the noise variance σ^2:

$$P_{SD}(\omega) = \sigma^2 \left| \Theta\left(e^{-i\omega}\right) \right|^2$$

that is, the power spectral density of an MA(1) process is the squared modulus of the lag-polynomial computed at $e^{-i\omega}$.

Exercise 7.5 With reference to Exercise 7.4, it can be demonstrated that the relationship between the lag-polynomial and the power spectral density P_{SD} holds for any MA(n) process.

(1) Justify the following assertion:

For a causal AR(1) process, P_{SD} can be expressed as the modulus of a polynomial $\Psi(e^{-i\omega})$:

$$P_{SD}(\omega) = \sigma^2 \left| \Psi\left(e^{-i\omega}\right) \right|^2$$

(2) What is the Ψ polynomial?

Exercise 7.6 The relationship of Exercise 7.4 can be further generalized for any causal ARMA process. For such a process $\Phi(L)X_t = \Theta(L)\epsilon_t$, with noise variance σ^2, we can write the power spectral density P_{SD} in rational form:

$$P_{SD}(\omega) = \sigma^2 \frac{\left|\Theta\left(e^{-i\omega}\right)\right|^2}{\left|\Phi\left(e^{-i\omega}\right)\right|^2}$$

Using the above rational formula, write the explicit expression of $P_{SD}(\omega)$ for an ARMA(1,1) process of coefficients ϕ and θ.

Exercise 7.7 Using the result of Exercise 7.6 write an R code to compute the $P_{SD}(\omega)$ of an ARMA(1,1) process with $\phi = 0.5$, $\theta = 0.5$ and $\sigma^2 = 1$. Plot it for $0 \le \omega \le \pi$.

Exercise 7.8 Following Section 7.4 and the example codes developed there, perform a singular spectrum analysis on the R dataset airmiles. Write an R code and obtain a plot like Fig. 7.26.

What can we say about the trend of that time series?

Exercise 7.9 Do the same analysis of Exercise 7.8 to the closing prices of the S&P 500 stock index recorded between 1950 and 2019. The datafile, available on the website of the book, can be read with the following code.

```
# change directory to that containing the data
setwd("C:/RPA/code/spectrum_analysis/data")
# read data
dat <- read.csv("snp500.csv")
dat.date <- as.POSIXct(dat$Date, format = "%Y-%m-%d", tz = "GMT")
dat.close <- dat$Close
# plot it
plot(dat.date,dat.close,t="l")
```

Exercise 7.10 The file `bike_rent_daily.csv` on the book website contains the daily number of bikes rent in Washington D.C. from January 1st 2011 to December 31st 2012 (Fanaee-T and Gama, 2013). Figure 7.36 shows the result of the SSA analysis.

Write an R code to obtain that figure, and discuss the meaning of the three plots below that of the original time series.

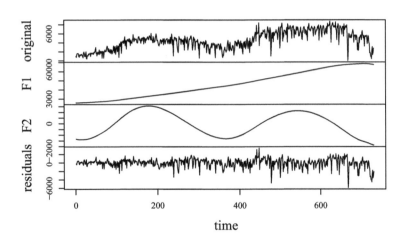

Fig. 7.36 SSA reconstruction of the bike rent data series

8
Markov Chain Monte Carlo

In a Monte Carlo problem the experimenter has complete control of his sampling procedure.
If for example he wanted a green-eyed pig with curly hair and six toes and this event had
a non zero probability, then the Monte Carlo experimenter, unlike the agriculturist,
could immediately produce the animal.

Herman Kahn, Use of Different Monte Carlo Sampling Techniques

8.1   Mother Nature's minimization algorithm

During the night of October 31 to November 1, 1952, on the Pacific Atoll Eniwetok,
the first thermonuclear reaction took place. On March 1, 1954, on the Bikini Atoll in
the Marshall Islands, the first H-bomb was detonated. Among the numerous scientists
who took part in the construction of the bomb, were Nicholas Constantine Metropolis,
Arianna Wright Rosembluth, Marshall Nicholas Rosembluth, Augusta Maria Harkanyi
Teller and Edward Teller. On March 6, 1953, the journal *The Journal of Chemical
Physics* received from these scientists the article 'Equation of State Calculations by
Fast Computing Machines', in which they put forward an algorithm, almost certainly
the most important algorithm of the last century (Metropolis *et al.*, 1953). This algo-
rithm is now called the *Metropolis algorithm*, or the $M(RT)^2$ algorithm with reference
to the initial letters of the authors' names.

The article begins with these words: 'The purpose of this paper is to describe a general method, suitable for fast computing machines, of calculating the properties of any
substance which may be considered as composed of interacting individual molecules.'
By 'fast computing machines' they refer to MANIAC (*Mathematical Analyser Numerical Integrator and Computer*), as it is written in the *Abstract*: 'Results for the
two-dimensional rigid-sphere system have been obtained on the Los Alamos MANIAC
and are presented here.' The 'general method' is described in their words:

So the method we employ is actually a modified Monte Carlo scheme, where instead of choosing configurations randomly, then weighting them with $\exp(-E/kT)$, we choose configurations
with a probability $\exp(-E/kT)$ and weight them evenly.

where T is the absolute temperature and k is Boltzmann's constant, relating temperature to energy. To realize this intent, that is to weight configurations according to
Boltzmann's probability, we have to rely on Mother Nature, to encode the minimization algorithm used by nature itself. A system may (albeit, with small probability)
exit from a local energy minimum in favour of reaching a better one.

Suppose the system is in the state $\mathbf{x}_i$, with energy E_i. It may try to reach the state
$\mathbf{x}_j$ having energy E_j. If $E_j \leqslant E_i$, the state $\mathbf{x}_j$ is certainly the new state of the system,
while if $E_j > E_i$, the system will be in the state $\mathbf{x}_j$ with probability $\exp(-\Delta E/kT)$,

where $\Delta E = E_j - E_i$. Encoding such a principle, our scientists achieve a Markov chain, which they prove to be irreducible and ergodic. As the length of the sequence tends to infinity, that is at stationarity, it visits the most probable states, the states having the lowest energy. The time average on those states is an estimate of the expected value of the process, in other words, one estimates an integral by an arithmetic mean computed only on the states whose contribution to the integrand is high, or 'important'. Let us see in more detail how the algorithm works.

---

1. Choose a starting microstate.
2. Choose the '*move*'. The move provides the change of the system, that is it determines how to go from state $\mathbf{x}_i$ with energy E_i to state $\mathbf{x}_j$ with energy E_j.
3. Compute the difference between the energies $\Delta E = E_j - E_i$.
4. If $\Delta E \leqslant 0$, state $\mathbf{x}_j$ is accepted and go to item 8. Indeed, in this case, the move led the system to a state with lower energy, which means an energy gain.
5. If $\Delta E > 0$, compute the transition probability:

$$p_{ij} = \exp(-\Delta E / kT).$$

6. Generate a random number u uniformly distributed in $[0, 1]$.
7. If $u \leqslant p_{ij}$, the state $\mathbf{x}_j$ is accepted, otherwise the system remains still in state $\mathbf{x}_i$. This means that the system goes in $\mathbf{x}_j$ with probability p_{ij}. If $u > p_{ij}$ the state $\mathbf{x}_i$ must be considered anyhow as a 'new' state, that is $\mathbf{x}_j \equiv \mathbf{x}_i$.
8. Compute the interested physical quantity $A(\mathbf{x}_j)$.
9. Repeat steps 1 to 8, in order to construct a chain of states

$$\mathbf{x}_1, \mathbf{x}_2, \ldots, \mathbf{x}, \ldots, \mathbf{x}_M$$

from which the values below are derived:

$$A(\mathbf{x}_1), A(\mathbf{x}_2), \ldots, A(\mathbf{x}_m), \ldots, A(\mathbf{x}_M).$$

10. After a suitable 'burn-in', time necessary to reach the invariant distribution, compute the time average on A_m, $m = 1, \ldots, M$.

It is apparent that the random numbers are employed to choose between different alternatives, which means that the procedure is based on the Monte Carlo method. It is likewise apparent that the procedure is devised to construct a Markov chain:

$$\text{Monte Carlo} + \text{Markov Chain} = \text{MCMC}$$

Some comments are in order. Curiously, Metropolis *et al.*, the scientists who conceived and realized the algorithm, did not actually apply this 'game of chance' – as they called it – in their experiments on two-dimensional rigid-sphere system, since, in that case, the difference $E_j - E_i$ was either zero or infinity.

In the present book, the reader has already encountered, and will encounter in the following, the term 'Monte Carlo'. In a broad sense, this method is to try the solution of a problem, representing it as a parameter of a hypothetical population and estimate such a parameter through a sample of the population obtained by random number sequences. For instance, suppose I is a definite integral that we cannot compute analytically, but we can regard it as the expected value of a random variable X. In this case, the Monte Carlo method consists of estimating I by generating N value of X with the use of random numbers. The mean of this obtained sample is an estimate of I. Note that with the Monte Carlo method we '*estimate I*', not numerically or approximately compute I, but we *make inference* from a sample to a population to estimate the parameter of interest.

Traditionally, the solution of a 'probabilistic' problem may be found through differential or integro-differential equations, with the Monte Carlo method; 'deterministic' problems are faced by finding a probabilistic analogue and acquiring an estimated solution through sampling procedures.

An example of Monte Carlo estimation of a numerical constant is the well-known 'Buffon's needle' problem, discussed in Chapter 3. Some consider Buffon's needle the first example of a Monte Carlo application, needing neither computer codes nor pencil and paper calculations. Georges Louis LeClerc, Comte de Buffon (1707−1788) used repeated needle tosses onto a lined background to estimate π.

The Monte Carlo method, as we know it today, was first applied during the Manhattan Project to construct atomic weapons, in particular to solve problems related to neutron travelling through fissile material. Those scientists, knowing the probabilities of the microscopic events occurring to a neutron (absorption, fission, etc.) devised a *stochastic process* following the 'story' (trajectory) of a single neutron. Constructing many neutron trajectories different from one another, that is a *sample* of trajectories, they were able to estimate the quantities of interest: how many neutrons were absorbed, how many neutrons were generated, etc.

The name 'Monte Carlo' is attributed to Metropolis, with a clear reference to the Monte Carlo Casino. Metropolis himself wrote (Metropolis, 1987):

It was at that time [in 1947] that I suggested an obvious name for the statistical method − a suggestion not unrelated to the fact that Stan [Stanislaw Ulam] had an uncle who would borrow money from relatives because he 'just had to go to Monte Carlo.' The name seems to have endured.

It is evocative that with the name 'Monte Carlo', the concept of probability, that is at foundations of the method, returns to gaming houses, which is where it was born.

8.2 From physical birth to statistical development

The history of science teaches us that great ideas go far beyond the immediate intents of those who had them. Now the applications of Monte Carlo methods cover a variety of areas, besides physics, ranging from medicine to economics, from archaeology to genetics, etc. In 1970, Keith Hastings, more in statistical than in physical style, wrote the paper 'Monte Carlo Sampling Methods Using Markov Chains and Their Applications', in which '... potential applications of the [Monte Carlo] methods in numerical problems arising in statistics, are discussed' (Hastings, 1970).

The essential idea of Markov Chain Monte Carlo (MCMC) is to generate realizations of a random variable X from an irreducible Markov chain *converging* to (i.e. having as its stationary distribution) the desired distribution of X. It is important to keep in mind that MCMC generated values are not independent, as in the 'ordinary' Monte Carlo method, rather they are often strongly correlated.

Let us enter a Bayesian environment, anticipating concepts that will be treated in the following chapter. In the Bayesian framework, inference is performed by calculating the posterior probability density function. For continuous distributions, we have for the final distribution:

$$p(\theta|D) = \frac{p(\theta)\,p(D|\theta)}{\displaystyle\int p(\theta)\,p(D|\theta)\,d\theta}$$

where D denotes the observed data, θ model parameters and missing data, $p(\theta)$ the prior distribution, and $p(D|\theta)$ the likelihood. The prior distribution expresses our knowledge (or uncertainty) about θ before seeing the data. The posterior distribution expresses our knowledge about θ after seeing the data. When the posterior expectation of a function of θ is the quantity of interest, we have

$$\mathbf{E}\left[f(\theta)|D\right] = \int f(\theta)\,p(\theta|D)\,d\theta$$

The basic idea is to draw samples from the posterior distribution and use those samples to estimate the properties of the posterior distribution (mean, variance, quantiles, etc.). For this aim we have to construct a Markov chain such that its equilibrium probability distribution is the required posterior density. The chain has to be stationary and ergodic, that is the stationarity can be reached, whatever the starting point at which the chain begins. This implies that the chain is irreducible and aperiodic.

As least in principle, with this method one is able to generate observations from any distribution. As already said above by Metropolis *et al.* (1953), MCMC was introduced as '... a general method, suitable for fast computing machines, of calculating the properties of any substance which may be considered as composed of interacting individual molecules'. Now this method has become 'a miraculous tool of Bayesian analysis' (Geyer, 1996) and the flag of what has been called, with a touch of irony, the 'model liberation movement' (Smith, 1992). On these topics, see the fine review by Robert and Casella (2011), which includes a rich list of references.

The Comprehensive R Archive Network (CRAN), the public clearing house for R packages, contains a considerable number of packages for the application of MCMC to various problems. Packages are: MCMCpack (Martin *et al.*, 2018), MCMCglmm (Hadfield, 2019), mcmc (Geyer, 2019). Here we will present a very simple code, showing step by step how to construct a Markov chain such that its invariant distribution is the standard normal.

In statistics, the terminology is a little different from physics. Dealing with MCMC, we talk about the 'Metropolis-Hastings (or Hastings-Metropolis) algorithm', which is more general than the original one. The basic formula is:

$$\alpha(x, y) = \min\left[1 + \frac{\pi(y)q(y, x)}{\pi(x)q(x, y)}\right] \tag{8.1}$$

where:

- y is the *candidate state*. That is, if at time t the system (process) is in the state $X_t = x$, y is the value assumed by X_{t+1}, which may be accepted or not. The value y is generated by the density $q(x, y)$.
- $q(x, y)$ is the *proposal distribution*, since it is the distribution that proposes the candidate. The proposal distribution is also written as $q(\cdot|X_t)$.
- $\pi(\cdot)$ is the invariant distribution, called the *target density*, that is the distribution of interest, from which a certain number of realizations are generated and their time average is estimated.
- $\alpha(x, y)$ is the *probability of move* (or *probability of acceptance*), since if $\alpha(x, y) \leqslant u$, then $X_{t+1} = y$, otherwise $X_{t+1} = x$. As usual, u is a random number, uniformly distributed in $[0, 1]$.

With this symbology, the transition probability of the chain $p(x, y)$ is:

$$p(x, y) = q(x, y)\, \alpha(x, y), \quad x \neq y \tag{8.2}$$

however, it is not complete at all, since it does not consider when the transition to y does not occur, that is when $X_{t+1} = x$. Let $r(x)$ be the probability that the system remains in x, that is:

$$r(x) = \mathsf{P}\left\{X_{t+1} = x | X_t = x\right\}$$

we have:

$$p(x, y) = q(x, y)\, \alpha(x, y) + r(x) \Bbbk_x(y)$$

where $\Bbbk_x(y)$ is equal to 0 or 1 depending on whether $x \neq y$ or $x = y$.

Note that the distribution of interest appears in the definition of $p(x, y)$ (8.2) through $\alpha(x, y)$, that is as the target density ratio $\pi(y)/\pi(x)$. This means that it is sufficient to know $\pi(x)$ up to a constant. In Bayesian statistics, we do not need to actually compute the normalization constant, and write simply the posterior distribution as the product between the prior distribution and the likelihood.

A possible choice of the proposal density is when it is symmetric $q(x, y) = q(y, x)$, as in the original Metropolis algorithm. More specifically it is:

$$q(x, y) = q(|x - y|)$$
$$q(y, x) = q(|y - x|)$$

This choice is equivalent to a random walk, since the new value y equals the old x + 'something', for instance a 'noise'. So we can write $y = x + z$, with z a random perturbation independently realized from x.

A further choice is $q(x, y) = q(y)$, that is the candidate y is independent from x. In other words, the value assumed by X_{t+1} does not depend on the current state $X_t = x$, that is to say that the candidates do not depend on the time instant at which the process is. Note, however, that the realizations, also in this case, are not independent, since the probability of accepting y depends on x. In this case, eqn (8.1) becomes:

$$\alpha(x, y) = \min\left[1 + \frac{\pi(y)q(x)}{\pi(x)q(y)}\right] = \min\left[1 + \frac{w(y)}{w(x)}\right]$$

with $w(x) = \pi(x)/q(x)$.

There are several choices for $q(x, y)$. It is desirable that:

1. $q(x, y)$ generates a chain rapidly converging to the invariant distribution.
2. $q(x, y)$ generates a chain that, after the burn-in period, has a 'good mixing'. The property of mixing refers to how fast the states of the chain acquire new values. Poor mixing means that the candidate state X_{t+1} is still
   $$= X_t = X_{t-1} = \dots$$
3. $q(x, y)$ generates the candidates in an easy way.

Let us now see how to construct a Markov chain having the standard normal $\mathcal{N}(0, 1)$ as invariant distribution. Of course, in R we can generate n realizations from a standard normal using the instruction `rnorm(n)`. We employ here the Metropolis-Hastings algorithm, following it step by step. The *target density* $\pi(x)$ is:

$$\pi(x) = \frac{1}{\sqrt{2\pi}} \exp(-x^2/2)$$

As *proposal distribution* $q(x, y)$ we choose:

$$q(x, y) = q(y, x) = q(|x - y|)$$

this means that the *candidate* y derives from x through $y = x + z$, where z is a realization of the random variable Z with a certain distribution. We choose for Z the uniform distribution in the interval $[-\delta, +\delta]$ with $\delta = 0.75$. Realizations of Z are then given by:

$$z = -0.75 + [0.75 - (-0.75)] u = -0.75 + 1.5u$$

where u is a uniform random number in $(0, 1)$. In R the instruction is simply:

```
z<-runif(1,-0.75,0.75).
```

In the code below only five steps are considered to show in detail how the code works.

```
## Code_8_1.R
## M-H algorithm
## target density : Standard Normal
## only 5 steps

mh <- function(nsteps,x0,d,pi){
x<- numeric()
x[1]<- x0  # initial state of the chain.  Note:  i=1 corresponds to the time t=0
for(i in 2:nsteps)  {
# extraction of the candidate y as possible state at the generic time i
z<-  runif(1,min=-d,max=d)
print(i)      # some printed values to follow the progress of the chain
print(x[i-1])
print(z)
y<-x[i-1]+z   # actually this y is y[i-1]
print(y)

# will y be the new x[i]?
alpha<- min(pi(y)/pi(x[i-1]),1)  # alpha: probability of move
u<- runif(1)  # uniform random number in (0, 1) to check whether  u <= alpha
```

```
print(u)
print(pi(y))
print(pi(x[i-1]))
print(alpha)
if(u <= alpha) x[i]<-  y  else  x[i]<- x[i-1]
                    }           # ending loop on the iterations
return(x)
                              }  # end function
# target density : standard normal
mu<-0
sigma<-1
target<-function(x){
dnorm(x,mean=mu,sd=sigma)
}
set.seed(2)
nsteps<- 5
x0<- -10
delta<- 0.75
x<- mh(nsteps,x0,delta,target)
```

As we said, in principle any choice of the initial value $x(1)$ is correct, since at stationarity the process does non depend on the starting state. In practice, however, it is advisable to run the code with different initial values (better: over-dispersed initial values), and to compare the plots by eye. Such a somewhat naïve method is also used to assess convergence (see next section).

We enumerate the states starting from $i = 1$ and set $x(1) = -10$, i.e. the random variable X at $t = 0$, that is X_1, takes the value -10. Now we need z. Let for instance $z = -0.4727$ (we limit ourselves to writing four decimal significant digits), then $y(1) = -10 + (-0.4727) = -10.4727$. We say that $y(1)$ is the candidate at the iteration 1 ($t = 0$) proposed by $q(x, y)$. Will $y(1)$ be the new $x(2)$? That is, will it be $x(2) = y(1)$? Such a decision is up to the *probability of move* $\alpha(x, y)$. We have to compute:

$$\alpha[x(1), y(1)] = \frac{\pi[y(1)]}{\pi[x(1)]}$$

This is:

$$\pi[y(1)] = \frac{1}{\sqrt{2\pi}} \exp[-(-10.4727^2)/2] = 6.0934 \times 10^{-25}$$

$$\pi[x(1)] = \frac{1}{\sqrt{2\pi}} \exp[-(-10^2)/2] \qquad = 7.6946 \times 10^{-23}$$

then $\alpha[x(1), y(1)] = 0.007919$. At this point a further random number u is required, for instance $u = 0.7024$. Is $u < \alpha[x(1), y(1)] = 0.007919$? No. So the process remains in the old state $x(1)$, that is $x(2) = x(1) = -10$. The value -10 has to be regarded as the realization of the random variable X_2.

The code goes ahead with a further value of Z and, for instance, let $z(2) = 0.1100$, from which $y(2) = -9.8900$. Repeating the above calculations results in:

$$\alpha[x(2), y(2)] = \frac{\exp[-(-9.8900^2)/2]}{\exp[-(-10^2)/2]} = 2.9857$$

The new random number u will be in any case < 2.9857. Therefore, the new state at $i = 2$ is surely accepted, that is $x(2) = y(1) = -9.8900$. The process continues up to the fixed number of iterations.

The next code is almost the same as `Code_8_1.R`, but without the printed values, and δ is varied to show the effect of the $q(x, y)$ on mixing. For completeness the whole code is reported.

```
## Code_8_2.R
## M-H algorithm
## target density : Standard Normal
## varying delta

mh <- function(nsteps,x0,d,pi){
x<- numeric()
x[1]<- x0  # initial state of the chain.  Note:  i=1 corresponds to the time t=0
for(i in 2:nsteps)  {
# extraction of the candidate y as possible state at the generic time i
z<-  runif(1,min=-d,max=d)
y<-x[i-1]+z    # actually this y is y[i-1]
# will y be the new x[i]?
alpha<- min(pi(y)/pi(x[i-1]),1)  # alpha: probability of move
u<- runif(1)  # uniform random number in (0, 1) to check whether  u <= alpha
if(u <= alpha) x[i]<-  y  else   x[i]<- x[i-1]
                          }      # ending loop on the iterations
return(x)
                          }  # end function
# target density : standard normal
mu<-0
sigma<-1
target<-function(x){
dnorm(x,mean=mu,sd=sigma)
}
set.seed(2)
nsteps<- 1000
x0<- -10
delta<- 0.75
x<- mh(nsteps,x0,delta,target)
par(mai=c(1.02,1.,0.82,0.42)+0.1)     # to control the margin size
plot(x[1:length(x)],type="l",main=paste("delta =",delta),
ylim=c(-10,5),xlab="iterations",ylab="x(i)",cex.lab=1.5,cex.main=1.1)
abline(h=0,lty=3,col="black",lwd=2)

##  2 chains with different initial states
nsteps<- 1000
delta<- 0.75
x0<- -10
x<- mh(nsteps,x0,delta,target)
plot(x[1:length(x)],type="l",main=paste("delta =",delta),
ylim=c(-10,+10),xlab="iterations",ylab="x(i)",cex.lab=1.7)
abline(h=0,lty=3,col="black",lwd=2)
x0<- +10
x<- mh(nsteps,x0,delta,target)
lines(x[1:length(x)],type="l",lty=2,col="black")

##   delta=0.1
set.seed(2)        # the seed is the same as at the beginning
nsteps<- 1000
delta<- 0.1
x0<- -10
```

```
x<- mh(nsteps,x0,delta,target)
plot(x[1:length(x)],type="l",main=paste("delta =",delta),
ylim=c(-10,5),xlab="iterations",ylab="x(i)",cex.lab=1.7)
abline(h=0,lty=3,col="black",lwd=2)


## delta=30
set.seed(2)        # the seed is the same as at the beginning
nsteps<- 1000
delta<- 30
x0<- -10
x<- mh(nsteps,x0,delta,target)
plot(x[1:length(x)],type="l",main=paste("delta =",delta),
ylim=c(-10,5),xlab="iterations",ylab="x(i)",cex.lab=1.7)
abline(h=0,lty=3,col="black",lwd=2)
```

In Fig. 8.1, it appears that the chain converges after about 200 steps and that there is a fast mixing around the line $x(i) = 0$. To sum up: we have generated 1000 values $x(i)$, which after $i \gtrsim 200$ are to be regarded as realizations of $\mathcal{N}(0,1)$.

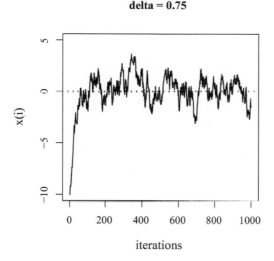

Fig. 8.1 Realizations of the Metropolis-Hastings algorithm with target density the standard normal $\mathcal{N}(0,1)$, and $\delta = 0.75$.

Figure 8.2, with the time series of the above figure, shows also the series with $x(0) = +10$, and the same convergence is apparent. Figures 8.3 and 8.4 show the effect of the choice of $q(x,y)$, with the same $x(0) = -10$. In Fig. 8.3, Z varies in the interval $[-0.1, 0.1]$, and it appears that after 1000 iterations, stationarity is not still reached, since variations induced in each state are too 'small' to produce substantial changes in the states within a reasonable period of time.

The opposite effect is shown in Fig. 8.4, in which Z varies in $[-30, 30]$. Even though stationarity is reached almost immediately, such a choice of δ is not a good choice, since the chain remains over or under the line $x(i) = 0$ for relative long period of time; in other word the mixing is very poor.

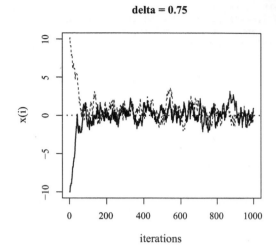

Fig. 8.2 As Figure 8.1, but with two different initial states

Fig. 8.3 As Figure 8.1, but $\delta = 0.1$

The estimate of the expected value $\mathbf{E}[X]$ of the random variable $X \sim \mathcal{N}(0,1)$ is given by:

$$\bar{x} = \frac{1}{n} \sum_{i=1}^{n} x_i$$

which is a time average, where x_1 is the first value after the chain has converged.

In `Code_8_2.R`, we increase the number of iterations `nsteps<- 11000` and introduce a burn-in period `burn.in<- 1000`. We add also the following lines to the section of `Code_8_2.R` relative to $\delta = 0.75$:

delta = 30

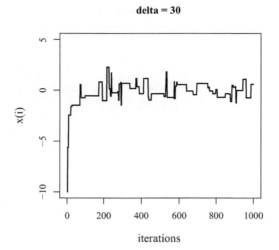

Fig. 8.4 As Figure 8.1, but $\delta = 30$

```
## CONTINUE Code_8_2.R

set.seed(11)
nsteps<- 11000
burn.in<- 1000
delta<- 0.75
x0<- -10
x<- mh(nsteps,x0,delta,target)
plot(x[1:length(x)],type="l",main=paste("delta =",delta),
ylim=c(-10,5),xlab="iterations",ylab="x(i)",
cex.lab=1.2,cex.main=1.1,
font.lab=3,lty=1,lwd=1.5)
abline(v=burn.in,lty=3,col="black",lwd=2)
ta<- burn.in
tb<- length(x)
lt<- length(x[ta:tb])

temp.av<-  mean(x[ta:(tb-1)] )
temp.av
se.temp<-  sqrt(var(x[ta:tb])/lt)
se.temp

# correlation
par(col="black",lwd=2,font.lab=3)
acf_x<-acf(x[ta:(tb-1)],lag.max=60)
#acf_x     # comment if you do not wish the acf values printed
```

Figure 8.5 reports the same time series of Fig. 8.1, but with 11000 iterations.

The code gives $\bar{x} = 0.059$ and the standard error of the mean $= 0.00979$, but this value makes no sense, since the realizations are strongly correlated as shown in Fig. 8.6.

The ACF already encountered in Chapters 6 and 7 measures the linear correlation at different time lags. The R function `acf` computes the ACF, and also automatically adds the horizontal dashed lines, corresponding to the 95% confidence intervals assuming a white noise input. This routine requires as input the value of the series x

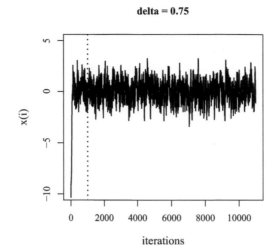

delta = 0.75

Fig. 8.5 As Fig. 8.1, but with 11000 iterations. The burn-in is located at the iteration = 1000.

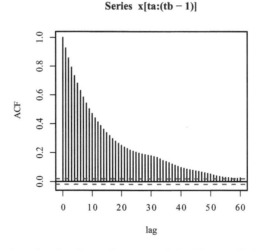

Series x[ta:(tb − 1)]

Fig. 8.6 Autocorrelation plot for the realizations of the Metropolis-Hastings algorithm with target density the standard normal $\mathcal{N}(0, 1)$, $\delta = 0.75$.

after the burn-in and a maximum delay, which we set at `lag.max=60`. If requested, the output `acf.x` can report numerical ACF values. Recall that the lag 0 autocorrelation is fixed at 1 by convention.

To end this section, we show as a further instance how to generate vectors: $\mathbf{x}_1, \mathbf{x}_2, \ldots, \mathbf{x}_n$. Let us consider vectors with two components, obtained by the bivariate normal distribution:

$$\mathcal{N}_2(\boldsymbol{\mu}, \Sigma)$$

where the mean vector μ is:

$$\mu = \begin{pmatrix} 1 \\ 2 \end{pmatrix}$$

and the covariance matrix Σ is:

$$\Sigma = \begin{pmatrix} 1.0 & 0.9 \\ 0.9 & 1.0 \end{pmatrix}$$

This distribution is 'cigar-shaped', because of very high correlation. Recall that if the random variables X_1 and X_2 have bivariate normal distributions, the joint density function is given by:

$$f(x_1, x_2) = \frac{1}{2\pi\sigma_1\sigma_2\sqrt{1-\rho^2}} \times$$

$$\times \exp\left\{-\frac{1}{2(1-\rho^2)}\left[\left(\frac{x_1-\mu_1}{\sigma_1}\right)^2 - \frac{2\rho(x_1-\mu_1)(x_2-\mu_2)}{\sigma_1\sigma_2} + \left(\frac{x_2-\mu_2}{\sigma_2}\right)^2\right]\right\}$$

(8.3)

where $-\infty < \mu_1, \mu_2 < +\infty$ and $\sigma_1^2, \sigma_2^2 > 0$ are the expected values and the variances of X_1 and X_2, respectively, while $-1 \leqslant \rho \leqslant +1$ is the correlation coefficient:

$$\rho = \frac{\text{Cov}\left[X_1, X_2\right]}{\sigma_1\sigma_2}$$

Turning to the example in which: $\sigma_1 = \sigma_2 = \sqrt{1} = 1$ and $\rho = 0.9/(\sigma_1\sigma_2) = 0.9$, the MCMC code to work on the bivariate normal distribution is reported below.

```
## Code_8_3.R
## M-H algorithm
## target density : Bivariate Normal

mh <- function(nsteps,x01,x02,d1,d2,g2){
x1<- numeric()
x1[1]<- x01  # initial state of the chain.  Note:  i=1 corresponds to the time t=0
x2<- numeric()
x2[1]<- x02  # initial state of the chain.  Note:  i=1 corresponds to the time t=0
for(i in 2:nsteps)  {                      # starting loop on the iterations
# extraction of the candidates y1 and y2
#  as possible state at the generic time i
z1<-   runif(1,min=-d1,max=d1)
z2<-   runif(1,min=-d2,max=d2)
y1<- x1[i-1]+z1
y2<- x2[i-1]+z2
g2num<- g2(y1,y2,mu1,mu2,sig1,sig2,ro)
g2den<- g2(x1[i-1],x2[i-1],mu1,mu2,sig1,sig2,ro)
alpha<- min(g2num/g2den,1)  #alpha: probability of move
u<- runif(1)  # uniform random number in (0, 1) to check whether  u <= alpha
if(u <= alpha) {
x1[i]<-   y1
x2[i]<-   y2      }
else  {
x1[i]<- x1[i-1]
x2[i]<- x2[i-1]
```

```
      }
                        }       # ending loop on the iterations
return(list(x1,x2))
                        }  # end function
# target density:
mu1<- 1
mu2<- 2
ssd11<- 1
ssd22<- 1
ssd12<- 0.9
ssd21<- 0.9
sig1<- sqrt(ssd11)
sig2<- sqrt(ssd22)
ro<- ssd12/(sig1*sig2)
g2<-function(x1,x2,mu1,mu2,sig1,sig2,ro){
  exp(-(1./(2.*(1.-ro^2)))  *
  (((x1-mu1)/sig1)^2 - (2.*ro*(x1-mu1)*(x2-mu2))/(sig1*sig2) + ((x2-mu2)/sig2)^2))
}
set.seed(2)
nsteps<- 10000
delta1<- 0.75
delta2<- 0.75
x01<- -4
x02<- +4
xx<- mh(nsteps,x01,x02,delta1,delta2,g2)
xx1<-as.numeric(unlist(xx[1]))
xx2<-as.numeric(unlist(xx[2]))
nplot<- 500          # to plot only nplot iterations
# xx1[1:nplot]        # uncomment to see x1(i) and x2(i)
# xx2[1:nplot]
par(mai=c(1.02,1.,0.82,0.42)+0.1)     # to control the margin size
plot(xx1[1:nplot],type="l",font.lab=3,
ylim=c(-5,5),xlab="iterations",ylab="x1(i) and x2(i)",
lwd=2,cex.lab=1.2)
lines(xx2[1:nplot],type="l",lty=2,col="black",lwd=2)
abline(h=mu1,lty=3,col="black",lwd=2)
abline(h=mu2,lty=3,col="black",lwd=2)
# to plot a different representation
plot(xx1[1:nplot],xx2[1:nplot],type="l",font.lab=3,
ylim=c(-2,5),xlab="x1(i)",ylab="x2(i)",
lwd=2,cex.lab=1.2)
# final check if desired:
burn.in<- 2000
x1m<- mean(xx1[burn.in:nsteps])
x1m
x2m<- mean(xx2[burn.in:nsteps])
x2m
```

The code is an extension of `Code_8_2.R`, where now the realizations $x_1, x_2, \ldots, x_n$ have to be generated. The candidate vector $\mathbf{y}$ is still chosen through:

$$\mathbf{y} = \mathbf{x} + \mathbf{z}$$

that is:

$$y_1 = x_1 + z_1$$
$$y_2 = x_2 + z_2$$

where z_1 and z_2 are the realizations of the uniform random variables Z_1 and Z_2, respectively, with Z_1 and Z_1 defined in $[-0.75, +0.75]$. The initial values are $x_1(0) = -4$ and $x_2(0) = +4$. Let it be, for instance, that $z_1 = -0.4727$ and $z_2 = 0.3036$, so $y_1 = -4.4727$ and $y_2 = 4.3036$, and $\alpha[\mathbf{x}(0)), \mathbf{y}(0)] = 3.0700 \times 10^{-13}$, computed by eqn (8.3). A random number u decides the change of state of $\mathbf{x}(0)$. A random number could be $u = 0.5733$, which is not less than $\alpha[\mathbf{x}(0)), \mathbf{y}(0)]$, therefore $\mathbf{x}(1)$ is still equal to $\mathbf{x}(0)$.

The statement `return(list(x1,x2))` of the function `mh` returns a list of multiple values, and the command `unlist`, repeated for `xx[1]` and `xx[2]`, converts the list into a single vector.

Figure 8.7 reports the values of x_1 and x_2 for the first 500 iterations (`nplot`). A

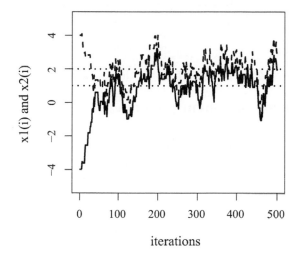

Fig. 8.7 Realizations of the Metropolis-Hastings algorithm with target density the bivariate normal distribution. The horizontal dotted lines mark the components $\mu_1 = 1$ and $\mu_2 = 2$ of the mean vector $\boldsymbol{\mu}$.

different representation is shown in Figure 8.8, plotting x_1 *vs* x_2. The last code lines are aimed to check if the computed time averages are approximately equal to $\mu_1 = 1$ and $\mu_2 = 2$, and indeed `x1m` and `x2m` result equal to 1.0155 and 1.99458, respectively, with 10000 iterations (`nsteps`) and a generous burn-in (`burn.in`) $= 2000$.

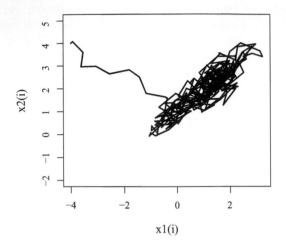

Fig. 8.8 As in in Fig. 8.7, but with a different representation.

8.3   The travelling salesman problem

In 1983, (Kirkpatrick *et al.*, 1983) proposed a computational method, that they called *simulated annealing*, to address the *travelling salesman problem* (TSP). The same method was proposed independently by Černỳ (1985). The name 'annealing' was inspired by the metallurgical process consisting of heating a solid material and allowing it to cool down very slowly, for the purpose of bringing the material closer to its equilibrium state. In this way, structural imperfections can be removed.

Suppose we have n cities in a certain geographical region, for instance in a whole nation. A salesman has to travel through all the cities and return to his original one. The problem is to find the shortest road route for his journey, taking into account that no city has to be visited twice. This is an NP-hard (non-deterministic polynomial-time hard) problem. A problem is NP-hard if it is as hard as or more than the hardest problem in NP, where in turn solutions of NP problems can be verified in polynomial time. For the salesman problem, we cannot be sure to find the shortest route within a reasonable computer time. If n are the cities including his own one, the salesman can choice the second city in $(n-1)$ ways, the third one in $(n-2)$ ways, and so on. Multiplying all these ways together gives:

$$(n-1) \cdot (n-2) \cdot (n-3) \cdots \cdots 3 \cdot 2 \cdot 1 = (n-1)!$$

therefore there are $(n-1)!$ possible routes, for instance, if $n = 10$, there are $362,880$. But if n is 20, the number of routes is $\approx 2^{18}$, using the Stirling approximation:

$$n! \approx \sqrt{2\pi n}\,(n/e)^n,\ n \to \infty$$

The problem is 'translated' into the Metropolis algorithm, discussed in the previous sections. In fact, there is an 'energy', there is a 'move', there is an 'acceptance probability', and there is a 'temperature'. But the 'energy' is a length, the length of the salesman's journey, the 'move' is how to change the order in which the cities are visited, the 'acceptance probability' is the probability of accepting a new route, the 'temperature' is also a length, measured, for instance, in kilometres, not of course with kelvin degrees. To explain, if the *energy* is the difference between the lengths associated with two consecutive 'moves', the initial *temperature* of the system is a value large compared to the largest energy we expect to encounter from move to move. We suggest (Applegate *et al.*, 2006), for the applications and theory of TSP and also for the history of the problem.

`Code_8_4.R` searches for a solution of the TSP with twenty cities, the main Italian provinces. Distances between cities, in kilometres, are read from a file (available on the book website). Figure 8.9 shows the distance table. The first city, where the salesman starts his journey, is L'Aquila, the wonderful town severely damaged by a terrible earthquake in 2009.

	AQ	PZ	CZ	NAP	BO	TS	RM	GE	MI	AN	CB	TO	BA	CA	PA	FI	TN	PG	AO	VE
AQ	0	412	629	244	394	690	137	546	606	189	190	725	424	710	940	319	620	173	790	542
PZ	412	0	320	159	658	954	368	841	871	454	199	989	166	948	631	614	884	512	1054	806
CZ	629	320	0	403	958	1258	612	1085	1155	774	480	1274	354	844	381	858	1169	756	1339	1110
NAP	244	159	403	0	573	873	227	700	770	418	155	889	261	806	714	473	784	370	954	725
BO	394	658	958	573	0	303	376	293	213	227	528	332	670	763	1272	121	227	254	397	155
TS	690	954	1258	873	303	0	673	548	413	522	823	551	965	1060	1569	418	360	551	592	161
RM	137	368	612	277	376	673	0	501	571	304	229	690	431	595	923	274	585	171	755	526
GE	546	841	1085	700	293	548	501	0	144	523	703	171	905	760	1396	245	350	378	248	396
MI	606	871	1155	770	213	413	571	144	0	443	744	144	885	885	1470	319	224	452	185	270
AN	189	454	774	418	227	522	304	523	443	0	324	558	466	808	1126	280	453	131	623	375
CB	190	199	480	155	528	823	229	703	744	324	0	859	224	809	793	476	754	373	924	676
TO	725	989	1274	889	332	551	690	171	144	558	853	0	1001	930	1586	434	357	567	114	402
BA	424	166	354	261	670	965	431	905	886	466	224	1001	0	1009	664	676	893	561	1063	815
CA	710	948	844	806	763	1060	595	760	885	808	809	930	1009	0	465	677	971	686	1006	912
PA	940	631	381	714	1272	1569	923	1396	1470	1126	793	1586	664	465	0	1170	1481	1067	1651	1421
FI	319	614	858	473	121	418	274	245	319	280	476	434	676	677	1170	0	328	151	498	269
TN	620	884	1196	784	227	360	585	350	224	453	754	357	893	971	1481	328	0	464	400	214
PG	173	512	756	370	254	551	171	378	452	131	373	567	561	686	1067	151	464	0	633	403
AO	790	1054	1339	954	397	592	755	248	185	623	924	114	1063	1006	1651	498	400	633	0	446
VE	542	806	1110	725	155	161	526	396	270	375	675	402	815	912	1421	269	214	403	446	0

Fig. 8.9 Distances between the main 20 provinces in Italy.

The initial visit order is chosen at random, for example by adopting the random city sequence corresponding to the city order of the table shown in Fig. 8.9.

```
# Code_8_4.R
# Read the province coordinates and mutual distances
setwd("C:/RPA/code/markov_chain_montecarlo")
city.coords <- read.table("data/provinces.txt",h=T)
cx <- as.matrix(city.coords[,3:4])
labs <- city.coords$Code
distances <- read.table("data/distances.txt",h=TRUE)
M_Cities <- data.matrix(distances)

# distance function
dista <- function(x) {
  ret <- M_Cities[x[1],x[2]]
  for (i in 2:20)
    ret <- ret + M_Cities[x[i],x[i+1]]
  as.numeric(ret)}

# Simulated Annealing (SA) code
r<-nrow(M_Cities) # number of raws of  M_Cities
# route starting from L'Aquila
s<-1:r
route <- c(s,1)
```

```
# route coming from SA
best.route <- route
# route corresponding to best minimum
very.best.route <- route
# Initial route length
L0 <- dista(route)
# Other parameters
# initial "temperature"
T0<-1000
# number of cooling steps
nT<-100
# number of tentative routes, for every T value
nL<-20
# slow-cooling parameter
a<-0.9
# route vector
L<-vector()
sL<-vector()
# best routes found during cooling
minL<-vector()
# Comment the following for an actual SA algorithm
set.seed(4321)

for (t in 1:nT)
{  # starting cooling loop
   T<-a^t*T0
   L[1]<-L0
   r_sample<-sample(s[2:20],replace = FALSE)
   sL<-r_sample
   route<- c(s[1],r_sample,s[1])
   for (l in 2:nL)
   {  # starting tour length loop
      route0<- route[2:20]
      L[l] <- dista(route)
      if(L[l] > L[l-1])           # worse tour
      {
         # energy
         DL <- L[l] - L[l-1]
         # transition probability
         P_tr   <- exp(-DL/T)
         u<- runif(1)
         if(u <= P_tr)
         {
            # Metropolis: accepted
            # generate new solution
            r_sample<- sample(sL,replace = FALSE)
            route<- c(s[1],r_sample,s[1])
         }
         else
         {
            # Metropolis: not accepted
            L[l] <- L[l-1]
            # generate new solution
            r_sample<- sample(route0,replace = FALSE)
            route<- c(s[1],r_sample,s[1])
         }
      } # ending worse tour
      else
      {
         # better tour
```

```
        best.route <- route
        if (dista(best.route)<dista(very.best.route)) very.best.route <- best.route
        # generate new solution
        r_sample<- sample(sL,replace = FALSE)      # better tour
        route<- c(s[1],r_sample,s[1])
        if (dista(route)<dista(very.best.route)) very.best.route <- route
    }
 }      # ending tour length loop
  minL[t]<- min(L)
}     # ending cooling loop

# Graphically show the route
# swap Long ang Lat
tmp <- cx
coords <- cx
coords[,1] <- tmp[,2]
coords[,2] <- tmp[,1]
xmin <- 38
xmax <- 46.5
ymin <- 7
ymax <- 17
# plot TSP
plot(seq(ymin,ymax,(ymax-ymin)/10),seq(xmin,xmax,(xmax-xmin)/10),
   type="n",xlab="Longitude",ylab="Latitude")
text(coords[s[1],2],coords[s[1],1],labels=labs[s[1]],col="black",cex=0.8)
text(coords[s[1],2],coords[s[1],1],labels="0",col="black",cex=2.5)
for (i in (2:20))
   text(coords[s[i],2],coords[s[i],1],labels=labs[s[i]],col="red",cex=0.8)
# best route found by simulated annealing
for (i in 1:19)
   arrows(coords[very.best.route[s[i]],2],coords[very.best.route[s[i]],1],
      coords[very.best.route[s[i+1]],2],
      coords[very.best.route[s[i+1]],1],col="blue",angle=20,length=0.1)
arrows(coords[very.best.route[s[20]],2],coords[very.best.route[s[20]],1],
   coords[very.best.route[s[1]],2],
   coords[very.best.route[s[1]],1],col="blue",angle=20,length=0.1)

# Show solution and route length
print(very.best.route)
print(dista(very.best.route))
```

Figure 8.10 shows the flow-chart of the simulated annealing (SA) procedure of
`Code_8_4.R`.

The simulated annealing (SA) code consists of two loops, one nested into the other.
The outer loop, defined 'cooling loop' in the above code, simulates the annealing
process. Starting with a temperature T_0, a slow cooling process is simulated, decreasing
the temperature down to a preset chosen limit. A cooling schedule, commonly used in
SA problems, prescribes a temperature decrease of the type $T_t = \alpha^t T_0$, with $\alpha = 0.8 -
0.9$. To propose a 'candidate' we need a 'move'. As we have said above, the move can be
a permutation of the cities configuration. Further strategies are proposed, for instance
a simple swap between two cities chosen at random. These type of permutations are
also called 'nearest permutations'.

Lin (1965) suggests the following procedure, known as 2-OPT. Let S_a be the configuration in which the salesman visits the cities in this order:

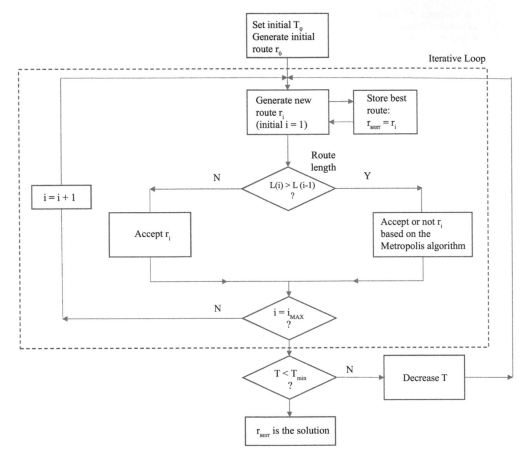

Fig. 8.10 Flow-chart of the Simulated Annealing solution to the TSP

$$C_{j-1}, C_j, \ldots, C_l, C_{l+1}$$

The new configuration S_b is:

$$C_{j-1}, C_l, \ldots, C_j, C_{l+1}$$

That is S_b is equal to S_a, except a stretch of road is travelled in the opposite direction. Let $\Delta L = L_l - L_{l-1}$ be the change of mileage, where L_l and L_{l-1} are the lengths of the existing and of the candidate configuration, respectively. If $\Delta L \leqslant 0$, that is L_l is less than L_l, we accept the candidate configuration as the new tour. If $\Delta L > 0$, that is L_l is greater than the existing tour, the proposed configuration still may be accepted, but with probability:

$$\mathsf{P}(\Delta L) = \exp\left(-\frac{\Delta L}{T}\right)$$

Then, when T is large, several configurations are accepted (to avoid getting trapped in a local minimum), while as T decreases fewer 'worse' candidates are accepted. That is the essence of the Metropolis algorithm, described by the flow-chart of Fig. 8.11, constituting the iterative loop nested in the cooling one.

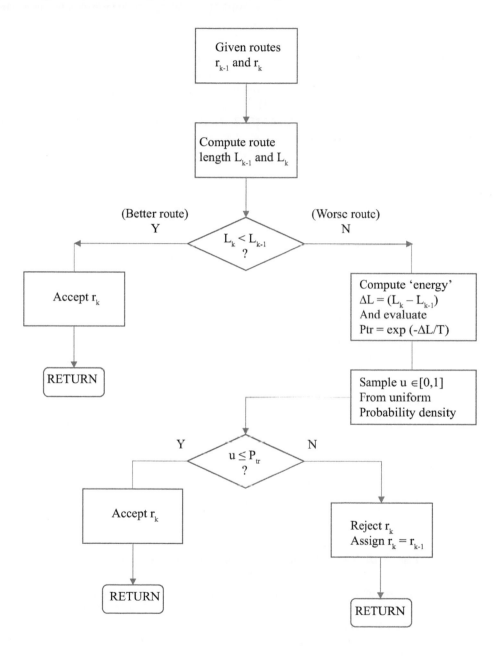

Fig. 8.11 Flow-chart of the Metropolis algorithm.

Concerning the annealing loop, two problems deserve attention: the initial temperature and the stopping criterion. Various suggestions can be found in the literature concerning the choice of T_0, see for example Ben-Ameur (2004) and references therein for a review of existing methods and for their own proposal. In the `Code_8_4.R` we assume for T_0 a value higher than the absolute value $|L_l - L_{l-1}|$, computed on some hundred permutations, namely $T_0 = 1000$. The algorithm of `Code_8_4` simply completes the cooling loop, saving at every step the best solution encountered.

Figure 8.12 shows a 'best' route resulting from the SA algorithm. That route (remember: it is not the absolute best!) corresponds to the sequence [1 10 18 20 19 5 8 6 2 13 11 4 16 17 12 9 7 14 15 3 1] and it has a length of 7849 km, as the last two lines of the code show.

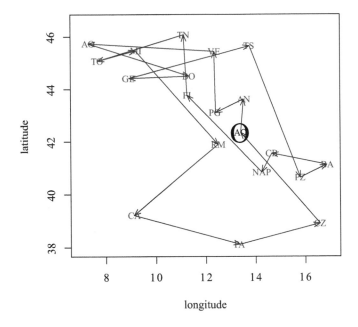

Fig. 8.12 TSP solution obtained by the simulated annealing algorithm.

Anticipating what will be presented in Chapter 12 in the framework of spatial analysis, we want to show the power of the R environment, Fig. 8.13 shows the same TSP route reported in Fig. 8.12, on a real geographical context. We will not go inside the `mapview` function. We only note, in passing, how a few lines of code, see `Code_8_5.R`, can accomplish a complex task!

The plot of Fig. 8.13 shows the first route-step, starting from L'Aquila, with a thick black line, while a thick grey line shows the last route-step coming back to that

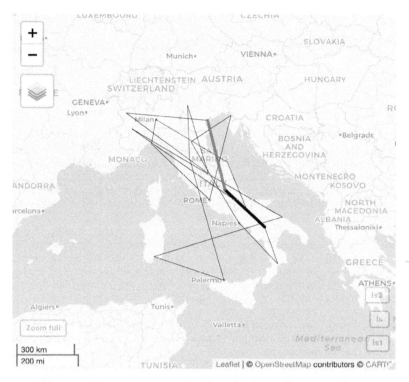

Fig. 8.13 TSP solution mapped on a true geographical context.

town. The remaining thin black lines are all other steps, corresponding to the arrows
in Fig. 8.12.

```
## Code_8_5.R
# Load required library
library(mapview)
# Coordinate projection on geo map
x = city.coords$Long
y = city.coords$Lat

xx = x[2:20]
yy = y[2:20]
pts = matrix(0, 19, 2)
pts[, 1] = xx
pts[, 2] = yy
ls = st_sfc(st_linestring(pts), crs = 4326)

x1= c(x[1],x[2])
y1 = c(y[1],y[2])
pts1 = matrix(0, 2, 2)
pts1[, 1] = x1
pts1[, 2] = y1
ls1 = st_sfc(st_linestring(pts1), crs = 4326)

x2 = c(x[20],x[1])
y2 = c(y[20],y[1])
pts2 = matrix(0, 2, 2)
```

```
pts2[, 1] = x2
pts2[, 2] = y2
ls2 = st_sfc(st_linestring(pts2), crs = 4326)

mapview(ls1,color="black",lwd=4) + mapview(ls,color="black",lwd=0.5) +
  mapview(ls2,color="gray",lwd=4)
```

8.4 Exercises

Exercise 8.1 Familiarize yourself with the Monte Carlo method in R. By means of `rnorm` sample N numbers (with N = 100, 1000, 10000) from a normal distribution with mean $\mu = 1.5$ and standard deviation $\sigma = 0.5$, compare the computed average and standard deviation with μ and σ. Use the density 'counterpart' `dnorm` of the random number generator to compute and plot the theoretical density, and superimpose it on the sample density (plotted by the `hist` command).

Exercise 8.2 The numerical computation of multidimensional integrals can be a quite demanding task. In addition to the computation time, which can become prohibitive, the geometrical definition of the boundaries can be extremely complicated if the domain of integration is not of simple geometry, but rather a complicated or irregular domain. A Monte Carlo approach greatly simplifies the task. Random numbers are generated within an area of simple boundaries (for instance a square in a two-dimensional domain, or a cube in a three-dimensional one) that contains the area with complicated boundaries. Then, a method is implemented to discriminate whether a generated random point of given coordinates is inside or outside the complicated region.

Write an R code using the Monte Carlo method to compute the volume of the sphere $x^2 + y^2 + z^2 = R^2$. Try to change the order of the discretization to see how that affects the accuracy.

Exercise 8.3 Modify the last section of `Code_8_1.R` to sample 5000 values from a gamma distribution with parameters: *shape = 2, scale = 0.5*.

(1) Try to reproduce the plot in Fig. 8.14, where the histogram of data coming from the MCMC simulation (excluding a burn-in of 100 values) is compared to the curve obtained by: `curve(dgamma(x,shape=2,scale=0.5),add=T)`
(2) Generate 5000 samples using the R function *rgamma* and compare the results.
(3) How does the number of discharged values (burn-in) affect the result?
Warning: pay attention to the starting value x(1), and remember the peculiar characteristics of the gamma distribution!

Exercise 8.4 Sampling by the MCMC algorithm is also allowed from a discrete distribution, with very simple changes to `Code_8_1.R` concerning the target distribution and the proposed one. Suppose, for example, you have to sample from a distribution of six integers (from 1 to 6) given by a table like the following:

1	2	3	4	5	6
0.15	0.30	0.20	0.10	0.17	0.08

Obtain 5000 samples for such a distribution and verify that the frequency of the obtained values is consistent with those of the above table.

Hint: change the target distribution to one based on the list `c(0.15,0.30,0.20,0.10,0.17,0.08)` *and use a 'fair' die roll for the proposed distribution.*

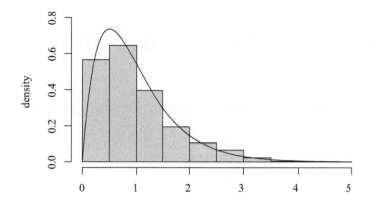

Fig. 8.14 MCMC simulation of a gamma distribution.

Exercise 8.5 Run the R code written to solve Exercise 8.4 with different numbers of steps: 100, 1000, 10000. Compute the relative frequencies of the six possible outcomes and compare them in a table with the values describing the discrete target distribution. How do the initial value and the burn-in period affect the results?

9

Bayesian Inference and Stochastic Processes

Every logical system must start somewhere,
and the question simply amounts to: where do we start?

Sir Harold Jeffreys, Scientific Inference

Statistical techniques based on the Bayesian paradigm can be applied in any flavour of data analysis (Gelman *et al.*, 2004) including the spectral analysis of time series (Broemeling, 2018) and for the study of stochastic processes (Insua *et al.*, 2012). On the other hand, stochastic processes (namely Markov chains) constitute an essential tool for Bayesian analysis.

We assume that the reader is familiar with the Bayes theorem, also known as the 'rule' of conditional probabilities, that is the theoretical foundation of Bayesian statistical methods (Bernardo and Smith, 1994). Essentially the Bayes theorem mathematically expresses the process of 'learning from observations', as it allows to quantify our state-of-knowledge about a hypothesis (a theory, a model) through the 'posterior' probability, computed in terms of the data (the likelihood) and of the previous state of knowledge, measured by the 'prior' probability (Robert, 1994).

Bayesian statistics has experienced a great development and diffusion in the scientific world mainly thanks to Sir Harold Jeffreys (Jeffreys, 1939) who re-discovered the Bayesian approach at the beginning of the past century, in a text developing a fundamental theory of scientific inference based on the Bayesian paradigm. Jeffreys '*who saw the truth and preserved it*', using the words of Jaynes (2003) in his book on the foundations of the theory of probability.

The Bayes theorem (for probability density) is formulated as follows. The (posterior) probability distribution for a parameter θ, given the observed dataset y is:

$$p(\theta|y) = \frac{p(y|\theta)p(\theta)}{\int_\Theta p(y|\theta)p(\theta)d\theta} \tag{9.1}$$

The terms on the right-hand side are: the prior probability density $p(\theta)$ and the likelihood $p(y|\theta)$ (sometimes called 'sampling distribution'), i.e. the probability of the observed data y given the parameter θ. The denominator is essentially a normalizing constant, conceptually given by the integral of the numerator over all possible values of θ, which is the definition of the marginal probability $p(y)$ of the data. Following (Smith, 1991), the Bayesian inference scheme is formally quite simple: the uncertainty about a parameter θ after data y have been observed, and it is computed simply by specifying $p(y|\theta)$ (usually referred to as the likelihood $l(\theta, y)$ when viewed as a function of θ) and $p(\theta)$, and normalizing their product to make it a probability distribution.

The presence of the *prior* probability, sometimes criticized by 'classical' statisticians for its apparent subjectivity, is one of the strengths of the Bayesian approach. Indeed, the prior probability represents our current state of knowledge *before* a bunch of data is taken into consideration. Prior can be chosen so as to express 'ignorance' about the parameters of interest, or it can take into account what we really know about them (for example, as a consequence of theoretical considerations or in virtue of previous data).

Using a very simple probabilistic example, it can be easily demonstrated that data always wins over the prior knowledge, but that the greater is our confidence in a hypothesis, the greater must be the evidence in the data to change our mind. The example comes from (Sivia, 1996), and concerns the verification of the hypothesis of unfairness of a coin in coin-tossing game. Let H be a number between 0 and 1 representing the bias-weighting of the coin, with $H = 0.5$ indicating a fair coin, $H = 0$ and $H = 1$ representing a coin which always gives tail or head, respectively. Our inference about the fairness of the coin is given by the conditional probability $p(H|D)$, where D represents the results of coin-tossing. Bayes theorem tells us that such a *posterior* probability is given by:

$$p(H|D) = \frac{p(D|H)p(H)}{p(D)} \propto p(D|H)p(H) \tag{9.2}$$

In (9.2) the denominator can be omitted, because it is only a normalizing constant.

The likelihood $p(D|H)$ for D = 'k heads in N tosses' is given by the binomial distribution, therefore:

$$p(D|H) \propto H^k(1 - H)^{N-k}$$

Suppose the true value of H is 0.25, i.e. the coin is biased with 'head'. We can compare how different priors influence the posterior $p(H|D)$ by computing:

$$p(H)H^k(1 - H)^{N-k}$$

for different choices of $p(H)$. Figure 9.1 (top-left) compares three priors: (1) a uniform one (ignorance) $p(H) = 1$; (2) an assumption of probable fairness (usually coins are fair...) expressed by a distribution centred at $H = 0.5$, e.g. a normal one: $p(H) \sim N(0.5, 5 \times 10^{-4})$; (3) an assumption of quasi-certain unfairness, e.g. expressed by a beta distribution $p(H) \sim Be(0.5, 0.5)$.

The effect of the above priors on the posterior probability $p(H|D)$ after 10, 100 and 1000 coin tosses is shown in Fig. 9.1. What it clearly shows is that a wrong hypothesis (fair coin) is rejected by a sufficiently large amount of data but it also tells us what we intuitively know: the more we are convinced of the correctness of a hypothesis, the more data confuting it we would need to definitely reject it. If from a single measurement we obtain that the gravitational acceleration on the Earth surface is different from 9.81 m/s^2 we do not reject Newton's law, but if we were to obtain 9.7 in one million independent measurements, made by different people with well-calibrated instruments, we should conclude that such a law is no longer valid!

The following R code has been used to generate Fig. 9.1.

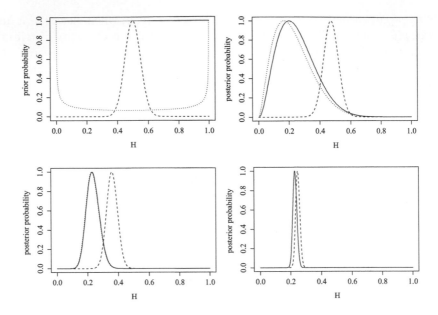

Fig. 9.1 Top-left: prior probabilities for H: uniform (solid line), fair-coin (dashed line), unfair-coin (dotted line). Top-right: posterior probabilities for H, after 10 coin tosses. Bottom-left and bottom right: posterior probabilities for H, after 100 and 1000 coin tosses, respectively

```
## Code_9_1.R
# Unfair coin tossing example
# PRIOR distributions
# true H value
h <- 0.25
H <- seq(0,1,0.001)
# uniform prior
prior1 <- rep(1,1001);
plot(H,prior1,t="l",lty=1,ylim=c(0,1),ylab="Prior probability")
# fair coin
prior2 <- exp(-(H-0.5)^2/5e-3)
lines(H,prior2,lty=2)
# unfair coin
prior3 <- dbeta(H,0.5,0.5)/10
# delete Inf values in H=0 and H=1
prior3[1] <- prior3[2]
prior3[1001] <- prior3[1000]
lines(H,prior3,lty=3)

# POSTERIOR distributions
# posterior after 10 coin tosses
k <- rbinom(1,10,0.25)
likelihood <- dbinom(k,10,H)
post1 <- prior1*likelihood/max(prior1*likelihood)
plot(H,post1,t="l",lty=1,ylim=c(0,1),ylab="Posterior probability")
post2 <- prior2*likelihood/max(prior2*likelihood)
post3 <- prior3*likelihood/max(prior3*likelihood)
lines(H,post2,lty=2)
lines(H,post3,lty=3)
# posterior after 100 coin tosses
```

```
k <- rbinom(1,100,0.25)
likelihood <- dbinom(k,100,H)
post1 <- prior1*likelihood/max(prior1*likelihood)
plot(H,post1,t="l",lty=1,ylim=c(0,1),ylab="Posterior probability")
post2 <- prior2*likelihood/max(prior2*likelihood)
post3 <- prior3*likelihood/max(prior3*likelihood)
lines(H,post2,lty=2)
lines(H,post3,lty=3)
# posterior after 1000 coin tosses
k <- rbinom(1,1000,0.25)
likelihood <- dbinom(k,1000,H)
post1 <- prior1*likelihood/max(prior1*likelihood)
plot(H,post1,t="l",lty=1,ylim=c(0,1),ylab="Posterior probability")
post2 <- prior2*likelihood/max(prior2*likelihood)
post3 <- prior3*likelihood/max(prior3*likelihood)
lines(H,post2,lty=2)
lines(H,post3,lty=3)
```

The power of the Bayesian approach is that of producing the posterior probability distribution, which in turn allows us to compute any needed statistics and confidence interval for them. Moreover, it can be easily demonstrated that well-established estimates, like maximum-likelihood and least-squares, are easily justified under the Bayesian paradigm (Sivia, 1996): maximizing the posterior probability with a uniform prior coincides with maximizing the likelihood function, while a posterior analysis using a uniform prior and a Gaussian likelihood with a least-squares estimate.

Bayesian analysis is increasingly used for studying random processes (Insua *et al.*, 2012; Broemeling, 2018), but also the reverse is true: stochastic process like MCMC (in particular, the Metropolis-Hasting and Gibbs algorithms) are largely employed for sampling from posterior probability distributions, which is the basis of the more widespread approach to numerical Bayes computation.

9.1 Application of MCMC in a regression problem with auto-correlated errors

Research activities in Antarctica, the greatest untouched 'natural laboratory' of the world, are of paramount importance for several sectors of science, from biology to geophysics. As an example, the study of the gas content trapped in ice cores drilled in the Antarctic pack, has allowed to prove the existence of climate cycles (first hypothesized in 1920 by the geophysicist Milutin Milankovitch).[1] Such a result, which is tightly connected with global climate change, suggests a causal relationship between temperature and carbon dioxide (CO_2) concentration in Antarctic ice as a function of age (Luethi *et al.*, 2008), which is directly correlated to the depth along the ice core.

Figure 9.2 compares the temperature anomaly (difference with respect to mean temperature of the last millennium, in °C) with the CO_2 content, in parts per million, relative to the last 800,000 years, obtained from ice cores drilled in the framework of the European project EPICA (European Project for Ice Coring in Antarctica). The drilling were conducted at Kohnen Station (75°009′060″S;00°049′ 040″E) and at

---

[1] A review of the Milankovitch theory and its experimental confirmations was published in (Berger *et al.*, 2015)

Concordia Station (Dome C; 75°069′040″S;123° 209′520″E), at a depth of up to 2,774 m and 3,270 m, respectively.

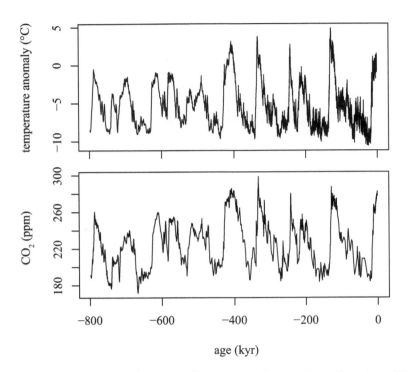

Fig. 9.2 Temperature anomaly (top) and CO_2 content (bottom) as a function of the age in thousands of years.

The relationship between two historical time series, one concerning the temperature and the other one the CO_2 concentration from about 800,000 years ago to the present time, appears to be approximately linear. Indeed, an ordinary linear regression analysis applied to the EPICA data obtained from *http://www.climatedata.info*, thinned out so as to have a time step of 2500 years (the original time step was 100 years), shows that such a linear relation exists, as Fig. 9.3 shows.

Figure 9.3 has been obtained by the following bunch of code, which uses the R function *lm* to perform the linear fitting.[2]

```
## Code_9_2.R
# Bayesian analysis of temperature and CO2
setwd("C:/RPA/code/bayesian")
ice.data <- read.table("data/Epica.txt",h=TRUE)
# assign column data to Year, T, CO2
Year <- ice.data$Year
T <- ice.data$T
CO2 <- ice.data$CO2
```

---

[2]The data file 'epica.txt' is available on the book website

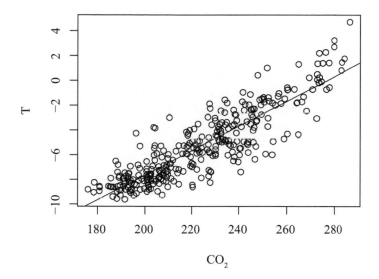

Fig. 9.3 Relationship between CO_2 content and temperature.

```
# Linear regression
Lfit <- lm(T~CO2)
plot(T~CO2)
abline(Lfit,col="red")
summary(Lfit)

# Linear regression
lm(formula = T ~ CO2)
```

The output produced from the last line in the code above shows that the linear regression coefficients obtained by the ordinary least squares (OLS) analysis appear to be statistically significant. The problem is that OLS, to be performed, requires the residual errors to be independent and identically distributed. Instead, if we compute the autocorrelation of the residuals, by means of the *acf* function in R, we verify that errors are correlated, as Fig. 9.4 shows.

```
lm(formula = T ~ CO2)
Residuals:
    Min      1Q  Median      3Q     Max
-3.7783 -0.7896  0.0081  0.8350  3.7894
Coefficients:
             Estimate Std. Error t value Pr(>|t|)
(Intercept) -27.54583    0.68712  -40.09   <2e-16 ***
CO2           0.09942    0.00305   32.60   <2e-16 ***
---
Signif. codes:  0 '***' 0.001 '**' 0.01 '*' 0.05 '.' 0.1 ' ' 1
Residual standard error: 1.394 on 318 degrees of freedom
Multiple R-squared:  0.7697,Adjusted R-squared:  0.769
```

F-statistic: 1063 on 1 and 318 DF,   p-value: < 2.2e-16

Figure 9.4 shows the presence of autocorrelation between data at consecutive times in the time series.[3] That is more clearly shown by computing the partial autocorrelation of the residuals, using `pacf(Lfit$residuals)`, which gives an autocorrelation of about 0.68 between consecutive data points (see Fig. 9.5). Such a correlation affects the reliability and robustness of the OLS regression analysis. When the Gauss-Markov assumptions (homoscedasticity and independence of errors) are violated, the variance of the regression parameters obtained by means of the OLS procedure is wrong and, as a consequence, also the confidence intervals of the parameters are incorrect.

Figure 9.5, showing a single dominant peak in the partial autocorrelation function, suggests a first-order autoregressive structure AR(1) for the residual error (see Chapter 6).

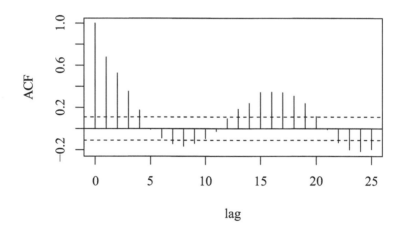

Fig. 9.4 Autocorrelation of residual errors versus time lag.

A robust Bayesian regression analysis on data with AR(1) errors can be performed based on Gibbs sampling (Chib, 1993) or, more generally, on MCMC sampling. We will briefly describe the rationale of the Bayesian analysis, schematically describing how the regression can be implemented in R, but instead of implementing it 'from scratch', we will use the R package *rjags* which works as an interface with the JAGS (Just Another Gibbs Sampler) program (Plummer, 2003). JAGS, which allows rather simple but efficient implementations of the Bayesian procedure to obtain a *posterior* probability distribution given a *prior* distribution and a *likelihood* function, is concisely described in Appendix B.

---

[3]A similar analysis has been conducted in (Hoff, 2009) for data coming from ice cores drilled at the Russian Vostok station in East Antarctica (78°S, 106°E) (Petit *et al.*, 1999).

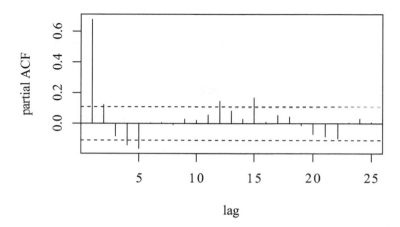

Fig. 9.5 Partial autocorrelation of residuals versus time lag.

We firstly need to understand how the presence of autocorrelation influences the regression analysis. Coming back to the uncorrelated-error case, the OLS procedure can be summarized in matrix form. Suppose we have $k = 1...n$ observations at k times of the variable $\mathbf{y} = (y_k)^T$ (the CO_2 concentration) corresponding to the variable $\mathbf{x} = (x_k)^T$, where the superscript 'T' denotes the transpose operation. Denoting by $\mathbf{X} = (\mathbf{x}\ \mathbf{1})$ the matrix of size $n \times 2$ having a column vector of ones as the rightmost column, by $\beta = (a\ b)^T$ the vector of the (unknown) linear regression coefficients, and by $\epsilon = (\epsilon_1...\epsilon_n)^T$ a vector of independent normally-distributed errors with standard deviation σ, the linear model can be written:

$$\mathbf{y} = \mathbf{X}\beta + \epsilon \tag{9.3}$$

In other words, each $y_k = ax_k + b$ has a normal distribution:

$$y_k \sim \mathcal{N}(ax_k + b, \sigma^2) \tag{9.4}$$

and therefore $\mathbf{y}$ has a multivariate normal distribution:

$$\mathbf{y} \sim \mathcal{MN}(\mathbf{X}\beta, \sigma^2\mathbf{I}) \tag{9.5}$$

where $\mathbf{I}$ is the identity matrix of rank n, and $\mathcal{MN}(\mu, \sigma^2\mathbf{I})$ is the product of n identical normal distributions.

Remark 9.1 We use the notation $\mathbf{Z} \sim \mathcal{MN}(\boldsymbol{\mu}, \boldsymbol{\Sigma})$ to mean that $\mathbf{Z} = (Z_1...Z_n)^T$ has a multivariate normal distribution, while $\mathcal{N}(\mu, \sigma^2)$ denotes an univariate normal distribution, when the variable $\mathbf{Z}$ is a vector. The general definition of a multivariate normal probability distribution is:

$$p(\mathbf{Z}) = \frac{1}{\sqrt{(2\pi)^n \det(\mathbf{\Sigma})}} \exp\left[-\frac{1}{2}(\mathbf{Z} - \boldsymbol{\mu})^T \mathbf{\Sigma}^{-1}(\mathbf{Z} - \boldsymbol{\mu})\right]$$

which reduces to (9.5), with $\mathbf{\Sigma} \equiv \sigma^2 \mathbf{I}$, if the Z_k are independent normally distributed variables with identical variance σ^2.

The OLS procedure consists of estimating a vector $\hat{\boldsymbol{\beta}}$ minimizing the sum of squared residuals (SSR):

$$\hat{\boldsymbol{\beta}} = \min_{\beta} SSR(\boldsymbol{\beta}) \tag{9.6}$$

It is easy to show that such a minimum corresponds to the solution of the system:

$$\hat{\boldsymbol{\beta}} = (\mathbf{X}^T \mathbf{X})^{-1} \mathbf{X}^T \mathbf{y} \tag{9.7}$$

provided $\mathbf{X}^T \mathbf{X}$ is not singular.

Proof Solving (9.3) with respect to ϵ, with $\hat{\boldsymbol{\beta}}$ in place of β, and multiplying to the left by its transpose, we obtain the sum of squared residuals:

$$SSR = \epsilon^T \epsilon = (\mathbf{y} - \mathbf{X}\hat{\boldsymbol{\beta}})^T (\mathbf{y} - \mathbf{X}\hat{\boldsymbol{\beta}}) \tag{9.8}$$

Developing the product:

$$SSR = \mathbf{y}^T \mathbf{y} - 2\hat{\boldsymbol{\beta}}^T \mathbf{X}^T \mathbf{y} - \hat{\boldsymbol{\beta}}^T \mathbf{X}^T \mathbf{X} \hat{\boldsymbol{\beta}} \tag{9.9}$$

and taking the derivative of SSR with respect to $\hat{\boldsymbol{\beta}}$ we obtain:

$$-2\mathbf{X}^T \mathbf{y} + 2\mathbf{X}^T \mathbf{X} \hat{\boldsymbol{\beta}} = 0 \tag{9.10}$$

which eventually gives (9.7). □

If, instead of being independent and identically distributed, errors are correlated, the matrix $\sigma^2 \mathbf{I}$ must be replaced by a covariance matrix $\mathbf{\Sigma}$ (see Remark 9.1). In the AR(1) case, errors are first-order autoregressive, i.e. error u_t at time t only depends upon the error at the previous time $t - 1$:

$$u_t = \rho u_{t-1} + \epsilon_t \tag{9.11}$$

with $\epsilon_t \sim \mathcal{N}(0, \sigma^2)$, i.e. are normal independent (uncorrelated) errors, identically distributed as required by the Gauss-Markov conditions necessary for the application of the ordinary least square procedure. In (9.11) ρ is the correlation among errors, in absolute value a number less than one. Modifying (9.3),the linear model with the serially correlated error $\mathbf{u}$ is:

$$\mathbf{y} = \mathbf{X}\beta + \mathbf{u} \tag{9.12}$$

The variance σ_u^2 of the true, serially correlated, error at time t is $E[u_t^2] = \sigma^2/(1-\rho^2)$ while the covariance between times t and $t - k$ is $E[u_t u_{t-k}] = \rho^k \sigma^2/(1 - \rho^2)$.

Proof Suppose there are N temporal steps. By applying (9.11) recursively, we obtain:

$$u_t = \rho^N t_{t-N} \sum_{i=0}^{N-1} \rho^i \epsilon_{t-i} = \sum_{i=0}^{\infty} \rho^i \epsilon_{t-i} \tag{9.13}$$

where the last passage holds for very large N (virtually infinite). The covariance γ_k for a time lag k is therefore:

$$\gamma_k = E[u_t u_{t-k}] = \sum_{i=0}^{\infty} \sum_{j=0}^{\infty} \rho^i \rho^j E[\epsilon_{t-i} \epsilon_{t-k-j}] \tag{9.14}$$

Remembering the assumptions, $E[\epsilon_i^2] = \sigma^2$ and $E[\epsilon_i \epsilon_j] = 0$ for $i \neq j$, we obtain:

$$\gamma_k = \sigma^2 \frac{\rho^k}{1 - \rho^2} \tag{9.15}$$

□

Therefore, in presence of AR(1) error, the covariance matrix can be computed in terms of the correlation coefficient ρ:

$$\Sigma = \sigma^2 \Omega = \frac{\sigma^2}{1 - \rho^2} \begin{bmatrix} 1 & \rho & \rho^2 & \cdots & \rho^{n-1} \\ \rho & 1 & \rho & \cdots & \rho^{n-2} \\ \vdots & \vdots & \vdots & & \vdots \\ \rho^{n-1} & \rho^{n-2} & \rho^{n-3} & \cdots & 1 \end{bmatrix} \tag{9.16}$$

If Ω is known, the OLS approach can be generalized (Sen and Srivastava, 1990) to give the required estimation of β:

$$\hat{\beta} = (\mathbf{X}^T \Omega^{-1} \mathbf{X})^{-1} \mathbf{X}^T \Omega^{-1} \mathbf{y} \tag{9.17}$$

All the above treatise requires is that we know the correlation coefficient ρ, but we don't!

9.1.1 MCMC implementation of Bayesian regression

The unknown parameters of the regression problem are four: the vector of regression coefficients $\beta = (\beta_0, \beta_1)$, the error variance σ^2 and the correlation coefficient ρ. From the Bayesian viewpoint we are not trying to obtain single estimates (and, possibly, variances or confidence intervals) for those parameters, but we are interested in evaluating their posterior probability distribution.

The procedure, as described by (9.2), specialized in (9.18) for the regression problem, requires us to write down a likelihood function and to assign a prior probability distribution to each parameter of interest:

$$p(\beta, \sigma^2, \rho | \mathbf{y}, \mathbf{X}) \propto p(\mathbf{y} | \beta, \sigma^2, \rho, \mathbf{X}) p(\beta, \sigma^2, \rho | \mathbf{X}) \tag{9.18}$$

The likelihood is no different from the joint probability distribution for the observation, supposing the regression parameters and the error correlation are known.

Therefore, using the AR(1) definition (9.11) in (9.12) it can be expressed (Chib, 1993) as a modified multivariate normal, given the first observation y_1. It comes out that the probability distribution of y_t depends on y_{t-1} as:

$$y_{t|t-1} \sim \mathcal{N}\left(\rho y_{t-1} + (\beta_0 x_t + \beta_1) - \rho(\beta_0 x_{t-1} + \beta_1), \sigma^2\right) \tag{9.19}$$

because, by subtracting ρy_{t-1} from y_t we obtain a linear model with normal, independent errors. As a consequence, the desired likelihood is given by:

$$p(y_2...y_n|y_1, \boldsymbol{\beta}, \rho, \boldsymbol{X}) \propto \frac{1}{\sigma^{n-1}} \exp\left(-\frac{1}{2\sigma^2} \sum_{t=2}^{n}(y_t - y_{t|t-1})\right) \tag{9.20}$$

The choice of the prior distribution is important, because the prior includes the *information* available (if any), but, as discussed at the beginning of this chapter, data always win. Therefore, a proper choice of the prior helps a faster convergence, but if no information is available, a 'diffuse' non-informative prior, assigning probability evenly over large regions of the parameter space, is a suitable choice. In the current regression problem, for example, we can assume that the prior distribution of $(\boldsymbol{\beta}, \sigma^2, \rho)$ is the product:

$$p(\boldsymbol{\beta}, \sigma^2, \rho) = p(\boldsymbol{\beta}|\sigma^2)p(\sigma^2)p(\rho) \tag{9.21}$$

meaning that $(\boldsymbol{\beta}, \sigma^2)$ is independent of ρ. A possible strategy is to choose for $p(\boldsymbol{\beta}|\sigma^2)$ and $p(\sigma^2)$ a pair of 'conjugate' priors.[4].

A possible choice for our problem (Chib, 1993; Hoff, 2009) is:

$$\begin{aligned} \boldsymbol{\beta}|\sigma^2 &\sim \mathcal{MN}(\boldsymbol{\beta_0}, \boldsymbol{\Sigma_0}) \\ \rho &\sim \mathcal{N}(\rho_0, \sigma_\rho^2) \\ \sigma^2 &\sim \mathcal{IG}(\tfrac{\nu_0}{2}, \tfrac{\delta_0}{2}) \end{aligned} \tag{9.22}$$

Different choices of the parameters $\boldsymbol{\beta_0}, \boldsymbol{\Sigma_0}, \rho_0, \sigma_\rho^2, \nu_0, \delta_0$ in (9.23) allow us to express the desired degree of knowledge about the regression parameters of interest. A diffuse prior is obtained posing $\boldsymbol{\beta_0} = \boldsymbol{0}$, $\rho_0 = 0$, $\nu_0 = 1$, and by assuming very large variances. The following R code implements the Bayesian estimation of the regression problem using JAGS for the MCMC sampling. Note that JAGS uses precisions $\tau = 1/\sigma^2$ instead of variances in the specification of the probability distributions. Therefore, a diffuse prior involves rather small precision, because of the inverse relationship between precision and variance.

```
## Code_9_3.R
# Linear regression using JAGS
library(rjags)

# Initialization
n <- length(T)
```

---

[4] A prior is conjugate to the likelihood function when it gives rise to a posterior belonging to the same distribution family (Gelman *et al.*, 2004)

```
jags.data = list("Y"=T,"N"=n,"X"=CO2)
jags.params=c("sigma","alpha","beta","rho")
jags.inits <- list("tau" = 1, "alpha" = 1, "rho" = 0)

# JAGS code
jags.model <- textConnection("model {
    alpha ~ dnorm(0, 0.01);
    tau ~ dgamma(1,0.001);
    sigma <- 1/sqrt(tau);
    beta ~ dnorm(0,0.001);
    rho ~ dnorm(0, 0.1);

    predY[1] <- Y[1];
    for(i in 2:N) {
        predY[i] <- alpha + beta*X[i] + rho * (Y[i-1] - alpha - beta*X[i-1]);
        Y[i] ~ dnorm(predY[i], tau);
    }
}")

# Evaluation
model <- jags.model(jags.model, data=jags.data, inits=jags.inits, n.chains=1)
# Burning in
update(model, n.iter=10000)
# Sampling
samples <- coda.samples(model, jags.params,30000)
# Transform into dataframe
out <- do.call(rbind.data.frame, samples)

# Plot histograms
hist(out$rho)
hist(out$beta)
hist(out$alpha)
hist(out$sigma)
```

The above code assigns suitable probability distributions to the parameters (α, β, ρ, τ) and defines the standard deviation σ as the reciprocal of the precision τ. Then it computes the likelihood function, or predictive distribution, based on the experimental data. JAGS evaluation consists of defining the model and running it for a suitable number of times. More details are given in Appendix B.

Figure 9.6 shows the posterior probability distributions of the four parameters. By means of the above analysis, we eventually obtain the desired regression model. Figure 9.7 compares the OLS linear fitting (dashed line) with the regression corrected for error autocorrelation (solid line).

```
## Code_9_4.R
# Plot results
plot(CO2,T)
abline(Lfit,col="red",lty=2)
interc <- mean(out$alpha)
slope <- mean(out$beta)
rho <- mean(out$rho)
abline(interc,slope,col="blue")
```

Let us take a look at the meaning of what we have found. Coming back to (9.19), that equation is simply an OLS regression involving 'lagged' variables. With reference to Chapter 6, the presence of an AR(1) error u_t means that the relationship among the dependent variable $y = T$ and the independent variable $x = CO_2$ is:

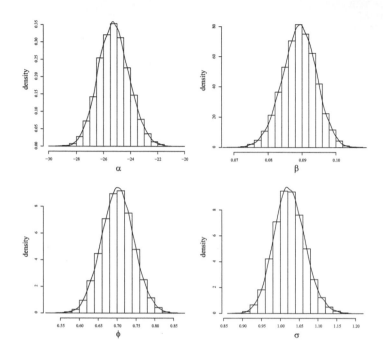

Fig. 9.6 From top-left to bottom-right, posterior probability distributions of α, β, ρ and σ.

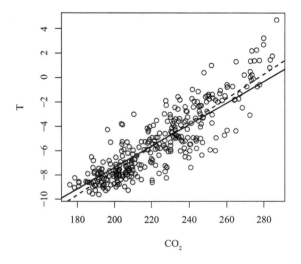

Fig. 9.7 Result of the Bayesian regression analysis.

$$y_t = \beta + \alpha x_t + u_t \tag{9.23}$$

By applying the lag operator L to the error term in (9.23), in terms of the computed correlation coefficient ρ, we transform it to:

$$y_t = \beta + \alpha x_t + (1 - \rho L)^{-1} \epsilon_t$$

where ϵ_t is white noise. Multiplying both members by $(1 - \rho L)$ gives:

$$(1 - \rho L)y_t = (1 - \rho L)\beta + \alpha(1 - \rho L)x_t + \epsilon_t$$

which defining $\tilde{y}_t = (1 - \rho L)y_t$, $\tilde{x}_t = (1 - \rho L)x_t$ and $\beta^* = (1 - \rho L)\beta$, gives an ordinary OLS equation:

$$\tilde{y}_t = \beta^* + \alpha \tilde{x}_t + \epsilon_t \tag{9.24}$$

i.e. the equation describing a correct linear regression (because ϵ_t is uncorrelated white noise) with the slope computed by the Bayesian algorithm and an intercept β^* that is scaled by a factor $(1 - \rho)$. The above procedure was firstly proposed by (Cochrane and Orcutt, 1949) and is known as Cochrane-Orcutt estimation.

Now, if we transform the original temperature and carbon dioxide data using the lag operator L and the correlation ρ, using the code below:

```
## Code_9_5.R
# define x and y as above
y <- T
x <- CO2
# redefine xx = x_tilde and yy = y_tilde
yy <- rep(0,n-1)
xx <- rep(0,n-1)
for (i in 2:n){
    yy[i-1] <- y[i] - rho*y[i-1]
    xx[i-1] <- x[i] - rho*x[i-1]
}
fit2 <- lm(yy~xx)
plot(xx,yy,
   xlab=expression(paste("(1-",rho,"L)CO"[2])),
   ylab=expression(paste("(1-",rho,"L)T")))
abline(fit2,col="red")
abline(interc*(1-rho),slope,col="green",lty=2)

# ACF
acf(fit2$residuals)
```

we verify that the OLS regression conducted on the lagged variables practically coincides with the line having the slope computed by the Bayesian procedure and the intercept transformed according to the correlation coefficient (Fig. 9.8).

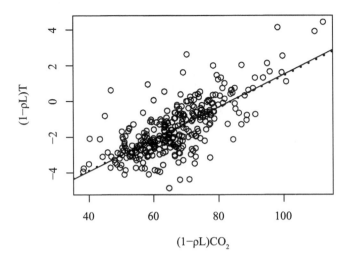

Fig. 9.8 OLS regression on the lagged variables. The solid line refers to the linear fitting, the dotted line to the parameters computed by the Bayesian procedure.

Finally, we can convince ourselves that the errors involved in the lagged regression are practically uncorrelated by looking at the autocorrelation function, computed in the last line of the above code, as Fig. 9.9 shows.

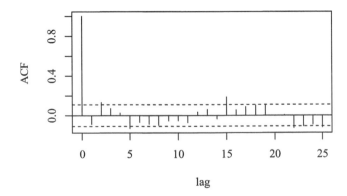

Fig. 9.9 OLS regression on the lagged variables: autocorrelation function.

Another powerful application of Bayesian analysis is the computation of the spectrum of a time series. In the following section we will show how Bayesian spectral analysis can help in the problem of target detection by a stepped frequency continuous wave (SFCW) RADAR.

9.2 Bayesian spectral analysis applied to RADAR target detection

Sometimes random phenomena obscure coherent signals. That happens when a deterministic signal is hidden by the presence of a stochastic signal (noise). That is exactly what happens when a RADAR operates in presence of intense rain, at wavelengths comparable to the size of the rain drops. Rain backscattering produces a stochastic signal having the characteristics of noise (Richards, 2014). Indeed, it has been demonstrated (Richards, 2014; Eaves and Reedy, 1987) that the motion of the clutter (rain droplets, in the present case) induces decorrelation in time and, therefore, also in space, which makes the clutter signal similar to white noise.

We will apply Bayesian analysis to the detection of an obstacle by means of a stepped-frequency continuous-wave RADAR in the presence of rain. The situation is depicted in Fig. 9.10.

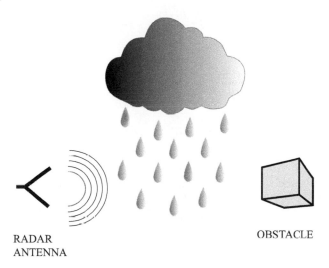

RADAR
ANTENNA

OBSTACLE

Fig. 9.10 Radar detection of an object through a rain curtain.

The SFCW radar is a practical implementation of the frequency modulated CW radar, where the modulating ramp actually consists of a stepwise digital signal making the radar output a succession of short CW pulses of increasing frequency. In the monostatic configuration, the transmitted radar signal is mixed with that received by the same antenna, reflected from an obstacle, giving rise to an oscillating low-frequency signal. In the case of a single reflector, this last signal depends harmonically on the path length, allowing the detection of the radar-to-target distance.

In general, if K reflectors are present and detectable (i.e. in the range of the radar, and having a radar cross section sufficient to produce an echo above the clutter/noise), the radar signal $s(f)$ consists of a sum of harmonics:

$$s(f) = \sum_{k=1}^{K} [A_k cos(wd_k) + B_k sin(wd_k)] \tag{9.25}$$

where w is a function of the signal frequency f. In particular, considering that the transmitted CW signal 'contains' N discrete frequencies:

$$f_i = f_0 + i\frac{BW}{N}, \quad i = 0...N \tag{9.26}$$

the received signal consists of the cosine of the phase difference between the echoed signal and the transmitted one:

$$s_i = cos(\phi_i)$$

with:

$$\phi_i = 2\pi\frac{2d}{c}\left(f_0 + i\frac{i}{N}BW\right) \tag{9.27}$$

In other words, the measured data belong to the 'frequency domain', and the elaboration (a discrete Fourier transform, in the simplest case) transforms the frequency domain into the 'distance domain'. A target at distance d_j gives rise to a signal:

$$s_j = cos(\phi_j) = cos\left(\frac{4\pi f}{c}d_j\right) \tag{9.28}$$

The distance domain is scanned by varying the index j in a suitable range. The probabilistic approach essentially consists of evaluating the *posterior* probability of the parameters of the problem $p(\delta|D, I)$ (δ denotes the set $\{d_k\}$ of the target distances) given the measured data D and all the available information I, starting from a *prior* probability $p(\delta|I)$ of those parameters and computing the likelihood $p(D|\delta, I)$ of the data given the assumed model. The Bayes theorem gives the following relation:

$$p(\delta|D, I) \propto p(\delta|I)p(D|\delta, I) \tag{9.29}$$

Assuming a Gaussian noise, the posterior probability can be demonstrated (Bretthorst, 1988) to have the form of a Student t-distribution:

$$p(\{d_k\}|D, I) \propto \left[1 - \frac{m\overline{h^2}}{N\overline{D^2}}\right]^{\frac{M-n}{2}} \tag{9.30}$$

where: $m = 2K$ is the number of model functions (two for a single reflection, four for two reflections, and so on), N is the number of data points, $\overline{D^2}$ is the squared mean of the data points, and $\overline{h^2}$ is the squared mean of the projections of the data on a set of orthogonal model functions derived from those in (9.25). This last function is given by:

$$\overline{h^2} = \frac{1}{m} \sum_{j=1}^{m} h_j^2 \tag{9.31}$$

In terms of a set of orthogonal functions H_k eqn (9.25) is written as:

$$s(f) = \sum_{k=1}^{m} C_k H_k(w, \{d_k\}) \tag{9.32}$$

the orthogonality property allows us to obtain a diagonal 'interaction matrix' $H \cdot H^T$ which simplifies the computation. Incidentally, $\overline{h^2}$ happens to be a generalization of the periodogram (Bretthorst, 1988) to arbitrary model functions, and thus it is a sufficient statistics for the set of investigated parameters $\delta = \{d_k\}$.

Referring to (Bretthorst, 1988) the computation procedure can be summarized as follows:

1. Assumed a 'model' for the main reflections ($m = 2$, in the usual case of 'separable' targets), for each possible value of the parameter of interest d_k in a guessed interval (d_{min}, d_{max}) derive the orthogonal model functions H_k, with a simple eigen-analysis of the initial harmonic functions (9.25).
2. For each distance value compute the statistics (9.31) and the relative probability value (9.30).

The mode (maximum value) of the posterior probability distribution gives the most probable reflector distance (or distances). The orthogonal model functions allow us to estimate the amplitude of the main reflection from the data and to evaluate the signal-to-noise ratio (SNR). The noise variance is easily computed from the data:

$$\overline{\sigma^2} = \frac{N}{N - m - 2} \left[\overline{d^2} - \frac{m}{N} \overline{h^2} \right] \tag{9.33}$$

The SNR is eventually computed by means of (9.31) and (9.33):

$$SNR = \frac{m \overline{h^2}}{N \overline{\sigma^2}} \tag{9.34}$$

In (9.34), the average signal power and the estimated noise variance are computed for the distance d_j giving the maximum probability value. In practice (as $m = 2$) the SNR is computed according to (9.35):

$$SNR = \frac{2 \, \max(\overline{h^2})}{N \min(\overline{\sigma^2})} \tag{9.35}$$

The posterior probability (9.30) would assume a simpler shape if the noise variance were known. In that case it would correspond to an exponential of the periodogram:

$$p(\{d_k\}|D,I) \propto exp\left[\frac{m\overline{h^2}}{2\sigma^2}\right] \tag{9.36}$$

Figure 9.11 shows the detection of a metallic cube placed at a distance of 7.6 m from the radar. The weather conditions were good (no rain) in that case. Figure 9.11 shows the periodogram $\overline{h^2}$ as a function of the distance from the radar.

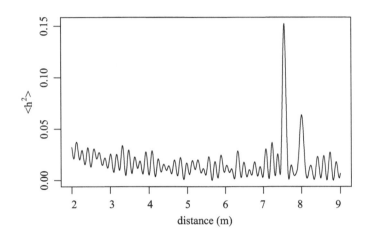

Fig. 9.11 Radar detection of an obstacle in dry air

Figure 9.12 shows the posterior probability computed from (9.36) by approximating the true variance with that computed from the data. Based on the posterior probability the distance detection is a sharp peak, as a consequence of the exponentiation of $\overline{h^2}$.

The problem with posterior probability is that a very intense echo makes lower reflections undetectable. In Fig. 9.12 the posterior probability related to Fig. 9.13 for a test reflector (a metallic target placed at a known distance as a reference) at a distance of 13.5 m from the radar appears as a sharp vertical line. No other obstacles are detected.

A plot of $\overline{h^2}$ as a function of distance, instead, suggests the presence of at least two small-reflections at about 10 and 11.8 m from the antenna, as Fig. 9.14 shows.

Now, let us see what happens during rain. The radar made a frequency sweep between 76 and 77 GHz, which means an average wavelength of 3.9 mm. Rain drops have a size of the same order of magnitude as the radar wavelength, being approximately distributed as a truncated gamma function with a mode around 0.5 mm (Fiser, 2010).

Moreover, their size is very small compared to the radar resolution, and of course they are (rapidly) moving objects, so that rain backscattering is a very dynamic process

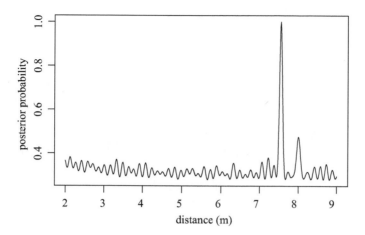

Fig. 9.12 Posterior probability for the detection of an obstacle in dry air.

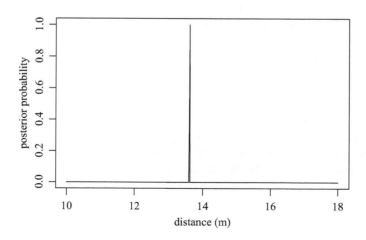

Fig. 9.13 Radar detection of a test reflector.

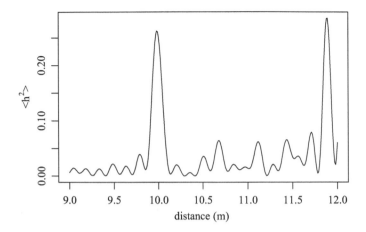

Fig. 9.14 Radar detection of obstacles close to a test reflector

involving a decorrelation time (decorrelation means, in the present context, a non-constant relationship between the phase of the transmitted signal and that of the echo) that decrease with increasing rain velocity. This is a very complex problem that can be summarized by asserting that rain is something like as Gaussian noise.

Figure 9.15 shows the disastrous tentative attempt at detection of the same obstacle as Fig. 9.11 in presence of rain. The metallic cube at 7.6 m is not detected at all!

It appears reasonable that averaging over successive measurements can bring a substantial gain in the 'signal-to-clutter' ratio, as successive measurements on immobile or very slow targets are very correlated whilst rain clutter is not. By averaging over five measurements, which span a time of the order of a couple of seconds (irrelevant for the purpose of detecting a still or slowly moving object, but long relative to the rain drops movement), the situation completely changes, allowing us to correctly detect the obstacle (Fig. 9.16).

Apart from the qualitative reasoning suggesting the averaging procedure to solve the detection problem in the presence of intense noise, that approach has an obvious statistical ground. Resorting to the approximate expression for posterior probability (9.36), observing that in N repeated measurements the posterior probability of the distance δ becomes:

$$p(\delta|\sigma, D, I) \propto exp \left[\sum_{k=1}^{N} \frac{C(\delta)_k}{\sigma^2} \right] \qquad (9.37)$$

where $C(\delta)_k$ is the k-th periodogram. The standard deviation is assumed known, as explicitly indicated in (9.37). It can be demonstrated that, if the spectrum amplitude

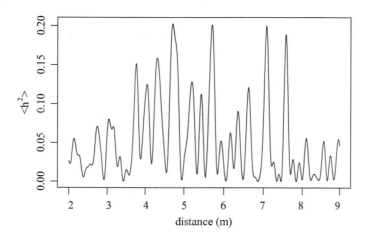

Fig. 9.15 Failure in detecting an obstacle in presence of rain.

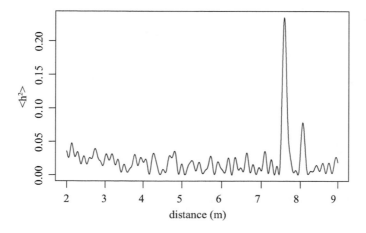

Fig. 9.16 Radar detection of an obstacle in presence of rain.

remains constant during the repetition, the estimation of δ becomes better by a factor $N^{1/2}$, as expected for the mean of a quantity (whose standard error is $\sigma/N^{1/2}$).

The following R codes can be used to generate the figures of this section. In particular: `Code_9_6.R` refers to a single measurement in dry weather, `Code_9_7.R` to a single measurement extracted from a set of multiple measurements (the number 40) and `Code_9_8.R` to the average of five measurements extracted from the same set. The data files are available on the book website.

```
## Code_9_6.R
# Radar Analysis
setwd("C:/RPA/code/bayesian")
# speed of EM signal in air
c <- 2.998e8
# read data: single measurement
data.dry <- readLines("data/single_dry.txt")
strsplit(data.dry,"SIGNAL,") -> tmp
header <- tmp[[1]][[1]]
dat <- tmp[[1]][[2]]
dat <- as.numeric(unlist(strsplit(dat, ",")))
# Elaboration
f1 <- 76e9
f2 <- 77e9
N <- length(dat)
deltaf <- (f2-f1)/(N-1)
f <- f1 + (0:(N-1))*deltaf
BW <- f2-f1
# normalized signal
dat1 <- dat-mean(dat)
si <- dat1/max(dat1)
# Bayesian spectrum analysis
d2m <- mean(si^2)
T <- round((N-1)/2)
T1 <- -T
T2 <- T-1
# choose distance interval
dd <- seq(2,9,0.01)
# uncomment to zoom a distance portion: Figure 9_14
# dd = seq(9,12,0.01)
# uncomment to see test reflector: Figure 9_13
# dd = seq(10,18,0.01)
Np <- length(dd)
# Needed matrices
Hj1 <- matrix(0,N,Np)
Hj2 <- matrix(0,N,Np)
G = matrix(0,2,N)
hj1 <- matrix(0,Np)
hj2 <- matrix(0,Np)
hh <- matrix(0,Np)
PP <- matrix(0,Np)
ss <- matrix(0,Np)
# Decompose in the orthogonal base sin/cos
k <- 1
m <- 2
for (d in dd)
{
  w <- 4*pi*d*BW/(N*c)

  C <- N/2 + sin(N*w)/(2*sin(w))
  S <- N/2 - sin(N*w)/(2*sin(w))
```

```
# Gjk matrix
for (i in 1:N)
{
  l <- -T + 2*T*(i-1)/(N-1)
  G[1,i] <- cos(w*l)
  G[2,i] <- sin(w*l)
}

# orthogonal functions
Hj1[,k] <- G[1,]/sqrt(C)
Hj2[,k] <- G[2,]/sqrt(S)

# project data on the orthogonal base
h1 <- sum(si*Hj1[,k]);
h2 <- sum(si*Hj2[,k]);

hj1[k] <- h1;
hj2[k] <- h2;
h2m = 0.5*(h1^2+h2^2)
hh[k] = h2m;

# Periodogram: equation (9.30)
P <- (1-m*h2m/(N*d2m))^((m-N)/2)
PP[k] <- P

# Periodogram: equation (9.36)
sigma2 <- N/(N-m-2)*(d2m - m*h2m/N)
#P <- exp(m*h2m/sigma2)
ss[k] <- sigma2
#PP[k] <- P

  k <- k+1
}

HHMAX <- max(hh)
SSMIN <- min(ss)
# Posterior probability: equation (9.30)
p <- PP/max(PP)

# Plots
plot(dd,p,t="l",xlab="Distance (m)",
ylab="Posterior Probability",mgp=c(2,0.8,0))

plot(dd,abs(hh),t="l",xlab="Distance(m)",
ylab=expression(paste("<",h^2,">")),mgp=c(2,0.8,0))


# Signal to Noise Ratio
snr <- (2*HHMAX/N)/SSMIN

## Code_9_7.R
setwd("C:/RPA/code/bayesian")
# speed of EM signal in air
c <- 2.998e8
# read data: multiple measurements
M <- readLines("data/multiple_rain.txt")
temp <- list()
vec <- c(list())
for (i in 1:length(M))
{
```

```
    one.line <- M[i]
    strsplit(one.line,"SIGNAL,") -> tmp
    header <- tmp[[1]][[1]]
    M.textdata <- paste(header,"SIGNAL,")
    dat <- tmp[[1]][[2]]
    dat <- as.numeric(unlist(strsplit(dat, ",")))
    M.data <- dat
    temp$data <- M.data
    temp$textdata <- M.textdata
    vec[[i]] <- temp
}
# vec[[1]][[1]], vec[[2]][[1]] contain data
# vec[[1]][[2]], vec[[2]][[2]] contain headers

choice <- 40
meas <- vec[[as.integer(choice)]][[1]];
# Elaboration
f1 <- 76e9
f2 <- 77e9
N <- length(meas)
deltaf <- (f2-f1)/(N-1)
f <- f1 + (0:(N-1))*deltaf
BW <- f2-f1
# normalized signal
meas <- meas-mean(meas)
si <- meas/max(meas)
# Bayesian spectrum analysis
d2m <- mean(si^2)
T <- round((N-1)/2)
T1 <- -T
T2 <- T-1
# choose distance interval
dd <- seq(2,9,0.01)
Np <- length(dd)
# Needed matrices
Hj1 <- matrix(0,N,Np)
Hj2 <- matrix(0,N,Np)
G = matrix(0,2,N)
hj1 <- matrix(0,Np)
hj2 <- matrix(0,Np)
hh <- matrix(0,Np)
PP <- matrix(0,Np)
ss <- matrix(0,Np)
# Decompose in the orthogonal base sin/cos
k <- 1
m <- 2
for (d in dd)
{
  w <- 4*pi*d*BW/(N*c)

  C <- N/2 + sin(N*w)/(2*sin(w))
  S <- N/2 - sin(N*w)/(2*sin(w))

  # Gjk matrix
  for (i in 1:N)
  {
    l <- -T + 2*T*(i-1)/(N-1)
    G[1,i] <- cos(w*l)
    G[2,i] <- sin(w*l)
  }
```

```
  # orthogonal functions
  Hj1[,k] <- G[1,]/sqrt(C)
  Hj2[,k] <- G[2,]/sqrt(S)

  # project data on the orthogonal base
  h1 <- sum(si*Hj1[,k]);
  h2 <- sum(si*Hj2[,k]);

  hj1[k] <- h1;
  hj2[k] <- h2;
  h2m = 0.5*(h1^2+h2^2)
  hh[k] = h2m;

  # Periodogram: equation (9.30)
  P <- (1-m*h2m/(N*d2m))^((m-N)/2)
  PP[k] <- P

  # Periodogram: equation (9.36)
  sigma2 <- N/(N-m-2)*(d2m - m*h2m/N)
  #P <- exp(m*h2m/sigma2)
  ss[k] <- sigma2
  #PP[k] <- P

  k <- k+1
}

HHMAX <- max(hh)
SSMIN <- min(ss)
# Posterior probability: equation (9.30)
p <- PP/max(PP)

# Plot periodogram
plot(dd,abs(hh),t="l",xlab="Distance
(m)",ylab=expression(paste("<",h^2,">")),mgp=c(2,0.8,0))

# Signal to Noise Ratio
snr <- (2*HHMAX/N)/SSMIN

## Code_9_8.R
setwd("C:/RPA/code/bayesian")
# speed of EM signal in air
c <- 2.998e8
# read data: multiple measurements
M <- readLines("data/multiple_rain.txt")
temp <- list()
vec <- c(list())
for (i in 1:length(M))
{
  one.line <- M[i]
  strsplit(one.line,"SIGNAL,") -> tmp
  header <- tmp[[1]][[1]]
  M.textdata <- paste(header,"SIGNAL,")
  dat <- tmp[[1]][[2]]
  dat <- as.numeric(unlist(strsplit(dat, ",")))
  M.data <- dat
  temp$data <- M.data
  temp$textdata <- M.textdata
  vec[[i]] <- temp
}
# vec[[1]][[1]], vec[[2]][[1]]
```

```
# vec[[1]][[2]], vec[[2]][[2]]

# average measurements
Nstart <- 40
Nave <- 5
meas = 0;
for (i in 1:Nave)
  meas <- meas + vec[[Nstart+i-1]][[1]]
meas <- meas/Nave
# Elaboration
f1 <- 76e9
f2 <- 77e9
N <- length(meas)
deltaf <- (f2-f1)/(N-1)
f <- f1 + (0:(N-1))*deltaf
BW <- f2-f1
# normalized signal
meas <- meas-mean(meas)
si <- meas/max(meas)
# Bayesian spectrum analysis
d2m <- mean(si^2)
T <- round((N-1)/2)
T1 <- -T
T2 <- T-1
# choose distance interval
dd <- seq(2,9,0.01)
Np <- length(dd)
# Needed matrices
Hj1 <- matrix(0,N,Np)
Hj2 <- matrix(0,N,Np)
G = matrix(0,2,N)
hj1 <- matrix(0,Np)
hj2 <- matrix(0,Np)
hh <- matrix(0,Np)
PP <- matrix(0,Np)
ss <- matrix(0,Np)
# Decompose in the orthogonal base sin/cos
k <- 1
m <- 2
for (d in dd)
{
  w <- 4*pi*d*BW/(N*c)

  C <- N/2 + sin(N*w)/(2*sin(w))
  S <- N/2 - sin(N*w)/(2*sin(w))

  # Gjk matrix
  for (i in 1:N)
  {
    l <- -T + 2*T*(i-1)/(N-1)
    G[1,i] <- cos(w*l)
    G[2,i] <- sin(w*l)
  }

  # orthogonal functions
  Hj1[,k] <- G[1,]/sqrt(C)
  Hj2[,k] <- G[2,]/sqrt(S)

  # project data on the orthogonal base
  h1 <- sum(si*Hj1[,k]);
```

```
h2 <- sum(si*Hj2[,k]);

hj1[k] <- h1;
hj2[k] <- h2;
h2m = 0.5*(h1^2+h2^2)
hh[k] = h2m;

# Periodogram: equation (9.30)
P <- (1-m*h2m/(N*d2m))^((m-N)/2)
PP[k] <- P

# Periodogram: equation (9.36)
sigma2 <- N/(N-m-2)*(d2m - m*h2m/N)
P <- exp(m*h2m/sigma2)
ss[k] <- sigma2
PP[k] <- P

 k <- k+1
}

HHMAX <- max(hh)
SSMIN <- min(ss)
# Posterior probability: equation (9.30)
p <- PP/max(PP)

# Plot periodogram
plot(dd,abs(hh),t="l",xlab="Distance
(m)",ylab=expression(paste("<",h^2,">")),mgp=c(2,0.8,0))

# Signal to Noise Ratio
snr <- (2*HHMAX/N)/SSMIN
```

9.3 Bayesian analysis of a Poisson process: the waiting-time paradox

The waiting-time paradox, also known as the inspection paradox, can be summarized with a question everyone has at some time asked: why is my bus always late? Apart from psychological considerations (waiting a time longer than expected is annoying, and our mind tends to record late bus arrivals more than early bus arrivals) the question is true, but bad luck should not be invoked to explain it. Actually, Mr Poisson is responsible of it.

Bus arrival is a Poisson process. Suppose three buses stop at a given station, identified by letters 'A', 'B' and 'C'. Suppose the average inter-arrival times, t_A, t_B and t_C, or their frequencies (the reciprocals) are known. You arrive at the bus stop at time T_0 to catch bus 'A' and, luckily, you don't miss it. Your chain of reasoning could be as follows: I did not miss it, therefore I should wait a time greater than zero minutes and lower than ten; in other words I expect it will arrive in five minutes, on the average.

Nothing more incorrect. On the average you will wait for ten minutes. This apparent paradox, see for example (Dobrow, 2016), was dealt with in the scientific literature about one century ago (Masuda and Hiraoka, 2020).

Suppose the average inter-arrival times are: $t_A = 10$ min, $t_B = 15$ min, $t_C = 20$ min, or which is the same, the bus frequencies are $\lambda_A = 1/10$, $\lambda_B = 1/15$, $\lambda_C = 1/20$. Nice! Your bus is the more frequent. But, a few minutes after arriving at the bus stop, you are disappointed to see that the first bus approaching the station is 'B'. You ask yourself: why is my bus always late? In the mathematical language: what is the probability that the first bus arriving at the bus stop is 'B', when I am actually waiting for bus 'A'?

If we know the values of λ_A, λ_B and λ_C there is no problem at all in computing the requested probability. With reference to Section 4.5.1, denoting by T_X the time before bus 'X' arrives (with X = A, B, C) the probability that the first bus to arrive is 'B' is simply given by:

$$P(min(T_A, T_B, T_C) = T_B) = \frac{\lambda_B}{\lambda_A + \lambda_B + \lambda_C}$$

which, with the given numbers, is 31%. Not so improbable, don't you think?

Moreover, the higher frequency of bus 'A' does not guarantee at all its early arrival. The following R code shows that it is perfectly plausible that bus 'B' and 'C' arrive before your bus, in spite of their lower frequencies.

```
## Code_9_9.R
# Arrival times
# Total observation time
t<-120
# Arrival times for bus A
# Frequency of bus A
lambda.A<-1/10
set.seed(456)
# Number of arrivals of bus A
N.A<-rpois(1,lambda.A*t)
unifs<-runif(N.A,0,t)
arrivals.A<-sort(unifs)
plot(arrivals.A,rep(1,length(arrivals.A)),yaxt="n",ylab="",
xlab="Bus arrival times",ylim=c(0,4),xlim=c(0,120),pch="A",cex=0.9)

# Arrival times for bus B
# Frequency of bus B
lambda.B<-1/15
set.seed(789)
# Number of arrivals of bus B
N.B<-rpois(1,lambda.B*t)
unifs<-runif(N.B,0,t)
arrivals.B<-sort(unifs)
points(arrivals.B,rep(2,length(arrivals.B)),pch="B",cex=0.9)

# Arrival times for bus C
# Frequency of bus C
lambda.C<-1/20
set.seed(123)
# Number of arrivals of bus C
N.C<-rpois(1,lambda.C*t)
unifs<-runif(N.C,0,t)
arrivals.C<-sort(unifs)
points(arrivals.C,rep(3,length(arrivals.C)),pch="C",cex=0.9)
```

Every Poisson process (A, B, C) is simulated by generating a number of arrivals (N.A, N.B, N.C) by the function `rpois`, then generating N.A, N.B and N.C arrivals having uniform distribution. The Poisson process results in sorting the arrivals in ascending order.

Figure 9.17 shows a possible distribution of arrivals. As you see, bad luck is not responsible of the two 'B' arrivals followed by a 'C' and, eventually, by an 'A' bus. It's randomness!

Of course, the explanation of the paradox is that the average arrival frequency is, indeed, the average of a random quantity. If bus 'A' arrives exactly every ten minutes, your average waiting time is five minutes. To say that a bus arrives 'on the average' every 10 minutes means that you will wait for it 10 minutes, on the average.

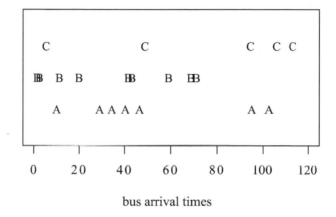

bus arrival times

Fig. 9.17 Simulation of the arrival times of three buses with frequencies $\lambda_A = 1/10$, $\lambda_B = 1/15$, $\lambda_C = 1/20$.

Now, what can I say if I don't know the bus frequencies? Well, I can spend a couple of hours recording the arrival times and I could be interested in estimating either (λ_A, λ_B and λ_C) and the probability that bus 'B' is the first to arrive.

Consider first a single bus line, for instance line 'A'. From Chapter 4, we know that the inter-arrival times $\{t_1, .., t_n\}$ are exponentially distributed. In the Bayesian framework, this means that the likelihood function (or sampling distribution) is given by:

$$p(\{t_1, .., t_n\}|\lambda_A) \propto (\lambda T)^n \exp(-\lambda T)$$

where $T = \sum_{k=1}^{n} t_k$. We can choose an improper prior probability $p(\lambda)$, for example an uniform distribution for λ, or a Jeffreys prior:

$$p(\lambda) \propto \frac{1}{\lambda}$$

corresponding to a non-informative prior suitable when observing times between events (Insua *et al.*, 2012). Both those choices are 'conjugate' to the likelihood and therefore they bring to similar posterior distributions for λ, for $Gam(n+1,T)$ and for $Gam(n,T)$.

We choose the last. Our aim is to estimate the (posterior) probability distributions of λ_A, λ_B and λ_C given the observations and, as a last step, to compute the probability that a bus 'B' arrives first. If the waiting times of the three bus lines are independent, with the above choice for prior and likelihood we have a closed-form expression for those probabilities:

$$p(\lambda_A|t_1^A..t_k^A) = Gam(n_A, T_A), \ n_A = k$$
$$p(\lambda_B|t_1^B..t_l^B) = Gam(n_B, T_B), \ n_B = l$$
$$p(\lambda_C|t_1^C..t_m^C) = Gam(n_C, T_C), \ n_C = m$$

The knowledge of analytical expression of the distribution, also allows the computation of the mean value and the variance of the expected λ's:

$$\mathrm{E}\,[\lambda_A] = \frac{N_A}{T_A} \qquad \mathrm{Var}\,[\lambda_A] = \frac{N_A}{T_A^2}$$
$$\mathrm{E}\,[\lambda_B] = \frac{N_B}{T_B} \qquad \mathrm{Var}\,[\lambda_B] = \frac{N_B}{T_B^2}$$
$$\mathrm{E}\,[\lambda_C] = \frac{N_C}{T_C} \qquad \mathrm{Var}\,[\lambda_C] = \frac{N_C}{T_C^2}$$

In order to compute

$$p\left(\frac{\lambda_B}{\lambda_A + \lambda_B + \lambda_C}\right)$$

we can resort to the Monte Carlo method. We extract N samples from each of the three distributions above, obtaining:

$$\lambda_{A_1}, \lambda_{A_2}, ..., \lambda_{A_N}$$
$$\lambda_{B_1}, \lambda_{B_2}, ..., \lambda_{B_N}$$
$$\lambda_{C_1}, \lambda_{C_2}, ..., \lambda_{C_N}$$

Now, for the assumed independence, the sequence $(\lambda_{A_1}, \lambda_{B_1}, \lambda_{C_1})...(\lambda_{A_N}, \lambda_{B_N}, \lambda_{C_N})$ consists of N independent samples from the joint distribution for $\lambda_A\lambda_B\lambda_C$. As a consequence of the law of large numbers, this last empirical distribution approximates the 'true' distribution, and we can estimate the required probability $P(min(T_A, T_B, T_C) = T_B)$ simply by making an average. The following code shows how to do it in R.

```
## Code_9_10.R
# Arrival times
# Total observation time

t <- 120
```

```
#Arrival times for bus A

lambda<-1/10
set.seed(123)
N<-rpois(1,lambda*t)
unifs<-runif(N,0,t)
(arrivals.A<-sort(unifs))
#Arrival times for bus B

lambda<-1/15
set.seed(987)
N<-rpois(1,lambda*t)
unifs<-runif(N,0,t)
(arrivals.B<-sort(unifs))
#Arrival times for bus C

lambda<-1/20
set.seed(543)
N<-rpois(1,lambda*t)
unifs<-runif(N,0,t)
arrivals.C<-sort(unifs)


N1 <- length(arrivals.A)
T1 <- arrivals.A[N1]
N2 <- length(arrivals.B)
T2 <- arrivals.B[N2]
N3 <- length(arrivals.C)
T3 <- arrivals.C[N3]

#Jeffreys (improper) Prior = 1/lambda
# Posterior = Gamma(n,T)
# yi = lambda_i

set.seed(111)
lam1 <- rgamma(50000,N1,T1)
set.seed(222)
lam2 <- rgamma(50000,N2,T2)
set.seed(333)
lam3 <- rgamma(50000,N3,T3)
hist(lam1,freq=FALSE,main="",xlab=expression(lambda[A]))
lines(seq(0,0.25,0.001),dgamma(seq(0,0.25,0.001),N1,T1))
hist(lam2,freq=FALSE,main="",xlab=expression(lambda[B]))
lines(seq(0,0.25,0.001),dgamma(seq(0,0.25,0.001),N2,T2))
hist(lam3,freq=FALSE,main="",xlab=expression(lambda[C]),ylim=c(0,15))
lines(seq(0,0.25,0.001),dgamma(seq(0,0.25,0.001),N3,T3))
c(N1/T1,mean(lam1))
c(N2/T2,mean(lam2))
c(N3/T3,mean(lam3))
p <- lam2/(lam1+lam2+lam3)
(h <- hist(p,freq=FALSE,main="",
xlab=expression(p(lambda[B]/paste(Sigma,lambda[i])))))
mean(p)
```

In the case implemented in the above code, the mode of the average probability distribution that bus B is the first to arrive at the bus stop is about 28%. The mode, which identifies the maximum value of the posterior probability, practically coincides with the mean value due to the shape of the probability distribution. Figures 9.18 show the posterior probability distributions of λ_A, λ_B and λ_C (superimposed on the empirical distributions) and the probability distribution of the required parameter $\lambda_B/(\lambda_A + \lambda_B + \lambda_C)$. Note that the focus of Bayesian analysis is on distributions, not just on point estimates.

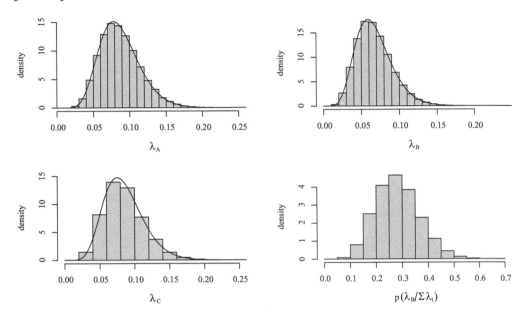

Fig. 9.18 From top-left to bottom-right, posterior probability distributions of λ_A, λ_B, λ_C and $\lambda_B/(\lambda_A + \lambda_B + \lambda_C)$.

The discussion above, as the Monte Carlo method itself, is grounded in the assumption that the law of large numbers (and the central limit theorem) is valid, a very common situation actually. We will consider a curious example where that is not true in the following section.

9.4   Bayesian analysis applied to a lighthouse

This example is taken and reworked from *Gull (1988) in 'Bayesian inductive inference and maximum entropy', in Maximum entropy and Bayesian methods in science and engineering, Kluwer.*

The example is remarkable for at least three reasons. (1) It shows how the Bayesian approach is a 'paradigm' that includes techniques normally used in a more or less uncritical way. (2) It shows how the mean is not **always** the best estimate of a random quantity, and how the solution exists even when the central limit theorem does not hold. (3) Finally, it shows how the information present in the data always wins in the end, if it is sufficient, and how the influence of the first choice of the a priori probability becomes negligible as the experimental data grows.

9.4.1 Description

A lighthouse is in a certain position on a straight stretch of coastline, at a position X_0 along the beach, measured from an arbitrarily chosen origin, and at a distance of Y_0 from the sea. The lighthouse is in constant rotation and emits short collimated flashes, at random time intervals (and therefore θ angles). Photo-detectors on the beach record the flash, but not the θ angle from which the beam is coming.

The experimental data is the set of $\{x_k\}$ positions of the photo-detectors that have been activated by a flash.

Suppose, for simplicity (just so as not to have to infer two parameters, but only one) we know the distance Y_0. What is the X_0 position? How can we estimate it from the data?

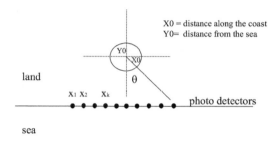

Fig. 9.19 Schematic of the problem.

9.4.2 Solution

Each x_k corresponds to an azimuth value θ_k. For the light beam to be visible, the angle must be between $-\pi/2$ and $\pi/2$ (extremes not included), i.e.:

$$\theta_k \in \left(-\frac{\pi}{2}, \frac{\pi}{2}\right)$$

Obviously, if the angle is exactly $\pm\pi/2$ the light beam does not hit the coast. Realistically we can assign a uniform probability density to the azimuth θ_k, that is to the k-th datum. If we denote by X the unknown position (while Y_0 is known):

$$p(\theta_k|X, Y_0) = \frac{1}{\pi}$$

The value $\frac{1}{\pi}$ comes from the integration of the uniform probability distribution:

$$\int_{-\infty}^{+\infty} p(\theta)d\theta = 1$$

In this case, the integral is definite:

$$p(\theta_k|X, Y_0) = \int_{-\pi/2}^{\pi/2} p(\theta)d\theta = \frac{1}{\pi/2 - (-\pi/2)} = \frac{1}{\pi}$$

Trigonometric considerations, when X_0 is known, allow us to say that:

$$Y_0 \tan(\theta_k) = x_k - X_0$$

By knowing X_0, Y_0 and x_k, the $\tan(\theta_k)$ is obtained and the angle is obtained with the arc tangent function.

The position of the photo-detector activated by the light beam is therefore:

$$x_k = X_0 + Y_0 \tan(\theta_k)$$

and the tangent of the angle is:

$$\tan(\theta_k) = \frac{x_k - X_0}{Y_0}$$

therefore the angle is:

$$\theta_k = \operatorname{atan}\left(\frac{x_k - X_0}{Y_0}\right)$$

We use the variable transformation from θ to x to get the probability density of x_k. If $x = x(\theta)$, for dx and $d\theta$ infinitesimal:

$$p(x)dx = p(\theta)d\theta$$

therefore:

$$p(x) = p(\theta)\left|\frac{d\theta}{dx}\right|$$

(the reason for the absolute value is that it must be a length ratio, i.e. always positive). The derivative $d\theta/dx$ is given by the derivative of the arc tangent with respect to x which is:

$$\frac{d\theta}{dx} = \frac{Y_0}{[Y_0^2 + (x_k - X_0)^2]}$$

while $p(\theta) = 1/\pi$ as shown above.
Finally:

$$p(x_k | X_0, Y_0) = \frac{Y_0}{\pi\,[Y_0^2 + (x_k - X_0)^2]}$$

In summary, if we know the (X_0, Y_0) position of the lighthouse, the probability of recording a flash at the x_k position has a Cauchy distribution. The Cauchy distribution is explained in detail in the next chapter about probability distributions. The Cauchy distribution is often used in statistics as an example of a 'pathological' distribution since both its expected value and its variance are undefined. The Cauchy distribution does not have finite moments of order greater than or equal to one.

Note: If we did not know the distance Y_0 of the lighthouse from the sea, we would have to estimate two parameters, a somewhat more complex problem (the solution of which risks overshadowing what we are interested in showing here).

To estimate (infer) the X parameter (the position of the light beam), we need to estimate the posterior probability of X, given Y_0 and the records $\{x_k\}$:

$$p(X|x_k, Y_0)$$

From Bayes' theorem:

$$p(X|\{x_k\}, Y_0) = \frac{p(\{x_k\}|X, Y0)p(X|Y_0)}{p(\{x_k\}|Y_0)}$$

The term in the denominator does not depend on the parameter sought. The first term of the numerator is what is called the 'likelihood' of the data, the second is the a priori probability of X. When we have no idea what a priori distribution a variable has, it is reasonable to take it uniform in a sensible range $[X_{min}, X_{max}]$ and zero outside:

$$p(X|Y_0) = p(X) = \begin{cases} \alpha, & X \in [X_{min}, X_{max}] \\ 0, & X \notin [X_{min}, X_{max}] \end{cases} \tag{9.38}$$

If the data x_k are independent, as is reasonable to assume, the probability $p(\{x_k\}|X, Y0)$ is the product of the probabilities of the single events x_k, therefore:

$$p(\{x_k\}|X, Y_0) - \prod_{k=1}^{N} p(x_k|X, Y_0)$$

Take the logarithm of the posterior probability $p(\{x_k\}|X, Y0)$:

$$L = \log(p(\{x_k\}|X, Y_0) = \beta - \sum_{k=1}^{N} \log(Y_0^2 + (x_k - X)^2)$$

where β includes everything that does not depend on the X parameter. The estimate of the position X_0 is obtained by looking for the maximum of the a posteriori distribution, that is, theoretically looking for the value of X which is a solution of:

$$\frac{dL}{dX} = 2 \sum_{k=1}^{N} \frac{x_k - X}{Y_0^2 + (x_k - X)^2} = 0 \tag{9.39}$$

9.4.3 Numerical procedure

The explicit solution of (9.39) is not analytically feasible. Instead of solving it numerically, it is more instructive to see how the posterior probability $\exp(L)$ behaves as the number of detections $\{x_k\}$ changes. This is what the following R code does, where we assume $Y_0 = 1$ km and the 'true' value of X_0 2 km. The code generates N angles θ_k and from these it calculates x_k, since the true value of X_0 is known.

```
## Code_9_11.R
## Where is the light ?

# distance from sea
Y0 <- 1
# distance along the coast
```

```
X0 <- 2
# possible values of X (positions of photo-detectors)
dx <- 0.05
X <- seq(-5,5,dx)
Nx <- length(X)

######################################################################
# number of measurements
# to change to observe the effect on the distribution a posteriori
N <- 10
######################################################################
# tetak <- runif(k,-pi/2,pi/2)
# instead of taking theta between -pi/2 and pi/2
# theta is selected such to determine "possible" x
# included between -x_max and x_max
x.max <- 50
# Here the angle is obtained using the arctan function, where x.max in an angle
tetak.max <- atan(x.max)
tetak <- runif(N,-tetak.max,tetak.max)
tetak
# compute the positions of the detectors activated by flash light
xk <- X0+Y0*tan(tetak)
# What is the distribution of the positions ?
hist(xk,main="")

L <- rep(0,Nx)
for (i in 1:Nx){
lk <- log(Y0^2+(xk-X[i])^2)
L[i] <- sum(lk)
}

hist(lk)

# posteriori probability
post <- exp(-L)
plot(X,dx*post/sum(post),t="l",ylab="p(X|x,Y0)")
abline(v=2,col="blue",lty=2)
abline(v=mean(xk),col="red")
```

9.4.4   Results

This is what happens (typically, the calculation is stochastic ...) as the number of detections changes.

The true value (dashed line) and the average value of x_k (solid line) are superimposed on the probability density curve. Due to the choice of a uniform distribution for θ (which is reflected in a Cauchy distribution for likelihood), and a uniform a priori probability, with little data the maximum posterior probability rarely hits the real position X_0.

For $N = 100$ detections, the maximum a posteriori probability starts hitting the true value (Fig. 9.22) almost always, but the average value of x_k can be very far from it.

At first sight this thing is surprising, because we are used to attribute almost 'magical' properties to the average value, by virtue of the central limit theorem, which asserts that for a sample $\{x_1...x_n\}$, collected from a distribution with mean μ and

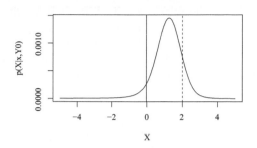

Fig. 9.20 N = 4.

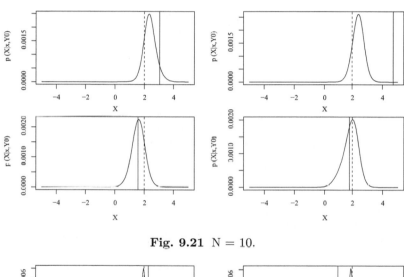

Fig. 9.21 N = 10.

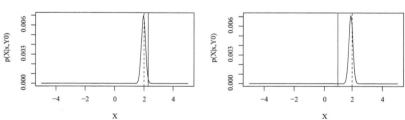

Fig. 9.22 N = 100.

variance σ^2 the distribution of the mean value $\bar{x}$ tends to a normal distribution with mean μ and variance σ^2/n, for $n \to \infty$.

The problem here is that the Cauchy distribution violates the validity conditions of this theorem, because it has large tails and as such gives a moment of order 2 of infinite value. From this example we can observe that although the central limit theorem is not valid, and therefore the average is not a sensible estimate of the position of the lighthouse, we can still calculate the a posteriori distribution, and the maximum of this gives us the correct position of the lighthouse. We note, incidentally, that this procedure coincides with the search for the maximum likelihood! In addition to shedding light on the latter, whose motivation is often unclear in any other way, it is evident that nothing prevents us from using information on the position of the lighthouse, if we have it, and using it by imposing a different form of a priori probability, and making the calculation of the true position correct and faster.

9.5 Exercises

Exercise 9.1 If H_1, H_2, ..., H_n are mutually exclusive hypotheses, the Bayes theorem asserts that the probability of a particular hypothesis H_i after the verification of an event A is given by:

$$P\{H_i|A\} = \frac{P\{H_i\}P\{A|H_i\}}{P\{A\}}$$

where:

$$P\{A\} = \sum_{i=1}^{n} P\{H_i\}P\{H_i|A\}$$

Demonstrate that the Bayes theorem is a corollary of the multiplication probability theorem, stating that the probability (or, also, the probability density) of an event C being the product of two events A and B is given by the product between the probability of A and that of B *conditioned* on A.

Hint: Use the multiplication theorem to express the probability of every event AH_i.

Exercise 9.2 In an archery club two archers A and B shoot an arrow each at the same target. Based on the previous performances of the two archers, we know that the probabilities of centring the target are 0.8 for archer A and 0.5 for archer B.

If only one arrow is stuck in the target board, what is the probability that it comes from archer B?

Hint: enumerate all possible mutually-exclusive hypotheses, for instance 'both archers hit the target' etc.

Exercise 9.3 A typical application of the Bayes formula is in medical diagnostics. Suppose a diagnostic test T for lycantrophe, executed on a blood sample, is 99% effective. This means that if you actually transform into a wolf on full moon nights, test T on your blood has 99% chance to result positive: $P\{T|H_L\} = 0.99$, where H_L denotes the hypothesis 'the tested individual is a lycanthrope'.

On the other side, suppose that the probability of a false positive test, $P\{T|\overline{H_L}\}$ is 10%: you are not a werewolf but the test is positive anyway. Knowing that lycanthropy is (luckily!) a rare condition, let say that its incidence in the population is 1%, if your brother Jack results positive to the test T estimate the probability that you better stay away from him especially on full moon nights.

Exercise 9.4 With reference to Exercise 9.3, what is the probability that Jack is actually a lycanthrope, if the test T on his blood gave a negative result?

Exercise 9.5 A bag contains an unknown number of dice, which you know can be of three types:

C: conventional, with faces enumerated from 1 to 6

E: even, with faces enumerated with even numbers from 2 to 12

R: repeated, two faces enumerated with 1, two with 2, two with 3

You blindly extract a die from the bag, and throw it four times obtaining the sequence 2 2 1 3. Assuming a uniform prior for the three different *hypotheses* $(P\{H_C\} = P\{H_E\} = P\{H_R\} = 1/3)$, and knowing the likelihood L (the probability of a given result):

(1) Fill a table like the following describing how the prior is updated after each result of the sequence or, which is the same, reporting the posterior probability after each outcome.

(2) Compute the posterior probability that the die is 'C' after the full sequence.

(3) Compute the posterior probability that the die is 'E' after the full sequence.

The table could be something like:

Hyp	Prior	outcome = "2" lik L	Post	outcome = "2" lik L	Post	outcome = "1" lik L	Post	outcome = "3" lik L	Post
H_C	1/3	1/6							
H_E	1/3	1/6							
H_R	1/3	1/3							

Hint: the likelihood of a particular outcome is the probability of that result, given the die, and of course it is zero if that outcome is impossible.

Exercise 9.6 How do the results of Exercise 9.5 change if you use a different prior? For example, suppose you are almost certain that dice of the 'E' type are not present in the bag, and someone has told you that there are presumably twice the number of conventional dice with respect to repeated ones.

Exercise 9.7 With reference to the bus waiting-time of Section 9.3, modify the R code to compute:

(1) The probability $p(\lambda_B > \lambda_A)$ for the given simulated data

(2) The quantiles of the empirical probability distribution of $\lambda_B / \sum \lambda_i$

Exercise 9.8 With reference to Exercise 9.7 how do things change assuming a uniform prior probability instead of $1/\lambda$?

Exercise 9.9 The JAGS language is a 'natural' extension of R for MCMC calculations. It is therefore useful to be acquainted with it. With reference to Section B.0.2 of Appendix B write an R code to compare normally-distributed random data generated in R with a sample extracted in JAGS, for example comparing the two histograms.

Note: despite the identical name, dnorm in JAGS is defined in terms of the precision instead of the standard deviation.

Exercise 9.10 We treated the problem of a fair coin at the beginning of this chapter. Suppose you want to decide about the fairness of a coin by an experiment consisting of counting the number of heads obtained in n flips. You know that the likelihood $P(\theta|h)$, the probability of obtaining h heads in n flips is the $Binomial(n, h)$ given the probability θ of a head:

$$P(\theta|h) = \binom{n}{h} \theta^h (1 - \theta)^{n-h}$$

Your prior expresses the information you have about the coin. If you have no reason to think the coin is fair, you can assume a uniform prior or, which is the same, a $Beta(1, 1)$ distribution. If you think the coin is fair, you assign a more 'informative' prior, such as $Beta(k, k)$ with k>1. The greater is k, the narrower is the prior distribution around the value $\theta = 0.5$.

Obtain the posterior distribution as a function of k and write down an R code to compute and plot likelihood, prior and posterior for $n = 100$ and $h = 40$. Maintaining the same ratio h/n, comment on how things change if $n = 10$ or $n = 1000$.

Exercise 9.11 Repeat Exercise 9.10 using JAGS. The JAGS model to be used in `rjags` is something like

```
H ~ dbin(theta,N)
theta ~ dbeta(k,k)
```

to be inserted into a `textConnection` command or to be read from a text file. Here H is the number of heads, whose distribution identifies the likelihood, and *theta* is the searched probability.

Experiment by changing the number of *burn-in* steps (using the `rjags` function `update`) and the number of extracted samples (via the `coda.samples` function).

Plot the posterior distribution and compare it to the analytical one obtained in the previous exercise.

Exercise 9.12 This problem concerns with a probability puzzle based on an (in)famous television game, known as Monty Hall. There are three doors A, B, C. Behind one door is a car, behind the others, goats. The contestant picks a door, say A, which remains closed. The host, who knows what's behind the doors, opens another one of the remaining doors, say C, where he knows there is a goat. The host asks the contestant 'Do you want to maintain your first choice A or change your mind and pick the door B?'. The answer is not indifferent, even though at a first sight it seems so. Prove via Bayes theorem that it is advantageous for the contestant to switch his choice. Try also to simulate the game.

10

Genetic algorithms: an evolutionary-based global random search

We are born by accident into a purely random universe.
Our lives are determined by entirely fortuitous combinations of genes.
Whatever happens happens by chance.

Robert Silverberg, The Stochastic Man

10.1 Introduction

In 1975 John H. Holland, professor of the University of Michigan, published the first edition of his book *Adaptation in natural and artificial systems*, summarising the result of his research activity on the application of the biological − namely, genetic − mechanisms to adaptive processes of different nature. As is well known, adaptive processes play a role in several fields, including those of artificial intelligence and computational mathematics. The book was essentially based on the following questions:

How does evolution produce increasingly fit organisms in environments which are highly uncertain for individual organisms? [...] How does an organism use its experience to modify its behaviour in beneficial ways (i.e. how does it learn or 'adapt under sensory guidance')?

And from the above questions, he asked himself:

How can computers be programmed so that problem-solving capabilities are built up by specifying '*what* is to be done' rather than '*how* to do it'?

His work was destined to become a milestone in this topic and to stimulate the development of one of the most smart and robust optimization techniques of the 20th century, the genetic algorithms (GA). GAs provide a stochastic method for the solution of problems of maximum/minimum search and of global optimization (Goldberg, 1989) based on the mechanics of natural selection and mimicking natural genetics. GAs belong to a more general class of heuristic methods inspired by biological evolutionism, known as **evolutionary algorithms (EA)** (Michalewicz, 1996). A common characteristics of EAs, as shown in the following, is the codification method: a tentative-solution for the problem is obtained in terms of a data structure that is treated as a chromosome and manipulated by means of genetic operators (e.g. crossover, mutation, inversion etc.) in order to reach the actual solution.

Leaving aside for the moment the genetic interpretation, a GA is a search model operating in a n-dimensional hyper-space where, after generating an initial distribution of hyperpoints, a solution is searched for by applying suitable operators to such a set of points to generate different configurations in the search space.

Turning back to the genetic interpretation, a GA simulates the Darwinian evolution process: it transforms a set of mathematical objects (chromosomes), each having associated a *fitness* property (measuring its 'distance' from the true solution), into a new set through a 'survival of the fittest' strategy.

10.2   Terminology and basics of GA

The solution of the problem must be representable in terms of a set of parameters (*genes*) whose union constitutes the *chromosome*. An objective function, called the *fitness function*, must be identified. The fitness function measures the goodness of a tentative-solution.

The GA begins with the generation of an initial *population* of tentative solutions, i.e. with a set of chromosomes. Such an initial population can be randomly generated (that is the typical case) or it can be built starting from a given number of individuals, whose closeness to the actual solution is postulated or known. It should be noted that usually the terms **individual** and **chromosome** have the same meaning. In other words, the GA individual is *haploid*. Models employing *diploid* individuals, i.e. possessing a couple of chromosomes, have been proposed too, but their use in problem solving is marginal.

Before discussing the function of the operators used in GA, a short definition of the biological terms that we will use in the following is in order.

10.2.1   Biological terms

Chromosome sequence of nucleotides (DNA) carrying all the information needed for cell growth, survival and reproduction.

Gene chromosome segment having functional autonomy. Through the synthesis of RNA, a gene controls protein synthesis. In other words, a gene is a functional unit regulating a series of operations in a living organism.

Genome the overall gene pool of an organism.

Haploid the condition of cells having a single series of chromosomes.

Diploid the condition of cells having a double series of chromosomes.

Alleles a couple representing the same gene (in diploid organisms) or a gene and its mutants (in haploid organisms).

Genotype the individuals seen as a gene 'carrier', i.e. as carrier of a given biological potentiality.

Phenotype the individual seen under the profile of his 'visible' properties.

Crossover genetic phenomenon consisting of the exchange of segments between *chromatids* (newly created chromosomes). After crossover, the final chromosome alternates segments coming from the original ones.

Mutation the modification of the nucleotide sequence in a gene.

Asexual reproduction a type of reproduction by which offspring arise from a single organism, and inherit the genes of that parent only

Sexual reproduction it happens through the meeting and fusion of two different reproductive cells (called gametes).

Inversion a chromosome segment is detached and re-attached in an inverted position.

Translocation a chromosome segment is detached and attached to a different chromosome.

10.2.2   Representation of the tentative solutions

The tentative-solution for a given problem usually consists of a certain number of real parameters. For example, in a problem where a function f of four real variables (X, Y, Z, T) is to be maximized, a tentative solution consists of a sequence of four real numbers. In order to conceptually simplify the treatment of the GA, such a sequence will be represented by four binary strings of fixed length. Each string represents a *gene*, whilst the binary string consisting of the concatenation of the four parameters is the *chromosome*.

The *population* we were talking about at the beginning of this section is therefore a set of binary strings. To obtain a satisfying solution, i.e. a string (X_s, Y_s, Z_s, T_s) maximizing the f function, requires the application of suitable 'genetic' operators to the individuals composing the population. Such operators can involve couples or single individuals. Before describing in detail the most commonly used operators, some specifications are in order concerning the *fitness* function.

In the simple case introduced above, the fitness function coincides with the function f. In more complex, realistic cases, the fitness function associates a number to a particular tentative solution, that number being increasingly higher as the guessed solution approaches the actual one. That association is not necessarily a function of the unknown variables. Indeed, GAs can be applied also to problems where the variables involved are not numerical.

10.2.3   Genetic operators

In this section the two most common operators are described: the **crossover** operator and **mutation**. Those operators always intervene in the *reproduction* process, that is the actual 'engine' of the search algorithm. The role of operators will be clear in the description of the simple GA, described in the following sections.

Crossover.   The crossover operator involves two chromosomes identifying different individuals (i.e. different tentative solutions). In the most common case, where the genome consists of a single chromosome, the function of the crossover operator is to exchange a chromosome section between two individuals.

Let us consider the situation depicted in Fig. 10.1.

The upper chromosomes (*PARENTS*) give rise to the lower ones (*OFFSPRING*) through the cross exchange of genetic material.

In that simple case, the crossover involves the following steps:

1. a cut point is established, identifying where to cut and reassemble the chromosome couple (*X-point* in Fig. 10.1);
2. the C_{1A} and C_{2A} sections are inherited without any alteration by the offspring;

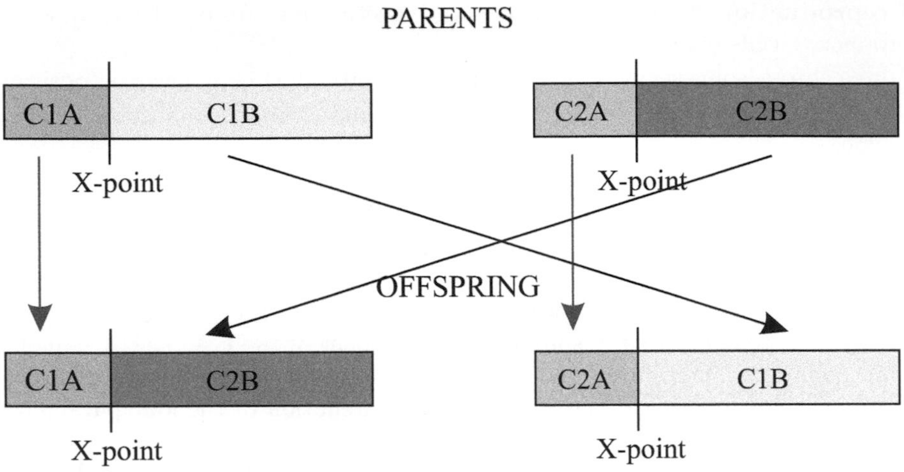

PARENTS

Fig. 10.1 Crossover between two individuals.

3. the C_{1B} section is chained to the C_{2A}, while C_{2B} attaches to the C_{1A} section.

The choice of the X-point is usually random.

If the chromosome C_1 is the binary string *10011001*, C_2 is **10111100**, and the *X-point* corresponds to the third bit, we obtain:

Parents: *10011001*   **10111100**
Offspring: *100***11100**   **101***11001*

The crossover operator has a key role in the search for the solution of a problem. GA's power in exploring the solution space largely depends on that operator, which allows us to obtain different individuals (tentative solutions) from the initial genetic material.

Mutation. The crossover operator fully exploits the information contained in the initial genetic material, but it is not able to 'create' new genetic material, i.e. material not belonging to one of the initial parents. We can understand this by means of a simple example. Suppose the parents are the following two 4-bit strings:

1100   1010

Starting from those parents, the following offspring couples can be generated:

1010   1100   if the X-point is between the first and the second bit
1110   1000   if the X-point is between the second and the third bit
1100   1010   if the X-point is between the third and the fourth bit

It should be noted that in the above case new individuals are obtained only when the cut point is in the middle of the chromosomes. Moreover, it should also be clear that, for example, a **1** in the fourth position cannot be generated, as well as the creation of a chromosome like **0010** is not possible.

These limitations are removed by the mutation operator, which can simply consist of the random complement of a bit in the chromosome, as shown in Fig. 10.2.

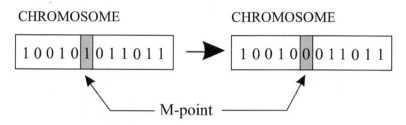

Fig. 10.2 Mutation in a chromosome.

As for the *X-point* the selection of the mutation point (*M-point*) is random.

10.3 Simple genetic algorithm

This section describes the most elemental genetic algorithm, known as *simple genetic algorithm* (SGA). The SGA consists of a given number of reproduction cycles applied to a population that remains constant in time. The reproduction process consists of:

1. Selection of a couple of individuals;
2. Mating (through the application of the crossover operator), optionally followed by mutation;
3. Computation of the fitness function for the offspring.

Steps 1 to 3 are executed on every couple extracted from the population, until all individuals are considered. Disregarding (for the moment) the modality used for selection and mating, the above sequence of steps is repeated until a predefined termination condition happens (e.g. an individual reaches a target fitness value, the allowed maximum number of generations is exceeded, the average fitness of the population falls below a given limit).

Figure 10.3 shows the flow diagram of SGA.

10.3.1 GA at work: selection and reproduction

One of the key points of the algorithm is the choice of the procedure used for selecting the individuals to be 'mated'. Such a procedure can be conducted in several ways, always bearing in mind that the purpose is to select the most suited (fit) individuals, that will be subsequently employed several times in the reproduction process to the detriment of the least fit ones, so as to propagate useful 'genetic material' to the future generations. A simple rank-based approach consists of evaluating the fitness of every member of the initial population and to sort it based on such a characteristic. Every 'potential' parent is associated a mating probability based on its rank.

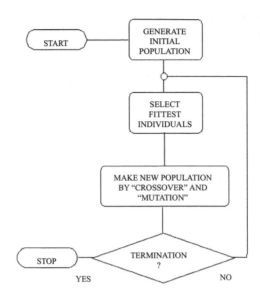

Fig. 10.3 simple genetic algorithm.

Another selection method, largely used in GA practice, is the so-called *roulette wheel selection* (RWS). In the RWS approach every individual 'occupies' a given number of roulette slots proportionally to its selection probability. If the initial population consists in four individuals A, B, C and D, respectively having fitness: $f_A = 50$, $f_B = 30$, $f_C = 15$, $f_D = 5$, their selection probabilities are: $p_A = 0.5$, $p_B = 0.3$, $p_C = 0.15$ and $p_D = 0.05$.

In the RWS, A is considered to occupy 50% of the slots, B 30% of them, and so on. At each roulette spin, A has 50% probability of being selected, B 30% and so on. At each spin any selected individual is stored to be successively mated, up to the achievement of the number of couple needed for generating a new population having the same numerosity of the parent population (eight times, in the considered example). The storing place is called the 'mating pool'. By excluding the possibility of an individual to mate with itself, in the process of selecting N couples, an individual having probability p will be chosen $p \cdot N$ on average. Coming back to the example, the A individual will be selected four times on average for the insertion in the mating pool. Likewise, B will be selected two times, C and D only once on average.

Note that neither religion nor moral commitments affect GAs individuals: they are allowed to mate with more than one partner, in the interests of the species! :)

Once the chromosomes to be reproduced have been identified, a couple of parents are randomly formed. The crossover operator is therefore applied with (usually high) probability p_X, as described in Section 10.2.3. In other words, the most part of the couples exchange genetic material between them, but nevertheless a non-zero number of individuals are transmitted unchanged to the successive generation. That could appear not too smart at first sight, as a chromosome not being a solution of the problem continues to be a wrong solution of the problem. But, if its fitness is sufficiently high to be placed in the mating pool, there is a chance that it contains useful genetic material, e.g. in the effort of finding five unknown parameters (genes) it could contain one correct parameter, the other four being completely wrong.

After mating (crossover), the individuals of the offspring generation are subjected to mutation with (usually low) probability p_M. The choice of p_M is dictated by a compromise: if $p_M = 0$, no new material is introduced, if $p_M = 1$ the GA transforms in a random search algorithm.

Let us see how the process works using a simple example.

10.3.2 An optimization problem: the Prof. Koza fast-food chain

This example is borrowed from (Koza, 1996). It is very meaningful, although very simple, and it should be useful for giving a clear idea of how GAs work.

A person is suddenly involved in the management of a chain of four fast-foods, without any previous experience. He is faced with the following binary choices, for maximizing the gain:

- Hamburger price (P): should it be high or low?
- Which beverage (B): coke or beer?
- Service (S): should it be slow and careful, or fast but hasty?

The manager only owns the weekly recording of the gains of the four fast-foods. The goal is to find the combination of choices bringing the highest total gain.

Implementation
The problem variables are three binary quantities: P, B and S. The chromosome consists of a 3-bit string. The search space consists of $2^3 = 8$ possible commercial strategies.

Representation
Each bit corresponds to a choice. The first bit is associated with the P variable, the second with B, the third with S. The following table shows the eight possible choices for a particular restaurant.

Price	Beverage	Service	Representation (PBS)
high	beer	careful	0 0 0
high	beer	fast	0 0 1
high	coke	careful	0 1 0
high	coke	fast	0 1 1
low	beer	careful	1 0 0
low	beer	fast	1 0 1
low	coke	careful	1 1 0
low	coke	fast	1 1 1

Table 10.1 Search space.

The chain manager has no idea which the most important variable is, and he/she does not possess any information concerning the 'gradients', i.e. how any variable influences the profit. Moreover, the manager cannot be sure that a step-by-step procedure is effective in maximizing the gain (e.g. optimize P, then optimize B and readjust P, etc.), because the three variables can be interrelated and the relation is not necessarily a linear one. What makes things more difficult is that the reason for the variation in the number of customer in a restaurant is unknown, and it cannot be known because clients do not motivate their choices and no warranty exists that their tastes are constant in time.

A possible strategy is to try, for a week, a different random strategy in the four fast-foods. This choice is such that the expected gain is the average available in the search space: by favouring the diversity, the probability of obtaining an 'average' profit is maximized. In the meanwhile the information that the manager can obtain in a week is maximized.

The above is exactly how GAs work. Indeed, a GA seeks for a solution in the search space by evaluating how much a number of individuals are 'compatible' with the environment. If no a priori information is available, such individuals are randomly generated to ensure the presence of the necessary diversity in the population. At every generation (iteration, as we should call it in *conventional* algorithms) the individuals are evaluated through the fitness function. In the example at hand, such a function measures the profit associated to a particular commercial strategy.

Suppose the initial population is that indicated in Table 10.2.

Fast-food	Price	Beverage	Service	Chromosome
1	high	coke	fast	0 1 1
2	high	beer	fast	0 0 1
3	low	coke	careful	1 1 0
4	high	coke	careful	0 1 0

Table 10.2 Initial population.

We arbitrarily assume the decimal value corresponding to binary strings of Table 10.1 as the profit (the value of the fitness function) associated to a given strategy (i.e. the gain that would come from the weekly takings). Of course, such a choice is meaningless; in a real situation the fitness function should be related to the actual earnings. With such an assumption the strings in Table 10.2 have the fitness values f shown in Table 10.3.

Note that the profit value for the different four strategies is the only information used for computing the results of Table 10.3.

The next step is the application of the *reproduction* operator to the initial population, as described in Section 10.3.1. The total fitness of generation 0 is 12, thus the four individuals of Table 10.3 have fitness $1/4$, $1/12$, $1/2$, $1/6$. At each selection, the best individual (110) has 50% probability of being chosen for mating, whilst the worse one (001) has about 8% probability. In terms of RWS (see Section 10.3.1), the fittest individual occupies half of the wheel slots, while the least fit occupies only $1/12$ of them.

i	PBM_i	Fitness $f(PBM_i)$
1	011	3
2	001	1
3	110	6
4	010	2
Total		12
Minimum		1
Average		3
Maximum		6

Table 10.3 Generation 0.

Selection takes place by spinning the roulette four times, to obtain the four individuals to mate. Given the selection probability of the four individuals, on the average we will obtain two times the third chromosome of Table 10.3 and one time the first and the fourth, as shown at the bottom of Table 10.4, where the individuals PBM_i are denoted by C_i.

Population at generation 0			
i	C_i	$f(C_i)$	$\frac{f(C_i)}{\sum f(C_i)}$
1	011	3	0.25
2	001	1	0.08
3	110	6	0.50
4	010	2	0.17
Individuals selected for mating			
i	C_i	$f(C_i)$	$\frac{f(C_i)}{\sum f(C_i)}$
1	011	3	0.25
2	110	1	0.50
3	110	6	0.50
4	010	2	0.17

Table 10.4 Selection.

Remember that GA is a stochastic method, thus Table 10.4 is only the most probable situation. The selection process could have produced four individuals having the lowest fitness, although with very low probability. If we compute the average and total fitness of the individuals at the bottom of Table 10.4 (pay attention, it not yet a population, because mating has not yet occurred) we see that the operation of selecting proportional to the fitness has the first effects: the total fitness rises from 12 to 17, the average fitness increases from 3 to 4.5, the minimum fitness from 1 to 2. The maximum fitness is unchanged, as no new genetic material has been introduced.

Recall that the main operator used to create new individuals is the *crossover*. To apply the crossover operator, couples of parents must be chosen from among the individuals listed at the bottom of Table 10.4. Since the crossover destroys the parents for giving rise to the offspring, its application to *all* the individuals selected for mating is not appropriate because that can push us away from the solution. Indeed, if one of the individuals at a given generation is close to **111**, which is the true solution of the problem, this individual should be given the chance to be conserved in the successive generation.

Having chosen two parents, for example 011 and 110, an X-point is randomly selected. Suppose it happens to be located between the second and third bits. The generated offspring are 010 and 111, respectively having fitness values of 2 and 7. Suppose the crossover operator to be applied to 50% of the cases, i.e. two individuals out of four selected for mating are transferred unchanged. If the mutation operator is not applied to any individual, the population at generation 1 is that shown in Table 10.5.

i	C_i	$f(C_i)$
1	111	7
2	010	2
3	110	6
4	010	2
Total		17
Minimum		2
Average		4.25
Maximum		7

Table 10.5 Population at generation 1.

The crossover has increased the maximum fitness from 6 to 7, producing an individual whose fitness is greater than those of its parents. In this simple example, two generations were sufficient to reach the best possible result. The best commercial strategy corresponds to the chromosome 111, i.e. it consists in selling the hamburgers at low price, serving coke as a beverage, and offering a fast service. In more complex, real problems obtaining the best solution is not possible, in general. To find the true solution will require a termination criterion to be established, and several generations will be needed.

10.3.3   Schemata theory. In other words, why genetic algorithms work

We have seen in the previous section *how* GAs work, but the question is: *why* do they work? Let us reconsider the example developed in Section 10.3.2. What did the manager learn after experimenting for a week? First of all, he learnt that the average fitness in the search space is 3. By analysing the four strategies employed during the week, he observes that the strategy *110* is 200% better than the average, *001* is 35% of the average and so on.

This analysis suggests to the manager several strategies for the following week:

1. He can use the same strategy as the first week, i.e. he can set up random commercial strategies in the four fast-foods. This choice clearly does not exploit at all the information gained in the first week and corresponds to a random search in the solution space (Rastrigin, 1963). This kind of search strategy quickly becomes unrealizable with increasing size of the search space. The chromosome (i.e. the

string of unknown parameters) of the simple example was three bits long, and it was based on an alphabet of length 2. In the general case, when the chromosome consists of a string of length L built with an alphabet of size K, the search space contains K^L points. Assuming we can analyse 10^9 points/second, one year of elaboration allows processing about $3 \cdot 10^{16}$, corresponding to binary strings ($K = 2$) of length $L = 55$. That length corresponds to 14 single precision real numbers, or to 7 double precision real parameters, i.e. to a rather small problem size.

2. He can try to fully exploit the obtained information, by applying to all the fast-foods the best strategy (110) resulting from the first week of experimentation. With this approach the obtained profit is doubled with respect to the average gain, but it does not allow us to establish whether a more productive strategy exists. If the manger's objective is to maximize the gain, this second strategy is also unsatisfactory.

3. He can adopt a search strategy based on the consideration that if a profit '6' is possible, perhaps a greater one is also possible.

The knowledge of the existence of a strategy giving an outcome of 6 poses a practical problem. When, during the experimentation, the manager obtains a profit 5, he feels he has lost a unit of gain. The strategy of point 3 is a compromise between the 'explorative' will, whose purpose is that of determining the *best* strategy (but that can occasionally produce gains lower than 6), and the 'conservative' will, which exploiting the current knowledge would push to a sub-optimal choice.

We can try to understand why the strategy 110 gives an over-average profit. Table 10.6 shows the possible explanations. A character # in the binary strings on the right means that the associated parameter is non-influential.

Explanation binary string			
It is the low price	1	#	#
It is coke	#	1	#
It is the careful service	#	#	0
It's the combination between low price and coke	1	1	#
It's the combination between low price and careful service	1	#	0
It's the combination between coke and careful service	#	1	0
It's the combination between low price, coke and careful service	1	1	0

Table 10.6 Explaining the goodness of the 110 strategy.

The strings in Table 10.6 are called *schemata*. Formally, in a problem based on strings of length L and consisting of characters belonging to an alphabet of size K, a schemata is a string whose characters belong to an alphabet of size $K + 1$ (K characters from the original alphabet, plus the # symbol). Let us consider, for example, the schemata ##0 referring to the hypothesis that the goodness of the 110 strategy depends only on the careful service and it is independent on the values of the remaining variables. That string clearly contains, as elements, the strings of the subset {000, 010, 100, 110}. All the strings belonging to a schemata share a number of characteristics, in the above example S = 'careful service'.

Geometrically, a schemata describes a subset of points in the search space exhibiting some kind of similarity. By considering the three bits P, B, S composing the chromosome as spatial coordinates in a 3D reference frame, the strings belonging to the schemata ##0 identify points on the plane S = 0. More generally, an L-order schemata (the *order* is the number of non-# characters) identifies a point, a schemata of order $L-1$ identifies a hyperplane of co-dimension 1, a schemata of order $L-2$ represents a hyperplane of co-dimension 2, etc. The geometrical representation of the solution space is a hypercube, the schemata being a partition of it.

A string of length L belongs to a schemata if its j-th character corresponds to the j-th character of the schemata, or such a position contains a #, for $j = 1...L$. We said that there exist K^L different strings of length L composed with characters from an alphabet of size K. Remembering that the size of the schemata alphabet is $K+1$, the number of different schemata is $(K+1)^L$. The number of different schemata in the fast-food example can be easily verified to be 27.

Let us have a look at Table 10.7. Each individual of the population belongs to K^L schemata, eight in the above example. Schemata have a key role in the genetic search for the solution of a problem. Table 10.6 showed the possible explanations for the above-average behaviour of the 110 strategy. Let us write down a similar table for the strategy 010, corresponding to a low profit, to understand what makes the 110 strategy better than the 010.

Schemata order	Hyperplane size	Geometry	Number of individuals in the schemata	Schemata number
3	0	point	1	8
2	1	line	2	12
1	2	plane	4	6
0	3	cube	8	1

Table 10.7 Schemata and their geometrical representation for the Koza Fast-food problem.

Explanation	binary string
It is the high price	0 # #
It is coke	# 1 #
It is the careful (but slow) service	# # 0
It's the combination between high price and coke	0 1 #
It's the combination between high price and slow service	0 # 0
It's the combination between coke and slow service	# 1 0
It's the combination between high price, coke and slow service	0 1 0

Table 10.8 Explaining the below-average outcome of the 010 strategy.

By comparing Tables 10.6 and 10.8, it is clear that the second, third and sixth row are contradictory because they give the same explanation to opposite phenomena. Those three hypotheses must therefore be discarded. If we were to repeat the procedure bringing to Tables 10.6 and 10.8 for the remaining strings (011 and 001) we would discard some more hypotheses, eventually getting the following explanation for the goodness of the 110 strategy: it is due to the low price (`1##`) alone or in combination with coke (`11#`) or with slow, careful service (`1#0`).

Following the point of view of a pioneer of GA, John Holland, schemata can be seen as competing explanations (Holland, 1992). Each schemata has a fitness value associated, that is the average fitness of the strings belonging to it. The simultaneous presence of high and low fitness strings inside a schemata eventually brings us to discarding such a schemata, as happens for the configurations (`#1#`) and (`#10`) in Tables 10.6 and 10.8. Conversely, the high fitness value associated with a particular string *implicitly* attributes such a fitness to all K^L schemata associated to such a string. This feature is known as *implicit parallelism* of the search algorithm, and it is a peculiar advantage of GA. The genetic algorithm, through the fitness-based selection, directs the search procedure in 'promising' regions of the search space, i.e. it leads towards schemata having above-average fitness. Crossover and mutation operate so as to maintain the population diversity, implicitly taking into consideration that the best current solution at a given generation can be wrong (false minimum/maximum). Intuitively, the rejection of a particular string (through the assignment of a low reproduction probability) acts as a partial, but temporary, rejection of the schemata to which it belongs.

A rigorous mathematical theory of GAs has not been fully developed, with the exception of the schemata theorem, described in the following section, that in some ideal conditions explains the power of GA. Anyway, apart from the partial theoretical demonstrations or intuitive justifications, what is actually convincing is the amazing effectiveness of the results obtained in various search/optimization problems.

10.3.4 The schemata theorem

Let $A_1 \ldots A_N$ be a population of strings at time t, and let us denote by S a generic schemata whose the subset $A_1 \ldots A_m$ belongs, i.e.:

$$S = \{A_i, i = 1 \ldots m\} \tag{10.1}$$

The fitness of the string A_i will be denoted by f_i, while f_t is the total fitness of the population at time (generation) t.

$$f_t = \sum_{i=1}^{N} f_i \tag{10.2}$$

Selecting proportionally to their fitness value the individuals to be inserted in the mating pool (i.e. to be propagated to generation $t + 1$), the selection probability for a generic individual A_k is f_k/f_t. The expected number of chromosomes A_k at time $t + 1$ is therefore $N f_k/f_t$.

From the definition of schemata (10.1), the expected number of strings belonging to the S schemata is:

$$N \frac{\sum_{i=1}^{m} f_i}{\sum_{j=1}^{N} f_j} \tag{10.3}$$

Denoting by $f(S)$ the average fitness of the schemata S at generation t:

$$f(S) = \frac{\sum_{i=1}^{m} f_i}{m} \tag{10.4}$$

we can eventually compute the expected number of strings belonging to the schemata S at generation $t + 1$:

$$n(S, t+1) = n(S, t)\frac{f(S)}{\bar{f}} \tag{10.5}$$

where $n(S, t) = m$ (number of strings belonging to the schemata S at time t) and $\bar{f} = f_t/N$ is the average fitness of the population al time t.

It should be noted that if a schemata S is above the average of a quantity $c\bar{f}$, (10.5) becomes:

$$n(S, t+1) = n(S, t)\frac{\bar{f} + c\bar{f}}{\bar{f}} = (1 + c)n(S, t) \tag{10.6}$$

and, in the simplifying hypothesis that c remains constant starting from the initial population:

$$n(S, t+1) = n(S, 0)(1 + c)^t \tag{10.7}$$

Equation (10.7) shows that, in the above hypothesis, reproduction allocates an exponentially increasing number of individuals to the above-average schemata.

Before analysing the effect of crossover and mutation on the time behaviour of a schemata, let us define two characteristics of schemata:

Defining length, $\delta(S)$, is the distance between the positions of the extreme defined values. For example, if $S = $ #011###1, $\delta(S) = 8 - 2 = 6$, distance between the first 0 and the last 1 of the string.

Order, $o(S)$, it is the number of specified positions (e.g. $o(S) = 4$ in the above defined S).

The crossover reduces the survival probability of a schemata. Limiting ourselves to the simple crossover introduced in 10.2.3, the survival probability of a schemata S of length L is:

$$p_s = 1 - \delta(S)/(L - 1) \tag{10.8}$$

because the schemata is destroyed if the crossover point, chosen among the $L - 1$ possible positions, is within the defining length. Denoting the crossover probability by p_c, the previous equation becomes:

$$p_s \geq 1 - p_c\frac{\delta(S)}{L - 1} \tag{10.9}$$

Assuming reproduction and crossover as independent from one another, at generation $t + 1$ the number of strings belonging to the schemata S is therefore:

$$n(S, t+1) = n(S, t)\frac{f(S)}{\bar{f}}p_s \tag{10.10}$$

By a similar reasoning, the probability of destroying a schemata due to mutation, given the mutation probability p_m and the schemata order $o(S)$, is:

$$\epsilon_m = (1 - p_m)^{o(S)} \tag{10.11}$$

Denoting by ϵ_c and ϵ_m the destruction probabilities respectively due to crossover and mutation, eqn (10.5) becomes:

$$n(S, t+1) \geq n(S, t)\frac{f(S)}{\bar{f}}(1 - \epsilon_m)(1 - \epsilon_c) \tag{10.12}$$

Equation (10.12) states that the expected number of strings belonging to a schemata at generation $t + 1$ is much greater the more the fitness of the schemata is above average. Given an equal average fitness, less specified schemata are favoured, i.e. short and low-order schemata.

Considering the usually low value of p_m, we obtain the following relation for the update of the number of strings in a schemata:

$$n(S, t+1) > n(S, t)\frac{f(S)}{\bar{f}}\left[1 - p_c\frac{\delta(S)}{L-1} - o(S)p_m\right] \tag{10.13}$$

Equation (10.13) is the symbolic expression of the **schemata theorem**, whose statement is the following:

In a genetic algorithm, short, low-order above-average schemata contain a number of strings exponentially increasing with increasing generations.

10.4 A simple application: non linear fitting

Let us apply GA to solve a simple, but realistic, problem. Suppose we have N couples of numbers (X_i, Y_i), coming from the measurement of a quantity Y dependent on another quantity X. The function $Y = f(X)$ will in general depend on a certain number of parameters that we can determine so as to minimize the mean squared error due to measurement uncertainty and to the various error sources:

$$\varepsilon = \sum_{i=1}^{N}(f(X_i) - Y_i)^2 \tag{10.14}$$

The couples (X, Y) are shown in Fig. 10.4.

The graph shape of Fig. 10.4 suggests an exponential function, with three unknown parameters:

$$Y = A \cdot e^{-BX} + C \tag{10.15}$$

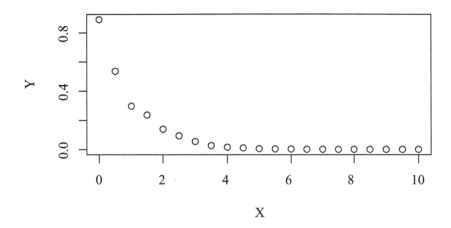

Fig. 10.4 Data for the non linear fitting problem.

10.4.1   Solution using a standard method

We can solve the fitting problem using the *nls* function of the *stats* R package:

```
## Code_10_1.R
# Non linear fitting with genetic algorithm
setwd("C:/RPA/code/genetic_algorithm")
dat <- read.table("data/exp_data.txt")
attach(dat)
plot(V1,V2,xlab="X",ylab="Y")
fit <- nls(V2~A*exp(-B*V1)+C,start=list(A=1,B=1,C=0.01))
summary(fit)
print(coef(fit))
X <- seq(min(V1),max(V1),length=101)
lines(X,predict(fit,list(V1=X)),col="red")
```

obtaining the following result:

```
        A            B            C
0.880675632 0.971162486 0.001310648
```

with 99.7% explained variance R^2. The fitting result is shown in Fig. 10.5.

10.4.2   Genetic solution

The R language has the awesome feature of possessing a huge number of application packages, due to its free nature. Therefore, it is not surprising that an excellent GA package exists, developed by (Scrucca, 2013; Scrucca, 2017). The GA package is loaded by:

```
library(GA)
```

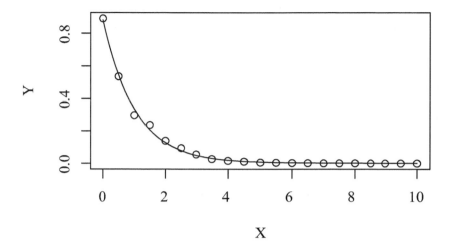

Fig. 10.5 Non linear fitting using *nls* in R.

The genetic solution of the above example can be obtained, for example, evolving a population of 100 individuals, each consisting of a binary chromosome of 60 bits. The chromosome codifies three real parameters (A,B,C), so there are 20 bits available for each parameter. The choice of a binary codification is mainly for didactic purposes. We will see that the chromosome can be efficiently coded directly as a real vector of parameters.

The correspondence between genes and real parameters is established in the present case by fixing the range of variability of each gene between $v_{min} = 0.0$ and $v_{max} = 1.0$. Given any minimum and maximum value, v_{min} and v_{max}, the binary representation of a gene G is such that the string $G_{min} = G_1 \ldots G_{20}$ is composed of all zeros and the string $G_{max} = G_1 \ldots G_{20}$ is composed of ones. Using the function `binary2decimal` of the GA package, shortened to $b2d$ in the following, a generic binary string G between $G_{min} = 00000000000000000000$ and $G_{max} = 11111111111111111111$ is decoded as:

$$v = v_{min} + \frac{v_{max} - v_{min}}{d_{max}} \text{b2d}(G) \qquad (10.16)$$

where $d_{max} = b2d(G_{max})$.

The reciprocal of the sum of squared residuals can be assumed as the fitness function:

$$fitness = \left(\sum_{i=1\ldots N} (y_i - \hat{y}_i)^2 \right)^{-1} \qquad (10.17)$$

where y_i are the experimental data and $\hat{y}_i$ the values expected from (10.15).

The first GA will adopt a crossover probability 0.7 and a mutation probability 0.1. In other words, 70% of the individuals are crossed during the reproduction phase (or, 30% of the individuals selected in the mating pool pass unaltered to the successive generation), while 10% of bits is flipped (mutation). The following table summarizes the parameters of the algorithm.

SGA parameters:

Total population size	100
Chromosome length (lchrom)	60
Maximum # of generations (maxgen)	100
Crossover probability (pcross)	0.7
Mutation probability (pmutation)	0.1

Table 10.9 GA parameters for the non-linear fitting problem.

The following code shows the GA implementation of the non-linear fitting.

```
## Code_10_2.R
# Non linear fitting with GA
# Load GA library
library(GA)
# Initialize
pop.size<- 100
chrome.len <- 60
max.gen <- 100
xover.prob <- 0.7
mut.prob <- 0.1
# Read data
setwd("C:\RPA\code\genetic_algorithm\data")
dat <- read.table("exp_data.txt")
X <- dat$V1
Y <- dat$V2

# Define fitness function
f <- function(x)  {
  chrom.a <- x[1:20]
  a <- a1+(a2-a1)*binary2decimal(chrom.a)/dmax
  chrom.b <- x[21:40]
  b <- b1+(b2-b1)*binary2decimal(chrom.b)/dmax
  chrom.c <- x[41:60]
  c <- c1+(c2-c1)*binary2decimal(chrom.c)/dmax
  N <- length(X)
  ret <- 0
  for (i in 1:N)
    ret <- ret + (Y[i] - a*exp(-b*X[i])+c)^2
  1/ret
}

# Exec GA
a1 <- 0
a2 <- 1.0
b1 <- 0
```

```
b2 <- 1.0
c1 <- 0
c2 <- 1.0
dmax = binary2decimal(c(1,1,1,1,1,1,1,1,1,1,1,1,1,1,1,1,1,1,1,1))
GA <- ga(type = "binary", fitness = f, popSize = pop.size,
    nBits = chrome.len, maxiter = max.gen, pmutation = mut.prob,
    pcrossover = xover.prob,keepBest = TRUE)
result <- unname(GA@solution)
a <- a1+(a2-a1)*binary2decimal(result[1:20])/dmax
b <- b1+(b2-b1)*binary2decimal(result[21:40])/dmax
c <- c1+(c2-c1)*binary2decimal(result[41:60])/dmax
param <- c(a,b,c)

# Output
max(GA@fitness)
min(GA@fitness)
mean(GA@fitness)
param

# Plot
# Exponential function
fexp <- function(x,p)
{
  a <- p[1]
  b <- p[2]
  c <- p[3]

  a*exp(-b*x)+c
}
plot(X,Y)
curve(fexp(x,param),from=0,to=10,add=TRUE,col="red",lty=2)

# Diagnostics
plot(GA@summary[,1],t="l",xlab="Generation",ylab="Fitness")
```

Table 10.10 shows an excerpt of the initial population, randomly generated. The real number on the right of each binary chromosome is the computed fitness.

1)	110010011011011110000001011100101010011111111000011001101111	0.14237
2)	011000011101101100110001000011100011010011100010011101001011	0.07571
3)	001000001110110110011011111001011111010110001010110010011001101	0.10940
4)	011100010001011001010001111100001100011001000110010100010111	1.14938
5)	011000111111010110111110100010000000010110110110100101001010	0.18846
6)	111001011100100100000110011011101011110010110000101111000010	0.14303
7)	111100101000000010010011010010011000010001111011110110100110	1.32572
8)	110011111100000010100011000111110010010100110000111110110101	2.54225
9)	000110111110110110110011000100110100010100110010111110001110	0.44466
10)	100101100001100111001101011011000111111110000100011001010111	0.15706
⋮	⋮	⋮
100)	010010000011110000110101010110100000011011001111111101101001	0.19736

Table 10.10 Initial population.

The initial population characteristics are summarized in Table 10.11.

min = 0.04359   max = 6.40820   avg = 0.762564
Global Best Individual so far, Generation 0:
10111110000110010110 10000111110101001100 00011010100000001000
A = 0.7425754   B = 0.5305906   C = 0.1035234 Fitness = 6.40820

Table 10.11 Initial population characteristics.

After ten generations, i.e. after applying ten times the sequence selection-reproduction-(crossover, mutation), the situation is that shown in Table 10.12. Table 10.13 summarizes the situation of Table 10.12.

1)	01010111001100011000111100101111010011010010100001000100010	0.69597
2)	10110110101000000011100110110111110110011000010001000011101000	16.44704
3)	01011011001100011000111100101111010011010010100010001110111	0.715674
4)	00001110000110000000000010101111111101000000110001110100000101	0.80283
5)	11100110000000000000000010101001101110110001100011101100000101	0.15559
6)	01011101001000110111010111110000110011000010101010100100011110	0.98354
7)	11110010010000011000000010110110010011011000010101000110000	1.29843
8)	11001100110011110111001001001110001010001000100011101110000010	0.73859
9)	11111011001100011000110011001110001010110000100011000110101000	28.45917
10)	01101011001100011000111101100010111010100100100011000000101	0.36534
$\vdots$	$\vdots$	$\vdots$
100)	11001100110011110111001011000111010011000011100010101000000101	2.56668

Table 10.12 Population at generation 10.

min = 0.05260   max = 67.75427   avg = 6.23810
Global Best Individual so far, Generation 10:
11001101100101111001 11001100010100100111 00000011100000100000
A = 0.80309468   B = 0.79813366   C = 0.01370241
Fitness = 67.75427

Table 10.13 Characteristics of the population at generation 10.

Note that the maximum fitness increases from 6.4 to 67.8 in ten generations, i.e. the error is reduced to one tenth. Figure 10.6 shows the behaviour of the maximum fitness as a function of the generation number. Observe that evolution is not linear, rather it proceeds in jumps (e.g. see the sharp jumps around generation 20 and 30) as a consequence of the stochastic nature of GA. Around generation 40 the algorithm reaches the minimum error (maximum fitness) compatible with the binary representation. Indeed, due to the 20-bit code, the available real numbers are multiples of $v_{min} + (v_{max} - v_{min})/(2^{20} - 1)$.

Table 10.14 shows the final population after 100 generations. Its characteristics are summarized in Table 10.15. Note that the final value of the maximum fitness is about 60 times the initial one. The average fitness of the population is high as well, about 500 times the initial one, denoting a global convergence of the population towards the

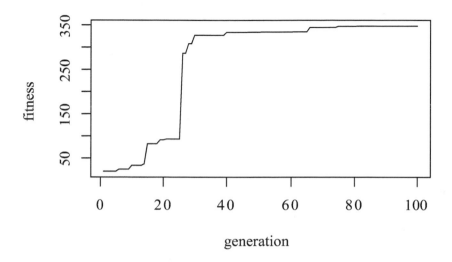

Fig. 10.6 Fitness value vs the number of generations

solution.

The fitting parameters A, B and C are very close to those obtained with *nls*. The result of the GA fitting is shown in Fig. 10.7.

1)	111000010100001101001111011011110110100100000000000000011101	347.0354
2)	111000010100001101011111011011110010100100000000000000011100	347.0414
3)	111000010100001110001111010111110111001000000000000000011100	346.3811
4)	111000011100001111001111010111110110100100000000000001111111101	342.1954
5)	111000010100001110001111011011100010100100000000000000001101	347.1046
6)	111000011100001110001111011011100110100100000000000110011101	345.6851
7)	111000010100001111001111010111100110100100000000000000001100	346.3201
8)	111000011100001101001111011011111001010010000000000000001101	347.2116
9)	111000011100001111001111011111100110100100000000000000011101	347.1570
10)	111000011100001101001111011011110110100100000000000000011101	347.1757
⋮		⋮
100)	111000010100101110001111010111110111001000000000000000011100	346.3827

Table 10.14 Population at generation 100.

A word of caution is necessary at this stage. The intrinsic randomness of GAs is evident from the plot of Fig. 10.8, showing the fitness behaviour in different runs.

min = 9.55632   max = 347.2292   avg = 318.6861
Global Best Individual so far, Generation 100:
11100001110000111100 11110111111001101001 00000000000000001101
A = 8.818940e-01   B = 9.683628e-01   C = 1.239778e-05
Fitness = 347.2292

Table 10.15 Population characteristics at generation 100.

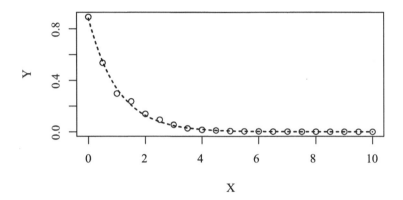

Fig. 10.7 GA solution of the non-linear fitting problem.

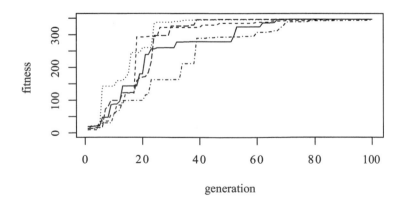

Fig. 10.8 GA fitness behaviour in several runs.

The realization of the maximum fitness is not guaranteed by a single run of the algorithm. In the example at hand, if we had set to 50 the maximum number of generations, we would have obtained a sub-optimal solution in three cases out of five. The GA approach therefore requires us to run the algorithm several times and to choose the best solution.

10.5   Advanced genetic algorithms

We introduce a new GA operator, the *elitism*. Elitism allows obtaining a faster convergence in several problems. Other criteria for reaching a solution will also be described, for example the external supply of genetic material during the creation of the initial population, or the recourse to a population of variable size.

10.5.1   Elitism

We saw that the simplest GA procedure is exposed to the possibility of a disappearance (temporary or definitive) of an individual having a relative maximum value. For example, the individual **k** at generation **i** can be selected for crossover and as a consequence it is no longer present in the population at generation **i+1**. On the one hand, that guarantees the variability of the population and it therefore helps in avoiding being trapped into a false minimum; on the other hand, the lost of the best individual of a generation can slow down the convergence to the solution.

To preserve the best individual of a generation, it can be propagated to the following one, giving it another chance to contribute with its genetic material to the future generations. If the total number of individuals must be conserved (constant population) this operation requires an individual in the following generation to be suppressed, usually that presenting the lowest fitness value.

The *elitism* operation can have side effects. In a minimum search problem, the objective function can exhibit a deep local minimum and the best individual at generation **i** can be located there. Propagating that individual to the generation **i+1** increases the probability of being trapped in a local minimum, i.e. elitism increases the probability of a premature (false) convergence.

In the GA library, by default 5% of the best fitness individuals survive in the next generation (i.e. the elitism operator was already present, although undeclared, in the previous analysis). The exact number of survivors is fixed by setting the parameter *elitism* in the *ga* function.

We can investigate the efficiency of elitism by applying it to the previous non-linear fitting example. Figure 10.9 compares the fitness behaviour with elitism percentages of 1% (solid line), 2%, 5% and 10%. The increasing convergence speed is manifest: although the final fitness value is comparable in the four cases, the elitism operator significantly reduces the computation time, usually reducing the number of jumps too.

A GA can also work without elitism (i.e., by setting *elism* = 0), but in this case a greater number of generations and a larger population size is usually needed. Moreover, if the best-fitness individual is not propagated to the next generation, the maximum fitness does not necessarily increase with each generation. As an example Fig. 10.10 shows the results for a GA with *elitism* = 0, having a population size of 500 and 200

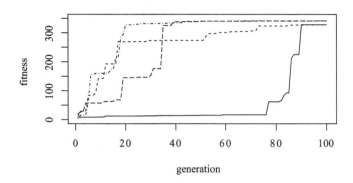

Fig. 10.9 Influence of elitism on the GA convergence.

generations. Although the correct fitting parameters are obtained (Fig. 10.10 (bottom)), the maximum fitness behaviour is erratic. Incidentally, in the particular case of Fig. 10.10 the maximum fitness is obtained at generation 37.

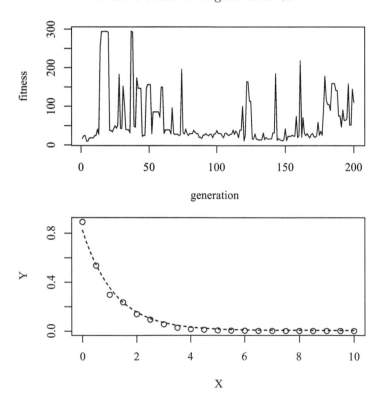

Fig. 10.10 Fitness behaviour (top) and fitting result (bottom) for a GA without elitism.

10.5.2   Inseminated and variable-size populations

Up to now, the initial population has been randomly generated. Sometimes, an a priori knowledge is available of the region of the solution space containing the true solution (e.g. the solution is a real positive number between 0 and π). In such a case, the initial population can also be forced to contain a small number of individuals with assigned characteristics, in other words it can be artificially *inseminated*. This action corresponds to giving a 'seed' in other algorithms.

Clearly, assigning a string having a particular bit combination initially forces the production of individuals containing that genetic material, at least partially. Therefore, as discussed for elitism, the artificial insemination can greatly increase the convergence speed but it also increases the probability of an untimely convergence to a false solution.

Note that all populations treated so far were constant-size. This constraint can be removed, allowing the number of individuals to vary in time. It is sometimes useful to start with a small population, allowing it to grow with increasing generations. A theorem has been developed by Goldberg concerning the 'implicit parallelism' (Goldberg, 1989): a genetic algorithm processes in parallel $O(N^3)$ schemata per generation, implying that doubling the population size eventually halves the total computation time. This means that the computational cost due to the increasing size of the population is at least partially compensated by the increase in the number of schemata processed by the GA.

10.5.3   Other genetic operators

Only the imagination can limit the operators suitable for application in a GA. For example, *inversion* and *translocation* are operators directly borrowed from biology (see Section 10.2.1) and sometimes used in GAs.

As an example, the inversion of the chromosome $\mathbf{C}=C_1 \dots C_N$ produces the chromosome $C^i=C_N \dots C_1$. Inversion can be applied in any GA. Translocation, instead, requires variable-length chromosomes. Indeed, this operation consists in detaching from $\mathbf{C}$ a section $\mathbf{S_c}=C_i \dots C_j$ and in attaching it to another chromosome $\mathbf{C}$', producing $C_1' \dots C_N' \mathbf{S_c}$ or $\mathbf{S_c} C_1' \dots C_N'$.

Finally, a note about crossover. The simple crossover introduced and used throughout this text is neither the only possible one nor the most used. At least two other crossover modalities are widely used in the literature: the two-point crossover and *uniform* crossover.

The former requires chromosomes $A_1 \dots A_N$ and $B1 \dots B_N$ to be cut in two points i e j (randomly chosen), to obtain:

$$A1 \dots A_{i-1} B_i \dots B_j A_{j+1} \dots A_N$$

and

$$B_1 \dots B_{i-1} A_i \dots A_j B_{j+1} \dots B_N$$

In the latter case, a crossover 'mask' is established, and subsequently the alleles of the parent A corresponding to 1's in the mask are copied in the offspring C together with the alleles 0's of the parent B coinciding with the mask. For example, if the parents are $A = A_1 A_2 A_3 A_4$ and $B = B_1 B_2 B_3 B_4$ and the chosen mask is 1011, we obtain $C = A_1 B_2 A_3 A_4$.

10.5.4   Real coded GA

The binary coding is historically motivated and, above all, it makes the algorithm easier to understand. Nevertheless, a direct real coding is perfectly equivalent and at least equally efficient.

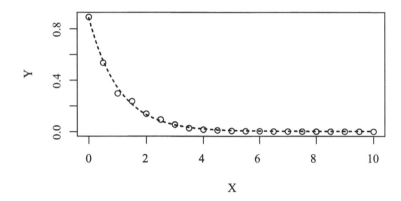

Fig. 10.11 Exponential fitting with real-coded GA.

The GA package by default uses real chromosomes. Figure 10.11 shows the result of the exponential fitting example using a real-coded GA with a population size of 100 and 200 generations. The fitness behaviour with increasing generations is shown in Fig. 10.12. The following R code shows the details of the computation.

```
## Code_10_3.R
# Real coded GA
setwd("C:/RPA/code/genetic_algorithm")
dat <- read.table("data/exp_data.txt")
X <- dat$V1
Y <- dat$V2
# Initialization
pop.size<- 100
max.gen <- 200
xover.prob <- 0.7
mut.prob <- 0.1
# fitness function
f <- function(x)
{
  a <- x[1]
```

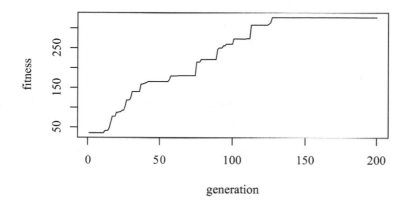

Fig. 10.12 Fitness behaviour for exponential fitting using real-coded GA.

```
  b <- x[2]
  c <- x[3]
  N <- length(X)
  ret <- 0
  for (i in 1:N)
    ret <- ret + (Y[i] - a*exp(-b*X[i])+c)^2
  1/ret
}
# Exec GA
library(GA)
GA <- ga(type = "real-valued", fitness = f, lower = c(0.,0.,0.),
    upper = c(1,1,1),maxiter=max.gen,popSize = pop.size)
result <- unname(GA@solution)
print(result)
# Plot
plot(X,Y)
curve(fexp(x,result),from=0,to=10,add=TRUE,col="red",lty=2)
# Diagnostics
plot(GA@summary[,1],t="l",xlab="Generation",ylab="Fitness")
```

10.6   Parameter estimation of ARMA models

We have met autoregressive moving-average processes in Chapter 6. An ARMA(p, q) model is used to analyse a stationary time series having an autoregressive component of order p and a moving-average component of order q.

AR(p) models, a subset of them where the moving-average component is negligibly small, were met in Section 9.1, where the linear relation between temperature anomaly and carbon dioxide concentration in Antarctic ice has been shown to be affected by autocorrelation. We have seen that the residual error of the linear regression is a time series corresponding to an AR(1) process, whose regression parameters (vector of regression coefficients $\boldsymbol{\beta} = (\beta_0, \beta_1)$, error variance σ^2 and correlation coefficient ρ) were estimated by the MCMC algorithm. Such a problem could be faced as well by means of a genetic algorithm, choosing a *chromosome* consisting of the unknown parameters and obtaining the best individual (solution) by maximizing the likelihood which would play the role of the fitness function.

As a complex example of GA, we will try to obtain the best-fit parameters of an ARMA(p, q) model describing the annual runoff of the Loire river, determined to be an ARMA(1,1) in Section 6.6.1 by a "manual" procedure involving the `arima` function of the R base package `stats`. The problem is the following. We have a set of data, the standardized annual runoff of Loire from 1860 to 1980, and we want to fit an ARMA(p, q) model, finding the order (p, q) and the values of the autoregressive coefficients $\phi_1, ..., \phi_p$ and of the moving-average coefficients $\theta_1, ..., \theta_q$. We will discuss what the fitting function is in a while. Before that, we have to understand what is the chromosome.

The chromosome naturally appears to be the sequence of unknown parameters, e.g. something like:

$$p \; q \; \phi_1\phi_2...\phi_p \; \theta_1\theta_2...\theta_q$$

Such a definition is not easy to implement, because of the variable number of genes. A simple solution is to define a maximum value for p and q, for instance $p = q = 4$, to implement a fixed-length chromosome as:

$$p \; q \; \phi_1\phi_2\phi_3\phi_4 \; \theta_1\theta_2\theta_3\theta_4$$

and to discard the ϕ_i, θ_k genes exceeding the best-fit p and q values maximizing the fitness function, i.e for $i > p$ and $k > q$.

Now, a short digression is in order concerning the fitness function. In Section 6.6.1 we searched for the minimum value of the Akaike's information criterion (AIC) function, expressed by (6.19), here repeated for convenience:

$$AIC = 2(p + q + 1) - 2\log(L)$$

which essentially corrects the log-likelihood for the number of the parameters in the model. The reciprocal of AIC is a good candidate for the fitness function.

The log-likelihood of an ARMA model can be obtained by the definition of the process $\{X_t\}$ (remember the arbitrary sign of ϕ's and θ's coefficients):

$$X_t = \phi_1 X_{t-1} + ... + \phi_p X_{t-p} + \epsilon_t + \theta_1\epsilon_{t-1} + ... + \theta_q\epsilon_{t-q} \qquad (10.18)$$

Following (Madsen, 2008), for a given set of parameters:

$$\boldsymbol{\Theta}^T = (\phi_1, .., \phi_p, \theta_1, ..., \theta_q)$$

the expected value of X at time t if the process at time $t - 1$ is known, i.e. knowing $\boldsymbol{X}_{t-1}^T = (X_{t-1}, X_{t-2}, ..., X_1)$, is:[1]

---

[1] That is usually called 'one-step prediction'.

$$\hat{X}_{t|t-1}(\Theta) \equiv \mathrm{E}\left[X_t | \Theta, \boldsymbol{X_{t-1}}\right] = \sum_{i=1}^{p} \phi_i X_{t-i} + \sum_{k=1}^{q} \theta_k \epsilon_{t-k}(\Theta)$$

where the dependence on the parameter vector Θ and conditioning on the previous time step have been explicitly indicated. The white noise terms ϵ_t here play the role of residuals, or 'prediction errors' and, of course, depend on the fitting parameters Θ. The prediction error at time t for a given choice of the parameters is:

$$\epsilon_t(\Theta) = X_t - \hat{X}_{t|t-1}(\Theta)$$

Assuming zero residuals for times $1 \ldots \max(p,q)$, i.e. $\epsilon_1 = \epsilon_2 = \ldots = \epsilon_{max(p,q)} = 0$ (this introduces a negligible error in the procedure when the number of observations is very much larger than p and q, as usually happens), we can compute the errors ϵ_t at any time by a recursive procedure. Explicitly, for the simple case $p = 1$, $q = 1$ and N observations:

$$\begin{aligned}
\epsilon_1 &= 0 & &\rightarrow & \hat{X}_{2|1} &= \phi X_1 \\
\epsilon_2 &= X_2 - \hat{X}_{2|1} & &\rightarrow & \hat{X}_{3|2} &= \phi X_2 + \theta \epsilon_2
\end{aligned}$$

...

$$\begin{aligned}
\epsilon_{N-1} &= X_{N-1} - \hat{X}_{N-1|N-2} & &\rightarrow & \hat{X}_{N|N-1} &= \phi X_{N-1} + \theta \epsilon_{N-1} \\
\epsilon_N &= X_N - \hat{X}_{N|N-1}
\end{aligned}$$

In order to get an explicit expression for the AIC we need the likelihood function. We will assume, as for maximum likelihood method, that $\{\epsilon_t\}$ has a normal distribution and variance $\mathrm{Var}\left[\epsilon_t\right] = \sigma_\epsilon^2$. The likelihood is the joint probability distribution for the observations for given values of the parameters Θ and σ_ϵ^2. Its explicit dependence from the parameters is not simple (Box *et al.*, 1994), with the exception of low-order cases ($p = 1$ or $q = 1$). In general, the log-likelihood is an expression like:

$$\log(L(\boldsymbol{X}, \Theta)) = -\frac{1}{2}\left[N\log(2\pi) + \log|\Gamma(\Theta)| + \boldsymbol{X}^T \Gamma(\Theta)^{-1} \boldsymbol{X}\right]$$

where Γ is the covariance matrix of the process, depending on the parameter vector Θ. If we solve (10.18) for ϵ_t, since they have been assumed as independent and normally distributed, the joint probability density function of $\boldsymbol{\epsilon}^T = (\epsilon_1, \epsilon_2, ..., \epsilon_N)$ is normal:

$$p(\boldsymbol{\epsilon}|\Theta, \sigma_\epsilon^2) = (2\pi\sigma_\epsilon^2)^{-N/2}\exp\left[-\frac{1}{2\sigma_\epsilon^2}\sum_{t=1}^{N}\epsilon_t^2\right]$$

and the likelihood of the ARMA model, which is the joint probability density of $\{X_t\}$, is something like (Abraham and Ledolter, 1983):

$$L(\Theta, \sigma_\epsilon^2|\boldsymbol{X}) = f(\Theta, \sigma_\epsilon^2)\exp\left[-\frac{1}{2\sigma_\epsilon^2}\sum_{t=1}^{N}\hat{\epsilon}_t^2\right] \tag{10.19}$$

where the function f depends on the orders p and q, and the expected value of ϵ_t can be computed with the recursive scheme discussed above. Thus, an approximate expression of the log-likelihood can be computed starting from the above one-step prediction errors (Madsen, 2008), sometimes called 'innovations' (Shumway and Stoffer, 2006), ignoring the function f (which is assumed equal to 1).

$$\log(L(\Theta, \sigma_\epsilon^2)) \approx -\frac{N}{2}\left[\log(\sigma_\epsilon^2) + \log(2\pi)\right] - \frac{1}{2\sigma_\epsilon^2}\sum_{t=p+1}^{N}\epsilon_t^2(\Theta) \qquad (10.20)$$

where the conditioning on the observed data $\boldsymbol{X}$ has been dropped from the notation. We have seen that what the best solution does is to exhibit the minimum AIC (6.19) rather than the maximum likelihood value. Therefore, the solution will correspond to a minimum of the sum of squared one-step prediction errors, $S(\Theta) = \sum \epsilon_t^2(\Theta)$, corrected for the orders p and q. A convenient fitness function $\mathcal{F}(p, q, \Theta)$ (to be maximized) is minus this last expression, that is:

$$\mathcal{F}(p, q, \Theta) = p + q + 1 - \sum_{t=p+1}^{N}\epsilon_t^2(\Theta) \qquad (10.21)$$

To summarize, if we neglect the influence of the covariance matrix on the true likelihood, we can implement the fitness function in the GA as a quantity mainly related to the sum of the squared residuals. The p and q orders of the ARMA, and of course the coefficients constituting Θ, enter the game through the estimates $\hat{X}_{t|t-1}$ and directly in the expression of the AIC-like function (10.21).

That is the approach adopted in the following R code. The first code section, coming from Chapter 6, computes the standardized annual runoff.

```
## Code_10_4.R
# GA for ARIMA optimization
setwd("C:/RPA/code/genetic_algorithm")
dat <- read.table("data/loire_runoff.txt",h=TRUE)
names(dat)
x <- ts(dat$DISCHRG)
# annual average
Y <- dat$YEAR
N <- length(dat$YEAR)
years <- Y[1]:Y[N]
n <- Y[N]-Y[1]+1
X <- rep(0,n)
for (i in 1:n){
  X[i] <- sum(x[(12*(i-1)+1):(12*(i-1)+12)])
}
X[113] <- 0.5*(X[112]+X[114])
# Standardized annual runoff
X.m <- mean(X)
X.sd <- sd(X)
X <- ts((X-X.m)/X.sd,start=1863)
plot(X,xlab="Year",ylab=expression(paste("Annual runoff (m"^"3","/s)")),
    mgp=c(2,0.7,0))
```

The fitness function (10.21) is computed as follows.

```
# CONTINUE Code_10_4.R

# Count the number of observations
n <- length(X)
# Initialize error
```

```
e <- rep(1, n)
# Fitness function
fitness.fun <- function(crom){
  g <- numeric

  mu <- crom[1]
  P <- round(crom[2])
  Q <- round(crom[3])
  if (P>Q)
    for (i in 1:P) e[i] <- 0
  else
    e[1] <- 0

  # Take into account a non-zero mean value
  XX <- X-mu

  fi <- c()
  teta <- c()
  if (P>0)
    for (i in 1:P) fi = cbind(fi,crom[i+3])
  if (Q>0)
    for (j in 1:Q) teta = cbind(teta,crom[j+P+3])

  if (P==0)
    n_start <- 2
  else
    n_start <- P+1

  for (t in (n_start : n)){
    X_hat <- 0
    if (P>0){
      for (i in 1:P)
        X_hat = X_hat+fi[i]*XX[t-i]
    }
    if (Q>0 & Q<=P){
      for (i in 1:Q)
        X_hat = X_hat+teta[i]*e[t-i]
    }
    e[t] <- (XX[t] - X_hat)
  }
  g <-  e%*%e
  return(-g-(P+Q+1))
}
```

Finally, the GA procedure using the GA package is the following.

```
# CONTINUE Code_10_4.R

# Load GA library
library(GA)
# Initialize GA
pop.size<- 30
max.gen <- 300
xover.prob <- 0.7
mut.prob <- 0.1
# Compute
GA <- ga(type = "real-valued", fitness = fitness.fun,
    lower = c(-0.01,0,0,-0.9,-0.9,-0.9,-0.9),
    upper=c(0.01,2,2,0.9,0.9,0.9,0.9),
    maxiter=max.gen,popSize = pop.size,elitism=4)
# Show solution
```

```
sol <- GA@solution[1,]
ARMA.params(sol)
# Plot diagnostics
plot(GA)
```

Remember that p and q are integers, so the **round** value of the relevant 'gene' must be taken. That's true always in the interpretation of the resulting best solution; the function **ARMA.params** is defined as follows:

```
# CONTINUE Code_10_4.R

ARMA.params <- function(crom){
  mu <- crom[1]
  p <- round(crom[2])
  q <- round(crom[3])
  fi <- c()
  teta <- c()
  if (p>0)
    for (i in 1:p) fi = cbind(fi,crom[i+3])
  if (q>0)
    for (j in 1:q) teta = cbind(teta,crom[j+p+3])

  cat("mean",mu,"\n")
  cat("AR",p,fi,"\n")
  cat("MA",q,teta,"\n")
}
```

Figure 10.13 shows the fitness result in ten different runs.

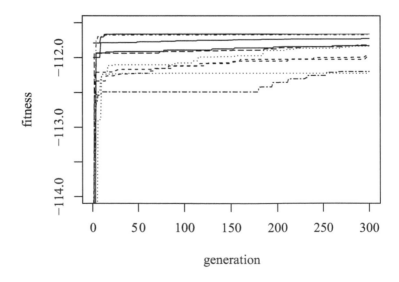

Fig. 10.13 Fitness behaviour for GA analysis of the annual Loire runoff data.

The best solution has the following parameters:

```
mean 0.001330264
AR 1 -0.4853239
MA 1 0.7153637
```

very close to those computed by `arima(X,order=c(1,0,1))`:

```
Coefficients:
         ar1      ma1  intercept
      -0.4535  0.6841     0.0016
```

Figure 10.14 shows the diagnostics of the best solution described above. The solid line is the fitness of the best 'individual', while the dots connected by dashed lines are the average fitness values at each generation.

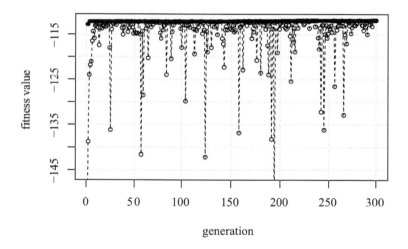

Fig. 10.14 Fitness behaviour for the best GA solution.

10.7   Solving the travelling salesman problem

We have encountered the travelling salesman problem (TSP) in Section 8.3 where it was approached by means of simulated annealing, based on the Metropolis algorithm. The TSP would ask for the **best** possible route, but the problem itself has stimulated the development of heuristic-search (HS) algorithms and it has nonetheless served as a benchmark of such algorithms. What is such an algorithm aimed to do in a problem like TSP, actually? Essentially it tries to find a *good* solution, possibly sub-optimal instead of the best solution.

GA are algorithms of the HS family. The peculiarity of the GA solution to TSP, with respect to the example discussed above, is in the chromosome structure. In all problems treated so far the genes (parameters) composing the chromosome were always quantitative or, in other words, they have a *ratio* measurement scale. In the TSP the natural definition of the chromosome is a succession of labels identifying the cities to be visited. Such labels could possibly be represented by numbers, but without any numeric meaning, i.e. the genes have a *nominal* scale.

While the fitness function, for example consisting of the inverse of the total distance travelled by the salesman, is a numeric value as it should be; the peculiar coding of the chromosome does not allow to define the main GA operators, crossover and mutation, as in Section 10.2.3. An example should clarify why.

Suppose, in a TSP involving six cities, labelled by the first alphabetic letters, you have generated two possible routes: A B C D E F and B D A E F C. A crossover point could be randomly chosen between the second and third letters:

```
A B | C D E F
B D | A E F C
```

With the rules of Section 10.2.3, mating between those parents would generate the following offspring:

```
A B | A E F C
B D | C D E F
```

which are clearly not valid routes, because cities cannot be visited more than one time.

A reasonable solution for the crossover is a procedure known as *ordered crossover*, which consist of choosing two points, for example second and fourth, and generating an offspring by a scheme like the following:

Parents:
```
A | b c | D E F
B | D A | E F C
```

Offspring:
```
D | b c | A E F
```

where the lower-case portion of the chromosome is the genetic material coming from the first parent, and the remaining are labels chosen *in order* from the second parent. Such a solution generates proper permutations of the tentative solutions, and that is what we need.

Similar considerations hold for the mutation operator: here we have not any bit to flip from 0 to 1 or vice versa The solution close to bit flipping is to exchange two adjacent sites, for a randomly chosen position (e.g. the second):

Original: A B C D E F   Mutated: A C B D E F

Ordered crossover is the default method for chromosomes of the `permutation` type in the GA package (Scrucca, 2013). Indeed, such a package can be usefully employed for finding an heuristic solution to TSP. The following code shows how to do it.

```
## Code_10_5.R
# Travelling salesman with GA
library(GA)

set.seed(12345)
# N cities: 1 2 .. N
N <- 15

# city coordinates
coords <- matrix(nrow=N,ncol=2)
test <- TRUE
while (test) {
  coords[,1] <- trunc(10*runif(N))
  coords[,2] <- trunc(10*runif(N))
  # avoid coincident cities
  test <- anyDuplicated(coords)
}

distances <- matrix(rep( 0, len=N*N), nrow = N)
for (i in 1:N){
  for (j in 1:N){
    distances[i,j] <- sqrt((coords[i,1]-coords[j,1])^2+(coords[i,2]-coords[j,2])^2)
  }
}

# fitness function
inverse_distance <- function(x) {
    N <- length(x)
    dist.total <- distances[1,x[1]]
    for (i in 1:(N-1))
      dist.total <- dist.total + distances[x[i],x[i+1]]
    dist.total <- dist.total + distances[x[N],1]
    1/dist.total
}

TSP.ga <- ga(type = "permutation", fitness = inverse_distance, lower = 2, upper = N,
    elitism = 1, maxiter = 500, popSize = 100)

# show GA parameters
summary(TSP.ga)

route.best <- unname(TSP.ga@solution)
# choose the first one
route.best <- c(1, route.best[1,])

# plot TSP
plot(0:10,0:10,type="n",xlab="",ylab="")
text(coords[1,1],coords[1,2],labels=1,col="black")
text(coords[1,1],coords[1,2],labels="0",col="black",cex=2)
for (i in (2:N))
  text(coords[i,1],coords[i,2],labels=i,col="red")
# best route find by GA
for (i in 1:(N-1))
```

```
arrows(coords[route.best[i],1],coords[route.best[i],2],coords[route.best[i+1],1],
    coords[route.best[i+1],2],col="blue",angle=20,length=0.1)
arrows(coords[route.best[N],1],coords[route.best[N],2],coords[route.best[1],1],
    coords[route.best[1],2],col="blue",angle=20,length=0.1)

# plot algorithm convergence
plot(TSP.ga)
```

The code above randomly generates the position of 15 cities in the domain $(0, 10) \times (0, 10)$, in arbitrary units. The cities are labelled with numbers from 1 to 15, city '1' being the home of the salesman where he should come back to after visiting the other 14 cities, as required by the TSP. As the first city is fixed, the number of possible permutation is 14 factorial, instead of 15 factorial. Then, the number of possible paths is *only* 8.7×10^{10} instead of 1.3×10^{12}. An exhaustive approach, i.e computing all routes and choosing the shortest, is out of the question!

Figure 10.15 shows the route compute by the above code, and Fig. 10.16 the relative fitness behaviour.

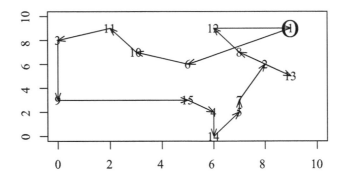

Fig. 10.15 GA solution of a TSP involving 15 cities.

The following `Code_10_6.R` implements the GA solution for the TSP of the 20 Italians provinces.

```
# Code_10_6.R
# load GA library
library(GA)
# Read the province coordinates and mutual distances
setwd("C:\RPA\code\genetic_algorithm\data")
city.coords <- read.table("provinces.txt",h=T)
cx <- as.matrix(city.coords[,3:4])
labs <- city.coords$Code
distances <- read.table("distances.txt",h=TRUE)
distances <- data.matrix(distances)
N <- 20
```

```
# distance function
dista <- function(x) {
  ret <- M_Cities[x[1],x[2]]
  for (i in 2:20)
    ret <- ret + M_Cities[x[i],x[i+1]]
  as.numeric(ret)}

# fitness function
inverse_distance <- function(x) {
  N <- length(x)
  dist.total <- distances[1,x[1]]
  for (i in 1:(N-1))
    dist.total <- dist.total + distances[x[i],x[i+1]]
  dist.total <- dist.total + distances[x[N],1]
  1/dist.total
}

TSP.ga <- ga(type = "permutation", fitness = inverse_distance, lower = 2, upper = N,
             elitism = 1, maxiter = 500, popSize = 100)

# show GA parameters
summary(TSP.ga)

route.best <- unname(TSP.ga@solution)
# choose the first one
route.best <- c(1, route.best[1,],1)

# plot TSP
route <- route.best
xmin <- 38
xmax <- 46.5
ymin <- 7
ymax <- 17
# plot TSP
plot(seq(ymin,ymax,(ymax-ymin)/10),seq(xmin,xmax,(xmax-xmin)/10),
type="n",xlab="Longitude",ylab="Latitude")
text(coords[1,2],coords[1,1],labels=labs[1],col="black",cex=0.8)
text(coords[1,2],coords[1,1],labels="0",col="black",cex=2.5)
for (i in (2:20))
  text(coords[i,2],coords[i,1],labels=labs[i],col="red",cex=0.8)
# best route find by GA
for (i in 1:19)
  arrows(coords[route[i],2],coords[route[i],1],coords[route[i+1],2],
         coords[route[i+1],1],col="blue",angle=20,length=0.1)
arrows(coords[route[20],2],coords[route[20],1],coords[route[1],2],
       coords[route[1],1],col="blue",angle=20,length=0.1)
# Uncomment to plot algorithm convergence
# plot(TSP.ga)

# Show solution and route length
print(route.best)
print(dista(route.best))
```

As a matter of comparison between the solution outlined in Section 8.3, we can apply the genetic approach to the problem of finding the best route between the 20 regional 'capitals' of Italy. The GA solution appears usually better than that based on SA: the route length of the travel in Fig. 10.17 is 6318 km, i.e. about 1500 km shorter.

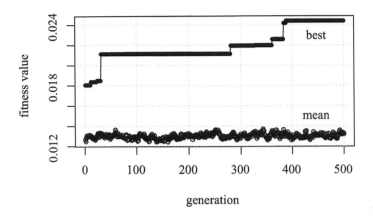

Fig. 10.16 Fitness behavior for the GA solution of the TSP.

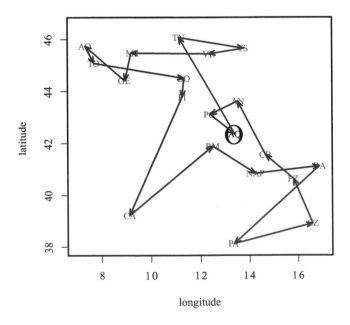

Fig. 10.17 GA solution of the TSP involving the 20 main Italian provinces.

10.8 Concluding remarks

Genetic algorithms work amazingly well, in spite of a non-fully developed theoretical framework. The ease of implementation and their conceptual clarity, joined with the power of the biological paradigm, makes them eligible for the solution of search and minimization problems in several fields.

GA does not require rigorous assumptions about probability distributions, i.e. they are intrinsically non-parametric, and they do not request an inversion procedure, working in a forward direction. It is only necessary to define a function to be maximized (or minimized), and to generate solutions almost randomly.

The simplicity of the method should not be a surprise. After all, it is simply a matter of generating and manipulating strings!

10.9 Exercises

Exercise 10.1 The following is a toy-problem, obviously not needing GA to be solved. Suppose you want to find the positive root of the polynomial:

$$f(x) = x^2 - 4x - 12$$

i.e. the value $x = 6$.

Your population consists of binary strings of four bits. The solution is, of course, the string 0110. Applying by hand a very rough GA, without mutation and with fixed crossover point in the middle of the chromosome, suppose you have obtained 0101 as the fittest individual.

Remembering the mating procedure introduced in Section 10.3 which, given two individuals $X_1X_2X_3X_4$ and $Y_1Y_2Y_3Y_4$, generates the new chromosomes $X_1X_2Y_3Y_4$ and $Y_1Y_2X_3X_4$, how many chromosomes (among the 16 possible) would give rise to the proper solution through crossover and mating with 0101?

Exercise 10.2 With reference to Exercise 10.1, define a fitness function suitable to be maximized such, for instance:

$$fitness(x) = 1/(10 + abs(x^2 - 4x - 12))$$

Write an R code to compute the fitness value of all possible chromosomes, and plot it.

Exercise 10.3 With reference to Exercise 10.2, implement a real-coded genetic algorithm based on the GA R library, taking the example in Section 10.5.4 as a starting point. Begin with a population of 10 individuals, 100 generations, crossover and mutation probabilities of 0.7 and 0.1 respectively, then play with numbers to see how the values of the four parameters affect the result. Compare the fitness histories for 10, 100 and 1000 individuals with a proportional number of generations, using `GA@summary[,1]`.

Exercise 10.4 Download the data for the example in Section 10.5.4 and modify the fitness function. Does the result significantly change if fitness is computed in terms of the absolute residuals instead of the residual sum of squares? How is convergence affected, if at all, by the choice of the fitness function?

Exercise 10.5 Following the lines of the previous exercises, write a GA code for fitting data generated by the following:

```
# Generate random signal, composition of sine and cosine
t <- seq(0,10,1/10)
A1 <- runif(1)
A2 <- runif(1)
T1 <- 1
T2 <- 0.5
f1 <- runif(1)/T1
f2 <- runif(1)/T2

s <- A1*sin(2*pi*f1*t)+A2*cos(2*pi*f2*t)
plot(t,s,t="l",xlab="t (s)",ylab="s(t)")
```

where the functional form of the signal s is supposed to be known:

$$s(t) = A_1\sin(2\pi f_1 t) + A2\cos(2\pi f_2 t)$$

but A_1, A_2, f_1, f_2 are unknown.

Write down the fitness function, and try to obtain a solution with a population of 100 chromosomes, with 100 generations. Repeat the whole code several times, to study how well different sine/cosine compositions can be managed.

Hint: we have four parameters, A_1, A_2, f_1, f_2, instead of the three in the code of Section 10.5.4. The fitness function must be modified accordingly.

Exercise 10.6 Section 10.7 shows how a quasi-optimal solution can be obtained for 15 cities. In a simpler case, say five cities including the starting one, the possible routes can be enumerated and all the relative distances can be computed. Modify the code to generate only five cities and compare the result of the heuristic algorithm to the true best solution. Write a simple R code to make that work.

11

The Problem of Accuracy

Accuracy and clarity of statement are mutually exclusive.

Niels Bohr

11.1 Estimating accuracy

In Chapter 3, we introduced some features of stochastic processes and their realizations, as stationarity, ergodicity, time and ensemble averages, convergence and so on. In the following, we show how such properties are realized in simulated series. We stressed that the convergence is *conditio sine qua non* to assign accuracy to estimates. In fact, only if the chain has achieved stationarity, do average and variance estimates make sense. In other words, we have to assess that the piece of chain we have generated, and on which we base our estimates, is representative of the underlying stationary distribution of the whole Markov chain.

As stressed by Cowles and Carlin (1996, p. 883), such a notion of convergence refers to something which is not 'a single number or even a distribution, but rather a *sample* from a distribution' (authors' italics). The problem of convergence will be discussed in a later section. In the following, we will deal with the problem of assigning an accuracy to the estimated parameters when the series has (presumably) reached stationarity.

Consider the discrete parameter stationary time series $\mathbf{X}_t = (X_1, X_2, \ldots, X_n)$. Let $\mu, \sigma_0^2, \gamma_k$ and ρ_k $(k = 0, \ldots, n-1)$ be the mean, variance, covariance and autocorrelation function of $\mathbf{X}_t$, respectively. Note that $\gamma_0 = \sigma_0^2$, and $\rho_k = \gamma_k/\gamma_0$. The variance of the estimator $\bar{X}$ of μ, is given by:

$$
\begin{aligned}
\sigma^2 = \mathrm{Var}\left[\bar{X}\right] &= \frac{\sigma_0^2}{n} + 2 \sum_{k=1}^{n-1} \frac{(n-k)}{n} \gamma_k \\
&= \sigma_0^2 \left[\frac{1}{n} + 2 \sum_{k=1}^{n-1} \frac{(n-k)}{n} \rho_k \right]
\end{aligned}
\tag{11.1}
$$

This equation is well known in the time series literature. For the general problem of estimating the standard error σ one can see, e.g. (Ripley, 1987), where various approaches are discussed and where it is stressed that finding a reliable value for σ^2 is not always a simple matter. In statistical mechanics σ^2 is usually written as (see e.g. Binder, 1992):

$$\sigma^2 = \sigma_0^2 \left(1 + \frac{2\tau}{\delta t}\right) \tag{11.2}$$

where τ is the 'integrated correlation time':

$$\tau = \int_0^\infty \rho(t) dt$$

and δt is the time interval between two successive observations. This is a quite important parameter, giving the whole information on the correlation structure of the observed data. However, in general the time dependence of the correlation function is not explicitly known, so that we have to rely on an estimate of τ which could be:

$$\hat{\tau} = \sum_{k=0}^\infty \hat{\rho}_k$$

with a suitable cutoff in the summation.

A plugged-in estimate of ρ_k is given by $\hat{\rho}_k = \hat{\gamma}_k / \hat{\gamma}_0$, where $\hat{\gamma}_k$ is the sample auto-covariance function:

$$\hat{\gamma}_k = \frac{1}{n} \sum_{i=1}^{n-k} [(X_i - \hat{\mu})(X_{i+k} - \hat{\mu})], \quad k = 0, \ldots, n-1 \tag{11.3}$$

The estimate $\hat{\rho}_k$ requires in turn the evaluation of (11.3), but we know (Priestley, 1989) that the above estimator $\hat{\gamma}_k$ must be used with caution, since it is not consistent even with divisor $n - k$ rather than n.

In the field of Markov chains, if the chain is *reversible* the autocovariance function γ_k declines smoothly to zero and is positive for all k, so one might think of cutting the sequence $\hat{\gamma}_k$ off when it becomes negative. This procedure, however, is only successful if the sample autocovariance function also goes smoothly to zero. But this is not always the case. So also in this context, the estimate of τ is a problematic task.

Other strategies have been proposed, for instance, Geyer (1992) suggested exploiting the function $\Gamma_k = \gamma_{2k} + \gamma_{2k+1}$, which is a non-negative, non-increasing convex function of k, proposing three different estimators in order to smooth the sample autocovariance function, if bumps are present. However, also in this case, the method and its success, depend on the particular form of the sample autocovariance function. In conclusion, the estimate of τ requires in any case the computations of intermediate estimates and, moreover, sometimes rather arbitrary approximations are involved.

11.2 Averaging time series

Consider now the following time series (written with three decimal places):

```
[1]      0.000 0.185 1.754 0.448 0.323 0.423 1.089 0.740 2.651 2.247 ...
..............................................................
[1191]  2.197 2.405 1.533 0.287 0.439 0.099 0.698 -0.195 0.418 0.991
```

That is a realization of the stochastic process:

$$y_i = y_{i-1} \phi + z_i \tag{11.4}$$

where $z_i \sim \mathcal{N}(0, \sigma)$. This kind of process, named AR(1), was described in Chapter 6, where we saw that an AR(1) process is stationary if and only if $|\phi| < 1$ or

$-1 < \phi < 1$. The case where $\phi = 1$ corresponds to a random walk process with a zero drift (see Chapter 5). The above realization of the process is obtained with the R code `Code_11_1.R` shown below (with $z_i \sim \mathcal{N}(0,1)$).

```
## Code_11_1.R
# Averaging time series
set.seed(2)
nsteps<- 1200          # number of iterations
phi<- 0.90
z<- rnorm(nsteps,mean=0,sd=1)
y<- numeric()
y[1]<- 0 # initial state of the process.
for(i in 2:nsteps)    {
y[i]<- y[i-1]*phi+z[i]
                }
#y   # comment if you do not wish the y values printed
plot(1:nsteps,y,type="l",xlab="steps",ylab="y",cex.lab=1.3,lty=1,
font.lab=3,lwd=1.5)
```

Figure 11.1 shows a time series plot of one realization of the eqn (11.4) AR(1) process.

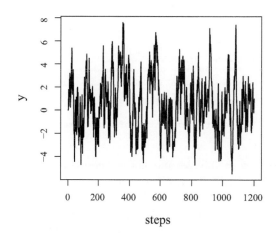

Fig. 11.1 One realization of the eqn (11.4) AR(1) process.

The plot shows typical features of a stationary series:

i) it extends roughly along the horizontal direction.

ii) the variance apparently remains almost constant.

iii) no patterns are evident in relative long time periods over or under the line $y = 0$.

The autocorrelation function for time series can be estimated using the `acf` R function, that we have frequently encountered in the previous chapters.

```
corr<-acf(y,type="correlation",lag.max=50,plot=TRUE,main=" ",
xlab="lag",ylab="ACF",cex.lab=1.2,cex.axis=1.2,font.lab=3,lwd=1.5)
#corr       # comment if you do not wish the acf values printed
```

Figure 11.2 shows the extension of the autocorrelation in our series. The ACF data exceeds the confidence bands (dashed horizontal lines) for the first 20 time lags, indicating a strong autocorrelation of the time series.

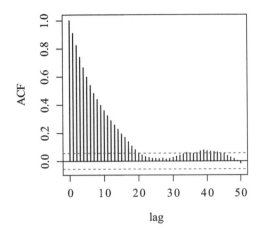

Fig. 11.2 Autocorrelation plot for the AR(1) series. Dashed lines: 95% confidence intervals assuming white noise input.

The lines:

```
ta<-1
tb<-1200
lt<- length(y[ta:tb])
m.temp<-  mean(y)    # time average (sample mean)
m.temp
se.temp<-  sd(y)/sqrt(lt)  # standard error of the m.temp
se.temp
```

give the time average $\bar{y} = 0.5928$ and the standard error of the time average $\hat{\sigma}_t = 0.06977$. Note that $\bar{y} = 0.5928$ is a *time average* (that is the reason for the subscript 't'), the arithmetic average of the i–th outcome of the process, in which the y^i's form an *i.i.d.* sample. Another realization of the process can be obtained with another **seed** in the command **set.seed(.)**. A different time average will result. For instance, with **set.seed(3)** it results $\bar{y} = -0.1053$ and $\hat{\sigma}_t = 0.06490$.

We also recall that, for an AR(1) process, the standard error can be analytically obtained:

$$\mathrm{Var}\,[y_i] = \frac{sd^2}{1 - \phi^2}$$

in our case is $sd = 1$, so the standard error is $\sigma_t = \mathrm{Var}\,[y_i]\,/\sqrt{l} = 0.06623$ (σ_t is the 'theoretical' quantity, while $\hat{\sigma}_t$ is its estimate), in good agreement with $\hat{\sigma}_t$.

The following example can be given because we know the generator process. In this manner, we can estimate the *ensemble average* executing a number of different trajectories and performing the average at a certain time. We add the following instructions to the code **Code_11_1.R**:

```
## Code_11_2.R
```

```
# Ensemble averages
set.seed(2)     #   reset random numbers if desiderata
nsteps<- 1200 #   nsteps  can be redefined
nhists<- 100   #   number of trajectories
ta<-1           #   ta and tb can be redefined
tb<-1200
lt<- length(y[ta:tb])
y<-numeric()
matr.history<-matrix(,nhists,nsteps+1) # to save each trajectory
ymed<-numeric()   # to obtain the 'mean trajectory'
mstep<- 600   # compute the ensemble average at step = mstep
# to plot only splot histories out of all nhists
splot<- 8
# to prepare the plot
plot(c(1,1), type="n",ylim=c(-10,10),xlim=c(0,nsteps),cex.lab=1.3,
xlab="t",ylab="y",font.lab=3,lwd=1.5)

for(l in 1:nhists){      # starting loop on histories
y[1]<- 0
for(i in 2:nsteps){      # starting loop on steps
y[i]<- y[i-1]*phi+rnorm(1)
matr.history[l,i]<- y[i]
                }        # ending loop on steps
              }          # ending loop on histories
for(l in 1:nhists){          # to add i=1
matr.history[l,1]<- 0     }

for(l in 1:nhists)      {    # to plot histories from 1 to splot
if(l<=splot)lines(matr.history[l,],lty=3,lwd=1)    }
for(i in 1:nsteps){
ymed[i]<- mean(matr.history[,i])       # mean trajectory
            }
lines(ymed,lwd=1,col="black")
abline(v=mstep,lwd=1.5,lty=4)
# print(matr.history[,mstep]) # to print the y_{mstep}^l (for each history)
m.ens<- mean(matr.history[,mstep])     # ensemble average
m.ens
se.ens<- sd(matr.history[,mstep])/sqrt(nhists)  # standard error
se.ens
```

Note the matrix `matr.history` where all the `nhists` trajectories (number of rows) and `mstep` (number of columns) are stored. The last parameter is the step at which the averages $\bar{y}$ are computed. In Fig. 11.3 only eight trajectories (dotted lines) out of all 100 are shown for clarity. The 'mean trajectory' on the eight trajectories (solid line) and the step $t = 600$ (dot-dashed line) are also reported.

Using the data coming from the AR(1) process simulation, eqn (11.4), we can compare these averages. Considering, for instance, the k-th trajectory, the time average is given by:

$$(y_1^k + y_2^k + y_3^k + \cdots + y_{1200}^k)/1200$$

for example:

$$(0.0000 + (-0.0.2925 + (-0.0045) + ... + (-3.1627))/1200 = -0.1053$$

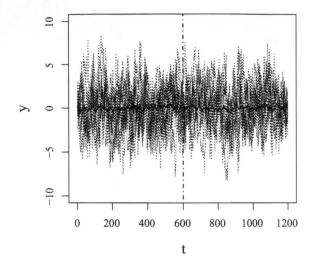

Fig. 11.3 Realizations of the eqn (11.4) AR(1) process. The plotted trajectories are eight (pointed lines). The continuous line represents the 'mean trajectory' on the eight trajectories. The vertical dot-dashed line indicates the time point ($t = 600$) at which the average is performed.

The ensemble average computation can be schematized as below (the subscript explicitly refers to the time step):

1 $y_{600} = -2.2638$.
2 $y_{600} = 0.2798$.
3 $y_{600} = -4.3918$.
... $y_{600} = \ldots$.
100 $y_{600} = 0.2230$.

Bold numbers $1, 2, 3, \ldots 100$ enumerate the 100 trajectories, at the step 600 used for the average computation:

$$(y^1_{600} + y^2_{600} + y^3_{600} + \cdots + y^{100}_{600})/100$$

for example:

$$-2.2638 + 0.2798 + (-4.3918) + \ldots + 0.2230\,)/100 = -0.22555$$

Time averages are averages in the 'horizontal direction' (see Chapter 3), and give synthetic information on the time dependence of the process. Ensemble averages are averages in the 'vertical direction' (see Chapter 3), and give an *estimate* $\hat{\mu}$ of the mean μ of the process. We remark that $\hat{\mu}$ is an 'estimate', since it is supposed that the 100 considered values are a sample of all the possible realization of the y_{600}'s. So it results in the ensemble average: $\hat{\mu} = -0.22555$ and $\sigma_e = 0.2549$, the subscript e referring to an 'ensemble average'.

We have seen that if the successive realizations of the process are in the form of a finite time series of *correlated* data, the accuracy associated to the estimate $\hat{\mu}$ of the mean of the process (and to other characteristic features) cannot be computed it can when we deal with an *i.i.d.* process. Several methods are proposed to overcome such a problem, two of them will be discussed in the following.

11.3 The batch means method

The idea underlying the batch means method is to divide the time series realized by a random process in pieces, or *batches* of the same length, to compute the mean of each batch, and to compute the global mean of these means. The batch length has to be large enough to ensure that realizations belonging to different batches are nearly statistical independent, while inside each batch the correlation is retained.

Let Y_n be a realization of length n of a random process:

$$Y_n : \{y_1, y_2, y_3, \ldots, y_n\} \tag{11.5}$$

The sequence Y_n is divided into h adjacent *non-overlapping* batches, each of length l. For convenience, we assume that n is a multiple of h, that is $n = hl$. Denote the batches as: $Y_i^{bm}, i = 1, \ldots, h$. In general, the i-th batch Y_i^{bm} with starting point y_j contains l elements, i.e.:

$$Y_i^{bm} \equiv (y_j, y_{j+1}, \ldots, y_{j+l-1})$$

with $1 \le i \le h$ and $il - (l-1) \le j \le il$.

For instance, assume $n = 18, l = 3$ and $i = 5$, so we have $j = 5 \times 3 - (3-1) = 13$, and $j + 1 = 14$, $j + (3-1) = 15$, then $Y_5^{bm} = (y_{13}, y_{14}, y_{15})$.

Figure 11.4 schematically shows how the sequence Y_n $(n = 18)$ is divided into $h = 6$ adjacent non-overlapping batches $Y_i^{bm}, i = 1, \ldots, 6$.

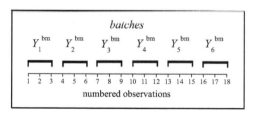

Fig. 11.4 Example of a sequence $Y_n, n = 18$ divided into $h = 6$ adjacent non-overlapping batches $Y_i^{bm}, i = 1, \ldots, 6$.

Denote the sample means of the h batches as:

$$\hat{\mu}_1, \hat{\mu}_2, \hat{\mu}_3, \ldots, \hat{\mu}_h$$

where:

$$\hat{\mu}_j = \frac{1}{l} \sum_{i=(j-1)l+1}^{jl} y_i, \quad j = 1, \ldots, h$$

An estimate $\hat{\mu}$ for the mean μ of the process is given by the average of these h 'coarse-grained' or 'time-smoothing' (Wood, 1968) observations $\hat{\mu}_j$. So $\hat{\mu}$ is given by the 'overall mean', or *grand mean* of the individual batch means:

$$\hat{\mu} = \frac{1}{h} \sum_{i=1}^{h} \hat{\mu}_i$$

An estimate of the variance of $\hat{\mu}$ is given by the sample variance $\hat{\sigma}^2$:

$$\hat{\sigma}^2 = \frac{1}{h-1} \sum_{i=1}^{h} (\hat{\mu}_i - \hat{\mu})^2 \tag{11.6}$$

Then the batch means estimate of the standard error is (the subscript bm stands for 'batch means'):

$$\hat{\sigma}_{bm} = \sqrt{\frac{\hat{\sigma}^2}{h}}$$

Every batch of size l has the same mean, equal to the sample mean of the entire process Y_n, while the standard error $\hat{\sigma}_{bm}$ depends on the length of the batches. The essential problem is to select an appropriate batch size l. If the batch size l is sufficiently large, the batch means $\hat{\mu}_i$, $i = 1, \ldots, h$ are approximately *i.i.d.* normal random variables with mean $\hat{\mu}$. In this case, we can apply 'classical' statistical methods and derive, for instance, confidence intervals for μ. Alternatively, if l is too small, the means $\hat{\mu}_i$, $i = 1, \ldots, h$ remain correlated, so the results are meaningless. On the other hand, if the number of terms in the summation eqn (11.6) becomes too small, the estimated results might be unreliable.

A small note, n not always is a multiple of h, that is not always $n = hl$ exactly. In this case, we use the function **floor()** that returns the largest integer not greater than the given number; for example, if $n = 12$ and $h = 5$, then floor(12/5) = 2. With this expedient, some observations are lost, but the remaining series is divided into batches having the same number of realizations.

The batch mean method is implemented in the R code Code_11_3.R below.

```
## Code_11_3.R
# Batch mean method
set.seed(2)
nsteps<- 1200          # number of iterations
phi<- 0.90
z<- rnorm(nsteps,mean=0,sd=1)   # sd: the standard deviation
y<- numeric()
y[1]<- 0 # initial state of the process.
for(i in 2:nsteps)   {
y[i]<- y[i-1]*phi+z[i]
                     }
# y   # comment if you do not wish the y values printed
```

```
plot(1:nsteps,y,type="l",xlab="steps",ylab="y",cex.lab=1.3,lty=1,
font.lab=3,lwd=1.5)
# autocorrelation function
par(mai=c(1.02,1.,0.82,0.42)+0.1)    # to control the margin size
corr<-acf(y,type="correlation",lag.max=50,plot=TRUE,main=" ",
xlab="lag",ylab="ACF",cex.lab=1.2,cex.axis=1.2,font.lab=3,lwd=1.5)
# corr  # comment if you do not wish the acf values printed
ta<-1
tb<-1200
lt<- length(y[ta:tb] )
# y[ta:tb]  # comment if you do not wish the y[ta:tb] values printed
m.temp<- mean(y)   # time average
m.temp
se.temp<- sd(y)/sqrt(lt)  # standard error of the m.temp (sample mean)
se.temp


# ....................................................

#  batch means method

set.seed(2)   #  reset random numbers if desired
nsteps<- 1200  #  nsteps can be  redefined
phi<- 0.90
z<- rnorm(nsteps,mean=0,sd=1)
y<- numeric()
y[1]<- 0 # initial state of the process.
for(i in 2:nsteps)   {
y[i]<- y[i-1]*phi+z[i]
                     }
ta<- 1      # ta, tb, redefined; the code can run  autonomously
tb<- nsteps
x<- numeric()
mbatch<- numeric()
se.batch<- numeric()
se.sebm<- numeric()
l<- numeric()
mj<- numeric()
n<- length(y[ta:tb] )      # n is the length of the series
m.temp<- mean(y)    # time average (reported again)
m.temp
se.temp<- sd(y)/sqrt(n)  # standard error of the m.temp (sample mean)
se.temp
## the interval [ta,tb] is divided in batches
## each batch has l observations, i.e., l = batch size
l<-  c(1,5,10,15,20,25,30,40,50,60,70,80,90,100,120)
# one can choose the initial value of l (lbmi) and the final one (lbmf)
lbmi<- 1
lbmf<- 15
h<- floor(n/l) # number of batches for each l
# starting most external loop on the batch size l
for (k in lbmi:lbmf)  {
j<- ta-l[k]
for (i in 1:h[k])       {      # starting loop inside each of the h batches
j<- j+l[k]
mj[i]<-mean(y[j:(j+(l[k]-1))])    # mean of each batch
                       }      #  ending loop inside each h batch
# print(mj[1:h[k]])  # to print the mean of each batch
mbatch[k]<-mean(mj[1:h[k]])
se.batch[k]<- sd(mj[1:h[k]])/sqrt(h[k])
se.sebm[k]<-  se.batch[k]/sqrt(2.*(h[k]-1))
```

```
                      }       # ending loop on the batch size l
# summary of quantities in [lbmi:lbmf]
l[lbmi:lbmf]              # batch size
h[lbmi:lbmf]              # corresponding number of batches
mbatch[lbmi:lbmf]        # check: they must be all equal
se.batch[lbmi:lbmf]
se.sebm[lbmi:lbmf]
par(mai=c(1.02,1.,0.82,0.42)+0.1)
plot(l[lbmi:lbmf],se.batch[lbmi:lbmf], type="b",ylim=c(0.05,0.42),
xlab="batch size",
ylab=expression(hat(sigma)[bm]),
font.lab=3,lwd=1.5,cex.lab=1.2)
arrows(l,se.batch, l,se.batch+se.sebm, length=0.05,angle=90)
arrows(l,se.batch, l,se.batch-se.sebm, length=0.05,angle=90)
```

The first part of the code, up to `batch means`, is a copy of `Code_11_1.R` reported for convenience. The time average of the series and its standard error are computed again. As we have seen, the time average $\bar{y}$ (`m.temp`) results in 0.5928, and the standard error σ_t (`se.temp`) is equal to 0.06977.

The part of the code that follows can run autonomously, if desired. The code is structured in two loops. The outer is on the number of batches h or, equivalently, on the size of the batches l. It is worth noticing that the code can vary the length of the batches to find the most appropriate one. Suppose, for instance, $h = 40$. In that case the series $Y_n, n = 1200$, is divided into 40 batches and each batch contains $l = 1200/40 = 30$ observations.

The inner loop is inside each batch h. For each batch with `l[k]`, the code computes the mean of each batch `mj[i]` ($\hat{\mu}_i$, $i = 1, \ldots, h$) and the standard error of the mean `se.batch[l]` ($\hat{\sigma}_{bm}$).

The accuracy of the estimate of $\hat{\sigma}_{bm}$, `se.sebm[hi:hf]` (s.e.($\hat{\sigma}_{bm}$)) can also be estimated. It is given by (see also Flyvbjerg and Petersen, 1989):

$$\text{s.e.}(\hat{\sigma}_{bm}) = \sqrt{\frac{D_4 - \hat{\sigma}_{bm}^4}{4(h-1)\hat{\sigma}_{bm}^2}} \tag{11.7}$$

where:

$$D_4 = \frac{1}{h}\sum_{i=1}^{h}(\hat{\mu}_i - \hat{\mu})^4$$

in the normal approximation:

$$\text{s.e.}(\hat{\sigma}_{bm}) = \hat{\sigma}_{bm} \times \left(\frac{1}{(2. * (h-1))}\right)^{1/2} \tag{11.8}$$

The final results are:

batch size `l[hi:hf]`

```
[1]    1    5   10   15   20   25   30   40   50   60   70   80   90  100  120
```

corresponding number of batches `h[hi:hf]`

```
[1] 1200   240   120    80    60    48    40    30    24    20    17    15    13    12    10
```

`mbatch[hi:hf]` mean of each batch with different l. Check: they must be all the same and equal to the time average `m.temp`.

```
 [1] 0.5928017 0.5928017 0.5928017 0.5928017 0.5928017 0.5928017
 [7] 0.5928017 0.5928017 0.5928017 0.5928017 0.5903281 0.5928017
[13] 0.5952046 0.5928017 0.5928017
```

standard error `se.batch[hi:hf]` $(\hat{\sigma}_{bm})$ of $\hat{\mu}$.

```
 [1] 0.06976656 0.14517286 0.18872174 0.21394611 0.23450259 0.25353158
 [7] 0.25251638 0.27594069 0.29530061 0.24951920 0.24837992 0.26547587
[13] 0.22170795 0.33959625 0.26176558
```

standard error `se.sebm[hi:hf]` (s.e.$(\hat{\sigma}_{bm})$) of the standard error $\hat{\sigma}_{bm}$, computed with eqn (11.8).

```
 [1] 0.001424698 0.006640052 0.012233014 0.017020639 0.021587714 0.026149785
 [7] 0.028591850 0.036232788 0.043539699 0.040477359 0.043907782 0.050170223
[13] 0.045255946 0.072402163 0.061698740
```

Figure 11.5 shows $\hat{\sigma}_{bm}$, the estimate of the standard error of $\hat{\mu}$, as a function of the size l of the batches. The error bars of these estimates are also added.

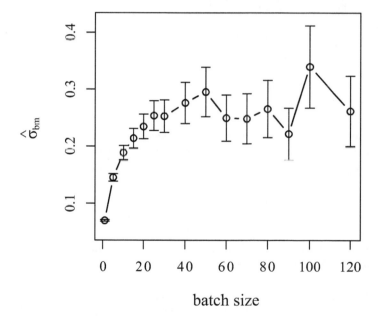

batch size

Fig. 11.5 Estimated standard errors $\hat{\sigma}_{bm}$ of the estimated mean $\hat{\mu}$ of eqn (11.4) AR(1) process, as a function of the size of the batches l. The error bars represent the standard error s.e.$(\hat{\sigma}_{bm})$, eqn (11.8) of $\hat{\sigma}_{bm}$.

It appears that $\hat{\sigma}_{bm}$ increases up to around $l = 25$ ($h = 48$), and it remains almost constant up to $l = 80$ ($h = 15$). It is said that $\hat{\sigma}_{bm}$ has reached a *plateau*.

The mean of the values of $\hat{\sigma}_{bm}$ in the interval $l = [25, 80]$ ($h = [48, 15]$) $= 0.2630$ can be assumed as the standard error of $\hat{\mu}$. The value $\hat{\sigma}_{bm} = 0.2630$, derived from one series only, appears in full agreement with $\sigma_e = 0.2549$ obtained from an ensemble of 100 simulated trajectories. Note, in passing, that with $l = 1$ the series remains as the original one, and obviously $\hat{\sigma}_{bm}$ ($l = 1$) is equal to $\sigma_t = 0.06977$.

Up to $l < 25$, the batch means $\hat{\mu}_i$, $i = 1, \ldots, 40$ are not independent, as is also shown in Figure 11.6 (left) with $l = 10$ ($h = 120$), where the large spike at lag 0 is followed by a decreasing wave that alternates ACF bars exceeding the positive and negative dashed lines. Instead, Figure 11.6 (right) shows that for $l = 60$ ($h = 20$) the $\hat{\mu}_i$'s become approximately independent, and only the lag 0 ACF is greater than the 95% confidence limits.

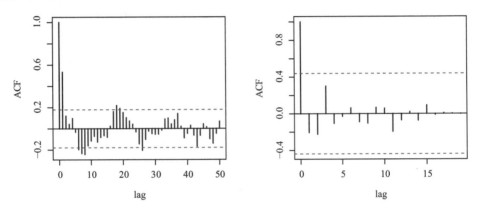

Fig. 11.6 Autocorrelation plots for the mean of each batch `mj`. Left: $h = 120$ ($l = 10$), right: $h = 20$ ($l = 60$). Dashed lines: 95% confidence intervals assuming white noise input.

As we have already said, the means of each batch, $\hat{\mu}_j = 0.5928$, are all equal, independently of the length of the batches, and equal to the 'grand mean' estimator $\hat{\mu}$. However, looking at the `mbatch[hi:hf]` values, we see that `mbatch[11]` and `mbatch[13]` are slightly different from 0.5928, due to the effect of the function **floor()**: for those batch means, it is not exactly $n = h\,l$.

This section ends with some historical and bibliographical references. The idea of breaking the original chain into independent statistical pieces was already exploited for a long time, for instance, Wood (1968), Friedberg and Cameron (1970), Landau (1976), Binder (1992), where the $\hat{\mu}_i$'s are named 'coarse-grained' or 'time-smoothing' values.

The batch means method was reintroduced in statistical literature by Carlstein (1986), so sometimes it is called with the name of 'Carlstein's method' or 'Carlstein's rule' (Hall *et al.*, 1995), or 'bootstrap with disjoint blocks' (Künsch, 2015).

A number of papers deal with the choice of the batch lengths, usually without the possibilities of varying l in a unique run, see among others Whitmer (1984) and Morales *et al.* (1990). An exception is in Flyvbjerg and Petersen (1989), where the batch sizes are automatically determined by halving them each time, rather than tentatively varying them in search of a *plateau*. We also point out that the behaviour approaching the *plateau* is an indirect indication of the strength of correlation.

Geyer (2011) applies the batch means method to an AR(1) process, Bennett *et al.* (1981) propose a variant of the batch means method by inserting spacers between the batches of observations. Pedrosa and Schmeiser (1993) use 'overlapping batch means', in which the batches have realizations in common. This variant is also the first step of the moving block bootstrap method discussed in the next section.

11.4   The moving block bootstrap method

The main difference with the 'classic', or *i.i.d.*, bootstrap is that in the moving block bootstrap (MBB), blocks of observations are resampled instead of single observations. The main difference with the batch means method is that in the MBB method the batches (or 'blocks') are overlapped.

As in the *i.i.d.* bootstrap, also in the MBB a number of replications are formed to compute the statistic of interest on each of them. As in the batches means method, also in the MBB the block size must be 'sufficiently' large such that the correlation between observations belonging to different blocks has decayed off. Reliable estimates cannot be obtained if the number of blocks is too small.

If the assumption of independent random variables is violated, as occurs when observations are serially correlated, the bootstrap is not applicable in the form sketched above, because this method ignores the dependence structure of the data. However, the problem can be overcome in a way which conserves the spirit of Efron's original idea, that is by resampling *blocks* of observations instead of individual observations. This improvement makes the original bootstrap more robust against serial dependence.

11.4.1   Introduction to the MBB

Theoretical foundations of the MBB are in Künsch (2015) and Liu and Singh (1992). A short account is reported in Mignani and Rosa (1995) where the MBB was applied for the first time in statistical mechanics in a study concerning the Ising model. Perhaps, the first idea to exploit such computer-intensive methods for Monte Carlo correlated data is found in a paper on lattice gauge theory (Gottlieb *et al.*, 1986).

Consider the discrete parameter stationary time series eqn (11.5)

$$Y_n : \{y_1, y_2, y_3, \ldots, y_n\}$$

Let $\hat{\mu}$ be an estimate of the mean of the process μ and $\hat{\sigma}_{bm}$ the standard error of $\hat{\mu}$, estimated via the batch means method. In the example presented in `Code_11_3.R`, we have $\hat{\mu} = 0.5928$ and $\hat{\sigma}_{bm} = 0.2630$.

In order to estimate σ through the MBB, the observed time series Y_n, is divided into *overlapping* blocks of l observations each and all possible contiguous blocks of length l are considered. Let us follow the example below.

We have seen above how to divide the realizations of the process $Y_n, n = 1200$ in $h = 40$ batches with $l = 30$ observations each. Consider now only $n = 18$ observations with $h = 6$

```
0.0000000, 0.1848492, 1.7542096, 0.4484130, 0.3233199, 0.4234082,
1.0890221, 0.7404219, 2.6508536, 2.2469813, 2.4399339, 3.1776933,
2.4672286, 1.1808367, 2.8449820, 0.2494147, 1.1030779, 1.0285768
```

and suppose we divide them into $h = 6$ batches with $l = 3$ (see Figure 11.4):

```
batch no. 1: 0.0000000, 0.1848492, 1.7542096
batch no. 2: 0.4484130, 0.3233199, 0.4234082
..........................................
batch no. 6: 0.2494147, 1.1030779, 1.0285768
```

With the MBB, we again divide the series into batches, now calling them *blocks*, with $l = 3$, but the blocks are overlapped and all possible contiguous blocks are considered. In this way, q blocks are obtained, with $q = n - l + 1$, here $q = 16$ (for a better choice of this number, see the following subsection). The 16 blocks are reported below with four significant decimal digits.

$$|0.0000, 0.1848, 1.7542|_1 \quad |0.1848, 1.7542, 0.4484|_2 \quad |1.7542, 0.4484, 0.3233|_3$$
$$|0.4484, 0.3233, 0.4234|_4 \quad |0.3233, 0.4234, 1.0890|_5 \quad |0.4234, 1.0890, 0.7404|_6$$
$$|1.0890, 0.7404, 2.6508|_7 \quad |0.7404, 2.6508, 2.2469|_8 \quad |2.6508, 2.2469, 2.4399|_9$$
$$|2.2469, 2.4399, 3.1776|_{10} \quad |2.4399, 3.1776, 2.4672|_{11} \quad |3.1776, 2.4672, 1.1808|_{12}$$
$$|2.4672, 1.1808, 2.8449|_{13} \quad |1.1808, 2.8449, 0.2494|_{14} \quad |2.8449, 0.2494, 1.1030|_{15}$$
$$|0.2494, 1.1030, 1.0285|_{16}$$

Call the blocks $Q_1, Q_2, \ldots, Q_q$. For example, $Q_3 = |1.7542, 0.4484, 0.3233|_3$. In general the i-th block Q_i with starting point y_i contains l elements, i.e.:

$$Q_i \equiv (y_i, y_{i+1}, \ldots, y_{i+l-1})$$

with $1 \leq i \leq q$.

From these q blocks Q_i $(i = 1, \ldots, q)$ we draw h $(hl = n)$ blocks at random *with replacement*. The *starting point* of each block is randomly selected from a uniform distribution of integers $(1, \ldots, n)$, so that all $\mathbf{Q}_i$'s are equally likely to be drawn. The h selected blocks, placed one after another, form the new full size series Q^*:

$$Q^* = Q_1^*, Q_2^*, \ldots, Q_h^*$$

where Q_1^* is one of the possible Q_i. In this sense, the blocks are 'moving': each Q_i may lie in any point of the new formed series Q^*.

In the example, a resampled series is formed with $h = 6$ blocks. If the random number i is equal to 4, $Q_4 \equiv (y_4, y_5, y_6)$, then $Q_1^* = Q_4 = 0.4484, 0.3233, 0.4234$. A further block is picked up, for instance $i = 10$, so $Q_2^* = Q_{10} = 2.2469, 2.4399, 3.1776$. This procedure is executed h times, so that the first *bootstrap sample* is formed, for instance: $Q^* = Q_4, Q_{10}, Q_1, Q_{10}, Q_7, Q_8$ that is:

```
0.4484, 0.3233, 0.4234, 2.2469, 2.4399, 3.1776,
0.0000, 0.1848, 1.7542, 2.2469, 2.4399, 3.1776,
1.0890, 0.7404, 2.6508, 0.7404, 2.6508, 2.2469
```

Note that some blocks may be extracted more than once, while others are never drawn.

The procedure is now perfectly analogous to that of the *i.i.d.* bootstrap. We are interested in the standard error of the estimate $\hat{\mu}$, that is computed by the arithmetic mean:

$$\hat{\mu} = \frac{1}{n}(y_1 + y_2, + \ldots, y_n)$$

in the same way, on the first bootstrap sample, the first *bootstrap replication* is computed:

$$\hat{\theta}_1^* = \frac{1}{h}\left(Q_1^* + Q_2^* + Q_3^*, + \ldots, Q_h^*\right)$$

Recall that, in general, the 'bootstrap replication' is the value of the statistics referred to the bootstrap sample (see Appendix A).

This drawing procedure is repeated many times, to obtain B ($B \approx 100 - 1000$) bootstrap replications:

$$\hat{\theta}_1^*, \hat{\theta}_2^*, \hat{\theta}_3^*, \ldots, \hat{\theta}_B^*$$

From the bootstrap replications, a bootstrap estimate of the standard error of the estimate $\hat{\mu}$ is derived. Call it $\hat{\sigma}^*$:

$$\hat{\sigma}^*(\hat{\theta}^*) = \left[\sum_{b=1}^{B} \frac{(\hat{\theta}_b^* - \bar{\theta}^*)^2}{B-1}\right]^{1/2} \tag{11.9}$$

where:

$$\bar{\theta}^* = \sum_{b=1}^{B} \frac{\hat{\theta}_b^*}{B} \tag{11.10}$$

As in the batch means method, we search for blocks sizes l for which observations belonging to different blocks are independent of one another. In practice, the plot of $\hat{\sigma}^*$ vs l shows that $\hat{\sigma}^*$ increases until it reaches a *plateau*, where it remains nearly constant. This is a sign that the blocks are actually *i.i.d.* random variables under the MBB scheme and, at the same time, inside each block the correlation is retained. For a review of variants of this method, see Politis (2003)

11.4.2 The MBB in R

An extremely schematic example allows us to further enter into the details of the MBB method, directly comparing it with the batch means method. Suppose we have an observed series of length $n = 12$, in which the data are numbered as:

$$Y_n : \{1, 2, 3, 4, 5, 6, 7, 8, 9, 10, 11, 12\}$$

If the batch means method is applied, the series is divided, for instance, in $h = 4$ batches, of length $l = 3$:

batch no. 1 $\frac{1+2+3}{3} = \bar{y}_1$
batch no. 2 $\frac{4+5+6}{3} = \bar{y}_2$
batch no. 3 $\frac{7+8+9}{3} = \bar{y}_3$
batch no. 4 $\frac{10+11+12}{3} = \bar{y}_4$

In general:

$$\bar{y}_i = \frac{1}{l} \sum_{j=il-(l-1)}^{il} y_j^i$$

where j refers to the j-th observation in the i-th batch. For example:

$$\bar{y}_2 = \frac{1}{3} \sum_{j=2\times3-(3-1)}^{2\times3} y_j^i = \frac{1}{3}\left(y_4^2 + y_5^2 + y_6^2\right)$$

The estimate $\hat{\mu}$ of the mean of the process μ is given by:

$$\hat{\mu} = \frac{1}{h}\sum_{i=1}^{h} \bar{y}_i = \frac{1}{lh}\sum_{i=1}^{h}\sum_{j=il-(l-1)}^{il} y_j^i$$

If the batch size is large enough, the sample means of the batches are approximately independent. Their sample variances are estimates of the variance of each $\bar{y}_i$, given by eqn (11.6).

Compare now the MBB procedure with the batch means method. Consider all possible contiguous blocks of $l = 3$ observations each, that is:

$$[1\,2\,3] \quad [2\,3\,4] \quad [3\,4\,5] \quad [4\,5\,6] \quad [5\,6\,7]$$
$$[6\,7\,8] \quad [7\,8\,9] \quad [8\,9\,10] \quad [9\,10\,11] \quad [10\,11\,12]$$

Now the number of blocks is no more $h = n/l = 4$ (still supposing $n = hl = 12$), but $q = n - l + 1$, in the example $q = 12 - 3 + 1 = 10$.

In the example, if only $q = 10$ blocks are considered, the observation 12 appears only in the block $[10\,11\,12]$, while all the observations y_i's must be present l times. To overcome this drawback, observations 1 and 2 are added. In general, to take into account all the observations, $l - 1$ initial observations have to be added, so the number of blocks becomes $q + (l - 1)$, here $= 12$ with the two added blocks $[11\,12\,1]$ and $[12\,1\,2]$. In this way, also the observations 1 and 2 are present $l = 3$ times.

As a further example, suppose $l = 5$, so $q = 12 - 5 + 1 = 8$, but $5 - 1$ observations must be added after the last one, and the number of blocks becomes $q = 12$.

$$[1\,2\,3\,4\,5] \quad [2\,3\,4\,5\,6]$$
$$[3\,4\,5\,6\,7] \quad [4\,5\,6\,7\,8]$$
$$[5\,6\,7\,8\,9] \quad [6\,7\,8\,9\,10]$$
$$[7\,8\,9\,10\,11] \quad [8\,9\,10\,11\,12]$$
adding
$$[9\,10\,11\,12\,1] \quad [10\,11\,12\,1\,2]$$
$$[11\,12\,1\,2\,3] \quad [12\,1\,2\,3\,4]$$

More formally, using the notion of congruence, it can be written:

$$y_{n+r} = y_{(n+r)(\bmod\,n)}$$

For instance, $(n = 12, l = 3)$ with $r = 2$:

$$y_{12+2} = y_{(12+2)(\bmod\,12)} = y_2$$

so in the place of the observation 14, which does not exist, the observation 2 is used to form the block $[12\,1\,2]$.

The idea of wrapping the observations around in a circle, before forming the blocks, is due to Politis and Romano (1992), who call the method 'circular block-resampling bootstrap'. By this method, the bootstrap distribution is centred around its mean instead of around the sample mean. This allows preventing the bias due to the reduced weight assigned to the observations y_i's not appearing l times in each block, that is if $i > n - l + 1$. Incidentally, similar procedures, involving data wrapping in a circle are exploited in statistical mechanics to reduce border effects. These kinds of 'tricks' are known as 'periodic boundary conditions'.

Figure 11.7 schematically shows how to look at the observations arranged in a circle. All of them are present l times in each block, represented by an arc.

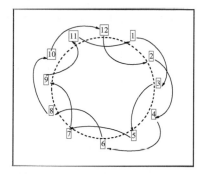

Fig. 11.7 Observations wrapped around in a circle to have all of them present in each block, depicted with arcs, an equal number of times ($n = 12, l = 3$).

Let us see how to construct a resampled series in the example with $l = 5$. The MBB method in implemented in the R code `Code_11_4.R`, where the function **ceiling()** is used, returning the smallest integer greater than the given number.

In that case, `ceiling(12/5) = 3`, so we need three blocks, and the resampled series will have 15 observations. The beginning of each block i is picked at random $1 \leq i \leq 12$). For instance:

$$i = 10 \quad \rightarrow \quad \text{block:} \quad [10\ 11\ 12\ 1\ 2]$$
$$i = 12 \quad \rightarrow \quad \text{block:} \quad [12\ 1\ 2\ 3\ 4]$$
$$i = 1 \quad \rightarrow \quad \text{block:} \quad [1\ 2\ 3\ 4\ 5]$$

The bootstrap sample is then:

$$(10\ 11\ 12\ 1\ 2\ 12\ 1\ 2\ 3\ 4\ 1\ 2\ 3\ 4\ 5)$$

On this bootstrap sample the bootstrap replication is computed. From here onwards, all proceeds analogously to the *i.i.d.* bootstrap. If the statistics of interest is the standard error of the mean, the mean of the first bootstrap sample above is computed. This computation is repeated B times to obtain the mean of the B bootstrap replications eqn (11.10) and a bootstrap estimate eqn (11.9) of the standard error of the mean $\hat{\mu}$.

```
## Code_11_4.R
# Bootstrap
set.seed(2)
nsteps<- 1200          # number of iterations
phi<- 0.90
z<- rnorm(nsteps,mean=0,sd=1)    # sd: the standard deviation
y<- numeric()
y[1]<- 0 # initial state of the process.
for(i in 2:nsteps)   {
y[i]<- y[i-1]*phi+z[i]
                   }
# y   # comment if you do not wish the y values printed
plot(1:nsteps,y,type="l",xlab="steps",ylab="y",cex.lab=1.3,lty=1,
font.lab=3,lwd=1.5)
# autocorrelation function
par(mai=c(1.02,1.,0.82,0.42)+0.1)    # to control the margin size
corr<-acf(y,type="correlation",lag.max=50,plot=TRUE,main=" ",
xlab="lag",ylab="ACF",cex.lab=1.2,cex.axis=1.2,font.lab=3,lwd=1.5)
# corr  # comment if you do not wish the acf values printed
ta<-1
tb<-1200
lt<- length(y[ta:tb] )
# y[ta:tb]  # comment if you do not wish the y[ta:tb] values printed
m.temp<- mean(y)   # time average
m.temp
se.temp<- sd(y)/sqrt(lt)  # standard error of the m.temp (sample mean)
se.temp


# .....................................................

#  Moving Block Bootstrap (MBB) method
set.seed(2)    #  reset random numbers if wanted
nsteps<- 1200  # nsteps can be  redefined
phi<- 0.90
z<- rnorm(nsteps,mean=0,sd=1)
y<- numeric()
y[1]<- 0 # initial state of the process.
for(i in 2:nsteps)   {
y[i]<- y[i-1]*phi+z[i]
                   }
ta<- 1      # ta, tb, redefined; the code can run  autonomously
tb<- nsteps
x<- numeric()
xb<-    numeric()
ntot<-  numeric()
mb<-    numeric()
nbl<-   numeric()
seb<-   numeric()
mmbb<-  numeric()
sembb<- numeric()
se.sembb<- numeric()

n<- length(y[ta:tb])       # n is the length of the series
x[1:n]<- y[ta:tb]          # x[1:n] are named the realizations
# x       # to see the realizations in [ta,tb]
m.temp<- mean(y)    # time average (reported again)
m.temp
```

```
se.temp<- sd(y)/sqrt(n)   # standard error of the m.temp (sample mean)
se.temp
#   l = block length, i.e., each block has l observations
l<-c(1,2,4,6,8,10,15,20,25,30,35,40,45,50,55,60,65)
llmb<- length(l)
# one can choose the initial value of lmb (lmbi) and the final one (lmbf)
lmbi<- 1
lmbf<- 17
h<- n/l          # number of blocks
q<- n-l+1        # number of overlapped blocks
# for each block, B replications are computed
B<- 400   # number of replications
# starting most external loop on the block length
for (k in lmbi:lmbf)  {
# ... periodic boundary conditions:
# ... each x[i] has to be appeared l times
# ... adding  the first l-1 observations
# total observations, after adding  the first l-1 observations
ntot[k]<- n+(l[k]-1)
n1<- n+1
if(l[k]>1){       # if l[k]=1 the loop to form the "new series" is skipped
ii=0
for (i in n1:ntot[k])  {
  ii<- ii+1
  x[i]<- x[ii]
                }            # ending "if l[k]=1"
                  }
# x is the  "new series" = the original one + (l-1) initial realizations
# print(x)   # to see  the "new series"
nbl[k]<- ceiling(n/l[k])      # number of blocks to form the resampled series
#  # starting loop on the  B replications
for (b in 1:B) {    # loop on replications
        nn<- 0
        for (j in 1:nbl[k]) {  # loop  on the number of sampled blocks
        i<- sample(n,1,replace=TRUE)  # beginning of a block picked at random
   # a block with length l is formed  (initial 'i' included)
              for (ll in i:(i+(l[k]-1)))  {
              nn<- nn+1
              xb[nn]<- x[ll]
                                    } # the block is finished
                              } # ending loop on the number of sampled blocks
# print(xb)      # to print the replications
mb[b] <- mean(xb)
# print(mb[b])  # to print the mean of each replication
seb[b]<- sqrt(var(xb))
# print(seb[b])  # to print the s.e. of each replication
            }            # ending loop on the replications
        mmbb[k]<- mean(mb)
        sembb[k]<-sum((mb-mmbb[k])^2/(B-1))^(1/2)    # to plot
        se.sembb[k]<- sembb[k]/sqrt(2.*(nbl[k]-1))*sqrt(2./3.)
           }   # ending most external loop on the length of blocks
# summary of quantities in [lmbi:lmbf]
l[lmbi:lmbf]
h[lmbi:lmbf]
q[lmbi:lmbf]
ntot[lmbi:lmbf]
nbl[lmbi:lmbf]
mmbb[lmbi:lmbf]
sembb[lmbi:lmbf]
se.sembb[lmbi:lmbf]
```

```
par(mai=c(1.02,1.,0.82,0.42)+0.1)
plot(l[lmbi:lmbf],sembb[lmbi:lmbf], type="b",ylim=c(0.05,0.35),
xlab="block length",font.lab=3,lwd=1.5,cex.lab=1.3,
ylab=expression(hat(sigma)*"*"))
arrows(l,sembb, l,sembb+se.sembb, length=0.05, angle=90)
arrows(l,sembb, l,sembb-se.sembb, length=0.05, angle=90)
```

As in `Code_11_3.R` the first part of the above code, up to `MBB`, simulates the AR(1) process eqn (11.4). The following part of the code is similar to that of `Code_11_2.R`. The series is divided into h blocks, each containing $l = n/h$ data values, namely l is the 'block length' varying inside a loop going from `lmbi` to `lmbf` (in the code equal to 1 and 17, respectively), that is $l = 1, \ldots, 65$. If, for instance, $l = 40$, this means that the series $Y_n, n = 1200$ is divided into 30 blocks, each block containing $l = 1200/30 = 40$ observations.

The outer loop (`for (k in lmbi:lmbf)` ...) is on the number of the block lengths `l<- c(1,2,4,...,65)`, given at the beginning, so that the blocks are not yet over-lapped. The observations are prepared to be replicated with two operations. As a first step, the observations are wrapped, adding the first $l - 1$ observations: `ntot[k]<- n+(l[k]-1)`. Then the length of blocks to constitute the resampled series is updated: `nbl[k]<- ceiling(n/l[k])`. The loop on the replications (`for (b in 1:B)` ...) is executed to assemble the bootstrap samples. The beginning of each 'moving' block is chosen by a random sampling with replacement from the vector $1 \ldots n$: `i<- sample(n,1,replace=TRUE)`. Finally, the average of the B bootstrap replications (eqn 11.10) and a bootstrap estimate (eqn 11.9) of the standard error of each block h (`h[lmbi:lmbf]`) is computed. The result obtained with $B = 400$ is:

block length `l[lmbi:lmbf]`

```
[1]  1  2  4  6  8 10 15 20 25 30 35 40 45 50 55 60 65
```

corresponding number of non-overlapping blocks `h[lmbi:lmbf]`

```
 [1] 1200.00000  600.00000  300.00000  200.00000  150.00000  120.00000
 [7]   80.00000   60.00000   48.00000   40.00000   34.28571   30.00000
[13]   26.66667   24.00000   21.81818   20.00000   18.46154
```

number of overlapping blocks `q[lmbi:lmbf]` $(q = n - l + 1)$

```
 [1] 1200 1199 1197 1195 1193 1191 1186 1181 1176 1171 1166 1161 1156 1151 1146
[16] 1141 1136
```

length of the blocks `ntot[lmbi:lmbf]` after adding the first $l - 1$ observations

```
 [1] 1200 1201 1203 1205 1207 1209 1214 1219 1224 1229 1234 1239 1244 1249 1254
[16] 1259 1264
```

length of the blocks after the ceiling function `nbl[lmbi:lmbf]`

```
 [1] 1200  600  300  200  150  120   80   60   48   40   35   30   27   24   22
[16]   20   19
```

means of each bock with different l, `mmbb[lmbi:lmbf]`. These are close to each other and close to the time average `m.temp` $= 0.5928017$.

```
[1]  0.5884565 0.5991287 0.5898515 0.5980998 0.6150934 0.6047890 0.6034414
[8]  0.6026749 0.5775105 0.5785146 0.6138597 0.5456157 0.5511164 0.5808206
[15] 0.5775882 0.5929105 0.5837608
```

bootstrap estimate for the standard error of the estimate $\hat{\mu}$ `sembb[lmbi:lmbf]` $(\hat{\sigma}^*)$

```
[1]  0.07006526 0.09598453 0.13178510 0.15841488 0.17378288 0.19585105
[7]  0.21543579 0.23066210 0.25532195 0.24458536 0.26150826 0.26335080
[13] 0.26030662 0.25882132 0.28130114 0.26969193 0.27711156
```

```
se.sembb[lmbi:lmbf]
standard error \texttt{se.sembb[lmbi:lmbf]}($\mbox{s.e.} (\hat\sigma^*)$) of
the standard error $\hat\sigma^*$.
[1]  0.001168241 0.002264265 0.004400176 0.006483489 0.008219648 0.010365537
[7]  0.013994058 0.017337625 0.021501987 0.022611925 0.025893176 0.028234170
[13] 0.029473921 0.031158426 0.035440612 0.035721569 0.037710106
```

It was observed, (Künsch, 2015), (Hall *et al.*, 1995), that the variance of the MBB estimators is less than that of the batch means estimators, because in the MBB the blocks are overlapped. The reduction with respect to the batch means s.e.$(\hat{\sigma}_{bm})$ is about a factor $\sqrt{(2/3)}$.

Figure 11.8 shows $\hat{\sigma}^*$, bootstrap estimate of the standard error of the estimate $\hat{\mu}$, as a function of the block length l.

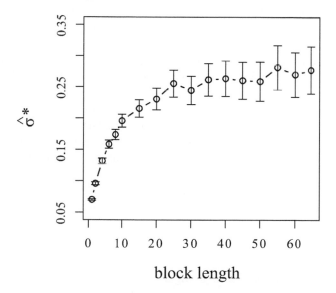

Fig. 11.8 Estimated standard errors $\hat{\sigma}^*$ of the estimated mean $\hat{\mu}$ of eqn (11.4) AR(1) process, as a function of the block length l. The error bars represent the standard error s.e.$(\hat{\sigma}^*)$ of $\hat{\sigma}^*$.

It appears that $\hat{\sigma}^*$ increases up to around $l = 25$, where it reaches a *plateau*, remaining nearly constant up to $l = 65$. As before, that is a sign that the mutual independence of the blocks has been achieved. The mean of the values of $\hat{\sigma}^*$ in the interval $l \in [25, 65]$ is 0.2636, which can be assumed as the standard error of $\hat{\mu}$, is almost the same value as $\hat{\sigma}_{bm} = 0.2630$ obtained by the batch means method.

In essence, in the batch means method the statistics of interest are recomputed in each of these smaller blocks of the type $(y_i, y_{i+1}, \ldots, y_{i+l-1})$, while in the MBB the statistics are recomputed in each of the new, *full size series* $(\mathbf{Q}_1^*, \mathbf{Q}_2^*, \ldots, \mathbf{Q}_h^*)$. In Mignani and Rosa (2001) it is shown, by means of computer experiments, that there are cases where the MBB outperforms other methods based on subseries, as the batch means method. This happens, for instance, with highly correlated states in relatively short chains.

For a recent review of variants of both the batch means method and the MBB, see Politis (2003).

11.5 Convergence diagnostic with the MBB method

We will show that the MBB method, used to assign accuracy to estimates in the presence of considerable correlations between successive realizations, can also help to shed light on the problem of convergence.

Since the beginnings of MCMC, it has been noticed that a 'formally' ergodic chain may nevertheless be 'computationally' non-ergodic or, in Wood and Parker's words, 'quasi-ergodic' (Wood and Parker, 1957). These authors warn against situations in which the state space may be divided into two (or more) 'pockets', formally belonging to the same ergodic class, but linked through an 'isthmus' of states of very small, but non-vanishing, probabilities. Suppose, for instance, that two simulations are run in parallel. If this quasi-ergodicity arises, two very different sequences might be obtained, not converging to a common sequence and consequently not yielding a final estimate. Individually each sequence might not reveal any flaws, appearing to have converged to its own stationary distribution. In such circumstances, the results from any one sequence would be unreliable. Examples can be found in Gelman and Rubin (1992a) and, by the same authors, in an article with a very eloquent title (Gelman and Rubin, 1992b). The notorious 'witch's hat' distribution (Matthews, 1989) serves as a further example giving the (erroneous) impression of convergence for the chain generated by the Gibbs algorithm, even though the sequence has not explored most of the target distribution.

A somewhat artificial example can be explanatory. Return to the code `Code_8_2.R` of Chapter 8 in which the 'target density' is now the function:

$$f(x) = \exp(-x^2)\, x^2 \tag{11.11}$$

reported in Fig. 11.9.

By applying the Metropolis-Hastings algorithm with

```
set.seed(2)
nsteps<- 10000
x0<- -10
delta<- 0.75
burn.in<- 2000
```

Fig. 11.10 is obtained. We see that the realizations are roughly around the lines $x(i) = 1$ and $x(i) = -1$, with sudden sign inversions. By changing random number sequences, different histories are obtained, but all show this kind of 'sign inversion', shown in Fig. 11.10. In such situations, speaking of convergence is obviously meaningless. An

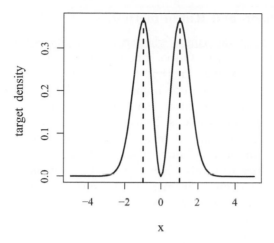

Fig. 11.9 Target density $f(x) = \exp(-x^2)\, x^2$. The dashed lines are at the two maximum values -1 and $+1$.

expedient, sometimes used in statistical mechanics, is to take the absolute values of all the realizations. Doing this, the computed time average results to be 1.145, in this example.

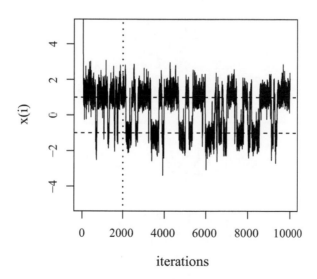

Fig. 11.10 Realizations of the Metropolis-Hastings algorithm with target density eqn (11.11), and $\delta = 0.75$. Dotted line at the burn-in $= 2000$ iterations, dashed line at the two maximum values -1 and $+1$ of the target density.

11.5.1 The Gelman and Rubin method

A variety of methods to determine whether stationarity is achieved (*convergence diagnostics*) have been proposed. Often, different authors give different suggestions, but all agree that there is no single method to deal with convergence, and each tool has its pros and cons. A rather common approach, e.g. (Fosdick, 1963; Wood, 1968; Ripley, 1987), consists in running more than one chain, all with the same parameters but different starting configurations. By comparing the results, one can detect possible dangers and try to circumvent them. This approach prescription was formalized and improved by Gelman and Rubin (G-R) and now it is one of the most popular convergence diagnostics methods. On the other hand, it has also been observed that sometimes many short chains appear to converge to a unique, but completely wrong, distribution (Geyer, 1992). So that it should be more advisable to trust results based on a single long run. For an extended overview of the efforts in this research area, the reader is referred to the articles by (Cowles and Carlin, 1996) and by (Brooks and Roberts, 1992), and to Chapter 8 of the book by (Robert and Casella, 1999), where almost all convergence diagnostics available in the literature are discussed and compared.

We use the basic version of the G-R method, as reported, for instance, in (Evans and Swartz, 2000). It consists essentially of the following steps. Generate m different chains at different starting values. Let the scalar $x_i^{(j)}$ be the i-th realization in the j-th chain, $(i = 1, \ldots, n; \ j = 1, \ldots, m)$ and $y_i^{(j)} = g(x_i^{(j)})$ be the quantity of interest. Define B the *between*-chain variance and W the *within*-chain variance, respectively, where:

$$B = \frac{n}{m-1} \sum_{j=1}^{m} \left(\overline{y}^{(j)} - \overline{y} \right)^2$$

and:

$$W = \frac{1}{m} \sum_{j=1}^{m} s_j^2$$

with:

$$\overline{y}^{(j)} = \frac{1}{n} \sum_{i=1}^{n} y_i^{(j)} \qquad \overline{y} = \frac{1}{m} \sum_{j=1}^{m} \overline{y}^{(j)}$$

and:

$$s_j^2 = \frac{1}{n-1} \sum_{i=1}^{n} \left(y_i^{(j)} - \overline{y}^{(j)} \right)^2$$

Consider the quantity:

$$R = \frac{B/n + [(n-1)/n]W}{W} \tag{11.12}$$

The idea is that for small n the numerator in (11.12) overestimates variability, since starting values are typically quite dispersed, whereas W underestimates variability because the different chains remain close to their initial values. Thus, we expect that R becomes close to 1 when each of the m chains approaches the target distribution.

Consider the following simple example, the simulation of the bivariate normal distribution $\mathcal{N}_2(\mu, \Sigma)$, where $\mu = (1, 2)'$ is the mean vector and Σ is the covariance matrix given by:

$$\Sigma = \begin{pmatrix} 1.0 & 0.4 \\ 0.4 & 1.0 \end{pmatrix}$$

As the candidate-generating density we choose the random walk generating density

$$y_i^{(u)} = x_i^{(u)} + z_i^{(u)} \qquad u = 1, 2$$

where $x_i^{(1)}$ and $x_i^{(2)}$ are the two components of the chain at the i-th step, and $z_i^{(u)}$ is the i-th realization of the bivariate uniform on the interval $[-\delta(u), +\delta(u)]$ (refer to `Code_8_4.R`). In this example, in order to have a relative slow convergence, we choose $\delta(1) = \delta(2) = 2$, and the starting points are $x_0^{(1)} = -10$ and $x_0^{(2)} = 10$. The length of the chain is $n = 4000$ moves. In the following we present the results only for the first component $x^{(1)}$, because those for the second one are pretty similar.

The results obtained by the MBB method, with 200 replications, are reported in Fig. 11.11. We see that a clear plateau is attainable only if the burn-in r is at least 500 moves. To compare the above results with those derived by the G-R diagnostics,

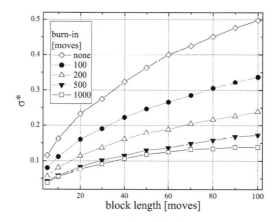

Fig. 11.11 Moving block bootstrap estimates $\hat{\sigma}^*$ of the standard errors as a function of the block length of the first component $x^{(1)}$ of the bivariate normal distribution defined in the text. The burn-in times for different runs are also reported.

we consider two approaches. In the first approach, 200 chains are generated starting at different initial points $x_0^{(1)}$, $x_0^{(2)}$ and the G-R method is applied to these multiple chains, with the burn-in time equal to zero. The results, shown in Fig. 11.12 (dotted lines), give the same indications of the MBB method, i.e. R reaches the value 1 after about 500 iterations.

We propose a second way to exploit the G-R diagnostics, consisting of using the MBB replications as 'virtual' multiple chains. To do so, we memorize all the 200 MBB replications used to obtain the results reported in Fig. 11.11 and the G-R method is applied to these series, regarded as multiple chains starting at different initial points.

The behaviour of R and W for such MBB replications is reported in Fig. 11.12 (solid lines), showing that, although in the initial region the graphs of both R and W are less smooth with respect to those derived from actual ('true') multiple chains (dotted lines), also in this case the plots suggest that the chain converges after about 500 iterations.

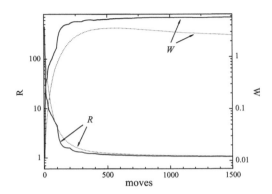

Fig. 11.12 Evolutions of R (scale on the left) and W (scale of the right) for the first component $x^{(1)}$ of the bivariate normal distribution defined in the text. Solid and dotted lines refer to MBB replications and actual multiple chains, respectively.

The conclusion (if any) may be summarized by the old idiom 'to kill two birds with one stone', namely, to assess accuracy and explore convergence with only the MBB method. The MBB is a very powerful non-parametric technique for handling statistical errors in MCMC estimates. So there are several reasons to rely on the MBB to estimate the simulation accuracy. When no plateau appears to be clearly distinguishable, it is not hard to search for a possible one by increasing the burn-in period. If we succeed in it, we are hopeful that the accuracy of estimates has been derived only after convergence has been reached. Eventually, having the MBB replications available, one can try to perform further checks, for example by means of the G-R method or with other diagnostic tools based on multiple chains, without needing to execute further runs.

11.6 Exercises

Exercise 11.1 Write a code to estimate, by means of the bootstrap, the standard error of the standard error both plug-in and correct.

Exercise 11.2 Demonstrate the expression in eqn (11.7).
Hint: use the operator notation for the variance of $\hat{\sigma}^2_{bm}$.

Exercise 11.3 Write a code to compare the distribution of means of a set of trajectories with that of a unique trajectory but obtained with bootstrap.

Exercise 11.4 Write a code to assign the standard error to the estimate by the batch means method and by the moving block bootstrap method. The target density of the MCMC algorithm is the standard normal.

Exercise 11.5 A the end of Section 11.5.1 we proposed to exploit the G-R diagnostics using the MBB replications as 'virtual' multiple chains. Consider the simulation of the bivariate normal distribution leading to Fig. 11.11 and implement a code in R using the MBB replications as described in the text.

12
Spatial Analysis

I think there's a great beauty to having problems. That's one of the ways we learn.

Herbie Hancock, Twitter

In the previous chapters the concept of the stochastic process was applied to time series. Many applications in the earth sciences and engineering, including geology, soil science, crop science, ecology, forestry and atmospheric sciences are dealing with models that compute the dependence of properties measured at different locations, therefore as a function of space. Geostatistics is the discipline that applies statistical concepts to spatial-dependent data. Geostatistics provides concepts and tools employed to derive information of spatial continuity common in natural phenomena and processes.

The application of random processes to problems of spatial analysis began in the early years of the twentieth century, when engineers and statisticians investigated methodologies for evaluating mineral resources in mining operations. Danie G. Krige, a South African statistician and mining engineer, was a pioneer in the new-born field of geostatistics (Krige, 1951). The *Kriging* technique was later formalized by the French mathematician and engineer Georges Matheron (Matheron, 2019), who also coined the term *Kriging*, in honour of Krige's contribution.

Many books are available on geostatistics, with both general concepts and applications to specific disciplines (Isaaks and Srivastava, 1989; Cressie, 1993; Kitanidis, 1997; Webster and Oliver, 2007; Leuangthong *et al.*, 2008; Chun and Griffith, 2013). The fundamental concepts of geostatistics are based on classical statistics including data distribution tools (mean, median, standard deviation) and bi-variate statistics (covariance, correlation coefficient, t-test and F-test), integrated into concepts of geometry such as distances, angles, lines, polygons and so forth.

Sources of spatial data are many, including satellite imagery, drone images, Geographical Positioning System (GPS), data acquired by direct sampling on the ground, satellite radar, digital representation of maps, contours, digital elevation models and many others. The integration of earth observation methods and computing techniques are studied by the field of geo-informatics, which deals with the science and technology of the structure of spatial information, classification, qualification, storage, modelling and others.

12.1 Geostatistical perspective

The fundamental notions of stochastic processes were described in Chapter 2. Random variables X_t's with probability space $(\Omega, \mathcal{F}, \mathsf{P})$ take given values in a measurable space, whose values are called *states*. All the values taken by a variable are called its *state space* and it was denoted as $\mathcal{S}$. In the examples presented, random variables changed as a function of time, leading to time series analysis. Clearly, a natural process such as precipitation or air temperature has also a spatial component, namely it changes with location.

The formal concept of a stochastic process applied to a spatial process is the one of *random fields*. A random field is defined as $Z(s)$, where s is a vector indicating a position in space, $s = (x, y, z)$. The vector is continuous in $\mathbb{R}$, $\mathbb{R}^2$ or $\mathbb{R}^3$. Indeed, the domain can be one-, two- or three-dimensional. If a transect for a given variable is measured the process is simply one-dimensional, if a variable changes on a surface (such as precipitation) the problem is two-dimensional. However, in geology, hydrogeology, seismology and soil sciences the domain is often three-dimensional, $\mathbb{R}^3$, since the variable of interest often displays a variation not only on the x, y coordinates (surface), but also with depth (z).

$Z(s)$ encompasses all the possible *states* and, analogously with the concepts applied to time series, all the values taken by a variable are called the *state space*. A probability density function of a random field is defined, encompassing all values in the space. Associated with the probability density function are statistical moments such as the expected value, variance and covariance functions.

Spatial random fields are random fields whose properties are determined by the position in space that are spatially correlated. A spatial random variable is then $\{Z_s; s \in \mathbb{T}\}$, where the set $\mathbb{T}$ is the *parametric space*. $\mathbb{T}$ is in $\mathbb{R}$ or its subsets. In analogy with the concepts applied to a *continuous*-time (-parameter) stochastic process, $\mathbb{T}$ can range from $\mathbb{T} = (-\infty, \infty)$, or $\mathbb{T} = [0, \infty)$, or $\mathbb{T} = [a, b)$, or $\mathbb{T} = [a, b]$. Mathematically, they are a set of a random variable Z_s tagged with a location s, and the location is part of a domain Ω:

$$\{Z(s) : s \in \Omega\} \tag{12.1}$$

where Z is the random variable over the locations s, where s is a spatial index.

In analogy with the definition of a random variable for time series, all the variables Z_s are defined in the same space Ω, then each Z_s is a random function of two arguments of different nature: the variable of probabilistic nature $\omega \in \Omega$ indicates the event, the variable of mathematical nature $s \in \mathbb{T}$ creates an order in the random variables family. For instance, the variable with probabilistic nature could be the concentration of a given chemical species in an agricultural field, and the variable with mathematical nature is its position in space.

When the concept of a stochastic process applied to time series was introduced in Chapter 2, time was defined as a physical process with an arrow, a direction, a before and an after, a past and a future. It is therefore expected that a realization x_t at time t of the random variable X_t is closer to observations x_{t-1} and x_{t+1}, rather than to those farther in time. In stochastic analysis of spatial processes the concept is conceptually analogous. It is expected that a realization z_s at location s of the random variable Z_s is closer to observations z_{s-h} and z_{s+h}, rather than to those farther away in space. In this case, the variable h is a distance that can span a few orders of magnitude depending on the observed process with incremental values of h, such as $k \times h$ and $(k = 1, ..., n)$.

The spatial stochastic process $\{Z_s; s \in \mathbb{D} \subset \mathbb{R}^d\}$, in a more complete manner, should be written as:

$$\{Z(\omega, s); \omega \in \Omega, s \in \mathbb{D}\}$$

to highlight the fact that the particular realization of the stochastic process at space s depends on the particular event $\omega \in \Omega$.

If the space is fixed, $s = \bar{s}$. Then $Z_{\bar{s}}(\omega) \equiv Z(\omega, \bar{s})$ is a random variable and, if the possible outcomes of the 'trial' are $\omega_1, \omega_2, \ldots$, the possible realizations of $Z_{\bar{s}}(\omega)$ are given by:

$$Z_{\bar{s}}(\omega_1) = z_1, Z_{\bar{s}}(\omega_2) = z_2, \ldots,$$

where the subscript i $(i = 1, 2, \ldots)$ of the z_i's numbers the different possible realizations of the same random variable $Z_s(\omega)$ at time $s = \bar{s}$. The same concepts can then be applied to $Z(s, \omega)$ as a function of s, by fixing $\omega = \bar{\omega}$ in Ω, so we have $Z_s(\bar{\omega})$.

Here, a particular outcome at spaces $s = s_1, s = s_2, \ldots$, that is:

$$Z_{s_1}(\bar{\omega}_1) = z_1, Z_{s_2}(\bar{\omega}_2) = z_2, \ldots, \tag{12.2}$$

in this case the subscripts i $(i = 1, 2, \ldots)$ of the z_i's, number the locations at location s_1 at which the event $\bar{\omega}_1$ has occurred, s_2 at which the event $\bar{\omega}_2$, has occurred, etc. For each fixed $\bar{\omega}$, the sequence $(z_1, z_2, \ldots)$ is also called spatial *realization* of the process, and the index is often referred to as spatial points with increments of s, s_1, s_2, s_3 and so forth, starting from an origin of a Cartesian system. Most commonly a georeferenced, x and y coordinate system is used where the point s_0 corresponds to the origin of a Cartesian system in geographic coordinates.

The state of a spatial random field can usually be decomposed into (Varouchakis, 2018):

$$Z(s) = Z'_\lambda(s) + m_Z(s) + e(s) \tag{12.3}$$

where $m_Z(s)$ is a deterministic trend, $Z'_\lambda(s)$ is a function expressing a spatial correlation and $e(s)$ is random noise. The trend is commonly used to develop models or regression of the mean value, while the correlation part is used to determine the covariance matrix for spatial correlation.

Overall, all the possible sample paths resulting from an experiment constitute an *ensemble*. Similarly to time series analysis where only a realization (time series) of a specific process is available, in geostatistical analysis often only one sample is collected at each point in space, therefore only one realization is observed (not multiple realizations). For practical reasons, observations are often limited and therefore only *spatial samples* are available, which are a *discrete measurement* of the variable. The observation set consists of sample locations at which the random variable is observed.

However, the purpose of the geostatistical analysis is to estimate a *continuous surface* from limited spatial points, therefore the stochastic process is continuous. The process is observed over a finite number of locations $\{s_1, s_2, ..., s_N\}$ and random variables are defined associated with those locations $\{Z(s_1), Z(s_2), ..., Z(s_N)\}$.

So what is the sampling process in a geostatistical analysis? How is it possible to treat a pattern (map) as if it is made by multiple realizations of a stochastic process? The idea is that each portion of a map is generated by different maps, therefore an infinite number of maps can be generated, corresponding to the spatial stochastic process. Clearly, to operate within this framework, the notion of *stationarity* is applied as introduced in Chapter 3.

12.1.1   Stationarity in spatial processes

In Chapter 3 the concept of stationarity was introduced. It was pointed out that it is difficult to make inferences about the process when, for instance, only a single realization of the process is available. If one time series is collected, it is indeed difficult to derive information about the stochastic process from a single realization. To obtain meaningful information, *stationarity* (restriction of heterogeneity) and *ergodicity* (independency) were described.

It is possible to conceptualize the stochastic process as a number generator that generates maps for the whole surface. All the generated maps have identical first moments (stationarity) but because of the their stochastic nature, they are also different. Therefore a surface is generated, a spatial subsample is taken and the mean is computed. Then another map is generated, another subset is selected and the mean computed. Since the mean is assumed to be the same, each measurement in space is visualized as realizations of different maps.

For instance, when a network of weather stations is represented on a map, each value for a given variable (for example, air temperature) represents a realization of all the infinite realizations of that variable on the map. Under the conditions of stationarity, if there are two hundred monitoring stations, each value of air temperature collected at each single station, is a realization of the same stochastic process.

Since the process is assumed to be the same (stationarity), it does not matter if the measurements are observed all at the same time or at different moments in time, since they are still different realizations of the stochastic process. Obviously, this is true if only the spatial process is analysed. If a multivariate, space-time stochastic process is analysed, the dependence of the variable with time becomes important.

As described above for time series, the assumption of stationary is a strong notion of equilibrium and it is a clear restriction upon the heterogeneity of the spatial data. In accordance with the description provided for time series, the application of stationarity to spatial data is to imagine dividing the surface into spacial chunks, where all the pieces are 'statistically similar', in the sense that statistical properties do not vary over space.

When a stochastic process is weakly stationary or simply stationary, the first and second moments exist and do not vary throughout space. The expected value is finite and constant, at all time points:

$$\mu_s = \mathrm{E}\,[Z_s] = \mu < \infty, \ \forall s$$

A relevant application for the expected value in geostatistics is the determination of large scale trends and spatial dependence patterns. For instance, it is obviously not realistic to assume that the mean air temperature has the same value (stationarity) over a large geographical area, and in particular where differences in altitude are present. Different kind of linear or non-linear models of the mean can be used.

Variance and autocovariance are finite and constant, at all time points:

$$\sigma_Z^2 = \mathrm{E}\left[(Z_s - \mu_s)^2\right]$$

The spatial autocovariance is written as:

$$\gamma(s_1, s_2) = \mathrm{E}\left[(Z_{s_1} - \mu_{s_1})(Z_{s_2} - \mu_{s_2})\right]$$

It is worth noting that under conditions of *spatial isotropy*, the direction in space for the calculation of geostatistical properties does not change. Therefore performing an analysis in north-south or south-north direction does not change the results. However, in real conditions isotropy is not common and is more often anisotropic, with the variables and the distributions assuming different values in different directions. In case of isotropic conditions covariance and correlation length provide information about the process. The application of correlation to spatial processes is described in the section below.

Overall, the assumption of stationarity (or weak stationarity) is that the mean value and the variance are constant. Strict stationarity imposes that the probability density function is the same at all spatial points, regardless of the transformations where the points are interchanged between each other, but preserving their position in space.

So after all, how can a process that has the same mean and the same variance be of interest? The informative part is related to the spatial correlation of the process. The variable of interest is the *covariance*. A few restrictions are also imposed on the covariance. In particular, the covariance is not a function of absolute location but only of spatial separation. As described in the example below the important aspect is always the distance between pairs where the observation was collected. If it is the same variable at two locations, then it is an *autocovariance*, but if there are multiple variables at multiple locations it is a *crosscovariance*.

In many cases error terms are assumed to be white noise, with no systematic trend (mean is zero) and no differences in variability (variance is constant). By exploiting the structure of the covariance it is possible to get better predictions (in a sense of higher precision). The idea again is to obtain information about a deterministic structure that appears as noise, but since it is not white noise, it provides information about the process. The two most common methods for obtaining information about spatial dependence are the *correlogram* and the *semivariogram*.

12.2 Correlation coefficient and correlogram

One of the most common statistical coefficients used to establish the strength of a correlation between two variables is the *correlation coefficient*. In Chapter 3, the autocovariance and autocorrelation functions were introduced. It is calculated as:

$$\rho = \frac{\frac{1}{n}\sum_{i=1}^{n}(X_i - \mu_X)(Y_i - \mu_Y)}{\sigma_X \sigma_Y} \tag{12.4}$$

where μ is the mean and σ is the standard deviation. The numerator is the covariance.

This well-known coefficient ranges between -1 and 1. It is a coefficient used for the interpretation of *scatterplots*. Scatterplots are plots where, on Cartesian coordinates, values for typically two variables are displayed. The correlation coefficient quantifies the distance of the data from a straight line (linear correlation). If $\rho = +1$, then the scatterplot will be a straight line with a positive slope; if $\rho = -1$, then the scatterplot will be a straight line with a negative slope. For $\rho < 1$ then the scatterplot appears as a cloud of points. The scattering of the points around the line increases at decreasing values of ρ. The concept is discussed here since it is important to understand spatial analysis tools such as the correlogram.

To describe the concept of *correlogram*, an example is presented. Measurements of soil's silt content along two transects were performed. Figure 12.1 shows the measurements for the two transects at incremental values (easting) of $h = 5$ metres. The top plate depicts transect 1 and the bottom plate shows transect 2.

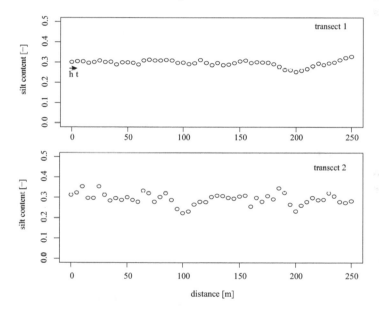

Fig. 12.1 Silt content values for two soil transects (Transect 1 top and Transect 2 bottom). The measurements was made at 5 metres increments (direction west to east). The arrow indicates the head (h) and tail (t), corresponding to measurements at z_s and z_{s+h}. In this case the one-dimensional position of the transect origin is at $s = 0$.

A first statistical analysis of the data showed that the following results: $\bar{Z}_1 = 0.29, \sigma^2_{Z_1} = 0.00022$ and $\sigma_{Z_1} = 0.015$ for transect 1 and $\bar{Z}_2 = 0.29, \sigma^2_{Z_2} = 0.0007$ and $\sigma_{Z_2} = 0.021$ for transect 2. The two datasets have the same mean and a slightly different standard deviation. Histogram plots of the two distributions (see Fig. 12.2) showed similar distributions, resembling a normal distribution. A first look at the summary statistics could mislead one to think that the two data sets have similar statistical properties.

A powerful approach to analysing spatial data is based on the idea of measuring correlation coefficients between pairs of data, selected at different spatial points s,

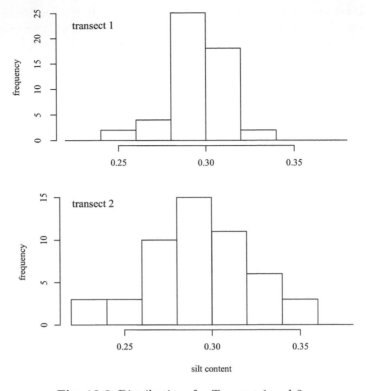

Fig. 12.2 Distributions for Transect 1 and 2.

separated by a distance h.

$$\rho = \frac{\frac{1}{n}\sum_{i=1}^{n}(Z_i - \mu_{Z_i})(Z_j - \mu_{Z_j})}{\sigma_{Z_i}\sigma_{Z_j}} \tag{12.5}$$

where Z_i and Z_j are measurements of the same variable at two positions (i and j), μ_{Z_i} and μ_{Z_j} are the means and σ_{Z_i}, σ_{Z_j} the standard deviations. The indexes i, j are locations separated by the distance h. Note that because of the lags, some numbers at the beginning of the series do not have a corresponding values as shown in the tables below. To compute a correct value of ρ, to have the same value of n for the two columns, the first pairs must be eliminated by the computation.

In Fig. 12.1 the measurement of a spatial random value Z at position s is indicated by Z_s for the *head* and the measurement at position Z_{s+h} is indicated with *tail*. A table can then be created with the values of Z_s at head and tail.

Distance	Tail	Head
x	t	h
0	0.301	NA
5	0.305	0.301
10	0.304	0.305
15	0.297	0.304
20	0.301	0.297
25	0.308	0.301
30	0.301	0.308
35	0.302	0.301
40	0.289	0.302
...	...	...

The indicated (NA) not available refers to the fact that no data are available before the point at $s = 0$. The concept is similar to the one of the 'lag' operator for time series as described in Chapter 6. The use of one spatial step (tail) ahead and the creation of this table is indeed called *Lag 1*. In this table, for each value of head there corresponds a value of tail that is the value at the previous space interval. The same example is shown for a *Lag 2* analysis:

Distance	Tail	Head
x	t	h
0	0.301	NA
5	0.305	NA
10	0.304	0.301
15	0.297	0.305
20	0.301	0.304
25	0.308	0.297
30	0.301	0.301
35	0.302	0.308
40	0.289	0.301
...	...	...

In this *Lag 2* table, for each value of head there is a value of tail corresponding to the value of two previous space intervals ($2h$). The procedure can be repeated for many incremental values of space intervals. In this case, since the Z_h value is related to the previous one, the lag operator is named 'backwards shift' or 'backshift' operator.

For each of these pairs a scatterplot is plotted for incremental values of lags (from *Lag1* to *Lag5*), corresponding to incremental distances of 5, 10, 15, 20 and 25 metres. Correlation coefficients were computed for each pairs at incremental distances.

The scatterplot for Transect 1 (Fig. 12.3) depicts a good correlation coefficient (0.832) for pairs 5 metres apart with decreasing R values at increasing distances. The points indicated in the plots are 5 metres apart in the upper left, 10 metres apart in the upper right, 20 metres apart in the lower left and 25 metres apart in the lower right panel. As the distances of pairs of sample increases, the correlation between the samples decreases (see Fig. 12.4).

On the other hand the correlation coefficients for the Transect 2 were very low, indicating no correlation between the data pairs regardless of the separation distance.

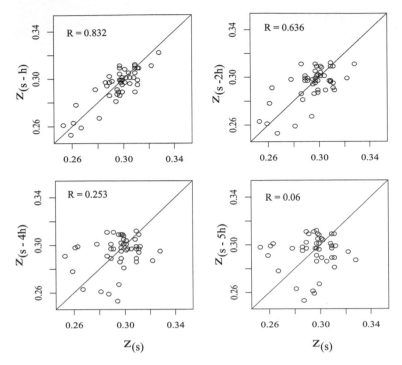

Fig. 12.3 Scatterplots for Transects 1 and 2. The value $z_{(s)}$ indicates the value of the variable (silt content) at different lags.

The correlation coefficients can be plotted against the incremental value of distances (Lag 1, Lag 2, Lag 3 and so forth) to create a *correlogram*. The correlogram provides information about the *spatial dependence*. It is function of distance between points and it estimates the correlation between points in space. Figure 12.5 depicts the two correlograms for the corresponding Transects 1 and 2.

Overall, the two correlograms are very different, indicating that the two samples, although they have the same mean, are very different. Since the lags were at increments of 5 metres, Transect 1 depicts a correlation of 0.83 at one lag (5 metres apart) and 0.6 at two lags (10 metres apart), while Transect 2 displayed a poor correlation 0.48 at Lag 1. In this example, a one-dimensional transect is presented, but in many cases the distribution of points in space are two- or three-dimensional; therefore the correlation can also be a function of direction (angle). The R code below shows the computations described above.

```
## Code_12_1.R
# Scatter plots and correlogram
library(sp)
library(gstat)
library(quantmod)

# ===========================================
setwd("C:/RPA/code/spatial_analysis")
# ===========================================
```

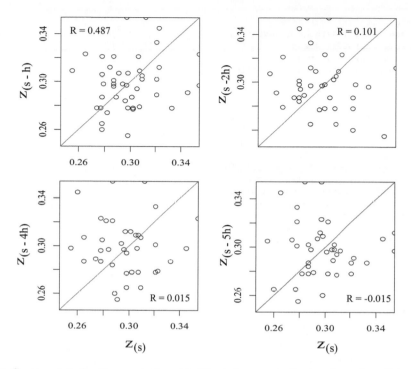

Fig. 12.4 Scatterplots for Transects 1 and 2. The value $z_{(s)}$ indicates the value of the variable (silt content) at different lags.

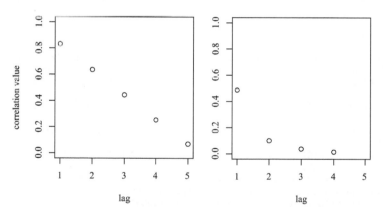

Fig. 12.5 Correlograms for Transects 1 (left) and 2 (right).

```
# load data
geodata1 = read.csv("data/transect_1.csv")
geodata2 = read.csv("data/transect_2.csv")
plot(geodata1$s,geodata1$z,xlab=("Distance [m]"),ylab=("Silt content [-]"),ylim=c(0,0.5))
plot(geodata2$s,geodata2$z,xlab=("Distance [m]"),ylab=("Silt content [-]"),ylim=c(0,0.5))
#edit(geodata)
#Compute basic statistics
mu1<- mean(geodata1$z) #mean
```

```
V1<- var(geodata1$z) #variance
S1<- sd(geodata1$z) #standard deviation
mu2<- mean(geodata2$z) #mean
V2<- var(geodata2$z) #variance
S2<- sd(geodata2$z) #standard deviation
mu1
V1
S1
##Compute lagged vectors and save as numerics into dataframe
geodata1$lag1<-Lag(geodata1$z, k = 1)
geodata1$lag1<-as.numeric(geodata1$lag1)
geodata1$lag2<-Lag(geodata1$z, k = 2)
geodata1$lag2<-as.numeric(geodata1$lag2)
geodata1$lag3<-Lag(geodata1$z, k = 3)
geodata1$lag3<-as.numeric(geodata1$lag3)
geodata1$lag4<-Lag(geodata1$z, k = 4)
geodata1$lag4<-as.numeric(geodata1$lag4)
geodata1$lag5<-Lag(geodata1$z, k = 5)
geodata1$lag5<-as.numeric(geodata1$lag5)
edit(geodata1)

##Compute lagged vectors and save as numerics into dataframe
geodata2$lag1<-Lag(geodata2$z, k = 1)
geodata2$lag1<-as.numeric(geodata2$lag1)
geodata2$lag2<-Lag(geodata2$z, k = 2)
geodata2$lag2<-as.numeric(geodata2$lag2)
geodata2$lag3<-Lag(geodata2$z, k = 3)
geodata2$lag3<-as.numeric(geodata2$lag3)
geodata2$lag4<-Lag(geodata2$z, k = 4)
geodata2$lag4<-as.numeric(geodata2$lag4)
geodata2$lag5<-Lag(geodata2$z, k = 5)
geodata2$lag5<-as.numeric(geodata2$lag5)
edit(geodata2)

##Plot scatter plots
str(geodata1)
plot(geodata1$z,geodata1$lag1,xlim=c(0.2,0.4),ylim=c(0.2,0.4),xlab=("x+h"),ylab=("x"))
abline(a=0,b=1,col="blue")
plot(geodata1$z,geodata1$lag2,xlim=c(0.2,0.4),ylim=c(0.2,0.4),xlab=("x+2h"),ylab=("x"))
abline(a=0,b=1,col="blue")
plot(geodata1$z,geodata1$lag3,xlim=c(0.2,0.4),ylim=c(0.2,0.4),xlab=("x+3h"),ylab=("x"))
abline(a=0,b=1,col="blue")
plot(geodata1$z,geodata1$lag4,xlim=c(0.2,0.4),ylim=c(0.2,0.4),xlab=("x+4h"),ylab=("x"))
abline(a=0,b=1,col="blue")
plot(geodata1$z,geodata1$lag5,xlim=c(0.2,0.4),ylim=c(0.2,0.4),xlab=("x+5h"),ylab=("x"))
abline(a=0,b=1,col="blue")
##=====================Correlations=========================
corlag1<-cor(geodata1$z,geodata1$lag1,use="complete.obs")
corlag2<-cor(geodata1$z,geodata1$lag2,use="complete.obs")
corlag3<-cor(geodata1$z,geodata1$lag3,use="complete.obs")
corlag4<-cor(geodata1$z,geodata1$lag4,use="complete.obs")
corlag5<-cor(geodata1$z,geodata1$lag5,use="complete.obs")
correlations1<-c(corlag1,corlag2,corlag3,corlag4,corlag5)
correlations1

##=========================================================
##Plot scatter plots
plot(geodata2$z,geodata2$lag1,xlim=c(0.2,0.4),ylim=c(0.2,0.4),xlab=("x+h"),ylab=("x"))
abline(a=0,b=1,col="blue")
plot(geodata2$z,geodata2$lag2,xlim=c(0.2,0.4),ylim=c(0.2,0.4),xlab=("x+2h"),ylab=("x"))
```

```
abline(a=0,b=1,col="blue")
plot(geodata2$z,geodata2$lag3,xlim=c(0.2,0.4),ylim=c(0.2,0.4),xlab=("x+3h"),ylab=("x"))
abline(a=0,b=1,col="blue")
plot(geodata2$z,geodata2$lag4,xlim=c(0.2,0.4),ylim=c(0.2,0.4),xlab=("x+4h"),ylab=("x"))
abline(a=0,b=1,col="blue")
plot(geodata2$z,geodata2$lag5,xlim=c(0.2,0.4),ylim=c(0.2,0.4),xlab=("x+5h"),ylab=("x"))
abline(a=0,b=1,col="blue")
##======================Correlations========================
corlag1<-cor(geodata2$z,geodata2$lag1,use="complete.obs")
corlag2<-cor(geodata2$z,geodata2$lag2,use="complete.obs")
corlag3<-cor(geodata2$z,geodata2$lag3,use="complete.obs")
corlag4<-cor(geodata2$z,geodata2$lag4,use="complete.obs")
corlag5<-cor(geodata2$z,geodata2$lag5,use="complete.obs")
correlations2<-c(corlag1,corlag2,corlag3,corlag4,corlag5)
correlations2
```

12.3 Semivariogram

The second most common method for quantifying spatial dependence is the *semivariogram* function. The variogram is the magnitude of the variance of the difference as a function of displacement:

$$2\gamma(s, h) = \text{Var}\left[Z_{s+h} - Z_s\right] \tag{12.6}$$

To compute the variance, the expected values must be computed. If the process Z_s is (second-order) stationary, the expected values are the same and their difference is zero:

$$\text{E}\left[Z_{s+h} - Z_s\right] = \text{E}\left[Z_{s+h}\right] - \text{E}\left[Z_s\right] = 0 \tag{12.7}$$

The variance of the difference between these two random variables is the expected value of their square minus the square of their expected value. The expected value is zero, therefore the only thing that remains is the first term:

$$\text{Var}\left[Z_{s+h} - Z_s\right] = \text{E}\left[(Z_{s+h} - Z_s)^2\right] - 0 \tag{12.8}$$

therefore the *semivariogram* is $\frac{1}{2}$ the expected value of the squared difference.

$$\gamma(s, h) = \frac{1}{2}\text{E}\left[(Z_{s+h} - Z_s)^2\right] \tag{12.9}$$

The value $\frac{1}{2}$ is obtained by the computation of the Euclidean distance between pairs of point.

The expected values can then be estimated. The semivariogram is therefore an average of squared differences for a given distance interval. To obtain a variogram, every observation relative to any other observation is computed. The process is to take the difference between the values of the same variable in each direction for each point and for each distance as shown in Fig. 12.6. It is the same concept described above for the transect, but the spatial relationship between pairs of numbers can be in two or three dimension and also depending on the angle (directional variograms).

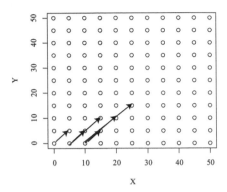

Fig. 12.6 Schematization for semivariogram computation.

The visualization of this calculation is called the variogram cloud plot, where the squared differences are plotted against the incremental displacement. We will see further ahead how to obtain a variogram cloud in R.

Because of the assumed stationarity, the semivariogram is a function of the displacement h only:

$$\mathrm{E}\left[(Z_{s+h} - Z_s)^2\right] = 2\mathrm{Var}\left[Z_s\right] - 2\mathrm{Cov}\left[Z_{s+h}, Z_s\right] \qquad (12.10)$$

Denoting by $C(h)$ the covariance function $\mathrm{Cov}\left[Z_{s+h}, Z_s\right]$, we obtain a notable relation between the semivariogram and covariance functions:

$$\gamma(h) = C(0) - C(h) \qquad (12.11)$$

Note that the variogram definition (12.6) does not require the process to be stationary. For example, for a non-stationary process whose first differences are stationary[1]

$$\mathrm{E}\left[Z_{s+h} - Z_s\right] = 0 \qquad (12.12)$$

the semivariogram is always defined while the correlation/covariance function is not necessarily:

$$\mathrm{Var}\left[Z_{s+h} - Z_s\right] = 2\gamma(s, h) \qquad (12.13)$$

The semivariogram can be empirically estimated by the 'classical' formula developed by its creator Georges Matheron (Cressie, 1993):

$$\hat{\gamma}(h) \equiv \frac{1}{2|N(h)|} \sum_{i,j \in N(h)} (Z(s_i) - Z(s_j))^2 \qquad (12.14)$$

for all couples s_i, s_j, where $|N(h)|$ is the number of distinct elements such that $s_i - s_j = h$.

---

[1]Such a process is defined as *intrinsic stationary*.

12.3.1   Variogram model

The main usefulness of variograms is connected to data interpolation (kriging) for obtaining continuous spatial maps. For that purpose, the empirical variogram (12.14) can be approximated by an analytical function interpolating it.[2]

Figure 12.7 shows the filled contour-plot of a Gaussian function with added white noise ϵ: $Z(X,Y) = 10 \cdot \exp(-(X-5)^2 - (Y-3)^2) + \epsilon$.

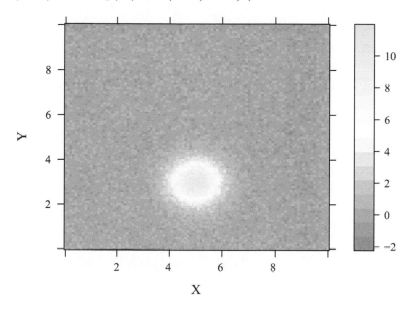

Fig. 12.7 Gaussian surface with noise.

For such a smooth function points close to one another are similar than distant points. That is confirmed by the empirical semivariogram (circle symbols) shown in Fig. 12.8. The continuous line is a Gaussian model $\gamma_M(h)$ fitting the experimental variogram:

$$\gamma_M(h) = A \left(1 - e^{-h^2/r^2} \right) + B \tag{12.15}$$

In the above case three parameters are sufficient to obtain a proper fitting: that's a usual situation with variograms of stationary spatial processes. For historical reasons related to mining, which incidentally gave the main acceleration to the development of geostatistics, the three parameters A, B and r are denoted 'sill', 'nugget' and 'range', respectively (Isaaks and Srivastava, 1989). In essence, the *sill* is the plateau reached by the variogram γ at large distances, the *range* is the distance at which γ reaches this plateau, while the *nugget* is a non-zero value at $h = 0$ due to sampling errors, short

---

[2]That also has the side-effect of avoiding to obtain an inconsistent semivariogram function, not respecting mathematical requirements such as the negative-definiteness of the variogram function itself (Cressie, 1993).

scale variability or other effects causing the sample variogram to be discontinuous in the origin (where it should be zero).

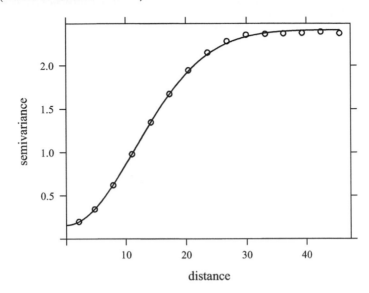

Fig. 12.8 Semivariogram of data in Fig. 12.7.

The following code shows how the above example is implemented in R.

```
## Code_12_2.R
# Semivariogram
# Load libraries

library(gstat)
library(sf)
library(reshape)
library(lattice)

# First part: generation of artificial data
# Generate Gaussian surface Z(X,Y) with added noise on a grid
X <- seq(0,10,0.1)
Y <- seq(0,10,0.1)
N <- length(X)
Noise <- matrix(rnorm(N*N,mean=0,sd=0.4),ncol=N)
exp(-(X-5)^2) -> eX
exp(-(Y-3)^2) -> eY
# create data matrix Z
Z <- 10*eX %*% t(eY)
Z <- Z+Noise
grid <- expand.grid(X=X, Y=Y)
levelplot(Z~X*Y,grid)

# Second part: semivariogram computation
# Convert Z to dataframe
Z.df <- melt(Z)
names(Z.df) <- c("X","Y","Z")
Z.df$X <- Z.df$X-1
Z.df$Y <- Z.df$Y-1
```

```
# Convert dataframe to sf class
Z.sf <- st_as_sf(Z.df,coords=c("X","Y"))
# compute empirical semivariogram
vgm <- variogram(Z~1, Z.sf)
plot(vgm)
# model fitting
vgm.model <- fit.variogram(vgm, vgm(sd(Z)^2, "Gau", sqrt(max(X)^2+max(Y)^2), 0.1))
plot(vgm,vgm.model)
```

The code above makes use of four R packages:

reshape to restructure and aggregate data by the `melt` function;

lattice for graphics, here to make a `levelplot` of the Gaussian data;

gstat for spatial and spatiotemporal geostatistical modelling, here used for computing the semivariogram;

sf providing classes for analysing spatial data.

The last two, in particular, will be extensively used in the analyses of the following sections. The **sf** package (Pebesma, 2018), for example, contains classes and functions operating on them (*methods*, in the object-oriented language) for manipulating spatial objects.

Looking at the R code above, the instruction `st_as_sf(Z.df,coords=c("X","Y"))` assigns the coordinates X and Y of the dataframe `Z.df` to the `sf` object `Z.sf` which 'rearranges' the information included in the former dataframe. Practically, this operation transforms the ordinary R dataframe object into a `sf` dataframe, a format more easily managed by the functions of the package `gstat`, like `variogram`. The **sf** package (see the related R help and pdf document) gives *support for simple features, a standardized way to encode spatial vector data. Binds to 'GDAL' for reading and writing data, to 'GEOS' for geometrical operations, and to 'PROJ' for projection conversions and datum transformations*, as the R manual literally explains.

By inspecting the object `vgm.model` we obtain nugget, sill and range of the Gaussian model:

```
vgm.model
```

```
  model     psill    range
1   Nug 0.1542854   0.0000
2   Gau 2.2321708  16.3435
```

i.e. nugget $= 0.1542854$, sill $= 2.2321708$ and range $= 16.3435$. Indeed, the plot of Fig. 12.8 could be obtained 'by hand' by means of the following lines of code.

```
plot(vgm$dist,vgm$gamma,ylab="Semivariogram",xlab="Distance")
dmax <- max(vgm$dist)
d <- seq(0,dmax,0.1)
Nugget <- vgm.model$psill[1]
Sill <- vgm.model$psill[2]
Range <- vgm.model$range[2]
Gam <- Nugget+Sill*(1-exp(-d^2/Range^2))
lines(d,Gam,col="blue")
```

A note is in order concerning the choice of the analytical function used for fitting the experimental variogram, which cannot be arbitrarily chosen. Among the admissible models, the most frequently used, having *sill s*, *nugget n* and *range a*, are the following:

Exponential

$$\gamma(h) = n + s\left[1 - \exp(-h/a)\right]$$

Gaussian

$$\gamma(h) = n + s\left[1 - \exp(-h^2/a^2)\right]$$

Spherical

$$\gamma(h) = n + s\left[1.5h/a - 0.5(h/a)^3\right], \text{if } |h| \le a, \gamma(h)= s \text{ otherwise}$$

A general-purpose sill-free model is the

Power function

$$\gamma(h) = n + \beta h^\theta$$

which includes, as a particular case, the frequently used linear model:

$$\gamma(h) = n + \beta h$$

R allows us to generate other variogram models by the function vgm,[3] in particular the Matérn model (Minasny and McBratney, 2005), very often used in soil analysis. The Matérn semivariance is expressed in terms of the modified Bessel function of the second kind. Its explicit expression can be found in the given reference.

12.3.2   Spatial prediction

Experimental spatial data are usually sparse (not equispaced) and limited in number. For example, Fig. 12.9 shows the spatial distribution of environmental monitoring stations in the Emilia-Romagna region in Northern Italy.

   Based on a discrete set of data, for example the measured maximum temperatures in a given day/week/month/year, the purpose is that of estimating the maximum temperature *everywhere* in the geographic domain of interest. Once the predictor variable of interest $Z(s)$ (e.g. the maximum temperature) has been individuated, we can explore its spatial correlation structure by computing the semivariogram cloud, i.e. plotting all possible squared differences $(Z(s_i) - Z(s_j))^2$ (actually, one half of them) as a function of the distance d_{ij} among every couple of stations. The resulting plot, like that in Fig. 12.10, is a graphical representation of dissimilarity between stations, in the specific case suggesting a linear increase of semivariance with the inter-point distance.

   The example above concerns spatial measurements of maximum temperatures in Emilia-Romagna (Antolini *et al.*, 2015) on November, 21, 2020.

   Indeed, computing the empirical (or sample) semivariogram by equation (12.14), we obtain the plot of Figure 12.11 evidencing what was suggested from the cloud.

   The sample semivariogram can depend on the direction, i.e. it can be not simply a function of the distance between spatial locations. In such a case, the parameters of the semivariogram model, in particular sill and range, depend on the direction. It is a frequently encountered situation in geostatistics, where quantities like temperatures,

---

[3]Note the difference between variogram and vgm, the former computing the sample semivariogram, the latter the semivariogram model.

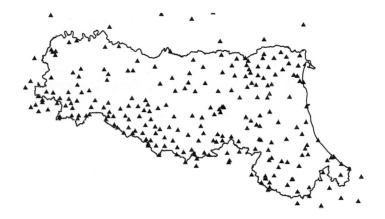

Fig. 12.9 Environmental monitoring stations in Emilia-Romagna, Italy.

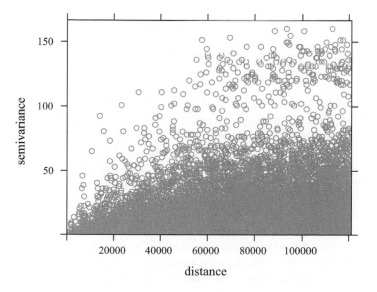

Fig. 12.10 Semivariogram cloud

precipitation etc. changes differently in different directions, i.e. with latitude and longitude. We can investigate the dependence on direction (anisotropy) by computing the directional variogram in R by specifying the parameter *alpha* in the `variogram` function. From the *help* of `variogram`, *alpha* specifies the *direction in plane (x,y), in positive degrees clockwise from positive y (north): alpha=0 for direction north (increasing y), alpha=90 for direction east (increasing x); optional a vector of directions in (x,y).*

Figure 12.12 shows the anisotropic semivariogram in directions increasing from 0, corresponding to north, to 135, corresponding to south-east. The influence of the cardinal direction on the daily maximum temperature is evident.

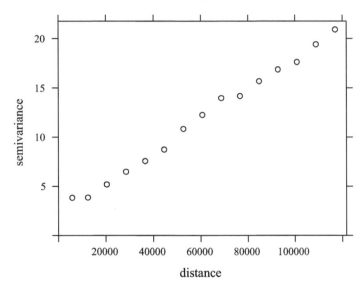

Fig. 12.11 Semivariogram of maximum temperature values in Emilia-Romagna on November, 21, 2020.

We will consider the variogram in the north direction for the analysis that follows. It is rather evident that the stochastic process 'maximum temperature in a given day on a given spatial extent', of which the data sample under consideration is a realization, is not stationary at all. Remembering the definition (12.3), it is reasonable that such a behaviour is due to a trend 'hidden' in the process. Altitude is highly probable to be at least in part responsible for it, together with other variables related to station locations. Figure 12.13 shows that the maximum temperature is linearly correlated to altitude ($R^2 = 0.9482$).

With reference to (12.3) we are interested in investigating the correlation structure of the residuals, i.e. of the de-trended stochastic process. Compare the semivariogram cloud of the residuals, Fig. 12.14, with that of Fig. 12.10: the linear dependence on distance has disappeared. Note also that the residuals are approximately normally distributed (Fig. 12.15).

Figures from 12.9 to 12.16 have been generated by the following code.

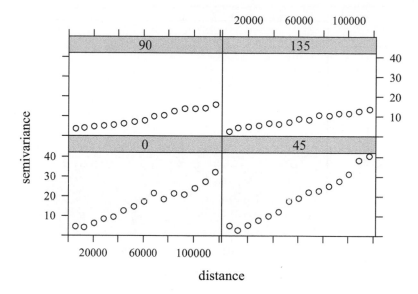

Fig. 12.12 Semivariogram of maximum temperatures in Emilia-Romagna on November 21 2020, in directions north (0 degrees), north-east (45 degrees), east (90 degrees) and south-east (135 degrees).

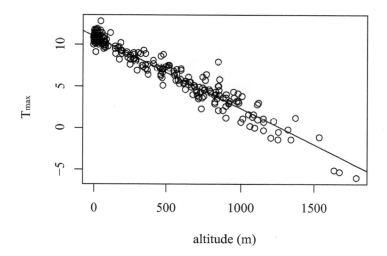

Fig. 12.13 Maximum temperature versus altitude in Emilia-Romagna on November, 21, 2020.

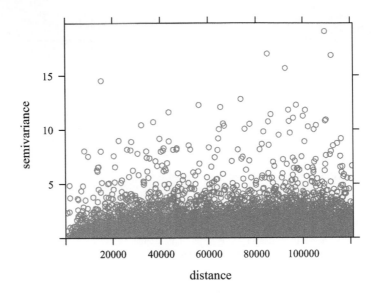

Fig. 12.14 Semivariogram cloud of residuals.

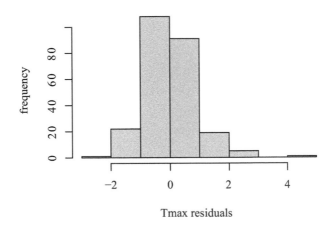

Fig. 12.15 Statistical distribution of residuals.

```
## Code_12_3.R
# Semivariogram with anisotropy
# Load libraries
library(gstat)
library(sf)

# Read data
setwd("C:/RPA/code/spatial_analysis")
```

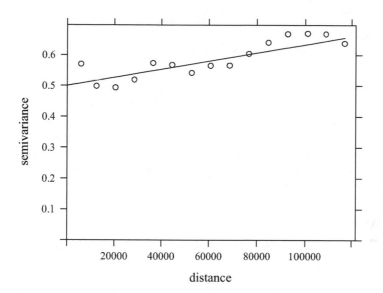

Fig. 12.16 Semivariogram of residuals, with superimposed model (solid line).

```
geo_data = read.csv("data/20201121_weather.csv")
xy <- geo_data[,c(1,2)]

# Read Emilia-Romagna shapefile and plots station locations
ER.shape <- st_read("RER.shp", quiet=TRUE)
plot(ER.shape$geometry)
points(xy,pch=17,col="red",cex=0.5)

# Semivariogram cloud
Tmax.vgm.cloud <- variogram(Tmax ~ 1, locations = ~UTM.X + UTM.Y, data = geo_data,
  cloud=TRUE)
plot(Tmax.vgm.cloud)

# Convert dataframe geo_data into object of the sf class
geo_data.sf <- st_as_sf(geo_data,coords=c("UTM.X","UTM.Y"),crs="EPSG:32632")
vgm <- variogram(Tmax~1,geo_data.sf)
plot(vgm)

# Anisotropy
vgm.aniso <- variogram(Tmax~1,geo_data.sp, alpha = c(0, 45, 90, 135))
plot(vgm.aniso)

# Linear dependence
Tmax.fit <- lm(Tmax~Alt,data=geo_data)
plot(Tmax~Alt,geo_data,xlab="Altitude (m)",ylab=expression(paste(T[max])))
abline(Tmax.fit,col="red")

# Work on residuals of linear model
# Semivariogram cloud
Tmax.res <- Tmax.fit$residuals
Tmax.vgm.cloud <- variogram(Tmax.res~1,locations = ~UTM.X + UTM.Y, data = geo_data,
cloud=TRUE)

plot(Tmax.vgm.cloud)
# Empirical semivariogram of residuals
```

```
Tmax.vgm <- variogram(Tmax.res~1,locations = ~UTM.X + UTM.Y, data = geo_data)
plot(Tmax.vgm)
Tmax.vgm.model <- vgm(0.2, "Lin", 150000, 0.5)
plot(Tmax.vgm,Tmax.vgm.model)
```

Semivariograms are computed in R by the function `variogram` of the `gstat` library. The second-last line of code uses `vgm` to define 'by eye' an analytical model fitting the experimental variogram, in this case a linear one. It must be clear that the choice of the model is constrained to the requirement of a definite-negative semivariogram function. A thorough discussion of this topic is out of the scope of the book. Readers interested to deepen their knowledge of that subject are encouraged to consult the cited books on geostatistics.

12.3.3   Kriging

The linear relationship between maximum temperature and altitude has been established on the covariates (Tmax, Alt) relative to the sparse irregular grid of environmental stations. But, luckily, altitude is known on very dense regular-grid maps known as DEM (digital elevation maps). Now, we are faced with the problem of how to interpolate temperatures at points between those on the irregular grid, taking into consideration the altitude.

Let us take a step backwards. Suppose our stochastic process $Z(s)$ is given by:

$$Z(s) = \mu + \epsilon(s) \tag{12.16}$$

where $\mu \in \mathbb{R}$ is a constant, possibly unknown, $s \in \mathbb{D} \subset \mathbb{R}^d$ and $\epsilon(s)$ includes the last two terms of (12.3), i.e. it is a second-order stationary process with zero mean and covariance function:

$$C(h) = \mathrm{E}\left[\epsilon(s)\epsilon(s+h)\right]$$

related to the semivariogram $\gamma(h)$ by (12.11).

One of the most diffused procedures for obtaining an interpolation over a grid is the 'kriging' (Cressie, 1993), which performs a spatial analysis of the data utilizing the variogram. We will just 'scratch' the surface of the theoretical basis of the kriging procedure, just to help our understanding of the importance of the covariance/correlation structure for the analysis of spatial data. The situation is pretty similar to what we encountered in the analysis of time series and ARMA processes, where almost all the relevant information about the process was contained in the correlation structure.

Reducing to the simplest case – a set of sampled points $(s_1, s_2, ..., s_n)$, one single point s_0 where producing the interpolation – the objective of kriging is to estimate $Z(s_0)$ given the values $z_1 = z(s_1), ..., z_n = z(s_n)$ via a linear predictor, so as to have that minimum variance. We write:

$$\hat{Z}(s_0) = \sum_{i=1}^{n} \lambda_i z_i \tag{12.17}$$

The problem is reduced to finding n weights λ_i whose sum is taken equal to one, to give an unbiased estimate: $\sum_i \lambda_i = 1$. If the process is stationary, the expected error is zero: $\mathrm{E}\left[\hat{Z}(s_0) - z(s_0)\right] = 0$, while the variance of the estimate is:

$$\mathrm{Var}\left[\hat{Z}(s_0) - z(s_0)\right] = \mathrm{E}\left[\left(\hat{Z}(s_0) - z(s_0)\right)^2\right] = 2\sum_{i=1}^{n}\lambda_i\gamma(s_i, s_0) - \sum_{i=1}^{n}\sum_{j=1}^{n}\lambda_i\lambda_j\gamma(s_i, s_j)$$

The procedure involving a Lagrange multiplier ψ is straightforward. We define a function $f(\lambda_i, \psi) = \mathrm{Var}\left[\hat{Z}(s_0) - z(s_0)\right] - 2\psi\left(\sum_{i=1}^{n}\lambda_i - 1\right)$ and find the values of λ_i and ψ minimizing f:

$$\begin{cases} \frac{\partial f(\lambda_i),\psi}{\lambda_i} = 0, i = 1,..,n \\ \frac{\partial f(\lambda_i),\psi}{\psi} = 0 \\ \sum \lambda_i = 1 \end{cases}$$

or:

$$\sum_{i=1}^{n}\lambda_i\gamma(s_i, s_j) + \psi = \gamma(s_j, s_0), \forall j$$

subject to $\sum_i \lambda_i = 1$.

Therefore, we see that the so-called 'ordinary' kriging, applicable to (second-order) stationary processes with unknown mean value[4] consists of a system of equations giving the weights λ_i in terms of the semivariances. Of course, that can be put in matrix form:

$$\mathbf{A}\underline{\lambda} = \underline{b} \tag{12.18}$$

where:

$$\mathbf{A} = \begin{bmatrix} \gamma(s_1, s_1) & \gamma(s_1, s_2) & \cdots & \gamma(s_1, s_n) & 1 \\ \gamma(s_2, s_1) & \gamma(s_2, s_2) & \cdots & \gamma(s_2, s_n) & 1 \\ \vdots & \vdots & \ddots & \vdots \\ \gamma(s_n, s_1) & \gamma(s_n, s_2) & \cdots & \gamma(s_n, s_n) & 1 \\ 1 & 1 & \cdots 1 & 0 \end{bmatrix}$$

$$\underline{\lambda} = \begin{bmatrix} \lambda_1 \\ \lambda_2 \\ \vdots \\ \lambda_n \\ \psi \end{bmatrix}$$

[4]The trivial case where the mean is known is called 'simple' kriging.

$$
\underline{b} = \begin{bmatrix} \gamma(s_1, s_0) \\ \gamma(s_2, s_0) \\ \vdots \\ \gamma(s_n, s_0) \\ 1 \end{bmatrix}
$$

The solution is formally: $\underline{\lambda} = \mathbf{A}^{-1}\underline{b}$. The advantage of the kriging procedure, with respect to other interpolation techniques, is that it automatically provides an estimate of the interpolation error, the kriging variance $\hat{\sigma}^2(s_0) = \underline{b}^t \underline{\lambda}$.

When we interpolate over a grid of points (a block) $B \subset \mathbb{D}$, i.e. in a subset of the domain $\mathbb{D}$ of the process instead of a single point, the procedure is formally the same although in that case the average semivariogram between sample points and B substitutes the values $\gamma(s_i, s_0)$ and an additional term $\bar{\gamma}(B, B)$ involving the within-block average variance comes into the play (Webster and Oliver, 2007):

$$
\bar{\gamma}(B, B) = \frac{1}{|B|^2} \int_B \int_B \gamma(s, s') ds ds'
$$

We are now able to come back to the example concerning the maximum temperature in Emilia-Romagna at a given date. Besides simple and ordinary kriging, a more general situation is when the variable of interest (Tmax in the present case) depends on one or more external variables (e.g. altitude). In those cases, the extension of ordinary kriging is called 'universal kriging', or sometimes 'kriging with trend' or, when the main variable Z is linearly related to an external variable Y, 'regression kriging'. In essence, regression kriging consists of regressing Z over Y, determining the correlation structure of the regression residuals, and performing the kriging procedure on such residuals. We will see this in a while.

The data file **20201121_weather.csv** contains maximum temperature, minimum temperature and precipitation for the Emilia-Romagna region in Northern Italy (Antolini *et al.*, 2015) on November, 21, 2020. We used the column containing the maximum temperatures for illustrating the computation of a semivariogram in Section 12.3.2. After loading it into a dataframe, the function **head** is useful to have an idea of its content, displaying the first lines of the file.

```
setwd("C:/RPA/code/spatial_analysis")
geo_data = read.csv("data/20201121_weather.csv")
head(geo_data)
```

The result:

```
    UTM.X   UTM.Y  Alt Tmax Tmin Prec
1 655965 4960223   23 10.5 -2.1 12.2
2 763298 4857077  629  4.6  1.5 28.0
3 800020 4874061    5 10.8  9.0 24.2
4 776616 4870711  680  4.0  0.9 17.4
5 643237 4944226   54 11.4 -1.3 17.0
6 825967 4860576    7 10.9  9.3 19.8
```

shows that the first two columns are the universal transverse mercator (UTM) coordinates of the environmental stations, the other four columns being the altitude (in metres), maximum and minimum temperature (in Celsius) and precipitation (in mm).

We are ready to apply regression kriging to the maximum temperature. We have computed the semivariogram of the residuals of `Tmax ~ Alt`, and quantified the parameters of a reasonable model (linear, see Fig. 12.16). We need a regular grid of points, covering the region of interest (all Emilia-Romagna) and including the altitude of every point of that grid: indeed, kriging mathematically is a function having the sample space as a domain, and the grid $B \subset \mathbb{D}$ as a codomain. The grid B can be obtained by `tif` files containing the DEMs of geographical regions. In R, we will use the function `read_stars` from the `stars` package:

```
## Code_12_4.R
# Kriging
# Load the stars package
library(stars)
library(ggplot2)
# import DEM data of Emilia-Romagna
setwd("C:/RPA/code/spatial_analysis")
ER.DEM <- read_stars("data/dem450_ER.tif")
# give the name Alt to the data column
names(ER.DEM) <- 'Alt'
# Set the CRS
st_crs(ER.DEM) <- st_crs(geo_data.sf)
# Show Altitude map
ggplot() + geom_stars(data = ER.DEM) +
  coord_equal()+
  theme_void() +
  scale_fill_steps(n.breaks=8,low="lightyellow2",high="red",na.value="white") +
  scale_x_discrete(expand=c(0,0))+
  scale_y_discrete(expand=c(0,0))+
  labs(fill="Alt") +
  theme(panel.border = element_rect(linetype = "solid", fill = NA),
        legend.key.size = unit(0.95, 'cm'))+
  theme(plot.margin=unit(c(1,1,1,1),"cm"))
```

Altitudes are distributed as shown in Fig. 12.17.

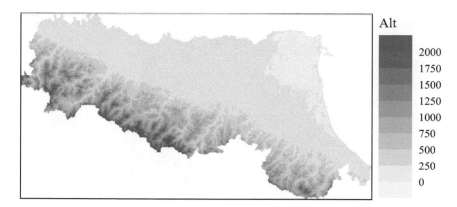

Fig. 12.17 Elevation Map of Emilia-Romagna.

The characteristics of the downloaded grid can be inspected by simply typing the name of the variable `ER.DEM`:

```
stars object with 2 dimensions and 1 attribute
attribute(s):
        Min. 1st Qu. Median    Mean 3rd Qu. Max.   NA's
Alt  -6.84   16.68  107.7 296.9984  510.16 2048 101857
dimension(s):
  from  to offset delta                   refsys point values x/y
x    1 635 515776    450 +proj=utm +zone=1 +datum=... FALSE   NULL [x]
y    1 337 4998818  -450 +proj=utm +zone=1 +datum=... FALSE   NULL [y]
```

The function `st_crs` of the `sf` package allows us to view the coordinate reference system (CRS) of the stars object ER.DEM:

```
ER.crs <- st_crs(ER.DEM)
ER.crs$input
```

ER.crs$input gives:

```
> ER.crs$input
[1] "+proj=utm +zone=1 +datum=WGS84"
```

It is mandatory for the regular grid (this last) and the sample grid to share the same CRS. Using the functions of the `gstat` package, ordinary kriging on residuals is performed by means of the function `krige`.

```
# Kriging
res.kriged <- krige(Tmax.res~1,geo_data.sf,ER.DEM,model=Tmax.vgm.model)
# show kriging results (predicted residuals)
ggplot() + geom_stars(data = res.kriged) +
  coord_equal()+
  theme_void() +
  scale_fill_steps(n.breaks=8,low="blue",high="orange",na.value="white") +
  scale_x_discrete(expand=c(0,0))+
  scale_y_discrete(expand=c(0,0))+
  labs(fill="Pred") +
  theme(panel.border = element_rect(linetype = "solid", fill = NA),
        legend.key.size = unit(0.95, 'cm'))+
  theme(plot.margin=unit(c(1,1,1,1),"cm"))
```

The result of the last line is shown in Fig. 12.18.

Fig. 12.18 Kriging of residuals.

The last step is to add up the residual map with a map obtained from the trend. That is what the last lines of code do. The final map is inserted into a geographical context: Figure 12.19 compares values and positions of the sample data (environmental stations, left) with the continuous interpolated Tmax over all the region (right).

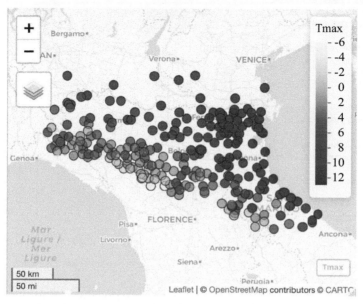

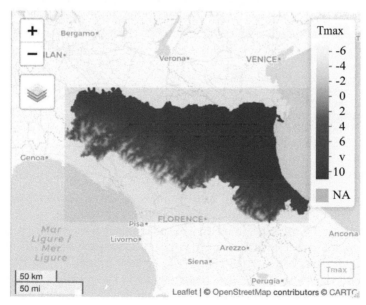

Fig. 12.19 Top: maximum temperatures at measurement stations. Bottom: maximum temperature over all the region.

This last passage requires the library `mapview`. Here is the R code.

```
## Code_12_5.R
predict(Tmax.fit,ER.DEM) -> Tmax.trend
```

```
matrix(unname(Tmax.trend),nrow=635,ncol=337) -> Tmax.trend1
Tmax.pred1 <- res.kriged
tmp <- res.kriged$var1.pred
# sum trend and residual kriging
Tmax.pred1$var1.pred <- tmp+Tmax.trend1
# plot
ggplot() + geom_stars(data = Tmax.pred1) +
  coord_equal()+
  theme_void() +
  scale_fill_steps(n.breaks=8,low="blue",high="orange",na.value="white") +
  scale_x_discrete(expand=c(0,0))+
  scale_y_discrete(expand=c(0,0))+
  labs(fill="Pred") +
  theme(panel.border = element_rect(linetype = "solid", fill = NA),
        legend.key.size = unit(0.95, 'cm'))+
  theme(plot.margin=unit(c(1,1,1,1),"cm"))

# load required libraries
library(mapview)
library(RColorBrewer)
pal <- colorRampPalette(brewer.pal(9, "YlOrRd"))
# regression kriging plot on geographic map
mapview(Tmax.pred1['var1.pred'],col.regions=pal,legend=TRUE,layer="Tmax")
# stations and measured values on geographic map
mapview(geo_data.sf[,2],legend=TRUE,layer="Tmax",col.regions=pal)
```

The procedure involving linear regression + kriging of residuals + recombination of maps can be automated in a single command using the *universal kriging* format in the `krige` function. In this case, the variogram is computed inserting the linear dependence of `Tmax` from `Alt` into the formula. The result is identical to that of Fig. 12.19 (right).

```
## Code_12_6.R
# Semivariogram
vgm1 <- variogram(Tmax~Alt,geo_data.sf)
vgm1.model =  vgm(0.2, "Lin", 150000, 0.5)
plot(vgm1,vgm1.model)
# Universal kriging
TTT <- krige(Tmax~Alt,geo_data.sf,ER.DEM,model=vgm1.model)
mapview(TTT['var1.pred'],col.regions=pal,legend=TRUE,layer="Tmax")
```

12.4   Spacetime analysis

We can only scratch the surface of the complex, intriguing topic of spatio-temporal analysis. We will do it by means of a simple real-world example.

The recording of a physical quantity like the ambient temperature in a space region clearly subtends a process that is stochastic both in space and in time. Indeed, we have already encountered random time series, whose properties are described by the autocorrelation structure (or the spectrum), while in the previous section we analysed spatial stochastic processes in terms of their semivariogram, very closely related to their covariance structure.

Limiting ourselves to a two-dimensional space domain $\mathbb{D} \subset \mathbb{R}^2$, the obvious fact that spatial characteristics change in time, naturally leads - at least conceptually - to adding a third dimension (the time) to the spatial ones and to consider the stochastic process in a domain $\mathbb{D} \times \mathbb{T}$, where $\mathbb{T} \subset \mathbb{R}$. If $z = \{z(s_k, t_k), k = 1..n\}$ is a sample of observed values of a stochastic variable Z, the purpose of a spatiotemporal analysis is usually that of predicting Z at unobserved time/space locations or that of interpolating it, as we did by means of kriging in the analysis of purely spatial phenomena.

We could naturally be tempted to treat time simply as an additional 'space' dimension, and to apply the spatial analysis to a problem defined on a domain $\mathbb{D}' \subset \mathbb{R}^3$ instead of $\mathbb{D} \times \mathbb{T} \subset \mathbb{R}^2 \times \mathbb{R}$. Apart from computational matters (one more dimension greatly increases the computational load), it must be understood that time is intrinsically different from space, because it is 'directional' and, moreover, time grids are usually more regular and dense than space grids: a measurement station can record data every minute/hour/day for very long periods but, of course, measurement stations cannot be placed everywhere!

Formally, a spacetime process $Z(s,t)$ is defined as:

$$\{Z(s,t) : s \in \mathbb{D}(t) \subset \mathbb{R}^2, t \in \mathbb{T} \subset \mathbb{R}\}$$

where both the spatial and temporal components can be discrete or continuous.

Suppose $Z(s,t)$ has mean $\mu(s,t)$. The covariance between to spacetime random variables is defined by:

$$\mathrm{Cov}\left[Z(s_1,t_1), Z(s_2,t_2)\right] = \mathrm{E}\left[(Z(s_1,t_1) - \mu(s_1,t_1))(Z(s_2,t_2) - \mu(s_2,t_2))\right]$$

Second-order stationarity requires that the following relations hold for any location s and time t:

$$\mathrm{E}\left[Z(s,t)\right] = \mu$$

$$\mathrm{Cov}\left[Z(s+h,t+u), Z(s,t)\right] = \mathrm{Cov}\left[Z(h,u), Z(0,0)\right] \equiv C(h,u)$$

for any spatial shift h and temporal shift u.

As in the spatial case, we understand that some kind of stationarity is needed (or, at least, desirable) to estimate the spacetime covariance structure. If the process is second-order stationary, we obtain the familiar expression for the semivariogram:

$$\gamma(h,u) = \frac{1}{2}\mathrm{Var}\left[Z(s+h,t+u) - Z(s,t)\right]$$

as in the purely spatial case. Furthermore, as in the spatial case, in order to predict values of the process Z at any location/time we will need:

1. to estimate the experimental (sample) semivariogram;
2. to individuate a suitable analytical function (model) to fit the sample semivariogram.

Spatiotemporal semivariogram models are usually not simply inherited from the spatial semivariogram by adding one more dimension. We cannot enter the jungle of spacetime models, a discipline still a research matter. We will only give a taste of that topic by developing an example on the time dependence of the same data used in the previous section, concerning the daily recording of maximum temperatures in the Emilia-Romagna region of Italy.

We will see that a reasonable model for a spacetime semivariogram is the 'separable' one.[5] Remaining in the familiar territory of second-order stationary processes, a separable semivariogram comes out when the spacetime covariance decomposes into separate spatial and temporal covariances, or:

$$C(h, u) = C_s(h)C_t(u)$$

where the space and time components usually have different parameters to allow space or time anisotropy. As a consequence, the semivariogram of a separable spacetime process is given by an expression like:

$$\gamma = S_{st}\left(\gamma_s(h) + \gamma_t(u) - \gamma_s(h)\gamma_t(u)\right) \tag{12.19}$$

where S_{st} is a 'joint' sill, and the spatial and temporal parts have their own sills (S), nuggets (N) and ranges (A):

$$\gamma_s(h) = N_s + S_s\left[1 - \exp(-h/A_s)\right]$$

$$\gamma_t(u) = N_t + S_t\left[1 - \exp(-u/A_t)\right]$$

when both the spatial and temporal part are assumed to be exponential.

Before studying a practical spatiotemporal example problem, some words are in order about data representation. We will use the `gstat` R library, whose functions for spatiotemporal analysis work well with data organized in 'long format' dataframes, as explained in the document of the `spacetime` package. In a long format dataframe every row refers to a single time and space position (Wikle, Zammit-Mangion and Cressie, 2019), as for example:

```
   Year Month Day Station    UTM.X      UTM.Y Alt Tmax proc        date
2  2019     1   1      10 659.314 4929.813 100  4.2 Tmax 2019-01-01
3  2019     1   2      10 659.314 4929.813 100  2.3 Tmax 2019-01-02
4  2019     1   3      10 659.314 4929.813 100  6.4 Tmax 2019-01-03
5  2019     1   4      10 659.314 4929.813 100  6.9 Tmax 2019-01-04
6  2019     1   5      10 659.314 4929.813 100  5.6 Tmax 2019-01-05
7  2019     1   6      10 659.314 4929.813 100 14.0 Tmax 2019-01-06
8  2019     1   7      10 659.314 4929.813 100  7.5 Tmax 2019-01-07
9  2019     1   8      10 659.314 4929.813 100  3.0 Tmax 2019-01-08
10 2019     1   9      10 659.314 4929.813 100  9.0 Tmax 2019-01-09
11 2019     1  10      10 659.314 4929.813 100  8.9 Tmax 2019-01-10
```

A long-format table as the above can be easily generated from the maximum temperature data (containing the station codes) and from the station description file (containing the coordinates of the meteo stations). With the development of the `stars` package, a simpler alternative is to transform the data into a stars object. The following code has that purpose.

```
## Code_12_7.R
# Spacetime
# required libraries
```

---

[5]Non-separable models are more complicated and they can take into account of complex spacetime interactions.

```
library(stars)
library(zoo)
library(xts)
library(mapview)

setwd("C:/RPA/code/spatial_analysis")
# Read maximum daily temperatures in years 2019-2020 and station coordinates
Tmax_data <- read.csv("data/ER_Tmax_2019-2020.csv",h=TRUE,check.names = FALSE)
# Read info about measurement stations
Stations <- read.csv("data/ER_stations.csv",h=TRUE)

cod.station <- names(Tmax_data)
cod.station[-1] -> cod.station
subset(Stations,Stations$ID_station%in%cod.station) -> Stations.sel
# Convert into a stars object:
# Step 1: Read in a zoo object
Tmax.zoo <- read.zoo(Tmax_data,index.column=1,format = "%d/%m/%Y")
str(Tmax.zoo)
# Step 2: Convert into an xts (time series)
xts(Tmax.zoo) -> Tmax.xts
str(Tmax.xts)
# Step 3: select stations having Tmax measurement, and obtain geometry
Stations.dat <- subset(Stations.sel,select=c("ID_station","Altitude","UTM_X","UTM_Y"))
space_part <- st_as_sf(Stations.dat,coords=c("UTM_X","UTM_Y"),crs = "EPSG:32632")
# Show measurement stations
space_part.geom <- st_geometry(space_part)
mapview(space_part.geom)
# Step 4: add Altitude data from file Stations
tmp <- subset(Stations.sel,select=c("ID_station","Altitude"))
Alt_data <- Tmax_data
nnn <- as.character(tmp$ID_station)
for (i in 1:length(nnn))
  Alt_data[nnn[i]] <- tmp$Altitude[i]
# Read altitudesa in a zoo object
Alt.zoo <- read.zoo(Alt_data,index.column=1,format = "%d/%m/%Y")
# Convert altitudes into xts
xts(Alt.zoo) -> Alt.xts
# Step 5: Make a stars object having two attributes: TMAX and ALT
st_as_stars(list(TMAX = as.matrix(Tmax.xts),ALT = as.matrix(Alt.xts))) %>%
  st_set_dimensions(names = c("time", "station")) %>%  # nomi colonne
  st_set_dimensions("time", index(Tmax.xts)) %>%  # trasforma time da char a POSIXct
  st_set_dimensions("station", space_part.geom) -> dat1.st
# Verify object
class(dat1.st)
dat1.st
```

We are now ready for spacetime (ST) analysis, using the same **gstat** package employed for spatial kriging. Data frames, although used for data storing, cannot be used directly for ST analyses which require to recast data into objects of the **stars** class. The object *dat1.st* built with the above code contains all data we need:

```
stars object with 2 dimensions and 2 attributes
attribute(s):
      Min. 1st Qu. Median     Mean 3rd Qu.    Max.
TMAX  -7.8   10.60   16.7 17.39916    24.2    40.8
ALT   -1.0   67.75  487.5 509.28889   832.5  1637.0
dimension(s):
         from  to   offset  delta                   refsys point
time        1 731 2019-01-01 1 days                    Date    NA
station     1 180           NA      NA  WGS 84 / UTM zone 32N  TRUE
```

```
                                            values
time                                        NULL
station POINT (655965 4960223),...,POINT (686582 4929444)
```

Stars objects are lists of arrays with a metadata table describing dimensions. We see from the above that *dat1.st* has two attributes (TMAX and ALT) and two dimensions (time and station). The latter consists in the 180 selected stations, while the former includes the 731 days from 2019 January 1st to 2020 December 31st. The construction of the `stars` object passes through a conversion ff the dataframe into a `zoo` object (class of ordered observations) and, successively, into an extensible time series (class `xts`) extending `zoo`.

An ST analysis on the full dataframe relative to two years of daily recordings from 282 stations (206,142 numbers) is computationally expensive and, however, the ST variogram computed on the full dataset (two years of daily recordings) cannot be modelled with 'standard' functions (exponential, spherical, Gaussian etc.). The time dependence on a yearly base could be described by a quasi-deterministic function taking into account the 'seasonal' behaviour of the temperature.

The analysis of a shorter period, e.g. one month, could be of more interest. Indeed, as the *gstat* authors quote, *local kriging is an attractive alternative* to a kriging on the complete data set (Graeler, Pebesma and Heuvelink, 2016). The left plot in Fig. 12.20 shows the sample semivariogram computed on the subset of the full data matrix corresponding to June 2020, obtained by the following lines of code:

```
## Code_12_8.R
# Variogram for space time
# load required libraries
library(tidyverse)
# Subset June 2020
dat1.st %>% filter(time>="2020-06-01",time<="2020-06-30") -> dat1_june.st
# sample semivariogram
v1.st <- variogramST(TMAX~1+ALT, dat1_june.st, width=10000, cutoff=300000, tlags = 1:10)
plot(v1.st)
# no map
plot(v1.st,map=FALSE)
```

We see from the code above that the sample semivariogram is computed, as without time, by taking *Alt* as a natural covariate, because the maximum temperature at a given time value is known to be almost linearly related to altitude. Before carrying out the analysis, it is useful to transform the coordinates into kilometres.

```
## Code_12_9.R
# Transform coordinates
# Latitude and longitude in km
st_as_stars(list(TMAX = as.matrix(Tmax.xts),ALT = as.matrix(Alt.xts))) %>%
  st_set_dimensions(names = c("time", "station")) %>%  # column names
  st_set_dimensions("time", index(Tmax.xts)) %>%  # transform time da char a POSIXct
  st_set_dimensions("station", space_part.geom/1000) -> dat2.st

dat2.st %>% filter(time>="2020-06-01",time<="2020-06-30") -> dat2_june.st
# verify geometry
plot(st_geometry(dat2_june.st))
# sample semivariogram
v2.st <- variogramST(TMAX~1+ALT, dat2_june.st, width=10, cutoff=300, tlags = 1:10)
plot(v2.st)
# no map
```

```
plot(v2.st,map=FALSE)
```

The procedure for computing the ST semivariogram model is pretty similar to that introduced in Section 12.3.2 for spatial data, the main difference being the 'guessed' model function which, this time, is two-dimensional. In particular, as mentioned before, we have to deal with a separable model (12.19). The model is specified in R as follows.

```
## Code_12_10.R
# Define model and starting parameters
v2.sepVgm <- vgmST(stModel = "separable",
                   space = vgm(1, "Exp",50, nugget = 0.1),
                   time = vgm(1, "Exp",1, nugget = 0.1),
                   sill = 10)
# Compute model parameters
(v2.sepVgm <- fit.StVariogram(v2.st, v2.sepVgm))
# Plot fitted semivariogram model
plot(v2.st,v2.sepVgm)
```

The first line specifies the guessed spatial and temporal parts, both as being described by an exponential function. The `vgmST` function defines the model and guesses the parameters, and `fit.StVariogram` computes the best fitting values taking into account the sample variogram. The right plot of Fig. 12.20 compares the model variogram to the experimental one. The agreement is not perfect but the general behaviour with distance and time lag appears to be reasonable.

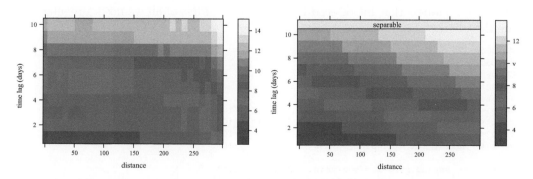

Fig. 12.20 Sample spacetime semivariogram (left) and separable model (right).

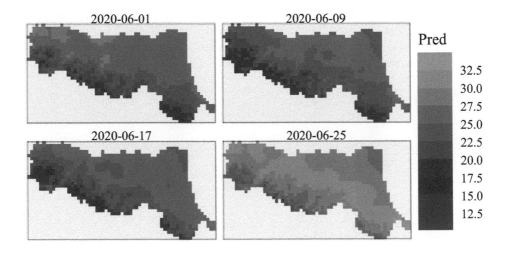

Fig. 12.21 Spatio-temporal kriging for June 2020.

One could question if the parameters of the fitting — nugget, sill and range of the separated components of equation (12.19) and the common sill value — are really the best one, for the assumed function (12.19). Actually they are not. A genetic algorithm can be used to find a slightly better set of values (see the discussion and code at the end of the chapter), but the difference is not a substantial one, and the result of the successive ST kriging is negligibly affected by that.

As a conclusion of the current example, we will see how the procedure for ST kriging closely remembers that introduced for spatial data. First of all we need a regular grid for the interpolation.

```
## Code_12_11.R
# Load DEM of Emilia-Romagna region
ER.DEM <- read_stars("dem450_ER.tif")
names(ER.DEM) <- "ALT"
# make a grid in km
xmin <- min(st_get_dimension_values(ER.DEM,"x"))/1000
xmax <- max(st_get_dimension_values(ER.DEM,"x"))/1000
ymin <- min(st_get_dimension_values(ER.DEM,"y"))/1000
ymax <- max(st_get_dimension_values(ER.DEM,"y"))/1000

newgrid <- expand.grid(x = seq(xmin, xmax, by = 4.5),
                       y = seq(ymin, ymax, by = 4.5))
# Transform to dataframe for successive elaborations
```

```
as.data.frame(newgrid) -> newgrid.df
as.data.frame(ER.DEM) -> ER.DEM.df
# Convert coordinates in kilometres
ER.DEM1.df <- ER.DEM.df
ER.DEM1.df$x <- ER.DEM1.df$x/1000
ER.DEM1.df$y <- ER.DEM1.df$y/1000
# Add altitude to grid points
newgrid.df$ALT <- NA
for (i in 1:nrow(newgrid.df)){
  ER.DEM1.df$ALT[ER.DEM1.df$x>=(newgrid.df$x[i]-2) & ER.DEM1.df$x<(newgrid.df$x[i]+2) &
    ER.DEM1.df$y>=(newgrid.df$y[i]-2) & ER.DEM1.df$y<(newgrid.df$y[i]+2)] -> xxx
  newgrid.df$ALT[i] <- mean(xxx)
}
# Delete NA values, re-obtaining the region shape
na.omit(newgrid.df) -> newgrid.df
# Transform into sf class
st_as_sf(newgrid.df,coords=c("x","y")) -> newgrid.sf
newgrid.geom <- st_geometry(newgrid.sf)
# Verify geometry
plot(newgrid.geom)
# time part of the stars object
t = st_get_dimension_values(dat2_june.st, 1)
st_as_stars(list(ALT = matrix(newgrid.sf$ALT, length(t), length(newgrid.geom),
  byrow=TRUE))) %>%
    st_set_dimensions(names = c("time", "station")) %>%
    st_set_dimensions("time", t) %>%
    st_set_dimensions("station", newgrid.geom) -> newgrid.st
# verify geometry
plot(st_geometry(newgrid.st))
```

Now we have a spatial grid for kriging. The following code allows us to predict the maximum temperatures in the whole Emilia-Romagna region for all days of 2020 June 30. The results are stored in the variable *June*.

```
## Code_12_12.R
# Regression kriging
June <- krigeST(TMAX~1+ALT, data=dat2_june.st, newdata = newgrid.st,
              modelList = v2.sepVgm,
              progress = TRUE)
# show object details
June
```

June is a `stars` object having an attribute (the predicted temperature) in two dimensions (coordinates and time):

```
stars object with 2 dimensions and 1 attribute
attribute(s):
            Min.   1st Qu.   Median      Mean   3rd Qu.      Max.
var1.pred  8.660726 23.64098 26.64732 26.34693 29.16412 35.18578
dimension(s):
     from   to     offset  delta refsys point
sfc    1 1019         NA     NA     NA  TRUE
time   1   30 2020-06-01 1 days   Date    NA
                                                values
sfc  POINT (745.5006 4851.893),...,POINT (574.5006 4991.393)
time                                               NULL
```

We now define a time grid consisting of four dates for 'prediction', using the previously defined spatial grid and we perform the kriging procedure by the command `krigeST`.

```
## Code_12_13.R
# Space time kriging
t = as.Date("2020-06-01") + seq(0, 30, by=8)
st_as_stars(list(ALT = matrix(newgrid.sf$ALT, length(t), length(newgrid.geom),
  byrow=TRUE))) %>%
    st_set_dimensions(names = c("time", "station")) %>%
    st_set_dimensions("time", t) %>%
    st_set_dimensions("station", newgrid.geom) -> newgrid.st

# Regression kriging
June.4dates <- krigeST(TMAX~1+ALT, data=dat2_june.st, newdata = newgrid.st,
              modelList = v2.sepVgm,
              progress = TRUE)
# show object details
June.4dates
```

You can compare *June.4dates* to the previous kriging result to verify that it actually refers to four dates in June:

```
stars object with 2 dimensions and 1 attribute
attribute(s):
              Min.  1st Qu.  Median    Mean  3rd Qu.    Max.
var1.pred  12.35336 23.50748 25.37122 25.20052 27.04161 32.64812
dimension(s):
     from   to    offset  delta refsys point
sfc    1  1019       NA    NA     NA   TRUE
time   1     4 2020-06-01 8 days  Date   NA
                                                    values
sfc  POINT (745.5006 4851.893),...,POINT (574.5006 4991.393)
time                                                  NULL
```

Finally, the kriging result can be plotted by using `ggplot` as shown in the code below. The steps for transforming the krige stars object into another stars object explicitly including the x and y coordinates are described in the comments. The result is shown in Fig. 12.21.

```
## Code_12_14.R
# For the purpose of using ggplot() with geom_stars()
# Step 1: Obtain the geometry of June.4dates
st_geometry(June.4dates) -> June.4dates.geom
xy <- do.call(rbind, June.4dates.geom) %>% as_tibble() %>% setNames(c("x","y"))
x <- xy$x
y <- xy$y
t = as.Date("2020-06-01") + seq(0, 30, by=8)
# Step 2: Convert stars object into an array and add x and y columns to it
# Then, convert in dataframe and reorder columns to have x and y as first two
June.4dates %>% pull(1) -> June.4dates.arr
June.4dates.arr <- cbind(June.4dates.arr,x,y)
June.4dates.df <- data.frame(June.4dates.arr)
June.4dates.df[,c(5,6,1:4)] -> June.4dates.df
# Step 3: rebuild stars object
June.4dates.new <- st_as_stars(June.4dates.df,dims=c("x","y"))
merge(June.4dates.new) -> June.4dates.new
setNames(June.4dates.new,"Tmax.pred") -> June.4dates.new
June.4dates.new = st_set_dimensions(June.4dates.new, names=c("x","y","time"))
June.4dates.new = st_set_dimensions(June.4dates.new, "time", values = t, names = "time",
  point = TRUE)

# Plot results
ggplot() + geom_stars(data = June.4dates.new) +
    coord_equal() +
```

```
    facet_wrap(~time) +
    theme_void() +
    scale_fill_steps(n.breaks=8,low="blue",high="orange",na.value="grey95") +
    scale_x_discrete(expand=c(0,0))+
    scale_y_discrete(expand=c(0,0))+
    labs(fill="Pred") +
    theme(panel.border = element_rect(linetype = "solid", fill = NA),
        legend.key.size = unit(0.95, 'cm'))+
    theme(plot.margin=unit(c(1,1,1,1),"cm"))
```

Look in particular at line 15 of the above code:

```
merge(June.4dates.new) -> June.4dates.new.
```

The *merge* operation merges the four attributes originated by the code above, one for each of the four dates:

```
stars object with 2 dimensions and 4 attributes
attribute(s):
        Min.  1st Qu.  Median    Mean 3rd Qu.    Max. NA's
V1  14.43113 23.11631 25.64355 24.62988 26.54723 28.59462  933
V2  12.53825 21.95759 24.31657 23.27692 25.21789 26.52291  933
V3  12.35336 22.92491 24.98222 24.00450 25.70258 27.27618  933
V4  19.41741 27.79361 29.39283 28.89077 30.43315 32.64812  933
dimension(s):
    from to  offset delta refsys point values x/y
x      1 61 522.751   4.5     NA    NA   NULL [x]
y      1 32 4993.64  -4.5     NA    NA   NULL [y]
```

into a single attribute whose values are ordered by an additional dimension that a successive line of code renames *time*:

```
stars object with 3 dimensions and 1 attribute
attribute(s):
              Min.  1st Qu.  Median    Mean 3rd Qu.    Max. NA's
Tmax.pred 12.35336 23.50748 25.37122 25.20052 27.04161 32.64812 3732
dimension(s):
      from to    offset  delta refsys point values x/y
x        1 61   522.751    4.5     NA    NA   NULL [x]
y        1 32   4993.64   -4.5     NA    NA   NULL [y]
time     1  4 2020-06-01 8 days   Date  TRUE   NULL
```

The new `stars` object, having one attribute and three dimensions, can conveniently be plotted under `ggplot()` by means of `geom_stars()` by selecting the different times with `facet_wrap(~time)`.

12.5 On the optimization of the spatio-temporal variogram

We now offer a few words to show how genetic algorithms (GA) can be usefully employed for computing a better set of parameters for the semivariogram model of Fig. 12.20. Remembering what was presented in Chapter 10, we can define a *fitness function* (or *objective* function) as the sum of the squared differences between the sample semivariogram γ and the model semivariogram $\hat{\gamma}$ computed on the same grid $G_d \times G_t \subset \mathbb{R}^2$ (distances × time-lags). We can compute such a function as follows:

$$SSD = \sum_{d,t \in G_d \times G_t} (\gamma(d,t) - \hat{\gamma}(d,t))^2 \qquad (12.20)$$

This function also is the 'core' of the GA procedure, which will try to maximize the reciprocal of (12.20). The *chromosome* of the current problem is a sequence of seven 'genes': n_s, s_s, a_s, n_t, s_t, a_t, S_{st}, i.e. nugget, sill and range of both components and the cumulative sill. The distance and time grids are those of the object *v2.st*, the sample semivariogram computed by `variogram` in the previous section. It consists of 31 space points (from 0 to 295 km) and 10 time lags (from 1 to 10 days). In the following code, such a grid of 31 × 10 spacetime points is arranged in matrix form, suitable to be plotted as a filled contour plot by means of the function `levelplot` of the package `lattice`.

The sample variogram *v2.st* is an object of class 'StVariogram data.frame', something like an ordinary dataframe:

```
> class(v2.st)
[1] "StVariogram" "data.frame"
> head(v2.st)
     np      dist     gamma    id timelag spacelag   avgDist
1  5220  0.000000 3.381070 lag0       1        0  0.000000
2 10672  6.854489 3.946344 lag0       1        5  6.854489
3 29348 15.382297 4.008843 lag0       1       15 15.382297
4 45472 25.263122 4.076748 lag0       1       25 25.263122
5 48430 35.097244 4.123429 lag0       1       35 35.097244
6 58870 44.850224 4.135518 lag0       1       45 44.850224
```

The matrices are created as follows:

```
## Code_12_15.R
# Matrices creation for GA analysis
library(lattice)
# rearrange v2.st in matrix form
tmp <- v2.st
dd <- tmp$spacelag
tt <- tmp$timelag
gg <- tmp$gamma
XY <- expand.grid(X=dd,Y=tt)
MX <- matrix(XY$X,nrow=10,ncol=31,byrow=TRUE)
MY <- matrix(rep(1:10,31),10,31)
GG <- matrix(gg,10,31,byrow=TRUE)
levelplot(GG ~ MX + MY,col.regions=heat.colors,
          xlab="distance",ylab="time lag")

# Semivariogram model
Ns = 0
Ss = 1
As = 1e5
Nt = 0.002744451
St = 0.997255549
At = 1088.299
Sst = 889.53781502222

gam.t = Nt+St*(1-exp(-MY/At))
gam.s = Ns+Ss*MX/As
gam = Sst*(gam.s+gam.t-gam.s*gam.t)

levelplot(gam ~ MX + MY,col.regions=heat.colors,
          xlab="distance",ylab="time lag")

# plot the difference between sample and model semivariogram
dif <- abs(gam-GG)
levelplot(dif ~ MX + MY,col.regions=heat.colors,
```

```
        xlab="distance",ylab="time lag")
# Sum of squared differences
(SSD <- sum(dif^2))
```

To verify the correctness of the grid, we can reproduce a plot like that on the left of Fig. 12.20 by the function `levelplot` of the package `lattice` (or by `filled.contour`, included in the standard *graphics* library).

The result is shown in Fig. 12.22 (left).

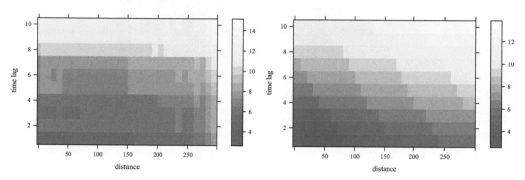

Fig. 12.22 Filled contour plot of the sample (left) and model (right) variogram recalculated on the $G_d \times G_t$ grid.

The model semivariogram $v2.sepVgm$ is an object of class 'StVariogramModel list':

```
> class(v2.sepVgm)
[1] "StVariogramModel" "list"
> v2.sepVgm
space component:
  model psill range
1   Nug     0 0e+00
2   Lin     1 1e+05
time component:
  model       psill    range
1   Nug 0.002744451    0.000
2   Exp 0.997255549 1088.299
sill: 889.53781502222
```

where linear spatial dependence and exponential time dependence have been assumed.

We can test the result of the fitting procedure by computing the two-dimensional semivariogram (12.19) and plotting the density plot, as shown in Fig. 12.22 (right). Figure 12.23 maps the difference inside the expression (12.20). Squaring and adding the plotted values we obtain $SSD = 229.55$.

The GA code is readily implemented in R by means of the functions of the library `GA`. The following code defines the *fitness* function, sets the GA parameters (crossover and mutation probability, number of individuals, maximum number of generations) and invokes the `ga` function to obtain a solution. Note that, as remarked in Chapter 10, the solution is itself stochastic, so the algorithm should be run a number of times (also with different choice of the GA parameters) to obtain the 'minimum' of the minima. We can obtain a solution like:

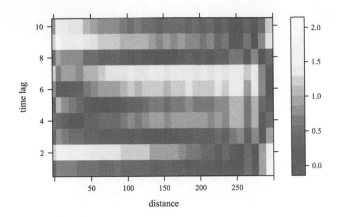

Fig. 12.23 Difference between sample and model semivariogram.

```
solution: 0.007663696 1.510145 69998.53 0.003812302 1.461451 467.5601 260.8526
space component:
Nugget 0.007663696
Sill 1.510145
Range 69998.53

time component:
Nugget 0.003812302
Sill 1.461451
Range 467.5601

cumulative Sill 260.8526
```

for which $SSD = 210.94$, i.e. a value about 8% lower than that obtained by means of `fit.StVariogram`. Of course, kriging is not heavily affected by such a difference but, anyway, Fig. 12.21 could be recalculated.

The GA code follows.

```
## Code_12_16.R
# Semivariogram with GA
library(GA)
# sample semivariogram
x <- MX[1,] # spacelag
y <- MY[,1] # timelag
z <- GG # gamma value

# fitness function
zeta <- function(crom){
  crom <- matrix(crom, ncol = 7)
  n_s <- crom[,1]
  s_s <- crom[,2]
  a_s <- crom[,3]
  n_t <- crom[,4]
  s_t <- crom[,5]
  a_t <- crom[,6]
  S_st <- crom[,7]

  g = 0;
  for (i in 1:31){
    gamma_s <- n_s + s_s*x[i]/a_s
```

```
    for (j in 1:10){
      gamma_t <- n_t + s_t*(1-exp(-y[j]/a_t))
      z.fit <- S_st*(gamma_s+gamma_t-gamma_s*gamma_t)
      Diff <- (z[j,i]-z.fit)
      if (!is.na(Diff)){
        g <- g + Diff^2
      }
    }
  }
  return (1/g)
}

# GA solution
pop.size<- 30
max.gen <- 1000
xovor.prob <- 0.7
mut.prob <- 0.1

# seed = 246 for reproducibility with the book results
# it should be removed in a real-world analysis
GA <- ga(type = "real-valued", fitness = zeta,
         lower = c(0,1,1e4,0,0,100,100),
         upper = c(0.05,2,1e5,0.05,2,2000,1000),
         maxiter=max.gen,popSize = pop.size,seed=246)

(solution <- unname(GA@solution))

# Post processing
n_s <- solution[1]
s_s <- solution[2]
a_s <- solution[3]
n_t <- solution[4]
s_t <- solution[5]
a_t <- solution[6]
S_st <- solution[7]

gam_t = n_t+s_t*(1-exp(-MY/a_t))
gam_s = n_s+s_s*MX/a_s
gam_st = S_st*(gam_s+gam_t-gam_s*gam_t)

# plot the solution
levelplot(gam_st ~ MX + MY,col.regions=heat.colors,xlab="distance",ylab="time lag")
# difference between sample and model semivariogram
D <- abs(gam_st-GG)
# plot the error
levelplot(abs(D) ~ MX + MY,col.regions=heat.colors,xlab="distance",ylab="time lag")
# Sum of squared differences
(SSD <- sum(D^2))
```

Note that a *seed* has been imposed in the **ga** function, such that the identical result presented here can be replicated. Of course, that is not useful in a real-world analysis: GA are stochastic algorithms, so a series of runs are usually necessary to find the 'best' solution, leaving the algorithm free to randomly generate the initial population of chromosomes for each run. For the particular case above, Fig. 12.24 shows the convergence of the genetic algorithm.

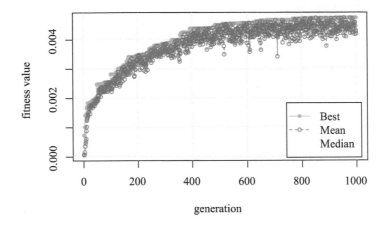

Fig. 12.24 Convergence of the genetic algorithm.

12.6   Exercises

Exercise 12.1 Demonstrate equation (12.11), i.e. that for a second-order stationary process the semivariogram is expressed in terms of the covariance $C(h)$ and the variance $C(0)$:

$$\gamma(h) = C(0) - C(h) \tag{12.21}$$

where:

$$C(h) = \mathrm{Cov}\,[Z_{s+h}, Z_s]$$

and:

$$C(0) = \mathrm{Var}\,[Z_s]$$

Exercise 12.2 If a variogram has a 'sill', what does this mean in terms of the covariance function (when it is defined, of course)?

Exercise 12.3 Replicate for T_{min} the analysis of Section 12.3.2 on the data contained into the file *20201121_weather.csv* conducted for T_{max}, limited to the computation of the sample and model variogram. Determine 'by eye' reasonable parameters for a linear semivariogram model.

Exercise 12.4 With reference to exercise 12.3, guess an exponential model instead of a linear one and use `fit.variogram` to refine its parameters.

Exercise 12.5 Perform the spatial kriging procedure on the T_{min} data of the last two exercises, and compare the interpolation results obtained by the linear and exponential variogram model.

Exercise 12.6 Using the precipitation data in *20201121_weather.csv* (variable *Prec*) try to conduct a variogram analysis like that developed for T_{max} in the chapter, or for T_{min}. What you can tell about the dependence on altitude?

Exercise 12.7 Use a linear variogram model in Exercise 12.6 and compute a spatial kriging. If **res.kriged** is the result of analysis on the residuals of linear dependence of *Prec* on *Alt*:

```
krige(Prec.res~1,geo_data.sf,ER.DEM,model=Prec.vgm.model) -> res.kriged
```

plot both 'prediction' and 'variance' as follows, and discuss what you see:

```
plot(res.kriged['var1.pred'])
plot(res.kriged['var1.var'])
```

13
How Random is a Random Process?

Le hasard est le plus grand de tous les artistes.

Honore' de Balzac, La vieille fille

13.1 Random hints about randomness

Let us consider the following sequence:

$$\textbf{seq}\,\pi \qquad 1\,4\,1\,\ldots\ldots\,6\,7\,9 \qquad \textit{1000 digits}$$

seq π reports the beginning and the end of the first 1000 decimal digits of π. Is such a sequence a random sequence? We would answer 'no', knowing that the sequence comes from the irrational number π. Indeed, with this knowledge we are able to say that the 1001-th digit is 3. However, for a person who does not know the decimal digits of π, probably she or he would think that the sequence was random. Such an opinion would be confirmed by submitting the sequence to statistical tests.

In the literature a great number of tests for randomness are reported that, if applied to the above sequence, would confirm that the sequence is random. Nevertheless, we know that the sequence is generated by a mathematical formula, therefore we say that **seq** π is not random, a claim however not supportable only on the basis of a statistical test. Now the difference is between the *mathematical formula* to generate a sequence and the *statistical properties* of such a sequence. A formula to compute π as:

$$\pi/2 = 2/1 \times 2/3 \times 4/3 \times 4/5 \times 6/5 \times 6/7 \times 8/9 \times \ldots$$

does not inform us about how many 0's followed by 1, by 2, etc. there are. Conversely, from any segment of the sequence, for instance 9 3 2 3 8 4 6 2 6 4, it is not possible to derive any formula to compute π.

This example of π seems to suggest that the randomness of a sequence is a notion relative to the mathematical generation rules. Is then randomness a subjective concept depending on knowing or not knowing how the sequence has been generated? If so, what is random for me, unaware of the generation method, it is not random for people who know it. This and similar questions will be discussed in the following.

TRN	0.8535	0.8038	0.6865 ...... 0.8276	0.8168	0.1555		*1000 digits*
PRN	0.2655	0.3721	0.5728 ...... 0.2821	0.1912	0.2655		*1000 digits*

Consider now the **TRN** and **PRN** sequences (written with four decimal digits). Both sequences pass statistical tests for randomness, but there is a profound difference between them. The numbers in the **TRN** sequence are *truly random numbers*, while those of **PRN** are *pseudo random numbers*. On the site `www.random.org` randomness is generated based on atmospheric noise, hence it derives from physical phenomena and it is transferred into a computer (do not try to exactly reproduce the **TRN** sequence written above, because all sequences downloaded from the site are always different from one another).

On the contrary, the **PRN** sequence is reproducible, since it is generated by an algorithm. In our opinion, it is very useful if we can obtain the same sequences when Monte Carlo methods are applied in computer experiments. This **PRN** sequence shown above is obtained with the R function **runif(.)**, combined with `set.seed(1)`, already used many times in the previous chapters. This function generates random deviates of the uniform distribution within defined intervals. The command ?RNG, or `help(RNG)`, can usefully be executed to list random number generators available in R.

Remark 13.1 In the 1970s, a number of generator were proposed, the most common called linear congruential methods. Although we have made use of functions already implemented in R, it is instructive to learn something about them. The name derives from the notion of *congruence*. Given two integer numbers a and b, it is said that a and b are *congruent modulo c*, written:

$$a \equiv b \pmod{c}$$

if $(a - b)$ is divisible by c, or equivalently if a and b have the same remainder when divided by c. In particular, the following generator is called a *mixed congruential* generator:

$$r_{i+1} \equiv (ar_i + b) \pmod{m}$$

where r_i, a, b, m are not negative integers. They are called:

a	*constant multiplier*
b	*increment*
m	*modulus*

r_0, the initial value, is called the *seed*. Random numbers between zero and one can be derived from $u_i = r_i/m$.

Congruential methods give rise to periodic sequences. The period of the sequences, as well as their statistical properties, greatly depend on the choice of a, b, m and r_0. As an example, let us choose:

$$r_0 = 2, \ a = 3, \ b = 1, \ m = 16$$

Then:

$$r_1 \equiv (3 \times 2 + 1) \pmod{16}$$

Now we divide 7 by 16 and take the remainder 7, so that: $r_1 = 7$, and so on. It results in the sequence:

$$2 \ 7 \ 6 \ 3 \ 10 \ 15 \ 14 \ 11 \ 2 \ 7 \ldots$$

after 11 the sequence starts over with 2, so the period is 8.

Not all generators proposed in the literature work equally well. Suppose we have generated a sequence. In order to establish its randomness we submit it to statistical tests. For example, are the numbers uniformly distributed? Are they independent? Does the sequence contain long monotonic segments? And, as we will discuss shorty, what features do these sequences, aggregated in couples, triples, n-ples, exhibit? Indeed, a 'good' random number generator should ensure that n-ples of numbers are random too, for n as large as possible, i.e. it would be desirable to find the same properties also in the n-dimensional space, for example a sequence grouped in n-ples should fill uniformly the space. Numbers generated by linear congruential methods tend, instead, to fall in planes.

An historical anecdote is worth be tolding. There was once the IBM infamous generator named RANDU. While for the pairs of generated numbers, it had a good statistical behaviour, the triplets lay on planes. Press *et al.* (1992) report that when one of the authors informed those concerned about the infelicitous performance of RANDU, he was assured 'We guarantee that each number is random individually, but we don't guarantee that more than one of them is random'. Unfortunately RANDU was widely used in many scientific works.

The default random number generator in R is the 'Mersenne-Twister', whose numbers are equi-distributed in 623 dimensions.

We just discussed about sequences generated by physical systems and mathematical algorithms. This leads, in turn, to considering physical randomness and mathematical randomness. Given the sequence $1\,0\,1\,0\,1\,0\ldots$ of a certain length, we feel that it is absolutely not random. But it could well be the results of the toss of a coin, realized by a tossing device. So, the distinction to be kept in mind is between the *generation* of random events and the *sequence* of random events.

RandProc.1	$-1\ -1\ +1\ldots\cdots-1\ -1\ -1$	*1000 digits*
RandProc.2	$+1\ +1\ +1\ldots\cdots+1\ +1\ +1$	*1000 digits*

The sequences **RandProc.1** and **RandProc.2** (written with four decimal digits) are *time series* of the random process described by the *i.i.d.* random variables:

$$X_1, X_2, X_3, \ldots, X_n, \ldots$$

such that:

$$\begin{cases} \mathsf{P}(X_i = +1) = p \\ \mathsf{P}(X_i = -1) = q = 1-p \end{cases}$$

Transitions from state i to state $i+1$ happen with probability p or to state $i-1$ with probability $q = 1-p$. This process is a random walk on a line, dealt with in Chapter 5. In the sequence **RandProc.1** it is $p = 0.5$, while $p = 0.7$ in the sequence **RandProc.2**. Both sequences are obtained by a random algorithm, now the question is: are the two sequences equally random? How can we measure their randomness?

13.2 Characterizing mathematical randomness

Does there exist a way to *objectively* define the randomness of a number sequence? A first difference, already seen before, is between *mathematical randomness*: what formal criteria a number sequence must meet to be random, and *physical randomness*: what *empirical* properties a sequence of events must have to be random.

Randomness and interpretation (or, better, conception) of probability are intimately related to one another. The von Mises' theory of probability (von Mises, 1928) (about von Mises see von Plato, 1994), is among the first and most well-known proposals characterizing mathematical randomness. We will briefly investigate the connection between randomness and the frequency conception of probability.

Randomness can be defined by means of two axioms applied to sequences with certain properties, called *collectives*. Using the words of von Mises (1928, p. 15):

a collective is a mass phenomenon or a repetitive event, or, simply, a long sequence of observations for which there are sufficient reason to believe that the relative frequency of the observed attribute would tend to a fixed limit if the observation were indefinitely continued. This limit will be called the probability of the attribute considered within the given collective.

Clearly, in von Mises' belief, it makes sense to talk about probability only if it refers to collectives.

Consider an *infinite* sequence of elements, e.g. the results of infinite tosses of an indestructible coin. Let $f = n/N$ the frequency of a certain attribute A, for instance the 'number of tails', determined on the basis of N observations (tosses). Von Mises *postulates* that the limit for $N \to \infty$ of f exists and it is finite. This limit is called *the probability* of the attribute A in the given sequence. Then the first axiom is:

1st Axiom *Existence of limits*. The relative frequency of that event A has a limit value.

However, the sequence must be structured in a specific way, to be defined as 'collective' on which to construct a theory of probability (*erst das Kollektiv, dann die Wahrscheinlichkeit*). The second axiom postulates randomness:

2sd Axiom *Insensitivity to place selection*. The limit value of the relative frequency in the sequence remains unchanged despite any possible selection that can applied. Moreover, the limit values of all the sub-sequences derived by place selections are equal to the limit of the original sequence.

This axiom is known as *Regellosigkeitaxiom*, literally 'without rules', i.e it is an 'axiom of irregularity'.

The von Mises definition of randomness is somehow related to the gambling house, and the concept of insensitive-to-place selection is also called *principle of impossibility of a successful gambling system*. Once again, games of chance, *roulettes*, coin tosses, and so on, are often evoked when defining randomness and they contribute, in a sense, to the foundation of probability. Perhaps it is because people effectively realize in such contexts, for example entering in a casino, how chance works in real life. Fyodor Dostoyevsky's gambler knows well those places, when he says:

[...] with a sort of fear, a sort of sinking in my heart, I could hear the cries of the croupiers– 'Trente et un, rouge, impair et passe', 'Quarte, noir, pair et manque'. How greedily I gazed upon the gaming-table, with its scattered louis d'or, ten-gulden pieces, and thalers;

According to von Mises' ideas, games of chance are indeed to be regarded as the paradigm of the notion of randomness. A sequence is called random when it exhibits the same characteristics as a game of chance, for instance the result of the roll of a die or of the spin of a roulette wheel. In games of chance racking the brain on strategies to improve the chance to win is pointless. That is the essence of randomness, after all: random is what you cannot forecast.

A very simple example. In an infinite sequence of this type:

$$1\,0\,1\,0\,1\,0\,1\,0\,1\,0\,1\,0\ldots$$

the limit to infinity of the relative frequency of 0's is $1/2$.

Applying the rule: select the elements in sites $2 \times i$, $i = 1, 2, 3, \ldots$ we obtain the sub-sequence:

$$0\,0\,0\,0\,0\,0\,0\,0\ldots$$

lefting the sequence

$$1\,1\,1\,1\,1\,1\,1\,1\ldots$$

In the former subsequence, the relative frequency of 0's has limit 1, while in the latter such a limit is 0. Both are different from $1/2$, the original limit. The original sequence, therefore, is *not* random, being not insensitive to place selection.

The von Mises' proposal aroused consensus and criticisms. In particular, it was argued that the application of a mathematical concept (existence of the limit of a series) to a non-mathematical property of the series itself, such as randomness (irregularity), can be contradictory. Waismann (1930, p. 7) writes:

[...] anyone who stipulates that a convergent series of numbers be constructed in an irregular way − and this is what is at issue − stipulates the impossible; neither the mathematical series nor the empirical one can fulfil these requirements. Against this it must be said emphatically that a statistical series does not have the properties of a mathematical one; that the mark of the accidental, unforeseeable, cannot be transferred from the empirical structure to the mathematical one without destroying the a priori necessity which is the characteristic mark of mathematics.

Now we briefly report some remarks by Popper (1934) to the von Mises' axioms, focusing on his proposed concept of randomness. The impossibility of applying a gambling system to infinite sequences, rules out any effective control of the randomness of the sequence. A sequence, even though regular 'at the beginning', may present itself as a collective, as long as 'at the end' it becomes irregular. A sequence of the type:

$$1\,0\,1\,0\,1\,0\,1\,0\,1\,0\,1\,0\ldots$$

consisting in, say, one million 0's and 1's, is 'at the beginning' regular, but if after those millions, 'at the end' it becomes insensitive to place selection, it will be legitimate to say it is a random sequence. Popper proposes a different type of random sequence, that is: a *finite* sequence such that, 'for each beginning segment, whether short or long, is as random as the length of the segment permits' (Appendix *VI, *On Objective Disorder or Randomness*). In this consists the 'objective characterization of disorder or randomness, as a type of order' (Appendix *VI).

Such sequences are called *ideally random sequences* (Appendix *VI) and are constructed by means of a *mathematical rule*, as described in Appendix *IV (*A Method of Constructing Models of Random Sequences*). Popper says that 'there are two tasks to be performed: the improvement of the axiom of randomness – mainly a mathematical problem; and the complete elimination of the axiom of convergence' (*ibid.*, p. 142). The axiom of randomness is replaced by a weaker requirement of irregularity. For that purpose Popper introduces the concept of *freedom from after effects* (*ibid.*, p. 159). Consider the sequence (or *alternative*):

$$(\alpha) \qquad 1\,1\,0\,0\,1\,1\,0\,0\,1\,1\,0\,0\,1\,1\,0\,0\,\ldots$$

with a thousand zeros and ones. Now 'select from α all terms with the neighbourhood-property β of immediately succeeding a one'. The selected subsequence will have the structure:

$$(\alpha.\beta) \qquad 1\,0\,1\,0\,1\,0\,1\,0\,1\,0\,\ldots$$

both the relative frequency of the ones and that of the zeros did not change, remaining equal to $1/2$, as in α. Then we may say that 'the alternative α is insensitive to selection according to the property β; or, more briefly, that α is insensitive to selection according to β'. In other words, Popper says that 'α is free from any after effect of single predecessors' or briefly, 'α is 1-free'. This notion of freedom from after effects, may be extended to **n**-free. For example, a 3-free alternative is obtained from the generating period

$$1\,0\,1\,1\,0\,0\,0\,0\,1\,1\,1\,1\,0\,1\,0\,0\,\ldots$$

Popper (*ibid.*, p. 151) writes:

It will be seen that the intuitive impression of being faced with an irregular sequence becomes stronger with the growth of the number **n** of its **n**-freedom.

A sequence **n**-free from after-effects for every **n**, is called 'absolutely free'. The concept of **n**-freedom assumes that of relative frequency. The insensitivity to selection according to certain predecessors is the relative frequency with which a property occurs.

Since the sequences are mathematically constructed, it makes sense to define limits of relative frequencies. With words by Popper: (*ibid.*, p. 155, author's italics):

The use of this concept [limit of relative frequencies] gives rise to no problem so long as we confine ourselves to reference-sequences which are constructed *according to some mathematical rule*. We can always determine for such sequences whether the corresponding sequence of relative frequencies is convergent or not. The idea of a limit of relative frequencies leads to trouble only in the case of sequences for which no mathematical rule is given, but only an empirical rule (linking, for example the sequence with tosses of a coin); for in these cases the concept of limit is not defined.

Randomness, as described by Popper, entails convergence in itself, with no need of a convergence axiom. The Appendix *IV of the cited work reports how to construct such sequences 'according to some mathematical rule'.

Using the notion of absolutely free sequences, Popper solves what he calls 'the fundamental problem of the theory of chance': 'the seemingly paradoxical inference from the unpredictability and irregularity of singular events to the applicability of the rules of the probability calculus to them is indeed valid' (*ibid.*, p. 180).

Physical randomness is derived from mathematical randomness: 'an empirical sequence is random to the extent to which tests show its statistical similarity to an ideal sequence' (Appendix *VI). It is also possible to establish the degree **n** of randomness on the basis of its n-freedom from after effect.

In summary, from the perspective of Popper, randomness should not be interpreted as our lack of knowledge. Mathematical randomness is rigorously defined for both finite and infinite sequences, since random sequences are the result of mathematical constructions. Physical randomness is assessed from comparison with mathematical randomness. Statistical tests evaluate how close a physical random sequence to a mathematical one.

Here we have limited ourselves to the notion of random sequence, as discussed in the *Logic*. Further philosophical problems are dealt with in the *Logic*. For Popper's philosophy see, among others, Keuth (2005). To mention the evolution of Popper's philosophy, in 1957 he proposed at the Ninth Symposium of the Colston Research Society a new interpretation of probability, called '*propensity interpretation*'. At the base of this proposal there is the belief that 'propensities' are physically real, like Newtonian forces, and that indeterminism is not a reflection of an epistemological state, but is inherent in physical systems, even *prima facie* deterministic. A sequence is random so far as it is the result of the propensities of the generating conditions (for this aspect of Popper's thought, see Galavotti, 2005).

As opposed to the above approach, there is a branch of the physical-mathematical thinking that has roots essentially founded on the Bayes theorem (whose applications were described in Chapter 9). Sir Harold Jeffreys, one of the eminent members of this branch (E. T. Jaynes dedicated his book to him (Jaynes, 2003) with the epigraph 'to Sir Harold Jeffreys: who saw the truth and preserved it.'), gives his perspective about that (Jeffreys, 1931), referring to statistical mechanics:

[...] the kinetic theory of gases deals essentially with fluctuating motions that we do not know in detail. It uses the principles of classical mechanics, but uses them to derive relations between statistical properties, and hence is essentially an application of probability theory.

Thus, he clearly states that randomness comes into the game due to a lack of knowledge. If we could know the initial position and velocity of the gas molecules we will not need to treat the problem in terms of probabilities.

Moreover, the so-called *subjective* approach to probability goes further. DeFinetti (1995) in his essay on the philosophy of probability discusses the randomness or not of a sequence like the following:

$$1\,1\,1\,0\,0\,0\,1\,1\,1\,1\,0\,0\,0\,0\,1\,1\,1\,1\,1\,0\,0\,0\,0\,0$$

He says (translated from Italian):

[...] a frequentist would say 'this sequence is not random [...] because it has been constructed taking three '1' followed by three '0', then four '1' followed by four '0' etc.'.

and he adds:

[...] if numbers were in a different order he would say: 'OK, this sequence appear as random.'

His conclusion is that both sequences may have been randomly chosen from the 2^{24} possible ones, the only actual difference being **our** perception of order/regularity or disorder/irregularity.

13.2.1   Randomness and complexity

Around 1965, Solomonoff (1964), Kolmogorov (1965) and Chaitin (1966) independently proposed a definition of the randomness of a sequence of numbers on the basis of *algorithmic complexity*. The basic idea is that the complexity of a sequence can be defined by the length of the shortest binary program for computing that sequence. Consider the sequences:

$$a)\ \ 1\ 0\ 1\ 0\ 1\ 0\ 1\ 0\ 1\ 0\ 1\ 0\ 1\ 0\ 1\ 0\ 1\ 0\ 1\ 0$$
$$b)\ \ 1\ 0\ 0\ 1\ 1\ 0\ 1\ 1\ 0\ 1\ 1\ 1\ 0\ 0\ 1\ 0\ 0\ 0\ 0\ 1$$

Suppose we have to input them into a computer. For the first one, it is enough to give the instruction 'print 1 0 ten times'. If we think of extending the sequence, this instruction will be 'print 1 0 N times', with N equal to one hundred thousand, one million, and so on. The amount of information (number of bits) in such a type of instructions is clearly much less than the number of bits of the sequences generated by them. Recall that a sequence of n binary digits holds n bits, since each digit conveys one bit of information by definition.

On the contrary, the same reasoning is no longer valid for the sequence *b*). To transmit it to a computer, the shortest instruction consists in rewriting the sequence as it is, through an instruction of the type: 'print 1 0 0 1 1 0 1 1 0 1 1 1 0 0 1 0 0 0 0 1'. Even if the sequence continued (without a rule of construction), the number of bits in the instruction would always be not less than the number of bits of the sequence. From this a new definition of random sequence can be derived: a number sequence is random, if the information contained inside cannot be compressed. More formally (Martin-Löf, 1966), consider a binary sequence of length n:

$$x = x_1, x_2, \ldots, x_n$$

Let A be a 'universal machine', e.g. a computer, and p a code (another binary string), with length $l(p)$, which gives $A(p, n) = x$ when executed by A, the elements of the sequence x constituting the data on which the function operates.

The Kolmogorov complexity of a sequence x relative to a 'machine' A is defined as:

$$K_A(x|n) = \min l(p)$$

that is, the Kolmogorov complexity of a string x with respect to the computer A is the minimum length of all possible programs read by the computer A which print out the string x. Note that we have assumed that $l(x)$, the length n of x, is known by the computer, otherwise we somehow have to inform the computer about the end of the sequence (Cover and Thomas, 1991).

If such a program p does not exist, we write:

$$K_A(x|n) = +\infty$$

In this case, we say that the sequence has the maximum complexity. It appears that the complexity measure depends on A, but Kolmogorov (1965) proved that a universal machine always exists that, at least asymptotically, is as efficient as any other machine. Therefore the index A can be omitted.

A sequence (finite or infinite) is defined as a *random sequence* when it has maximum complexity. Therefore a random sequence cannot be computed by means of a finite algorithm, i.e. the contained information cannot be encoded. The only way to state a random sequence is to rewrite the sequence itself.

Martin-Löf (1966) proved that random sequences, in the above specified sense, have the same properties required by the probability theory to characterize certain sequences as random. It is clear, for instance, that a maximum complexity sequence must pass all tests for randomness. Suppose we put all these tests together in a unique 'universal test' defined as follows. A test is 'universal' if, when a sequence passes it, it passes every other conceivable test. Martin-Löf (1966) shows that the maximum complexity sequences pass the universal test. Therefore, the random sequence definition agrees with the intuitive idea that a sequence is random if it is so unpredictable that no mathematical expedient, humanly conceivable, is able to reveal some property in it.

Come back to the two sequences $a)$ and $b)$. Both are *not* random, since an algorithm exists which can allow a computer to generate them. But the complexity degree of $a)$ is less than that of $b)$, since the algorithm to generate $a)$ is shorter than that needed to generate $b)$.

Some properties of the Kolmogorov complexity are reported in the following (Cover and Thomas, 1991). Let U be a universal computer. Then for any computer A we have:

$$K_U(x) \leqslant K_A(x) + c$$

for every string x and c independent of x. Then we always have:

$$\left| K_U(x) - K_A(x) \right| \leqslant c$$

The complexity $K_A(x|l(x))$ is less than the length $l(x)$.

A sequence of n 0's has a finite Kolmogorov complexity c (the computer knows n):

$$K(000\ldots0|n) = c, \forall n$$

Similarly for n decimal digits of π.

In the next section the Shannon entropy H will be introduced. We will see that, for a sequence with entropy H, the Kolmogorov complexity K is approximately equal to H.

The basic features discussed before, concerning the notion of randomness, are certainly useful for determining the degree of complexity of the sequence at hand, but they are not helpful if we intend to *generate* random numbers.

13.3   Entropy

The notion of entropy is thoroughly dealt with in Huffaker *et al.* (2017). The term 'entropy' was created in the mid 19th century by the German physicist Rudolf Clausius in the context of classical thermodynamics. Originally the thermodynamic entropy was defined as $\delta S = \delta Q/T$. The infinitesimal increase of the entropy δS is equal to the infinitesimal element δQ of heat, for a system in a *reversible* infinitesimal process at absolute temperature T. The second law of thermodynamics can be expressed saying that in an irreversible process, entropy always increases with time, so the change in entropy $S_2 - S_1$ is positive. Since, entropy increases with time it is implied that there is a direction of time, a 'time's arrow'. What is measured is the difference $S_2 - S_1$ between states 1 and 2 of a body. Heat and temperature are physical quantities experimentally measured.

The idea that increasing entropy means nothing more than loss of information, was expressed by scientists like Born, Rosenfeld, Brillouin and others. Born (1949, p. 72), for instance, thought that 'Irreversibility is a consequence of the explicit introduction of ignorance into the fundamental laws'.

In statistical mechanics, Ludwig Boltzmann linked the entropy to the concept of probability, through the well-known formula $S = k \log W$, i.e. the entropy S is proportional to the logarithm of the probability W (*thermodynamische Wahrscheinlichkeit*), and the constant k will be called Boltzmann constant.

This equation is based on the assumption that all the microstates of the system are equally probable, so W represents the number of real microstates corresponding to the same macrostate. In Chapter 3, we posed the example of two dice, and quoted Boltzmann about the extraction from an urn of black and white balls, arguing that the most probable result is that corresponding to the greatest number of permutations. Now, we can rephrase that concept by saying that, according to the above formula, the system spontaneously reaches the state of equilibrium, where the entropy S has its maximum value, because all other states are extremely unlikely. For an extended discussion on the concept of probability in classical statistical physics, see von Plato (1994).

13.3.1   Shannon's entropy

In 1948 Claude Shannon (1948) proposed a measure of the amount of information contained in a message sent along a transmission line. He, following some advice from von Neumann called this measure 'entropy'. Surely, the entropy defined in statistical mechanics and that defined in information theory have a common source in the probability context, but it does not imply they have the same meaning. Perhaps, the choice of a common name has not been the best one. We will return to this issue later.

Let X be a discrete random variable with support S_X, i.e. the set of all possible values of X. In the context of information theory the support is called an 'alphabet'. If H has to be a measure of information, it must satisfy the following requirements, as Shannon himself pointed out:

1. If the probabilities p_i are all the same (uniform distribution), that is $p_i = 1/n$, then H is a monotonically increasing function of n.
2. H must be a function of the probability distribution $\{p_1, \ldots, p_n\}$, independently of how events are gathered within this distribution.

For the latter point, consider the following example. Let X be a random variable with distribution:

$$\mathsf{P}\{X = a\} = 0.5, \ \mathsf{P}\{X = b\} = 0.2, \ \mathsf{P}\{X = c\} = 0.3$$

We can also say that the events $\{b\}$ and $\{c\}$ are realized half the time. When it happens, the event $\{b\}$ occurs with probability 0.4 and the event $\{c\}$ with probability

0.6. Then, the above distribution can also be written:

$$\mathsf{P}\{X = a\} = 0.5, \ \mathsf{P}\{X = Y\} = 0.5, \ \mathsf{P}\{Y = b\} = 0.4, \ \mathsf{P}\{Y = c\} = 0.6$$

H must be the same in the two cases.

Only the function H written below satisfies these requirements:

$$H = -k \sum_{i=1}^{n} p_i \log p_i$$

The constant k depends on the selected unit of measurement, that is on the base of the logarithm. If $k = 1$ and if the base of the logarithm is 2, Shannon called the measure

$$H = - \sum_{i=1}^{n} p_i \log_2 p_i$$

the *entropy* of the set of probabilities $\{p_1, p_2, \ldots, p_n\}$. With the chosen basis 2, the unit of measurements is expressed in the term *bit*. It should be understood that the notation $H(X)$ does not mean that X is an argument of the function H but, rather, that it is a function of the probability distribution $\{p_1, p_2, \ldots, p_n\}$ of X. This allows us to distinguish, for instance, the entropy of the random variable X, $H(X)$, from the entropy of the random variable Y, $H(Y)$.

Important properties of the Shannon's entropy are:

1. H is equal to 0 if and only if X is known with certainty, that is all the probabilities p_i are zero, except one which is 1:

$$\mathsf{P}\{X = j\} = 1 \quad \text{and} \quad \mathsf{P}\{X = i\} = 0, \ \forall i \neq j$$

   This means that H takes the value 0 only when the result is certain, otherwise is always $H > 0$.

2. Given a certain dimension n, H is maximal when all the probabilities p_i are equal, $p_i = 1/n$, uniform distribution, situation with the greatest uncertainty.

The quantity $-\log_2 \mathsf{P}\{X = i\}$, or $-\log_2 p(x_i)$, or $-\log_2 p_i$ is called also *surprisal*. When p_i is small, we are 'surprised' if the result is that associated to x_i. Then, if p_i is small $-\log_2 p_i$ is big. If p_i is big, the surprise is small, as $-\log_2 p_i$ has a small value. All this is in agreement with the interpretation of H as a measure of the amount of uncertainty: the more uncertain we are of an outcome, the more surprised we are if this result occurs.

Are there two worlds? The world of atoms, in which gas of molecules expands and the magnitude of the physical entropy increases, and the world of 'emotions' in which a quantity with the same name measures our surprise, uncertainty, lack of knowledge. Might the entropy be the bridge, as someone said, between the two worlds? That is, of course, a philosophical question, quite outside the scope of this book.

Statistical mechanics and information theory have undoubtedly influenced one another. The crucial point is that randomness in physical phenomena is interpreted as a lack of knowledge about the phenomena themselves. In this view, entropy is considered as a measure of incompleteness of knowledge. Starting in 1957, E. T. Jaynes developed an alternative foundation for statistical mechanics. Jaynes's articles are collected in Rosenkrantz (1983). The focus of Jaynes's thinking, in the interweaving of physics, philosophy and statistics, is the *principle of maximum entropy* (also written as MaxEnt). It asserts that entropy is maximized by the uniform distribution when no constraint is imposed on the probability distribution. This principle is just (or better: an extension of) the principle of indifference: equal probabilities have to be assign to each occurrence if there is no reason to think otherwise. Among all the possible probability distribution consisting with the data at hand, we have to choice the distribution $p_i = (p_1, \ldots, p_n)$ that maximized the function:

$$H(p_1, \ldots, p_n) = -\sum_{i=1}^{n} p_i \log p_i$$

In this equation, Shannon's entropy is easily recognizable as a measure of the amount of uncertainty accounted for by the probability distribution $(p_1, \ldots, p_n)$. Such a distribution, once obtained, can be exploited in statistical mechanics to derive the thermodynamical properties of the system or can be introduced into the Bayes-Laplace formula in the role of an a priori probability. As a consequence, how to assign the a priori probabilities should be guaranteed on a sound basis, not by more or less arbitrary evaluations. Moreover, the *a priori* probabilities are updated when new information is received. So we understand the intimate connection between entropy and Bayes' theorem.

In his view of statistical mechanics, concerning irreversibility, Jaynes, as Born quoted before (Born, 1949), expressed the belief (Jaynes, 1957, p. 171): '[...] it is not the physical process that is irreversible, but rather our ability to follow it'.

Concerning entropy, Jaynes supports the idea that entropy has to be interpreted in a subjective sense, as a measure of 'our degree of ignorance as to the unknown microstate when the only information we have consists of the macroscopic thermodynamic parameters' (Jaynes, 1965, p. 396). Perhaps with a bit of provocation, he also wrote (Jaynes, 1965, p. 398) (author's italics):

[...] *Even at the purely phenomenological level, entropy is an anthropomorphic concept. For it is a property, not of the physical system, but of the particular experiment you or I choose to perform on it.*

As expected, Jaynes's conception aroused both consent and criticism. For a critical analysis of Jaynes's viewpoint see, among others, Denbigh and Denbigh (1985).

Going back to Shannon's entropy, the second property introduced before becomes:

2. Let a time series of a random process $x_1, x_2, \ldots, x_n$, and p_i the probability of each x_i. It is:

$$H \leqslant \log n$$

the equal sign holding when $p_i = 1/n, \forall i$, that is the x_i are uniformly distributed, the situation with the maximum uncertainty and, consequently, with maximum entropy H.

Consider an alphabet with two symbols: **H** (head) and **T** (tail). We describe the tosses of the coin with a random walk on a line. The binary entropy function is the function:

$$H(p) = -p \log_2 p - (1-p) \log_2 (1-p)$$

If $p = 0.5$ (symmetric random walk), it is $H(0.5) = 1$, while if $p = 0.75$, $H(0.7) = 0.8813$. Of course, if p is equal to 0 or 1, $H(p) = 0$. It is like saying that there is no randomness if the coin has two tails or two heads, and the most random outcomes are obtained with a fair coin. This example answers the question put at the beginning of this chapter, concerning the sequences **RandProc.1** and **RandProc.2**. They are not equally random and $H(p)$ may be taken as a measure of their randomness.

Some variants of H are reported in the following. The *joint entropy* of two discrete random variables X and Y is defined as:

$$H(X,Y) = \sum_i \sum_j p(x_i, y_j) \log_2 p(x_i, y_j)$$

The *conditional entropy* is defined as:

$$H(X|Y) = \sum_i \sum_j p(x_i, y_j) \log_2 p(x_i|y_j)$$

$H(X|Y)$ measures the mean uncertainty of X, if the Y value is known. It is also:

$$H(X,Y) = H(X) + H(Y|X)$$

As a consequence:

$$H(Y|X) = H(X,Y) - H(X$$

that is, conditioning reduces entropy.

Consider now the *mutual information*. Let X and Y be two discrete random variables. The mutual information between them $I(X;Y)$ is defined as:

$$I(X;Y) = \sum_i \sum_j p(x_i, y_j) \log_2 \frac{p(x_i, y_j)}{p(x_i)\, p(y_j)}$$

where $p(x_i, y_j)$ is the joint probability distribution function of X and Y, and $p(x_i)$ and $p(y_j)$ are the marginal probability distribution functions of X and Y, respectively. Note that $I(X;Y)$ is in *bits*.

Introducing the entropy $H(X)$, $I(X;Y)$ can be written:

$$\begin{aligned} I(X;Y) &= H(X) + H(Y) - H(X,Y) \\ &= H(X) - H(X|Y) \\ &= H(Y) - H(Y|X) \end{aligned}$$

The mutual information is a measure of how much the knowledge of X (Y) reduces uncertainty about Y (X). If the knowledge of Y reduces our uncertainty about X, then we say that Y carries information about X. This implies that $I(X;Y)$ is 0, if and only if X and Y are independent random variables, that is $p(x_i, y_j) = p(x_i)\,p(y_j)$, or if one of the two variables has zero entropy. This measure is symmetric, that is $I(X;Y) = I(Y;X)$, and it is always non-negative.

We compute now the entropy for the sequence **PRN** generated by the function **runif(.)**, with the code `Code_13_1.R`

```
## Code_13_1.R
# Shannon's entropy of pseudo random numbers sequence
cell<-numeric()
entr<-numeric()
probk<- numeric()
ncell<-100        # number of cells
n<-1000           # number if iterations
set.seed(1)
u<-runif(n,0,1)      # generate n uniform random number
cell[1:ncell]<-0
entr<-0
for(j in 1:n)        { # starting loop on the iterations
y<- u[j]
k<-trunc(ncell*y)+1
cell[k]<-cell[k]+1
                  } # ending loop on the iterations
for(k in 1:ncell){          # loop to compute entropy
prob<-cell[k]/n
probk[k]<- prob
if(prob>0){entr<- entr-prob*log2(prob)}
          }                 # ending loop to compute entropy
# cell      # uncomment to print cell
# hist(cell, xlim=c(0,20), freq=FALSE) uncomment to see the distribution of the cells
entr
entr_max<- log2(ncell) #  analytically computed entropy
entr_max
```

Very simply, the interval $[0,1]$ is divided into N (`ncells`) subintervals I_k:

$$I_k = \big[(k-1)/N, k/N\big], \quad k = 1, \ldots, N$$

and the number of iterates in each interval is computed. The probability p_k (`prob`) that an iterate falls in the I_k interval is estimated by the frequency of occurrence in this interval (`prob<-cell[k]/n`). H is maximal, `entr_max<- log2(ncell)`, if all the intervals are equally probable.

The result is $H = 6.5862$, and the maximum entropy (`entr_max`) $= 6.6439$. The small difference is due to the fact that the cells are not equally filled up. We have to remember that results from numerical computations are approximations of analytical ones, and such approximations may be very sensitive to the dimension of the generated sequences. In effect, by executing 5000 runs (`ncases`), every time with sequences 10000 iterations long (`n`) and `ncells` = 500, the mean of 5000 computed entropies results to be 8.9295, standard error 0.0023, closer to the maximum entropy 8.9658, with respect to the previous result.

The R package `Entropy` (Hausser and Strimmer, 2021) implements various estimators of entropy for discrete random variables. Moreover other functions, as Kullback-Leibler divergence, mutual information, and others, can be estimated. See Hausser and Strimmer (2019), for a detailed statistical comparison of the estimators available in the package.

```
## Code_13_2.R
# Shannon's entropy of pseudo random numbers sequence
estimate with the function `entropy'
#install.packages("entropy", dep=TRUE) # to install the package if not present
library("entropy")
### 1D example ####
# sample from continuous uniform distribution
set.seed(1)
n<- 1000
x = runif(n)
# hist(x, xlim=c(0,1), freq=FALSE) # uncomment to see the x distribution
# discretize into  (n/numBins) categories
numBins<-100
y = discretize(x, numBins, r=c(0,1))
#y  # # uncomment to print the intervals
# compute entropy from counts
entropy(y,unit="log2") # empirical estimate near theoretical maximum
log2(numBins) #   analytically computed entropy
```

Note 'unit= ' to choose the unit in which entropy is measured, and with `unit="log2"` it is in *bits*. The results are the same as the ones obtained by `PRN_entr` code, as it must be.

13.3.2 S_ρ Entropy

Another metric entropy measure is due to Granger, Maasoumi and Racine (Granger *et al.*, 2004) and references therein, often denoted as S_ρ. It is defined as:

$$ S_\rho(k) = \frac{1}{2} \int_{-\infty}^{+\infty} \int_{-\infty}^{+\infty} \left(\sqrt{f_{(X_t, X_{t+k})}(x_1, x_2)} - \sqrt{f_{X_t}(x_1) f_{X_{t+k}}(x_2)} \right)^2 dx_1 dx_2 $$

where $f_{X_t}(\cdot)$ and $f_{(X_t, X_{t+k})}(\cdot, \cdot)$ are the probability density function of X_t and of the vector (X_t, X_{t+k}), respectively. This quantity is a metric measure of the dependence between two series or between elements of the same series. In other words, it can been interpreted as a nonlinear autocorrelation function, that is, if S_ρ exceeds the confidence band at lag k, then there is a significant correlation between X_t and X_{t+k}, which are distant k steps in the sequence. Similarly, if two series X_t and Y_t are independent, $S_\rho = 0$. This entropy S_ρ is a normalized version of the Bhattacharya–Hellinger–Matusita distance (Maasoumi and Racine, 2002) and meets the following six properties:

- It is defined for both continuous and discrete variables. In the case of binary series the measure becomes:

$$ S_\rho(k) = \frac{1}{2} \sum_{i=0}^{1} \sum_{j=0}^{1} \left(\sqrt{\mathsf{P}\{X_t = i, X_{t+k} = j\}} - \sqrt{\mathsf{P}\{X_t = i\}\, \mathsf{P}\{X_{t+k} = j\}} \right)^2 $$

- It is normalized and varies between 0 and 1. It takes the value 0 if X_t and X_{t+k} are independent.
- It takes the value 1 if there is a measurable exact (non-linear) relationship, say $Y = m(X)$, between the variables.
- It reduces to a function of the linear correlation coefficient in the case of a bivariate normal distribution.

- It is a metric measure, not only a divergence measure; indeed, it obeys the triangular inequality and is a commutative operator.
- It is invariant with respect to continuous, strictly increasing transformations $\alpha(.)$. This is useful since X and Y are independent if and only if $\alpha(X)$ and $\alpha(Y)$ are independent. Invariance is important since otherwise clever or inadvertent transformations would produce different levels of dependence.

The function `Srho.test` in the R package **tseriesEntropy** (Giannerini, 2017) implements the entropy measure S_ρ of serial and cross dependence for integer or categorical data. A simple example is to compute S_ρ of the **RandProc.1**. The code is:

```
## Code_13_3.R
## S entropy
#install.packages("tseriesEntropy", dep = TRUE)
# to install the package if not present
library(tseriesEntropy)
x <- rbinom(n=1000,size=1,prob=0.5)
x<- as.integer(2*x-1)
S<- Srho(x,lag.max=6)
```

and the output is:

```
Srho computed on 6 lags
----------------------------------------------------------------------
         1            2            3            4            5            6
0.0000215000 0.0003135686 0.0002334794 0.0001630541 0.0001552996 0.0006956368
----------------------------------------------------------------------
Data type          : integer-categorical
Stationary version : TRUE
```

very close to 0, X_t and X_{t+k} being independent.

In Huffaker *et al.* (2017), we used the package **tseriesEntropy** to construct tests for non-linear serial dependence for continuous and categorical time series.

13.3.3 Approximated entropy

With the *approximated entropy* (*ApEn*) we look at the patterns of a sequence. We can say that a sequence is random if there is no repetition of patterns, while it is somewhat predictable if there are present patterns repeating themselves throughout the series. High *ApEn* levels indicate randomness and unpredictability, low levels mean the existence of repeated patterns. The *ApEn* was proposed by Steve Pincus in the final decade of the last century, originally in the field of medicine, in particular in the framework of heart rate time series (Pincus, 1991; Pincus *et al.*, 1991; Pincus and Singer, 1996; Pincus and Kalman, 1997), but it has been expanded to further different fields from psychology to finance. The *ApEn* is characterized by the authors (Pincus and Singer, 1996, p. 2083):

ApEn$(m, r, N)(u)$ measures the logarithmic frequency with which blocks of length m that are close together remain close together for blocks augmented by one position. Thus, small values of *ApEn* imply strong regularity, or persistence, in a sequence u. Alternatively, large values of *ApEn* imply substantial fluctuation, or irregularity, in u.

The symbol m, r, N, u will be defined below. It is remarkable that the *ApEn* can potentially distinguish order in series generated by both stochastic and deterministic processes, that is time series deriving from mathematical algorithms, from random processes and from periodic and chaotic systems.

The *ApEn* is defined as follows (Pincus and Singer, 1996).

Given quantities

1. N: length of the sequence
2. m $(m < N)$: length of the compared patterns
3. r: margin of tolerance
4. sequence of real numbers $u = (u(1), u(2), ...u(N))$

Defined quantities

1. distance $d(x_i, x_j)$ between two blocks x_i and x_j:

$$d(x_i, x_j) = \max_{k=1,2,...,m} \left| u(i+k-1) - u(j+k-1) \right|$$

where:

$$x(i) = (u(i), u(i+1) \ldots, u(i+m-1)) \qquad \text{and}$$
$$x(j) = (u(j), u(j+1) \ldots, u(j+m-1))$$

Two blocks are 'close together' if $d(x_i, x_j) \leqslant r$. Note $d(x_i, x_j)$ is the maximum distance of the scalar components.

2. $C_i^m(r) = \dfrac{\text{number of } j < N - m + 1 \text{ such that } d(x_i, x_j) \leqslant r}{N - m + 1}$

The numerator of $C_i^m(r)$ counts, within resolution r, the number of blocks of length m that are approximately the same as a given block acting as template. Appropriately $C_i^m(r)$ is also called *correlation sum* because it acts like a measure of the summed correlation of vector (block) x_i with all other vectors.

3. $\Phi^m(r) = \dfrac{1}{N - m + 1} \sum_i^{N-m+1} \log C_i^m(r)$

Then the *ApEn* with parameters m and r for a sequence of N elements is given by:

$$ApEn(m, r, N)(u) = \Phi^m(r) - \Phi^{m+1}(r), m \geq 1$$
$$ApEn(0, r, N)(u) = -\Phi^1(r)$$

We have to keep in mind that the choice of m and r are quite critical. Different choices might potentially lead to opposite results. The authors in their works suggest m about 2-3, and r between 0.1 and 0.25 standard deviations of the sequence. If the data are affected by noise, r has to be larger than most of the noise.

To see how the *ApEn* works, let us limit ourselves to binary sequences of 0's and 1's. In this case, the resolution r must be $r < 1$. Therefore in binary sequences, to set $r < 1$ is equivalent to considering 'close together' two elements if they are equal. To set $r \geqslant 1$ would be meaningless, since the inequality $d(x_i, x_j) \leqslant r$ would always be fulfilled. So we control if any matches exist in the blocks x_i and x_j, that is whether:

$$\left| u(i+k-1) - u(j+k-1) \right| = 0 \text{ or } 1$$

It can be simply written $ApEn(m, N)(u)$, omitting r.

With $r = 0$, and for simplicity $m = 1$, $C_i^m(r)$ and $\Phi^m(r)$ become:

$$\Phi^1 = \frac{1}{N} \sum_{i=1}^{N} \log C_i^1$$

$$C_i^1 = \frac{\text{number of } j \leqslant N \text{ such that } d(x_i, x_j) = 0}{N}$$

Let k be the number of 1's in the sequences, if $x_i = 1$, the number of matches will be k, that is $d(x_i, x_j) = 0$ will be fulfilled k times, so that $C_i^1 = k/N$. If $x_i = 0$, the number of matches will be $N - k$, therefore Φ^1 is given by:

$$\Phi^1 = \frac{1}{N} \left[k \log \frac{k}{N} + (N-k) \log \frac{(N-k)}{N} \right]$$

then:

$$\Phi^1 = \frac{k}{N} \log \frac{k}{N} + \frac{N-k}{N} \log \frac{N-k}{N}$$

and by putting $k/N = p$ and $(N-k)/N = q$, we have:

$$ApEn(0, N) = -(p \log p + q \log q)$$

which is just the Shannon's entropy.

From infinite sequence $u = (u(1), u(2), ...)$ and $r < 1$, it is selected the sequence $u^{(N)} = (u(1), u(2), ..., u(N))$ of finite length N, and the $ApEn(m)(u)$ for the sequence $u(N)$, with $N \to \infty$ is defined as (Pincus, 2008):

$$ApEn(m)(u) = \lim_{N \to \infty} ApEn(m, N)(u^N)$$

For an infinite binary sequence u, the concept of *computationally random* sequence (C-random) is introduced. A sequence is C-random if $ApEn(m)(u)$ is equal to $\log 2$, $\forall m \geq 0$. Recall that Shannon's H is maximal for a uniform distribution. The extension to a sequence with an alphabet of k symbols is immediate. In this case, a sequence is C-random if $ApEn(m)(u)$ is equal to $\log k$, $\forall m \geq 0$.

The `Code_13_4.R` below computes the $ApEn$ in R. We compare two short ($N = 20$) binary sequences, to show how the $ApEn$ can identify randomness even in relatively short strings. The examined sequences were proposed by Chaitin (1975) and utilized as classic example of application of the $ApEn$ by other authors, for instance (Pincus and Kalman, 1997). The sequences are:

a) 0 1 0 1 0 1 0 1 0 1 0 1 0 1 0 1 0 1 0 1
b) 0 1 1 0 1 1 0 0 1 1 0 1 1 1 1 0 0 0 1 0

The two sequences are equiprobable, as they are all the 2^{-20} sequences.

```
## Code_13_4.R
# Approximated entropy
# binary sequence
m<- 1
r<- 0.1
N<- 20
u<- rep(c(0,1), 10)
nsum<- N-m
id<-rep(0,nsum)
ic<-rep(0,nsum)
i <-0
while(i < nsum){
i <- i+1
j <- 0
while(j < nsum){
j <- j+1
k <- 1
while(k <= m){
diff <- abs(u[i + k - 1] - u[j + k - 1])
if(diff > r) break
k <- k+1
        }
if(k > m){
ic[i] <- ic[i] + 1
diff <- abs(u[i + m] - u[j + m])
if(diff <= r) id[i] <- id[i] + 1
}
}
}
rate <- sum(log(id/ic))    # base = exp(1)
ApEn <- -1* rate/nsum
print(c(N,m,r,ApEn))
```

Sequence a) is clearly non random, indeed it results $ApEn(1,20) = 0$.

```
print(c(N,m,r,ApEn))
[1] 20.0  1.0  0.1  0.0
```

Note the parameters $m = 1$ and $r < 1$. As Pincus and Kalman (1997) point out, the result $ApEn(1,20) = 0$ means that there are no blocks $\{0,0\}$ or $\{1,1\}$ in a). On the contrary, the sequence b) shows as random since $ApEn(1,20) = 0.677453$, very close to the maximum value $\log 2 = 0.693147$.

Let us slightly 'perturb' the string a) by substituting a 0 with a 1. The result is now $ApEn(1,20) = 0.263369$ and with two substitutions $ApEn(1,20) = 0.379489$. These illustrative examples show how the $ApEn$ is able to detect irregularities in series and also to characterize a sequence on the basis of its irregularity level, from full order ($ApEn = 0$) up to the complete disorder when the $ApEn$ reaches its maximum value.

Consider now the periodic sequence of length $N = 1000$:

$$0\ 0\ 1\ 1\ 0\ 0\ 1\ 1...0\ 0\ 1\ 1$$

that is 250 blocks 0 0 1 1. In the code `Code_13_4.R`, the input string is now given by `u<- rep(c(0,0,1,1),250)` and with $m = 1$ the result is $ApEn(1,1000) = 0.693146$, that is the sequence is random. On the other hand, with $m = 2$, $ApEn(2,1000) = 0$, that is the periodicity is identified. As we have already said, different choices of m can potentially lead to opposite results.

An applicative, comparative and exhaustive study between several entropy-based indicators of independence and correlation, applied to series generated by complex deterministic processes, stochastic and chaotic processes, is in (Bertozzi, 2011).

13.4   A final note

Chapter 2 began with a section entitled *The Philosopher and the Gambler*. It is undeniable that everything revolving around the concepts of randomness, stochasticity, probability closely interlinks with 'chance' (in particular, games of chance) and philosophy. So, at the end of this book we 'close the circle' with a few words about different points of view related to what 'random' actually means.

Chaitin (1975) writes:

Although randomness can be precisely defined and can even be measured, a given number cannot be proved to be random.

To some that might seem obvious, but it is certainly not trivial. We have seen indeed that the notion of (mathematical) randomness makes sense if referred to a well-defined sequence, which could be a *Kollektive*, mathematically constructed sequences, sequences having maximum complexity, and so on.

Poincaré's conception of randomness, has a central role in the topic of chaos (Huffaker *et al.*, 2017) and his ideas have been rediscovered and enhanced much later. Poincaré does not believe in subjective randomness, i.e. due to lack of knowledge; rather he thinks that randomness derives from the intrinsic nature of certain phenomena. He analyses several sources of randomness, and among them he finds what we now call 'sensitivity to initial conditions' (Poincaré, 1908, Chapter 4):

A very small cause, which escapes us, determines a considerable effect, which we cannot ignore, and we then say that this effect is due to chance.

On the opposite philosophical side, there is the so-called 'subjective' definition of randomness, whose roots are based on the thinking of philosophers like David Hume − see, for example, (Hume, 2007) − who, in his well-known *Treatise of the Human Nature* writes:

On the other hand, as chance is nothing real in itself, and, properly speaking, is merely the negation of a cause, its influence on the mind is contrary to that of causation; and 'tis essential to it, to leave the imagination perfectly indifferent, either to consider the existence or non-existence of that object, which is regarded as contingent. A cause traces the way to our thought, and in a manner forces us to survey such certain objects, in such certain relations. Chance can only destroy this determination of the thought, and leave the mind in its native situation of indifference.

In other words, chance represents insufficient knowledge or indifference. By using a tautology, random is what the human mind 'feels' as random or, more precisely, what the lack of knowledge forces us to treat by means of a probabilistic approach.

In this sense, the work of Lorenz (1963) and Lorenz (1995) on dynamical systems that are highly sensitive to initial conditions, was indeed a deterministic description of an apparently random process. The deterministic process could be unveiled from the apparent randomness by using the appropriate conceptual and mathematical tools called deterministic chaos theory (Huffaker *et al.*, 2017).

The two different viewpoints on randomness essentially reflect themselves in the dispute between 'objective' and 'subjective' probability. One of the re-founding fathers of Bayesian probability, Jeffreys (1939), writes:

I should query whether any meaning can be attached to 'objective' without a previous analysis of the process of *finding out* what is objective If it is done from experience it must begin with sensations, which are peculiar to the individual [...] We must and do begin with the individual, and we never get rid of him, because every new 'objective' statement must be made by some individual and appreciated by other individuals.

Both viewpoints have strengths and weaknesses. We are not interested, here, in embracing one side or the other. Whatever our (and your) opinion is about the essence of randomness, when the phenomenon we are studying appears as a stochastic process — that is it does not fit into a fully deterministic framework — or we do not have access to the system parameters to solve it with deterministic tools, we need to treat it with probabilistic tools.

Appendix A
Bootstrap

The name bootstrap was introduced by Efron (1979) and described more fully in Efron and Tibshirani (1993). This name comes from the expression 'to pull oneself up by one's bootstrap', as the Baron Münchhausen did in the book written in 1785 by Rudolf Erich Raspe. It is a name that evokes something apparently 'miraculous', indeed in statistics a 'bootstrap estimate' is an estimate with no a priori assumptions, but that can only be obtained by using the data at hand.

The idea underlying the bootstrap method is as simple as it is powerful. Suppose we have a random sample of independent observations:

$$\mathbf{x} = (x_1, x_2, \ldots, x_n)$$

which are distributed according to some unknown probability distribution F. Let $\theta = t(F)$ be the unknown population parameter of interest, such as population mean. The letter 't' represents the procedure to be applied to F to obtain θ, t may be simple formulae or computer-based methods. Since F is unknown, we must exploit what we have to hand, the sample $\mathbf{x}$. Let $\hat{\theta}(\mathbf{x})$ be an estimate of θ based on the observed sample $\mathbf{x}$. We wish also to estimate how good is this estimation, for instance through the standard error $\sigma(\hat{\theta})$ of $\hat{\theta}(\mathbf{x})$. The bootstrap method gives a non-parametric solution for estimating a sampling distribution and assessing the statistical accuracy of an estimation. This method involves resampling the original data $\mathbf{x}$, and performing inference from the resampled data.

The bootstrap method prescribes what follows.

1. F is estimated by $\widehat{F}$, the *empirical distribution function*, constructed by putting probability mass $1/n$ on each x_i.

2. A *bootstrap sample* $\mathbf{x}^*$:

$$\mathbf{x}^* = (x_1^*, x_2^*, \ldots, x_n^*)$$

   is generated from $\widehat{F}$ by making independent random draws with replacement from the data.

3. A bootstrap replication $\hat{\theta}^*$:

$$\hat{\theta}^* = s(\mathbf{x}^*) = s(x_1^*, x_2^*, \ldots, x_n^*)$$

   is computed, s being the statistic of interest.

Let us remember some terminology. $\hat{\theta}$, named (sample) *statistic*, is a function that associates each n-tuple $(x_1, x_2, \ldots, x_n)$ to a real number. If the statistic $\hat{\theta}$ is used to estimate a parameter $\hat{\theta}$ it is called *estimator* or *summary statistics*. The value assumed by the estimator is named *estimate*. Now we put t in the place of s and $\widehat{F}$ in the place of F

$$\hat{\theta} = t(\widehat{F})$$

We estimate the function $\theta = t(F)$ of the probability distribution F with the same function of the empirical distribution $\widehat{F}$. This is the plug-in principle. Efron (1989) writes:

The bootstrap [...] is essentially the oldest idea in the statistical book: substitute the empirical distribution of the data for the (unknown) true distribution in anything you wish to estimate.

What does such principle say? It says: have you to estimate the population mean? Use the sample mean. Do you need the population variance? Compute the sample variance, and so on. The idea is old, but the instruments used to implement it really are not. As opposed to the evangelical parable discernment 'And no one pours new wine into old wineskins', now old wine is just put into new wineskins. The old wine is still the conceptual and technical statistical methodology developed in previous centuries, the new wineskins are the computational techniques able to handle enormous amount of data with great speed and little expense.

Replacing F with $\widehat{F}$ is correct thanks the law of large numbers. This law says: let $(X_1, X_2, \ldots, X_n)$ be n *i.i.d.* random variables. If, and only if, $\mathrm{E}\,[X_n] = \mu$, $\forall n$, it is:

$$\frac{1}{n} \sum_{i=1}^{n} X_i \xrightarrow[n \to \infty]{a.s.} \mu$$

That is the convergence is 'almost sure'.[1] In our case the $\{X_n\}$ are the $\widehat{F}(x)$ (they depend on n even though not explicitly noted) and μ is $F(x)$, therefore, from a certain $x \in \mathbb{R}$:

$$\widehat{F}(x) \xrightarrow[n \to \infty]{a.s.} F(x) \Leftrightarrow \widehat{F}(x) - F(x) \xrightarrow[n \to \infty]{a.s.} 0$$

Moreover, the Glivenko-Cantelli theorem ensures that such a convergence is not only punctual, but also uniform, which is a more strong property.

Formally, let $(X_1, X_2, \ldots, X_n)$ be n *i.i.d.* random variables, with cumulative distribution F, then:

$$\sup_{x \in \mathbb{R}} |\widehat{F}(x) - F(x)| \xrightarrow[n \to \infty]{a.s.} 0 \tag{A.1}$$

To compute the empirical distribution function $\widehat{F}$, in principle *all* the possible bootstrap replications are necessary, i.e. we would need all the bootstrap samples. The

---

[1] We encountered several types of convergence in Chapter 3. A succession $\{X_n\}$ *almost surely* converges to X if the probability $\mathrm{P}\,\{\lim_{n \to \infty} X_n = X\}$ is one.

number of bootstrap samples of length k that can be extracted from n observations is given by the number of combinations with repetition

$$C_{n,k}^R = \binom{n+k-1}{k}$$

In our case $k = n$, then, putting $C_{n,n}^R = m$, the number of distinct bootstrap samples is

$$m = \binom{2n-1}{n} = \frac{(2n-1)!}{n!(n-1)!}$$

If $n = 9$, a small enough value, m is 24310. But with $n = 20$, it is $m = 68923264410$. Therefore the computation of $se_{\widehat{F}}(\hat{\theta}^*)$ becomes impracticable, so it is rightly called an 'ideal bootstrap estimate'. The problem is solved by computing only a number B of bootstrap samples, obtaining the so-called '*Monte Carlo* bootstrap estimate'.

A.1   Bootstrap standard error

Recall some formulae. Let X be a random variable with probability distribution function F. Let $\mu_F = \mathrm{E}_F[X]$ be the expected value of X and σ_F^2 the variance:

$$\mathrm{Var}_F[X] = \sigma_F^2 = \mathrm{E}_F\big[(X - \mu_F)^2\big] \tag{A.2}$$

The *standard deviation* σ_F of X is defined as the square root of the variance:

$$\sigma_F = \sqrt{\mathrm{Var}_F[X]}$$

Define now the *standard error*. By 'standard error' $se(\cdot)$ is meant the standard deviation of a summary statistic (or estimator), and $se(\cdot)$ is used to give a measure of the statistical accuracy of an estimate.

Consider the random variable *sample mean* $\overline{X}$. The standard error $se_F(\overline{X})$ is given by:

$$se_F(\overline{X}) = \sqrt{\mathrm{Var}_F[\overline{X}]} = \sigma_F/\sqrt{n}$$

The central limit theorem (under quite general conditions on the F) states that for large sample sizes the sample mean $\overline{X}$ is approximately distributed as a normal with mean μ_F and variance σ_F^2/n:

$$\overline{X} \approx \mathcal{N}(\mu_F, \sigma_F^2/n)$$

Note that 'approximately' refers to normality, not to μ_F and σ_F^2/n, which are exact. From the properties of the normal, it follows:

$$\mathsf{P}\left\{|\bar{x} - \mu_F| < \sigma_F/\sqrt{n}\right\} = 0.683 \quad \text{and} \quad \mathsf{P}\left\{|\bar{x} - \mu_F| < 2\sigma_F/\sqrt{n}\right\} = 0.954$$

Let $\bar{x}$ be a realization of $\overline{X}$, that is the mean of the sample $\mathbf{x} = (x_1, x_2, \ldots, x_n)$. In other words, we can expect $\bar{x}$ to be within one standard error of the mean μ_F about 68% of the time and within two standard errors about 95% of the time.

In the case of the sample mean, the estimate $\hat{\sigma}_F^2$ of σ_F^2 is given by:

$$\hat{\sigma}_F^2 = \frac{1}{n-1} \sum_{i=1}^{n} (x_i - \bar{x})^2 \tag{A.3}$$

then, the estimate $\widehat{se}(\bar{x})$ of $se(\overline{X})$ is given by:

$$\widehat{se}(\bar{x}) = \left[\frac{1}{n(n-1)} \sum_{i=1}^{n} (x_i - \bar{x})^2 \right]^{1/2} \tag{A.4}$$

By applying the plug-in principle we replace $\mu_F = \mathrm{E}_F[X]$ with $\mu_{\hat{F}} = \mathrm{E}_{\hat{F}}[X]$, and we write $\mu_{\hat{F}} = \bar{x}$. Then we have:

$$\sigma_{\hat{F}}^2 = \mathrm{E}_{\hat{F}}\left[(X - \mu_{\hat{F}})^2\right] = \frac{1}{n} \sum_{i=1}^{n} (x_i - \bar{x})^2 \tag{A.5}$$

a bit different from eqn (A.3) for the denominator n in the place of $n-1$. Note that $\sigma_{\hat{F}}^2$ is a *distorted* estimator with respect to σ_F^2, while eqn (A.3) is a *correct* estimator. An estimator is called 'correct' if its mean is equal to the what is to be estimated. The random variable sample mean $\overline{X}$ has the same mean μ of the random variable X of the population, then it is a correct estimator. From eqn (A.5) we estimate the standard error of the variable $\overline{X}$, that is $se_F(\overline{X}) = \sigma_F/\sqrt{n}$, by putting in the place of σ_F its plug-in estimate $\sigma_{\hat{F}}$. Then the estimated standard error applying the plug-in principle is:

$$\widehat{se}(\bar{x}) = \sigma_{\hat{F}}/\sqrt{n} = \left[\frac{1}{n^2} \sum_{i=1}^{n} (x_i - \bar{x})^2 \right]^{1/2} \tag{A.6}$$

The difference between eqn (A.4) and the above eqn (A.6) is n^2 in the place of $n(n-1)$. Obviously, if n is not too small, the two estimates are the same. It might be noticed that the plug-in principle has been applied twice. The first was when μ_F was estimated with $\mu_{\hat{F}} = \bar{x}$, the second when $se_F(\overline{X})$ was estimated with $se_{\hat{F}}(\overline{X})$.

We estimate now the standard error using the bootstrap method. The prescriptions introduced before proceed as follows:

4. B bootstrap replications are computed, each corresponding to a bootstrap sample x_b^*, $b = 1, \ldots, B$:

$$\hat{\theta}_1^* = s(\mathbf{x}_1^*), \quad \hat{\theta}_2^* = s(\mathbf{x}_2^*), \ldots, \hat{\theta}_B^* = s(\mathbf{x}_B^*)$$

5. $se_{\hat{F}}(\hat{\theta}^*)$ is estimated by the estimator standard deviation of the B replications $(\hat{\theta}_1^*, \hat{\theta}_2^*, \ldots, \hat{\theta}_B^*)$:

$$se_B(\hat{\theta}^*) = \left[\sum_{b=1}^{B} \frac{\left[\hat{\theta}_b^* - \hat{\theta}^*(\cdot)\right]^2}{B-1} \right]^{1/2} \tag{A.7}$$

where:

$$\hat{\theta}^*(\cdot) = \frac{1}{B} \sum_{b=1}^{B} \hat{\theta}_b^*$$

We can look at the observed samples as a population, therefore it does not vary. What varies are the replications $\hat{\theta}_b^*$. They are *i.i.d.* random variables, then, for the strong law of large numbers, their sample mean converges almost certainly to the expected value of the population, that is to the mean of all the replications. To sum up: the standard error $\hat{\theta}$ of the statistics $se_F(\hat{\theta})$ is estimated by the standard deviation of *all* the replications $se_{\widehat{F}}(\hat{\theta}^*)$ that, in turn, is estimated by the standard deviation of B replications $se_B(\hat{\theta}^*)$.

There are two main packages for bootstrapping in R: `bootstrap` and `boot`. The package `bootstrap` is considered as help for the book Efron and Tibshirani (1993), Here we report the code `Code_1_Appendix_A.R` to show how the bootstrap works in a very simple case.

```
## Code_1_Appendix_A.R
# bootstrap standard error of the mean
set.seed(123)
thb<-numeric()   # vector of bootstrap replications
B<-200           # number of replications
x<- c(80,82,78,81,80,79,84)  # observed data
nx<-length(x)
x_bar<-mean(x)
x_bar
se_mean<- sd(x)/sqrt(length(x)) # correct standard error
# se_mean<-sum((x-x_bar)^2/(nx*(nx-1)))^(1/2) # explicit formula
# the plug-in standard error by the function:
sig2.pl<-function(x){var(x)*(longth(x)-1)/length(x)}
# distorted estimator to compute se plug-in
se_pl<-sqrt(sig2.pl(x)/length(x)) # se plug-in
for(b in 1:B)   {     # starting loop on replications
# sample(...) extracts a sample of size length(x)
#          from the elements of x, with replacement
xb<-sample(x,length(x),replace=TRUE)
thb[b]<- mean(xb)     # b-th bootstrap replication
                }     # ending loop on replications
mthb<- mean(thb)      # bootstrap mean of replications
mthb
seb<-sqrt(var(thb))   # bootstrap standard error
#seb<-sum((thb-mthb)^2/(B-1))^(1/2) # explicit formula
se_mean      # se correct
se_pl        # se plug-in
seb          # se bootstrap
quant<-quantile(thb, probs = c(0.025, 0.1587, 0.5, 0.8413, 0.975))
quant
half.68<- (quant[4]-quant[2])/2
half.68
shapiro.test(thb)
hist(thb,lty=1,main="",xlim=c(78,83),font.lab=3,freq=T,
cex.lab=1.2,xlab="replications",ylab="frequency")
abline(v=quant[2],lty=2,col="black",lwd=2)
abline(v=quant[4],lty=2,col="black",lwd=2)
abline(v=quant[3],lty=3,col="black",lwd=3)
```

Suppose the observed data are measures of one side of a rectangular piece of land, performed by different people. Figure A.1 shows the replications distribution `thb`, that is $\hat{\theta}_b^*, 1 \leqslant b \leqslant B$. The vertical lines mark the quantiles corresponding to 15.87%

and 84.13% quantiles (dashed lines), and 50.0% quantile (dotted line). The quantiles are:

```
    2.5%    15.87%        50%    84.13%      97.5%
79.14286 79.85714 80.42857 81.28571 81.85714
```

Note that the bootstrap mean of replications, `mthb` $= 80.505$, is very close to the mean of the observed data `x_bar` $= 80.571$. Half of the width of the interval $[16\%, 84\%]$ equal to 0.7143 (`half.68`) may be considered as the 68% bootstrap confidence interval, in this case close to the standard error `se_mean` $= 0.7514$, eqn (A.4). The Shapiro-Wilk normality test gives a p-value equal to 0.2765, therefore the normality of the distribution cannot be rejected, being greater than 0.05.

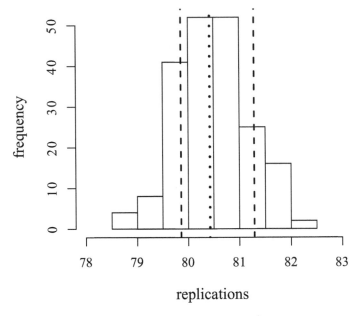

Fig. A.1 Distribution of $B = 200$ bootstrap replications $\hat{\theta}_b^*, 1 \leqslant b \leqslant B$. The dashed lines define the 15.87% and 84.13% quantiles, while the dotted line marks the 50% quantile.

A.2   Parametric bootstrap

Suppose we have only one measurement of the longer side of that piece of land, for instance $x = 81$ m. Suppose also one measurement has been done of the shorter side, $y = 58$ m. We compute the area $A = (81 \times 58) = 4698$ m^2. The example is somewhat artificial, as it is supposed that the two opposite sides are equal. The problem is to estimate the uncertainty on A. Each of the two measurements has its uncertainty Δx and Δy which is supposed to be $\Delta x = \Delta y = 2$ m. Uncertainty in measurement may be instrument sensitivity, small change in observations, or other quantities, as the standard deviation, determined once for all through appropriate experimental operations.

We can now apply the *error propagation law*. The measurement of n quantities has as a result $x_i \pm \Delta x_i, i = 1, \ldots, n$, where Δx_i are the uncertainties, assumed as known. From the measured quantities a new quantity has to be calculated. The procedure may be a simple analytical formula, as in this example $A = f(x, y)$, but also it may be a complex computer code. In this case, the measured quantities $x_i, i = 1, \ldots, n$, are the input of the code and the values of the derived quantities $q_j, j = 1, \ldots, m$ are the output. It may be $m = 1$, but in general it is $m \neq n$. The problem is that of the evaluation of the errors Δq_j of the derived quantities in terms of the combined effect of the errors of the direct measurements.

In our example the error propagation law is written:

$$A + \Delta A = (x + \Delta x)(y + \Delta y)$$
$$= xy + x\Delta y + y\Delta x + \Delta x \Delta y$$

Usually $\Delta x \ll x$ and $\Delta y \ll y$, therefore the last term can be neglected. Now $A = xy$, then:

$$\Delta A = x\Delta y + y\Delta x$$

and as relative errors it is written:

$$\frac{\Delta A}{A} = \frac{\Delta x}{x} + \frac{\Delta y}{y}$$

by replacing the standard deviations with the measurement uncertainties, the result is:

$$\frac{\Delta A}{A} = \sqrt{\left(\frac{\Delta x}{x}\right)^2 + \left(\frac{\Delta y}{y}\right)^2}$$

With the measured data, it is:

$$\frac{\Delta A}{4698} = \sqrt{\left(\frac{2}{81}\right)^2 + \left(\frac{2}{58}\right)^2} = 0.04241136$$

so that:

$$\Delta A = 0.04241136 \times 4698 = 199.2486$$

and the final result is written as $A = (4700 \pm 200)\, \text{cm}^2$, the uncertainty is rounded to three significant figures.

Let us approach the problem with 'the brute force application of the bootstrap' (Efron, 1979, p. 17). The parametric bootstrap differs from the 'non-parametric' version outlined above in the steps 1 and 2 in the bootstrap algorithm. At the step 1, instead of estimating F by the empirical distribution $\widehat{F}$, F is estimated by the distribution $\widehat{F}_{par}$ derived from a parametric model for the data. So, at the step 2, instead of sampling with replacement from the data, we generate B samples from $\widehat{F}_{par}$ After that, we proceed exactly as in steps 3, 4, and 5. The $\widehat{F}_{par}$ is the normal distribution with mean x (y) and standard deviation Δx (Δy). It is assumed that the data come from a normal distribution.

The code `Code_2_Appendix_A.R` below reports the estimate of the uncertainty of the area A in the above example with the parametric bootstrap.

```
## Code_2_Appendix_A.R
# parametric bootstrap

## Now I know only that the available data are:
x<-81
Dx<- 2
y<-58
Dy<-2
A<- 4698        # measured area 81*58
Ab<- numeric()
set.seed(1)
B<-400
for(b in 1:B)    {    # starting loop on replications
# generate one value of x and y under the normality hypothesis
xb<- rnorm(1,x,Dx)
yb<- rnorm(1,y,Dy)
Ab[b]<- xb*yb          # b-th bootstrap replication
                 }     # ending loop on replications
mA_b<- mean(Ab)        # bootstrap mean of replications
mA_b
DA_b<-sqrt(var(Ab))   # bootstrap standard error
DA_b
quant<-quantile(Ab, probs = c(0.025, 0.1587, 0.5, 0.8413, 0.975))
quant
half.68<- (quant[4]-quant[2])/2
half.68
hist(Ab,lty=1,main="",xlim=c(4000,5800),font.lab=3,freq=T,
cex.lab=1.2,xlab="replications",ylab="frequency",ylim=c(0,80))
abline(v=quant[2],lty=2,col="black",lwd=2)
abline(v=quant[4],lty=2,col="black",lwd=2)
abline(v=quant[3],lty=3,col="black",lwd=3)
```

Figure A.2 shows the replications distribution Ab. The vertical lines mark the quantiles corresponding to 15.87% and 84.13% quantiles (dashed lines), and 50.0% quantile (dotted line). The quantiles are:

```
   2.5%    15.87%       50%    84.13%      97.5%
4291.436 4491.453 4689.361 4902.506 5103.030
```

Note that, as expected, the bootstrap mean of replications mA_b $= 4696.941$ and the bootstrap accuracy DA_b $= 203.5017$ are almost equal to the corresponding quantities, $A = 4698$ and $\Delta A = 200$, derived analytically by the measured data. Half of the width of the interval $[16\%, 84\%]$ equal to 205.5264 (half.68) may be considered as the 68% bootstrap confidence interval, close to the DA_b. We conclude with a warning. If the observations can be no longer considered the realization of mutually independent random variables, the bootstrap is not applicable in the form outlined above, because the dependence structure of the data is disregarded. So the (non-parametric or parametric) bootstrap for independent and identically distributed variables has to be replaced by techniques based on *block resampling*, as discussed in Chapter 11.

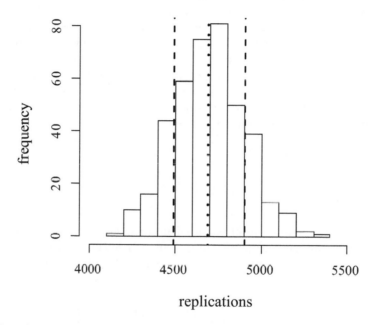

Fig. A.2 Distribution of $B = 400$ generated replications Ab. The dashed lines define the 15.87% and 84.13% quantiles, while the dotted line marks the 50% quantile.

Appendix B
JAGS

Just Another Gibbs Sampler, or JAGS, is a platform-independent program for simulation/analysis of Bayesian model using Markov Chain Monte Carlo (MCMC). The language, developed and implemented by Martyn Plummer (Plummer, 2003) has been developed to be compatible with BUGS (Bayesian inference Using Gibbs Sampling), a language for Bayesian inference developed at the end of the 1980 (Lunn *et al.*, 2012).

We have seen in Chapter 9 that Bayesian inference is based on determining the probability distribution of the unknown parameters involved in the problem. JAGS computes these probability distributions (and the summarizing parameters of interest such as their moments) by computer simulation based on (Gibbs) sampling and Monte Carlo integration. Suppose a random variable X has a probability distribution $p(x)$ and that we are able to generate a large number of values x_1, x_2, ..., x_N from that distribution. From the (strong) law of large numbers we know that we can approximate the expected value of X by:

$$E[X] \approx \frac{1}{N} \sum_{k=1}^{N} x_k$$

Any summary of $p(x)$ can be approximated in a similar way, i.e. by computing summaries of sampled values 'extracted' from that distribution. For example, the probability that X lies in a given interval (x_L, x_R) can be simply computed by a counting procedure:

$$P(x_L < X < x_R) \approx \frac{\text{number of values } x_k \in (x_L, x_R)}{N}$$

B.0.1   The JAGS language

The JAGS language is very similar to R. We will approach its syntax by means of an example: an example is worth a thousand words.

Suppose you are flipping a fair coin, and you are interested in computing the probability that you get no more than k heads in N tosses. We know from probability theory that the probability of getting **exactly** k heads is given by the binomial distribution:

$$P(k) = \binom{N}{k} \left(\frac{1}{2}\right)^k \left(\frac{1}{2}\right)^{N-k}$$

and, as a consequence, the required probability is:

$$Pr(x \le k) = \sum_{i=0}^{k} P(i)$$

The JAGS model for computing via MCMC simulation the probability of no more than three heads in ten flips is:

```
model {
    Y ~ dbin(0.5,10)
    Pr <- step(3.1 - Y)
}
```

In the preceding code, we define a variable Y having a binomial distribution $Binomial(N, p)$ with sample size $N = 10$ and $p = 0.5$. The required probability $Pr - Prob(Y \le 3)$ is computed by means of the function step, defined in JAGS by:

$$\text{step}(X) = \begin{cases} 1 & X \ge 0 \\ 0 & X < 0 \end{cases}$$

The JAGS model can be written in a file, but it is definitely better to include it in a string and use the R function textConnection to use the model inside R. The R interface to JAGS is the R package rjags. A complete R code for the coin tossing simulation is shown below.

Let us analyse every step of the listing. For not too complex (too long) models textConnection usefully avoids the need to write a model to a file, to be read by the jags.model function, which is the key-point function of rjags. The jags.model requires as input:

- a model file;
- a list or environment containing the data (possibly empty);
- a specification of initial values (optional), in form of list or function;
- the number of parallel chains.

In the example, we choose a single chain and we to leave JAGS the task of randomly generating the initial values. An empty data list completes the model definition, because we have no data here. The function coda.samples sets a monitor for the 'nodes' defined in the second argument (corresponding to *variable.names*, see help(coda.samples)), and outputs the results of the analysis in an MCMC object.

The remaining lines extract Y and P from the MCMC list and perform various summarizing procedures. The exact result is $Pr(x \le 3) = 0.171875$. The result of the JAGS computation is close to that value, but it obviously changes from one simulation to another and also depends on the number of iterations and other MCMC parameters. For example, it could result in $Pr(x \le 3) \equiv \text{sum(out\$}Pr == 1)/\text{length(out\$}Pr) = 0.1735$ when the distribution of 'heads' is that of Figure B.1

```
## Code_1_Appendix_B.R
# Use rjags library
library(rjags)
# Export JAGS model
model <- textConnection("model {
  Y ~ dbin(0.5,10)
  Pr <- step(3.1 - Y)
}")
```

```
# Init
jags.inits <- NULL
# Data
jags.data <- list()
# Perform Bayesian analysis using JAGS
model <- jags.model(model, data=jags.data, inits=jags.inits, n.chains=1)
update(model, n.iter=10000)
# Parameters
jags.params <- c("Y","Pr") # parameters to be monitored
samples <- coda.samples(model, jags.params,10000)
# Analysis of the simulation results
plot(samples)
summary(samples)
# extract P and Y
out <- do.call(rbind.data.frame, samples)
names(out)
# Probability
summary(out$Pr)
sum(out$Pr==0)/length(out$Pr)
sum(out$Pr==1)/length(out$Pr)
# Flipping results
summary(out$Y)
hist(out$Y,breaks=c(0,1,2,3,4,5,6,7,8,9,10),main="",freq=FALSE,xlab="Number of heads")
```

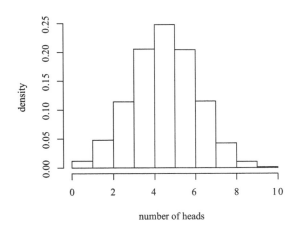

Fig. B.1 Distribution of heads.

B.0.2   Extracting samples from a distribution

JAGS computation is based on simulation using data extracted from a distribution. Indeed, JAGS can be used for sampling from one of the several distributions included in the language. The following R code shows how to draw 1000 samples from a *Poisson* distribution with parameter $\lambda = 2$. Of course, the same result can be obtained directly in R using a single line of code. Figure B.2 compares the samples extracted with the two methods.

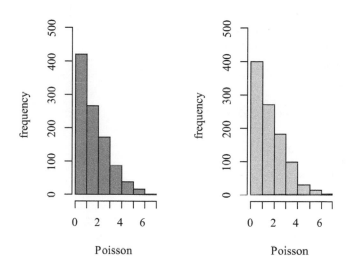

Fig. B.2 Samples extracted from a Poisson distribution, in R (dark grey) and JAGS (light gray).

```
## Code_2_Appendix_B.R
N <- 1000
lambda <- 2
# using R
Pois.R <- rpois(N,lambda)

# USING RJAGS
library(rjags)
# Model
model <- textConnection("model {
    Y ~ dpois(lambda);
}")
# Init
jags.inits <- NULL
# Data
jags.data <- list("lambda"=lambda)
# Run
model <- jags.model(model, data=jags.data, inits=jags.inits, n.chains=1)
update(model, n.iter=N)
# Parameters
jags.params <- c("Y") # parameters to be monitored
samples <- coda.samples(model, jags.params,N)
out <- do.call(rbind.data.frame, samples)
Pois.J <- out$Y
# Compare R and JAGS results
```

```
par(mfrow=c(1,2))
hist(Pois.R,breaks=max(Pois.R),ylim=c(0,500),col="red",main="",xlab="Poisson")
summary(Pois.R)
hist(Pois.J,breaks=max(Pois.J),col="blue",xlab="Poisson",ylim=c(0,500),main="")
```

B.0.3 Regression example

The following example, extracted from the 'JAGS User Manual', is a bit more complicated than the previous ones and it is perfect for explaining the JAGS syntax in some more detail. It implements a linear regression analysis:

```
## Code_3_Appendix_B.R
model {
    for (k in 1:N) {
        Y[k] ~ dnorm(mu[k], tau)
        mu[k] <- alpha + beta*(x[k] - x.bar)
    }
    x.bar <- mean(x)
    alpha ~ dnorm(0.0, 1.0E-4)
    beta ~ dnorm(0.0, 1.0E-4)
    sigma <- 1.0/sqrt(tau)
    tau ~ dgamma(1.0E-3, 1.0E-3)
}
```

Let us briefly analyse the content of the previous listing. The regression model is defined in terms of a set of relations, each one defining a *node* in the model.[1] Nodes on the left of a relation are defined in terms of nodes on the right-hand side. We see two kinds of relation:

- Deterministic relations, written in terms of the symbol '<-', like x.bar <- mean(x), whose meaning is identical to an assignment in R.
- Stochastic relations, written in terms of the symbol '~', like alpha ~ dnorm(0.0, 1.0E-4), which defines a random variable alpha as normally distributed with mean 0 and 1E-4, i.e. variance 1E4. Note that the coincidence of the name dnorm with the analogous R function can be misleading: the former is defined in terms of the *precision*, i.e. the reciprocal of variance, while the latter is in terms of the standard deviation. Therefore, we have the following equivalence:

$$\text{dnorm}_{\text{JAGS}}(m, p) \equiv \text{dnorm}_{\text{R}}(m, 1/\sqrt{p})$$

The for loop at the very beginning of the model definition means:

$$Y_1 \sim \text{dnorm}(\text{mu}_1, \text{tau})$$
$$Y_2 \sim \text{dnorm}(\text{mu}_2, \text{tau})$$
$$\text{...}$$
$$Y_N \sim \text{dnorm}(\text{mu}_N, \text{tau})$$

i.e. it allows us to generate N values Y_k having normal distribution with mean mu_k and precision tau, with the values mu defined by the subsequent deterministic relation. In

---

[1] Interested readers can deepen this relationship with the Bayesian framework and the 'directed graphical model' in (Lunn *et al.*, 2012).

such a model Y[k] and x[k] represent the vector of observed dependent and independent values, respectively. Assuming normally distributed residuals in the regression model: $\epsilon_k = Y_k - (\alpha + \beta x_k)$ the dependent variable is normal, and $(\alpha + \beta x_k)$ is its expected value, i.e. the mean.

The JAGS code for linear regression is inserted into a `textConnection` as before. We apply JAGS linear regression to data coming from the dataset *cars* included in the standard R Datasets Package (see `library(help = "datasets")`). Note from the R listing that the covariate (the X data) is 'centred', i.e. `mean(X)` is subtracted from each X value, because this reduces the posterior correlation between the angular coefficient and the intercept term, and high levels of posterior correlation are known to be problematic for Gibbs sampling.

Figure B.3 compares the regression line computed by JAGS (solid line) with the linear fitting in R (dashed line).

```
## Code_4_Appendix_B.R
# Read data
data(cars)
X <- cars$speed
X <- X-mean(X)
Y <- cars$dist
# Perform JAGS computation
N <- length(X) # number of data points

# Model
mod <- textConnection("model {
    mu ~ dnorm(0, 0.01);
    tau.pro ~ dgamma(0.001,0.001);
    sd.pro <- 1/sqrt(tau.pro);
    beta ~ dnorm(0,0.01)

    for(i in 1:N) {
       predY[i] <- mu + beta*X[i];
       Y[i] ~ dnorm(predY[i], tau.pro);
    }
}")
# Init and definitions
jags.data = list("Y"=Y,"N"=N,"X"=X)
jags.params=c("sd.pro","mu","beta")
jags.inits <- list("tau.pro" = 1, "mu" = 0, "beta" = 0)
# Run
model <- jags.model(mod, data=jags.data, inits=jags.inits, n.chains=1)
update(model, n.iter=10000)
samples <- coda.samples(model, jags.params,10000)
out <- do.call(rbind.data.frame, samples)
names(out)
# Comparison with R linear fitting
plot(X,Y)
abline(lm(Y~X),col="red")
abline(mean(out$mu),mean(out$beta),col="blue")
# Coefficients
coefficients(lm(Y~X))
c(mean(out$mu),mean(out$beta))
```

In these short notes we have only scratched the surface of Bayesian inference using JAGS. JAGS is very powerful and can do much more, like model comparison, analysis of time and spatial series, non-parametric Bayesian analysis, and more.

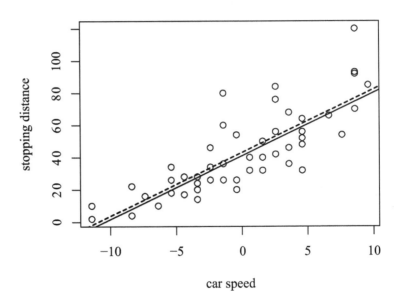

Fig. B.3 Linear regression on the *cars* data, in R (dashed line) and JAGS (solid line).

List of Symbols

Lowercase Latin symbols

a	partial autocorrelation
c	speed of light
e	random noise
f	frequency
g	gravitational acceleration
h	spatial distance
k_B	Boltzmann constant
i	imaginary number
m_Z	deterministic trend in spatial analysis
t	time
v	velocity
w	angular frequency
x	where referred to a random variable is a realization of X

Uppercase Latin symbols

A	amplitude
B	backshift operator
$Bern$	Bernoulli process
Bin	binomial distribution
D	diffusion coefficient
E	expected value
	energy
Exp	exponential distribution
F	distribution function
$\mathcal{F}$	family of subsets
$\mathscr{F}$	future set of events
Gam	gamma distribution
L	lag operator
P	probability
$\mathscr{P}$	past set of events
$Poiss$	Poisson distribution
R	autocorrelation of a deterministic signal
S	entropy
	Fourier transform
$\mathcal{S}$	state space of a random variable
$\mathbb{T}$	parametric space for random variables
$\mathcal{T}$	set of all transient states

Var	variance
Weib	Weibull distribution
X	random variable
Z	random fields

Lowercase Greek symbols

γ	autocovariance function
λ	intensity parameter in Poission distribution
μ	mean
ρ	autocorrelation function
σ	standard deviation
τ	integrated correlation time
ω	angular frequency

Upper case Greek symbols

Δ	difference operator
Θ	estimator
Ω	probability space [-]

List of R Codes

References

Abraham, B. and Ledolter, J. (1983). *Statistical Methods for Forecasting*. Wiley & Sons, USA.

Akaike, H. (1974). A new look at the statistical model identification. *IEEE Transactions on Automatic Control*, **19**, 716–723.

Antolini, G., Auteri, L., Pavan, V., Tomei, F., Tomozeiu, R., and Marletto, V. (2015). A daily high-resolution gridded climatic data set for Emilia-Romagna, Italy, during 1961-2010. *Journal of Climatology*, **08**.

Applegate, D. L., Bixby, R. E., Chvatàl, V., and Cook, W. J. (2006). *The Traveling Salesman Problem: A Computational Study*. Princeton University Press, Princeton.

Bachelier, L. (1900). Théorie dela spéculation. *Annales Scientifiques de l'École Normale Supérieure*, **27**, 21–86. English translation 'Random Character of Stock Market Prices', translated by A. J. Boness in P. H. Cootner (ed.), 1964, MIT Press, Cambridge.

Ben-Ameur, W. (2004). Computing the initial temperature of simulated annealing. *Computational Optimization and Applications*, **29**, 369–385.

Bennett, L. F., Goldsman, D., and Swain, J. J. (1981). Spaced batch means. *Operations Research Letters*, **10**, 255–263.

Berger, A., Crucifix, M., Hodell, D., Mangili, C., McManus, J. F., Otto-Blisner, B., POl, K., Raynaud, D., Skinner, L. C., Tzedakis, C., Wolff, E. W., Yin, Q., Abe-Ouchi, A., Barbante, C., Brovkin, V., Cacho, I., Ferretti, P., Ganopolski, A., Grimalt, J., and Riveiros, N. Vazquez (2015, 11). Interglacials of the last 800,000 years. *Reviews of Geophysics*, 162–219.

Bernardo, J. M. and Smith, A. F. M. (1994). *Bayesian Theory*. Wiley & Sons, USA.

Bertozzi, F. (2008). Relazione su i processi stocastici. Technical report, Universitá degli Studi di Bologna. In Italian.

Bertozzi, F. (2011). Indicatori di correlazione e di disordine basati sul concetto di entropia. Technical report, Università degli Studi di Bologna. Dissertation thesis. Dottorato di ricerca in metodologia statistica per la ricerca scientifica. In Italian.

Binder, K. (1992). Introduction. In *The Monte Carlo Method in Condensed Matter Physics* (ed. K. Binder), pp. 1–22. Springer, Germany.

Born, M. (1949). *Natural Philosophy of Cause and Change*. Oxford University Press, Oxford, UK.

Box, G. E. P., Jenkins, G. M., and Reinsel, G .C. (1994). *Time Series Analysis: Forecasting and Control*. Prentice Hall, New Jersey, USA.

Bretthorst, G. L. (1988). *Bayesian Spectrum Analysis and Parameter Estimation*. Springer, Germany.

Broemeling, L. D. (2018). *Bayesian Inference for Stochastic Processes*. CRC Press, USA.

Brooks, S. P. and Roberts, G. O. (1992). Convergence assessment techniques for Markov chain Monte Carlo. *Statistics and Computing*, **8**, 319–335.

Brown, R. (1828). A brief account of microscopical observations made in the months of June, July and August 1827, on the particles contained in the pollen of plants; and on the general existence of active molecules in organic and inorganic bodies. *The Philosophical Magazine*, **4**, 161–173.

Brush, S. G. (1964). *L. Boltzmann, Lecture on Gas Theory (Translated by S. G. Brush)*. University of California Press, Berkeley and Los Angeles.

Brush, S. G. (1968). A history of random processes: I. Brownian Movement from Brown to Perrin. *Archive for History of Exact Sciences*, **5**, 1–36.

Buffon, George LeClerc (1777). *Essai d'Arithmétique Moralc*. Supplemént a l'Histoire Naturelle, Volume 4.

Carlstein, E. (1986). The use of subseries values for estimating the variance of a general statistic from a stationary sequence. *The Annals of Statistics*, **14**, 1171–1179.

Černỳ, V. (1985). Thermodynamical approach to the traveling salesman problem: an efficient simulation algorithm. *Journal of Optimization Theory and Applications*, **45**, 41–51.

Chaitin, G. J. (1966). On the length of programs for computing binary sequences. *Journal of the Association for Computing Machinery*, **13**, 547–569.

Chaitin, G. J. (1975). Randomness and mathematical proof. *Scientific American*, **232**, 47–52.

Chib, S. (1993). Bayes regression with autoregressive errors: a Gibbs sampling approach. *Journal of Econometrics* (58), 275–294.

Chun, Y. and Griffith, D. A. (2013). *Spatial Statistics and Geostatistics*. SAGE, Los Angeles, USA.

Cochrane, D. and Orcutt, G. H. (1949). Application of least squares regression to relationships containing auto-correlated error terms. *Journal of the American Statistical Association*, **44**, 32–61.

Costantini, D. and Garibaldi, U. (2004). The Ehrenfest fleas: from model to theory. *Synthese*, **139**, 107–142.

Cover, T. M. and Thomas, J. A. (1991). *Elements of Information Theory*. Wiley & Sons, New York.

Cowles, M. K. and Carlin, B. P. (1996). Markov chain Monte Carlo convergence diagnostic: A comparative review. *Journal of the American Statistical Association*, **91**, 883–904.

Cressie, N. A. C. (1993). *Statistics for Spatial Data*. Wiley & Sons, New York.

DeFinetti, B. (1995). *Filosofia della probabilità*. Il Saggiatore, Milano.

Delft, D. van (2014). Paule Ehrenfest's final years. *Physics Today*, **67**, 41–47.

Denbigh, K. G. and Denbigh, J. S. (1985). *Entropy in Relation to Incomplete Knowledge*. Cambridge University Press, Cambridge, UK.

Dobrow, R. P. (2016). *Introduction to Stochastic Processes with R*. Wiley & Sons, USA.

Eaves, J. L. and Reedy, E. K. (1987). *Principles of Modern Radar*. Chapman & Hall, UK.

Efron, B. (1979). Bootstrap methods: Another look at the jackknife. *The Annals of Statistics*, **7**, 1–26.

Efron, B. (1989). Computer-intensive statistical inference. *Current Contents*, **29**, 16.

Efron, B. and Tibshirani, R. J. (1993). *An Introduction to the Bootstrap*. Chapman & Hall, New York.

Ehrenfest, P. and Ehrenfest, T. (1907). Über zwei bekannte Einwände gegen das Boltzmannsche H-Theorem. *Physikalische Zeitschrift*, **8**, 311–316.

Einstein, A. (1905). Investigations on the Theory of the Brownian Movement. *Annalen der Physik*, **17**, 549–560. Edited with notes by R. Fürth, translated by A. D. Cowper, 1956, Dover.

Einstein, A. (1925). Quantentheorie des einatomigen idealen gases. *Berliner Berichte*, **23**, 3–14.

Evans, M. and Swartz, T. (2000). *Approximating Integrals via Monte Carlo and Deterministic Methods*. Oxford University Press, Oxford, UK.

Fanaee-T, H. and Gama, J. (2013). Event labeling combining ensemble detectors and background knowledge. *Progress in Artificial Intelligence*, 1–15.

Feller, W. (1970). *An Introduction to Probability Theory and its Applications* (3rd edn). Wiley & Sons, New York.

Fiser, O. (2010). The Role of DSD and Radio Wave Scattering in Rain Attenuation. In *Geoscience and Remote Sensing, New Achievements* (ed. P. Imperatore and D. Riccio), Chapter 23, pp. 437–456. Intech Open, India.

Flyvbjerg, H. and Petersen, H. G. (1989). Error estimates on averages of correlated data. *Journal of Chemical Physics*, **91**, 461–466.

Fosdick, L. D. (1963). Monte Carlo computations on the Ising lattice. In *Methods in Computational Physics* (ed. B. Alder, S. Fernbach, and M. Rotenberg), pp. 245–280. Academic Press, New York.

Friedberg, R. and Cameron, J. E. (1970). Test of the Monte Carlo method: fast simulation of a small Ising lattice. *The Journal of Chemical Physics*, **52**, 6049–6058.

Galavotti, M. C. (2005). *Philosophical Introduction to Probability*. CSLI Pubblications, Stanford University.

Gelman, A., Carlin, J. B., Stern, H. S., and Rubin, D. B. (2004). *Bayesian Data Analysis*. Chapman & Hall/CRC, New York, USA.

Gelman, A. and Rubin, D. B. (1992a). Inference from iterative simulation using multiple sequences. *Statistical Science*, **7**, 457–511.

Gelman, A. and Rubin, D. B. (1992b). A single series from the Gibbs sampler provides a false sense of security. In *BBayesian of Statistics 4* (ed. J. M. Bernardo, J. O. Berger, A. P. Dawid, and A. F. M. Smith), pp. 625–632. Oxford University Press, Oxford, UK.

Geyer, C. J. (1992). Practical Markov chain Monte Carlo. *Statistical Science*, **7**, 473–511.

Geyer, C. J. (1996). Estimation and optimization of functions. In *Markov Chain Monte Carlo in Practice* (ed. W. R. Gilks, S. Richardson, and D. Spiegelhalter), pp. 241–258). Chapman and Hall, London.

Geyer, C. J. (2011). Introduction to Markov Chain Monte Carlo. In *Handbook of*

Markov Chain Monte Carlo (ed. S. Brooks, A. Gelman, G. Jones, and X.-L. Meng), pp. 3–48. CRC Press-Chapman and Hall/CRC.

Geyer, C. J. (2019). Package **mcmc**: Marco chain monte carlo.

Giannerini, S. (2017). Package **tseriesEntropy**: Entropy Based Analysis and Tests for Time Series.

Glaz, J. and Balakrishnan, N. (1999). *Scan Statistics and Applications*. Springer, Germany.

Goldberg, D. E (1989). *Genetic Algorithms in Search, Optimization, and Machine Learning*. Addison Wesley, Reading, Massachusetts.

Golyandina, N., Korobeynikov, A., and Zhigljavsky, A. (2018). *Singular Spectrum Analysis with R*. Springer, Germany.

Gottlieb, S., Mackenzie, P. B., Thacker, H. B., and Weingarten, D. (1986). Hadronic coupling constants in lattice gauge theory. *Nuclear Physics B*, **263**, 704–730.

Graeler, B., Pebesma, E., and Heuvelink, G. (2016). Spatio−temporal interpolation using gstat. *The R Journal*, **8**, 204–218.

Granger, C. W., Maasoumi, E., and Racine, J. (2004). A dependence metric for nonlinear time series. *Journal of Time Series Analysis*, **25**, 649–669.

Grimmett, G. (2018). *Probability on Graphs* (2nd edn). Cambridge University Press, Cambridge, UK.

Grimmett, G. and Stirzaker, D. (2001). *Probability and Random Processes* (3rd edn). Oxford University Press, Oxford, UK.

Hacking, I. (1975). *The Emergence of Probability*. Cambridge University Press, Cambridge, UK.

Hadfield, J. (2019). Package **MCMCglmm**: Mcmc generalised linear mixed models.

Hall, P., Horowitz, J. L., and Jing, B.-Y. (1995). On blocking rules for the bootstrap with dependent data. *Biometrika*, **8**, 561–574.

Hastings, K. (1970). Monte Carlo sampling methods using Markov chains and their applications. *Biometrika*, **57**, 97–109.

Hausser, J. and Strimmer, K. (2019). Entropy inference and the james-stein estimator, with application to nonlinear gene association networks. *Journal of Machine Learning Research*, **10**, 1469–1484.

Hausser, J. and Strimmer, K. (2021). Package **entropy**: Estimation of Entropy, Mutual Information and Related Quantities.

Hipel, K. W. and McLeod, A. I. (1994). *Time Series Modelling of Water Resources and Environmental Systems*. Elsevier, Amsterdam, The Netherlands.

Hoff, P. D. (2009). *A First Course in Bayesian Statistical Methods*. Springer, Germany.

Holland, J. H. (1992). *Adaptation in Natural and Artificial Systems*. MIT Press, Cambridge, UK.

Huffaker, R., Bittelli, M., and Rosa, R. (2017). *Non Linear Time Series Analysis with R* (1st edn). Oxford University Press, Oxford, UK.

Hulme, M. (1999). Global monthly precipitation, 1900−1999.

Hume, D. (2007). *A Treatise of Human Nature. A critical edition*. Oxford University Press, Oxford, UK.

Hyndman, R. J. and Fan, Y. (1996). Sample quantiles in statistical packages. *American Statistician*, **50**, 361–365.

Iacus, S. M. and Masarotto, G. (2003). *Laboratorio di Statistica con R.* McGraw-Hill, Italy.

Insua, D. R., Ruggeri, F., and Wiper, M. P. (2012). *Bayesian Analysis of Stochastic Process Models.* Wiley & Sons, USA.

Isaaks, E. H. and Srivastava, R. M. (1989). *Applied Geostatistics.* Oxford University Press, Oxford, UK.

Jaynes, E. T. (1957). Information theory and statistical mechanics II. *The Physical Review*, **108**, 171–190. Also in Rosenkrantz (1983), pp. 19-38.

Jaynes, E. T. (1965). Gibbs vs. Boltzmann entropies. *American Journal of Physics*, **33**, 391–398. Also in Rosenkrantz (1983), pp. 79-86.

Jaynes, E. T. (2003). *Probabillity Theory: the logic of science.* Cambridge University Press, Cambridge, UK.

Jeffreys, H. (1931). *Scientific Inference.* Cambridge University Press, Cambridge, UK.

Jeffreys, H. (1939). *Theory of Probability.* Oxford University Press, Oxford, UK.

Jenkins, G.M. and Watts, D.G. (1968). *Spectral analysis and its applications.* Holden-Day, San Francisco, U.S.A.

Jones, P. W. and Smith, P. (2018). *Stochastic Processes: An Introduction.* CRC Press, Boca Raton, Florida, USA.

Kac, M. (1959). *Probability and Related Topics in Physical Sciences.* Lectures in applied mathematics series, Vol. 1A. American Mathematical Society, Providence.

Keuth, H. (2005). *The Philosophy of Karl Popper.* Cambridge University Press, Cambridge, UK.

Kirkpatrick, S., Gelatt, C. D., and Vecchi, M. P. (1983). Optimization by simulated annealing. *Science*, **220**, 671–680.

Kitanidis, P.K. (1997). *Introduction to geostatistics. Applications to hydrogeology.* Cambridge University Press, Cambridge, UK.

Klein, M. J. (1985). *Paul Ehrenfest. The Making of a Theoretical Physicist Vol. 1.* Elsevier, Amsterdam, The Netherlands. Volume 2 has never been published.

Kolmogorov, A. (1965). Three approaches to the quantitative definition of information. *Problems of Information Transmission*, **1**, 1–7.

Kolmogorov, A. N. (1950). *Foundations of the theory of probability* (Translation of Grundbegriffe der Wahrscheinlchksrechnung, Springer, Berlin, 1933 edn). Chelsea, G.

Koza, J. R. (1996). *Genetic Programming.* MIT Press, Cambridge, UK.

Krige, D. G. (1951). A statistical approach to some basic mine valuation problems on the witwatersrand. *Journal of the Chemical, Metallurgic and Mining Society of South Africa*, **52**(6), 119–139.

Künsch, H. R. (2015). The jackknife and the bootstrap for general stationary observations. *The Annals of Statistics*, **17**, 1217–1241.

Landau, D. P. (1976). Finite-size behavior of the Ising square lattice. *Physical Review B*, **13**, 2997–3011.

Laplace, P.S. (1825). *Essai Philosophique sur les probabilités.* Cambridge University

Press, Cambridge, UK.

Lawler, G. F. (1995). *Introduction to Stochastic Processes*. Chapman and Hall/CRC, New York.

Lawler, G. F. (2006). *Introduction to Stochastic Processes* (2nd edn). Chapman and Hall.

Leuangthong, O., Khan, K. D., and Deutsch, C. V. (2008). *Solved Problems in Geostatistics*. Wiley & Sons, New Jersey, USA.

Lin, S. (1965). Computer solutions of the traveling salesman problem. *Bell System Technical Journal*, **44**, 2245–2269.

Liu, R. Y. and Singh, K. (1992). Moving blocks jackknife and bootstrap capture weak dependence. In *Exploring the Limits of Bootstrap* (ed. R. Lepage and L. Billard), pp. 3–43. Wiley & Sons, New York.

Lord Rayleigh (1880). On the resultant of a large number of vibrations of the same pitch and of arbitrary phase. *Philosophical Magazine and Journal of Science*, **10**, 73–78. Reprinted in Scientific Papers, Vol. I (1869-1881), p. 491.

Lord Rayleigh (1899). On James Bernoulli's theorem in probabilities. *Philosophical Magazine and Journal of Science*, **47**, 246–251. Reprinted in Scientific Papers, Vol. IV (1892-1901), p. 370.

Lord Rayleigh (1905). The problem of the random walk. *Nature*, **72**, 1476–4687. Reprinted in Scientific Papers, Vol. I (1869-1881), p. 491.

Lorenz, E. N. (1963). Deterministic nonperiodic flow. *Journal of the atmospheric sciences*, **20**, 130–141.

Lorenz, E. N. (1995). *The essence of chaos* (2nd edn). University of Washington Press, Seattle.

Luethi, D., Floch, M. Le, Bereiter, B., Blunier, T., Barnola, J., Siegenthaler, U., Raynaud, D., Jouzel, J., Fischer, H., Kawamura, K., and Stocker, T. F. (2008). High-resolution carbon dioxide concentration record 650,000-800,000 years before present. *Nature*, **453**, 379–382.

Lunn, D., Jackson, C., Best, N., Thomas, A., and Spiegelhalter, D. (2012). *The BUGS Book: A Practical Introduction to Bayesian Analysis*. Chapman & Hall/CRC Texts in Statistical Science. Taylor & Francis.

Maasoumi, E. and Racine, J. (2002). Entropy and predictability of stock market returns. *Journal of Econometrics*, **107**, 291–312.

Madsen, H. (2008). *Time Series Analysis*. Chapman and Hall, UK.

Martin, A. D., Quinn, K. M., and Park, J. H. (2018). Package **MCMCpack**: Markov chain Monte Carlo (mcmc) package.

Martin-Löf, P. (1966). The definition of random sequences. *Information and Control*, **9**, 602–619.

Masuda, N. and Hiraoka, T. (2020). Waiting-time paradox in 1922. *Northeast Journal of Complex Systems*, 1–19.

Matheron, G. (2019). *Matheron's Theory of Regionalised Variables*. (Ed. V. Pawlowsky-Glahn and J. Serra). Oxford University Press,, Oxford, UK.

Matthews, P. (1989). A slowly mixing Markov Chain with implications for Gibbs sampling. *Statistics and Probability Letters*, **17**, 231–236.

Medhi, J. (2003). *Stochastic Models in Queueing Theory* (2nd edn). Academic Press,

Amsterdam, The Netherlands.

Metropolis, N. (1987). The Beginning of the Monte Carlo Method. *Los Alamos Science Special Issue*, **15**, 125–130.

Metropolis, N., Rosembluth, A. W., Rosembluth, M. N., Teller, A. H., and Teller, E. (1953). Equation of state calculations by fast computing machines. *The Journal of Chemical Physics*, **21**, 1087–1092.

Michalewicz, Z. (1996). *Genetic Algorithms + Data Structures = Evolution Programs*. Springer, Berlin. Germany.

Mignani, S. and Rosa, R. (1995). The moving block bootstrap to assess the accuracy of statistical estimates in Ising model simulations. *Computer Physics Communications*, **92**, 203–213.

Mignani, S. and Rosa, R. (2001). Markov chain monte carlo in statistical mechanics: the problem of accuracy. *Technometrics*, **43**, 347–355.

Minasny, B. and McBratney, A. B. (2005). The matérn function as a general model for soil variograms. *Geoderma* (128), 192–207.

Morales, J. J., Nuevo, M. J., and Rull, L. F. (1990). Statistical error methods in computer simulations. *Journal of Computational Physics*, **89**, 432–438.

Øksendal, B. (2003). *Stochastic Differential Equations: An Introduction with Applications* (6th edn). Springer, Germany.

Ore, O. (1960). Pascal and the invention of probability theory. *The American Mathematical Monthly*, **67**, 409–419.

Pearson, K. (1905). The problem of the random walk. *Nature*, **72**, 294.

Pebesma, Edzer (2018). Simple Features for R: Standardized Support for Spatial Vector Data. *The R Journal*, **10**(1), 439–446.

Pedrosa, (A. C. and Schmeiser, B. W. (1993). Asymptotic and finite-sample correlations between obm estimators. In *Proceedings of the 1993 Winter Simulation Conference* (ed. G. W. Evans, M. Mollaghasemi, E. C. Russel, and W. E. Biles), pp. 481–488.

Peterson, S. C. and Noble, P. B. (1972). A two-dimensional random-walk analysis of human granulocyte movement. *Biophysical Journal*, **12**, 1048–1055.

Petit, J., Jouzel, J., Raynaud, D., Barkov, N., Barnola, J. M., Basile-Doelsch, I., Bender, M., Chappellaz, J., M. Davis, M, Delaygue, G., Delmotte, M., Kotlyakov, V. M., Legrand, M., Lipenkov, V., Lorius, C., Pepin, L., Ritz, C., Saltzman, E., and Stievenard, M. (1999, 06). Climate and atmospheric history of the past 420,000 years from the Vostok ice core, Antarctica. *Nature*, **399**, 429–436.

Pincus, S. (1991). Approximate entropy as a measure of system complexity. *Proceedings of the National Academy of Sciences USA*, **88**, 2297–2301.

Pincus, S. (2008). Approximate entropy as an irregularity measure for financial data. *Econometric Reviews*, **27**, 329–362.

Pincus, S., Gladstone, I. M., and Ehrenkranz, R. A. (1991). A regularity statistic for medical data analysis. *Journal of Clinical Monitoring and Computing*, **7**, 335–345.

Pincus, S. and Kalman, R. E. (1997). Not all (possibly) 'random' sequences are created equal. *Proceedings of the National Academy of Sciences USA*, **94**, 3513–3518.

Pincus, S. and Singer, B. H. (1996). Randomness and degrees of irregularity. *Proceedings of the National Academy of Sciences USA*, **93**, 2083–2088.

Plummer, M. (2003). JAGS: A program for analysis of Bayesian graphical models using Gibbs sampling. *3rd International Workshop on Distributed Statistical Computing (DSC 2003); Vienna, Austria*, **124**.

Poincaré, J. H. (1908). *Science et méthode*. Flammarion, Paris. English edition 'Science and Method', 1914, Thomas Nelson & Sons, London. Reprinted Thoemmes Press, Bristol 1996.

Poisson, S. D (1837). *Recherches sur la probabilité des jugements en matière criminelle et en matière civile*. Bachelier, Paris.

Politis, D. N. (2003). The impact of bootstrap methods on time series analysis. *Statistical Science*, **18**, 219–230.

Politis, N. D. and Romano, J. P. (1992). A circular block-resampling procedure for stationary data. In *Exploring the Limits of Bootstrap* (ed. R. Lepage and L. Billard), pp. 366–381. Wiley & Sons, New York.

Popper, K. R. (1934). *Logik der Forschung*. Springer, Wien. English enlarged edition 'The Logic of Scientific Discovery', 1959, Hutchinson, London, third revised edition 1968.

Poularikas, A. D. (1999). *The Handbook of Formulas and Tables for Signal Processing*. CRC Press, USA.

Preece, D. A., Ross, G. J. S., and Kirby, S. P. J. (1988). Bortkewitsch's horse-kicks and the generalised linear model. *Journal of the Royal Statistical Society: Series D (The Statistician)*, **37**, 313–318.

Press, H., Teukolsky, S. A., Vetterling, T., and Flannery, P. (1992). *Numerical Recipes in Fortran 77: the Art of Scientific Computing* (2nd edn). Cambridge University Press, Cambridge, UK.

Rastrigin, L. A. (1963). The convergence of the random search method in the extremal control of a many parameter system. *Automation and Remote Control*, **10**, 1337–1342.

Richards, A. M. (2014). *Fundamentals od Radar Signal Processing*. McGraw Hill, New York, USA.

Ripley, B. D. (1987). *Stochastic Simulation*. Wiley & Sons, New York.

Robert, C. and Casella, G. (2011). A history of markov chain monte carlo – subjective recollections from incomplete data. *Statistical Science*, **26**, 102–115.

Robert, C. P. (1994). *The Bayesian Choice: A Decision-Theoretic Motivation*. Oxford University Press, Oxford, UK.

Robert, C. P. and Casella, G. (1999). *Monte Carlo Statistical Methods*. Springer-Verlag, New York.

Rosa, R. (1993). At the birth of quantum statistics: Challenge and defence of the 'a priori' statistical counting. *Philosophia Naturalis*, **22**, 84–105.

Rosenkrantz, R. D. (1983). *E. T. Jaynes: Papers on Probability, Statistics and Statistical Physics*. Reidel, Dordrecht.

Ross, S.M. (2014). *Introduction to Probability Models* (11th edn). Academic Press, San Diego, CA, USA.

Ross, S. M. (2019). *Introduction to Probability Models* (12th edn). Elsevier, Amsterdam, Netherlands.

Salas, J.D. and Smith, R.A. (1981). Physical basics of stochastic models of annual flows. *Water Resource Research*, **17**, 428–430.

Scrucca, L. (2013). GA: A package for genetic algorithms in R. *Journal of Statistical Software*, **53**(4), 1–37.

Scrucca, L. (2017). On some extensions to GA package: hybrid optimisation, parallelisation and islands evolution. *The R Journal*, **9**(1), 187–206.

Sen, A. and Srivastava, M. (1990). *Regression Analysis: Theory Methods and Applications*. Springer, Germany.

Shannon, C. E. (1948). A mathematical theory of communication. *The Bell System Technical Journal*, **27**, 379–423.

Shumway, R.H. and Stoffer, D.S. (2006). *Time Series Analysis and Its Applications*. Springer, New York, U.S.A.

Sivia, D. S. (1996). *Data Analysis: a Bayesian Tutorial*. Oxford University Press, Oxford, UK.

Smith, A. F. M. (1991). Bayesian computational methods. *Philosophical Transactions: Physical Sciences and Engineering*, **337**, 369–386.

Smith, A. F. M. (1992). Discussion of C. J. Geyer and E. A. Thompson 'Constrained Monte Carlo Maximum Likelihood for Dependent Data'. *Journal of the Royal Statistical Society. Series B (Methodological)*, **54**, 657–699.

Smoluchowski, M. (1906). Zur kinetischen theorie der brownschen molekularbewegung und der suspensionen. *Annalen der Physik*, **21**, 756–780.

Solomonoff, J. (1964). A formal theory of inductive inference. Part I. *Information and Control*, **7**, 1–22.

Song, S. and Song, J. (2013). A note on the history of the Gambler's ruin problem. *Communications for Statistical Applications and Methods*, **20**, 1–12.

Stoica, P. and Moses, R. (2005). *Spectral Analysis of Signals*. Pearson Prentice Hall, New Jersey, USA.

Tijms, H. C. (2003). *A First Course in Stochastic Models*. Wiley & Sons, U.S.A.

Todhunter, I. (1865). *A History of the Mathematical theory of Probability*. Cambridge University Press, Cambridge, UK.

Varouchakis, E. A. (2018). Geostatistics: Mathematical and statistical basis. In *Spatiotemporal Analysis of Extreme Hydrological Events*. Elsevier, Amsterdam, The Netherlands.

von Mises, R. (1928). *Wahrscheinlickeit, Statistik und Wahrheit*. Springer, Wien. English edition 'Probability, Statistics and Truth', Allen and Unwin, 1939, London-New York. Reprinted, Dover, 1968, New York.

von Plato, J. (1994). *Creating Modern Probability*. Cambridge University Press, Cambridge, UK.

Vorosmarty, C. J., Fekete, B. M., and Tucker, B. A. (1998). Global River Discharge, 1807-1991, ORNL Distributed Active Archive Center.

Waismann, F. (1930). Logische analyse des wahrscheinlichkeitsbegriffs. *Erkentnis*, **1**, 228–248. English edition 'A Logical Analysis of the Concept of Probability', in 'Philosophical Papers', B. McGuinness (ed.), Reidel, 1977, Dordrecht, Holland.

Webster, R. and Oliver, M.A. (2007). *Geostatistics for Environmental Scientists*. Wiley & Sons, UK.

Whitmer, C. (1984). Over-relaxation methods for Monte Carlo simulations of quadratic and multiquadratic actions. *Physical Review D*, **29**, 306–311.

Wikle, C. K., Zammit-Mangion, A., and Cressie, N. (2019). *Spatio-Temporal Statistics with R*. Chapman and Hall/CRC, New York.

Wood, W. W. (1968). Monte Carlo studies of simple liquid models. In *The Physics of Simple Liquids* (ed. H. Temperley, J. Rowlinson, and G. Rushbrooke), pp. 115–230. North-Holland, Amsterdam.

Wood, W. W. and Parker, F. R. (1957). Monte Carlo equation of state of molecules interacting with the Lennard-Jones potential. I. Supercritical isotherm at about twice the critical temperature. *Journal of Chemical Physics*, **27**, 720–733.

Yates, R. D. and Goodman, D. J. (2015). *Probability and Stochastic Processes* (3rd edn). Wiley & Sons.

Ye, L., Hanson, L. S., Ding, P., Wang, D., and Vogel, R. M. (2018). The probability distribution of daily precipitation at the point and catchment scales in the United States. *Hydrology and Earth System Sciences*, **22**, 6519–6531.

Yule, G. U. (1927). On a method of investigating periodicities in disturbed series with special reference to Wolfer's sunspot numbers. *Philosophical Transactions*, **226**, 267–298.

Index

absorbing barrier, 147
  code, 148
Antarctica exmaple, 297
ApEn, 461
  code, 463, 464
  computationally random, 463
AR process, 196
ARIMA process, 209
ARMA process, 204
asymptotic transition matrices, 38
autocovariance
  function, 17
autoregressive process, 196

Bayes
  theorem, 294
Bayesian
  analysis, 300
  approach, 297
  paradigm, 294
  regression, 303
birth-death process, 73
boot(R package), 471
bootstrap
  empirical distribution function, 467
  error propagation, 473
  ideal estimate, 469
  in R, 471
  Monte Carlo estimate, 469
  origin of the name, 467
  plug-in principle, 468
  replication, 467
  sample, 467
bootstrap(R package), 471
Brownian motion
  drift parameter, 183
  first passage times, 179
    expected value, 180
  fractal dimension, 181
  independent increment, 173
  random set, 181
  reflection principle, 180
  standard, 174
  stationary increment, 173
  stochastic differential equation, 183
  stopping time, 179
  strong Markov property, 179
  time homogeneity, 174

Cantor set, 182

Hausdorff–Besicovitch dimension, 182
Cesaro
  convergence, 50
Chapman-Kolmogorov equation, 35, 42, 62
closed class, 44
closed set, 44
Communicability
  property, 42
complexity, 453
conditional entropy, 458
conditional probability, 294
counting process, 108

dead time, 138
detailed balance, 74
DFT, 232
difference equation, 153
distribution
  equilibrium, 38
  invariant, 38
  stationary, 38

entropy
  Shannon, 456
  Shannon , 456
  code, 459
  surprise, 456
equilibrium probability, *see* stationary
            probability distribution
Ergodic
  hypothesis, 25
  processes, 20
  theory, 25
Ergodicity
  ergodicity, 18
estimator, 83
expected duration, 157

fair game, 155
FFT, 234
first visit time, 51
Fourier transform, 232
  discrete, 232
  fast, 234
fractal dimension
  Hausdorff–Besicovitch, 181

gambler's ruin, 152
genetic algorihm, 335
genetic algorithm